INFORMATION SOURCES
FOR RESEARCH AND DEVELOPMENT

series under the General Editorship of

R. T. Bottle, B.Sc., Ph.D., F.R.I.C., F.L.A., M.I.Inf.Sc.
and
D. J. Foskett, M.A., F.L.A.

INFORMATION SOURCES
FOR RESEARCH AND DEVELOPMENT

Use of
Engineering Literature

Use of
Engineering Literature

Editor
K. W. Mildren, B.SC., A.L.A.,A.I. INF.SC.
Engineering Librarian, Portsmouth Polytechnic

BUTTERWORTHS
LONDON – BOSTON
Sydney - Wellington - Durban - Toronto

THE BUTTERWORTH GROUP

ENGLAND
Butterworth & Co. (Publishers) Ltd.
London: 88 Kingsway, WC2B 6AB

AUSTRALIA
Butterworths Pty Ltd.
Sydney: 586 Pacific Highway, NSW 2067
Melbourne: 343 Little Collins Street, 3000
Brisbane: Commonwealth Bank Building,
    King George Square, 4000

SOUTH AFRICA
Butterworth & Co. (South Africa) (Pty) Ltd.
Durban: 152–154 Gale Street

NEW ZEALAND
Butterworths of New Zealand Ltd.
Wellington: 26–28 Waring Taylor Street, 1

CANADA
Butterworth & Co. (Canada) Ltd.
Toronto: 2265 Midland Avenue,
    Scarborough, Ontario, M1P 4S1

USA
Butterworths (Publishers) Inc.
161 Ash Street,
Reading, Mass. 01867

First published 1976

ISBN 0 408 70714 3

© Butterworth & Co. (Publishers) Ltd., 1976

Printed in Great Britain by
Cox & Wyman Ltd, London, Fakenham and Reading.

Preface

As part of a series of guides to information sources, this book has been produced in an attempt to assist engineers, librarians and information officers in their awareness and use of published literature. It is hoped that it will be of value to others involved in the organisation of courses for undergraduate and postgraduate students in librarianship, information science and the various engineering disciplines.

Material included in the chapters has been chosen on a selective rather than a comprehensive basis and the items mentioned are those found to be of most use by the respective contributors. Each author naturally has his own style and this is reflected in the presentation of the chapters, which have not therefore been subjected to restrictive editing.

Chapter 1 provides an insight into the structure of the literature and the channels of communication. This is followed by a chapter on classification and indexing in engineering. Chapters 3–11 deal with the various forms of literature and other services. Chapter 12 gives a somewhat introductory guide to literature searching and Chapter 13 discusses the problems of personal indexes. The remaining chapters are concerned with the literature in the various engineering fields.

The contributions were completed during 1974 and minor alterations and additions have been made at proof stage. Owing to space limitations, the place of publication of books cited has been omitted except in isolated cases where the publisher is not, perhaps, well known.

I should like to express my thanks to the contributors and to the numerous colleagues who have made helpful suggestions and comments on parts of the manuscript, and supplied source material.

Special thanks are due to Mr D. Jackson, Mrs D. J. Northcott and Mrs F. M. Littlefield for their assistance.

I am also indebted to several organisations for allowing parts of their publications to be reproduced, namely GEC Power Engineering (*Thesaurofacet* in Chapter 2), Institute of Electrical and Electronics Engineers (*Table 3.1*), National Translations Center (*Translations Register-Index* in Chapter 4), HMSO (*About Patents: Patents as a Source of Technical Information* in Chapter 6), Institution of Electrical Engineers (*Electrical and Electronics Abstracts* in Chapter 16), and British Standards Institution and Institution of Mining and Metallurgy (UDC Summary as used in *IMM Abstracts*, based on *BS 1000, Figure 35.1*). Thanks are also due to Clive Bingley Ltd for providing *Figure 13.2*, as used in A. C. Foskett's *A Guide to Personal Indexes* (2nd edn, Bingley, 1970).

Portsmouth, 1975   K.W.M.

Contributors

G. K. ANDERSON, B.SC., PH.D., M.I.P.H.E., A.M.I.W.P.C., Department of Civil Engineering, University of Newcastle-Upon-Tyne

C. M. ARCHER, B.E., M.PHIL., C.ENG., M.I.C.E., Department of Civil Engineering, Portsmouth Polytechnic

J. B. ARNOLD, B.SC., D.I.C., D.M.S., C.ENG., M.I.MECH.E., A.M.B.I.M., Department of Mechanical Engineering and Naval Architecture, Portsmouth Polytechnic

V. J. BENNING, B.SC., Technology Reports Centre, Orpington

R. M. BIRSE, B.SC., C.ENG., M.I.C.E., M.I.H.E., Department of Civil Engineering and Building Science, University of Edinburgh

J. BROWN, C.ENG., F.I.MECH.E., Assistant Director, Technical, British Standards Institution, London

L. S. BROWN, C.ENG., M.I.MECH.E., M.I.MAR.E., M.INST.F., Department of Mechanical Engineering and Naval Architecture, Portsmouth Polytechnic

J. O. COOKSON, B.SC., The Machine Tool Industry Research Association, Macclesfield

L. B. COUSINS, A.R.I.C., A.I.INF.SC., Heat Transfer and Fluid Flow Service, A.E.R.E., Harwell

D. P. EASTON, B.COM., Technical Indexes Limited, Bracknell

A. C. FOSKETT, M.A., F.L.A., A.L.A.A., South Australian Institute of Technology, Adelaide

D. F. FRANCIS, B.A., A.L.A., The Library, Portsmouth Polytechnic

A. GOMERSALL, B.SC., M.SC., M.I.INF.SC., Head, Research Library, Greater London Council

D. W. H. HAMPSHIRE, B.SC., C.ENG., M.I.E.E., Department of Electrical and Electronic Engineering, Portsmouth Polytechnic

D. J. HARRIS, O.B.E., B.SC.(ENG.), PH.D., C.ENG., F.I.E.E., F.I.E.R.E., Professor and Head of Department of Electrical and Electronic Engineering, University of Wales Institute of Science and Technology, Cardiff

M. E. HORSLEY, B.SC., PH.D., C.ENG., F.INST.F., M.I.CHEM.E., Department of Mechanical Engineering and Naval Architecture, Portsmouth Polytechnic

D. JACKSON, B.SC., DIP.LIB., The Library, Portsmouth Polytechnic

W. J. D. JONES, B.SC., PH.D., D.I.C., A.R.S.M., A.I.M., Reader, Department of Mechanical Engineering, University College London

R. C. KAHLER, B.SC., M.I.INF.SC., British Ship Research Association, Wallsend

D. R. KING, B.SC., M.SC., The Library, Portsmouth Polytechnic

J. A. LEE, B.SC., M.SC., A.M.I.C.E., Department of Civil Engineering, University of Melbourne

G. M. LILLEY, M.SC., D.I.C., C.ENG., F.R.A.E.S., M.I.MECH.E., F.I.M.A., Professor and Head of Department of Aeronautics and Astronautics, University of Southampton

A. R. LUXMOORE, M.SC., Department of Civil Engineering, University of Wales, Swansea

J. MCFARLANE, B.SC., B.ENG., F.G.S., M.I.M.M., C.ENG., Department of Mining and Mineral Sciences, University of Leeds

N. G. MEADOWS, B.SC., PH.D., C.ENG., M.I.E.E., M.B.I.M., Depute Director, Glasgow College of Technology

P. MEINHARDT, DR. IUR., Barrister Inner Temple, London, Associate of the Chartered Institute of Patent Agents

K. W. MILDREN, B.SC., A.L.A., A.I.INF.SC., The Library, Portsmouth Polytechnic

C. A. MURFIN, D.I.C., C.ENG., M.I.MECH.E., Department of Mechanical Engineering and Naval Architecture, Portsmouth Polytechnic

C. A. O'FLAHERTY, B.E., M.S., PH.D., C.ENG., F.I.MUN.E., F.C.I.T., M.INST.H.E., M.I.E.I., Professor of Transport Engineering, Institute for Transport Studies, University of Leeds

A. M. PARKER, M.SC., PH.D., Department of Electrical Engineering, University of Adelaide

D. PETERSEN, B.SC., C.ENG., M.I.E.E., Department of Electrical and Electronic Engineering, Portsmouth Polytechnic

M. J. SHIELDS, M.I.INF.SC., formerly British Ship Research Association, Wallsend

N. E. SIMONS, PH.D., M.A., M.SC., F.I.C.E., F.G.S., Reader, Department of Civil Engineering, University of Surrey, Guildford

J. R. SMITH, A.R.I.C.S., F.R.G.S., Department of Civil Engineering, Portsmouth Polytechnic

W. G. STEVENSON, M.A., B.SC., formerly The Motor Industry Research Association, Nuneaton

D. N. WOOD, B.SC., PH.D., British Library, Lending Division, Boston Spa, Wetherby

Contents

1

Structure of the literature and channels of communication

D. Jackson

Many attempts have been made to define the term 'engineering', none of which would be acceptable to all engineers. However, one definition serves to show the wide range of human activities that engineering encompasses and to indicate the extent of the information needs of the engineer.

> Engineering is the profession in which a knowledge of the mathematical and natural sciences, gained by study, experience and practice, is applied with judgement to develop ways to utilise economically the materials and forces of nature for the benefit of mankind.
>
> Engineers' Council for Professional Development

The engineer, like the scientist, is totally dependent on his ability to acquire and subsequently retrieve information. It is necessary for him to keep aware of current progress in his own field. Specific items of information, or factual data, are continually required in his everyday work and occasionally it may be necessary to acquaint himself with work outside his own field. The information requirements of the engineer vary widely according to the nature of his work. The research engineer who is involved with theoretical information and the concepts of pure science has totally different needs from those of the design engineer, who applies existing materials, devices and systems to particular situations. As the volume of information is growing at an ever-increasing rate, the task of acquiring relevant information is becoming more and more difficult. A great

1

deal has been written on the communication problems of scientists and technologists and proposals made for their solution (e.g. Meadows, 1974; Passman, 1969; Committee on Scientific and Technical Information (SATCOM), 1969). Much less has been written on the specific requirements of engineers and the use they make of existing sources of information (e.g. Lufkin, 1966, Waldhart, 1974, Wolek, 1969). An international symposium, entitled *Information Systems for Designers*, was held in 1971 and the proceedings were published by the organisers, University of Southampton, Department of Mechanical Engineering. A useful source for further references on this subject is *Annual Review of Information Science and Technology*, currently published by the American Society for Information Science. A cumulative index to the first seven volumes was published in 1972.

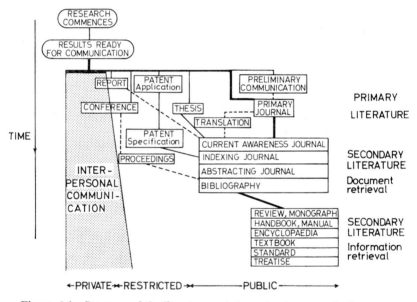

Figure. 1.1   Structure of the literature and channels of communication

A large number of formal and informal communication channels have evolved to satisfy the varying information requirements of scientists and engineers. In order to exploit them to the full, it is necessary to understand the relationships between the channels and the structure of the literature (Bottle, 1973). *Figure 1.1* illustrates the progress of an item of information through a variety of communication media from the time its discoverer commences his research until it is accepted into the body of recorded knowledge. Major information transfer pathways are indicated by heavy lines, while broken lines represent steps which have a lower probability of occurring, e.g. information communicated at a conference may progress into a formal publication, but there is a chance that it may not and the information may effectively be lost. The horizontal axis represents increasing availability of the information to the community and the vertical axis represents the relative time-lag until its appearance in each communication medium.

INTERPERSONAL COMMUNICATION

Informal interpersonal communication is probably the most important information channel for the majority of scientists and engineers. At its simplest level this may mean the engineer discussing his work with colleagues in his own organisation. Unless positive barriers are erected to preserve secrecy, novel information can be transmitted in this way to an ever-widening group.

From the point of view of the recipient of information, interpersonal communication has many advantages. If the source is selected carefully, information can be acquired rapidly and with little effort on the part of the enquirer. There is the possibility of a dialogue, which improves the efficiency of the channel and avoids misunderstandings. Information not usually carried in other channels, e.g. practical expertise, is often available. From a study of engineers in an industrial environment, it has been shown that certain individuals are recognised as good information sources and are relied upon as 'technological gatekeepers', bridging the gap between colleagues within the organisation and formal and informal channels outside (Allen, 1970).

THE PRIMARY LITERATURE

The first documentary record of new information is likely to be in the form of a **technical report**, a term which covers everything from a laboratory notebook recording day-to-day work and submitted to the research supervisor, to more formal documents, which have been produced as a result of a contractual obligation in return for research funds. There has been a great proliferation in the published and semi-published report literature since World War II, due to the increased sponsorship by government of industrial and academic research

(Brearley, 1973). By means of a technical report, information can be disseminated rapidly to a controlled population of users. The information content is usually very detailed, giving a full account of the research history with illustrations, tables and discussions of unsuccessful approaches. However, as an archival record reports have certain disadvantages. They may not be adequately refereed, so that false or misleading information may be included, and the difficulties of bibliographical control may mean that useful information is lost. Report literature is discussed in more detail in Chapter 5.

Conferences, held at local, national or international level, provide a further link between the informal and formal channels of communication. A great deal of information is communicated by personal contact between people with similar interests brought together by the conference. Equipment manufacturers take advantage of this and send representatives to conferences and organise displays of their products. The documentary record may take the form of duplicated sheets circulated to conference delegates or more formally, but at the expense of speed, the presented papers may be published as one-off volumes, as part of a series, or as articles in journals. The communication processes taking place at conferences have been analysed in depth by the Centre for Research in Scientific Communication at Johns Hopkins University, with the ultimate aim of providing guidance on conference presentation and control, to facilitate information flow (Garvey, 1970). A fuller description of conferences and the bibliographical control of their proceedings may be found in Chapter 3.

If a new discovery is likely to be of economic significance, it may be first communicated to the outside world as a **patent**. The patent system effectively guarantees the discoverer of a new invention or technique a temporary monopoly in the exploitation of his discovery in return for making the knowledge public as a patent specification. Although the main function of the patent system is to establish legal priority for inventions and discoveries, patent specifications are an important primary medium of communication. The national and international patent system and the use of patent specifications by the engineer are described in Chapter 6.

Information produced as the result of a research project in an academic institution may be published in the form of a **thesis** or **dissertation**. The thesis may be considered as a specialised form of report, presented to the awarding body in partial fulfilment of the requirements of a higher degree, and exhibits most of the characteristics of the technical report literature discussed earlier. The use of theses as an information source is considered in Chapter 3.

The **journal** has been the main medium for the announcement of new knowledge for the last 300 years and, at the same time, is expected to form the fundamental archival record of science and technology. The two functions, unfortunately, have different requirements as far as the publisher is concerned. The archival function, and the need to maintain the reputation of the journal, requires time-consuming and expensive refereeing of submitted articles by eminent workers in the field. The announcement function requires, above all, speed of publication and to this end a new breed of **preliminary communication journals** has arisen, e.g. *Electronics Letters*, in which the time-lag is reduced to a few weeks. The need for speed of publication and economic considerations have led some publishers to experiment with microfilm as the publication medium. It has also been suggested that conventional journal publication should be discontinued. Primary articles would be kept on microfilm in central stores from which hard copy enlargements would be available on demand. A conference on the *Future of Scientific and Technical Journals* was held in 1973 and the proceedings were published in the September, 1973 issue of *IEEE Transactions on Professional Communication*. The journal literature of engineering is analysed in Chapter 3.

THE SECONDARY LITERATURE

Abstracting and indexing journals provide a key to the documents which make up the primary literature. Approaching 2000 titles are now available in science and technology, not counting those produced for their own use by individual firms or institutions. They range from the comprehensive international, e.g. *Engineering Index*, to those limited to specific subject areas, e.g. *Steel Castings Abstracts*, or forms of literature, e.g. *Dissertation Abstracts International*.

The provision of deep indexing and informative abstracts is an expensive and time-consuming process, but is vital if the service is to provide efficient access to the archive of primary documents. Abstracting and indexing journals are also expected to aid the announcement function of the primary literature and with this aim are usually issued at frequent intervals, with cumulative editions covering longer periods. Again the two requirements are obviously incompatible, and, during the last 20 years, the functions have been separated by the publication of **current awareness journals**. These may take the form of reproductions of contents pages of primary journals, e.g. *Current Contents—Engineering, Technology & Applied Sciences*, or lists of articles arranged under broad subject headings, e.g. *Current Papers in Electrical and Electronics Engineering*. In some cases, rudimentary computer-produced indexes based on keywords in the title of the article are also provided. Current awareness journals are produced very rapidly, in some cases concurrently with the publication of the primary journal. They have no archival function and can be discarded when the material has been covered in the conventional abstracting and indexing journals.

The scanning of abstracting and indexing journals can be a time-consuming process and a **selective bibliography**, if available on the subject, is more convenient as a short cut to the primary literature. Bibliographies may be published as books, as pamphlets or in journals, but often they are not formally published and are only available to a limited group of users.

Computer-based information systems, e.g. COMPENDEX (Computerised Engineering Index), which can match articles with individual or group user interest profiles and produce current or retrospective bibliographies, have recently become available. Such services are very convenient as far as the user is concerned, but at the moment they are expensive and their efficiency is totally dependent on the quality of the indexing of the articles and the preparation of the user profile. There is little scope for the accidental discovery of interesting articles, which can be so rewarding when browsing through primary journals, abstracting journals or even current awareness lists. The major abstracting and indexing journals in engineering are reviewed in Chapter 11.

The forms of secondary literature so far discussed refer the user to documents only and carry limited information themselves. Each primary document is treated as an integral unit and no attempt is made to relate the information content to other articles on the same subject and the general framework of knowledge. The next stage is for the information to be evaluated and consolidated to give a state-of-the-art review which represents a synthesis of contemporary knowledge on the subject complete with an extensive bibliography of the literature on which it is based. **Review** articles may be published in journals or in series of the 'Annual Reviews in . . .' or 'Progress in . . .' type. Alternatively, covering a wider area, the review may be published separately as a **monograph** or as part of a multi-volume **treatise**.

The information may also be reprocessed into forms aimed at particular types of use. **Encyclopaedias**, designed for quick reference, may be thought of as collections of short reviews arranged in alphabetical order of subject. Textbooks contain the information in a form suitable for students and their teachers. The practising engineer is served by **handbooks** and **manuals**, which contain factual information in easily retrievable form as tables and diagrams. Practically orientated reviews and news of new methods and equipment may be found in **trade journals**, which are highly prized by the engineer, not least for the advertising material they contain. Product information is also to be found in **trade catalogues** and brochures, issued as publicity material by manufacturers. **Standards,** issued by national and international organisations, and covering a wide range of topics, include glossaries of terms, definitions and symbols; specifications for quality, safety, performance or dimensions; and codes of practice.

BARRIERS TO COMMUNICATION

The total literature/information system described has evolved over 300 years to attempt to satisfy man's search for comprehensive awareness and retrieval of information. However, the system has many inherent imperfections which act as barriers to effective communication, leading to delay and loss of valuable information (Basu, 1972).

Perhaps the greatest barrier is formed by the sheer size of the body of recorded knowledge from which information can be retrieved and the rate at which it is growing. This problem is made worse because the literature archive includes a large proportion of duplicated information. This may be a result of the 'publish or perish syndrome' which prevails in academic research, where an individual is often judged by the number of publications to his credit. Alternatively, it may be due to unconscious duplication of research effort, itself arising from a failure in communication. Such duplication, carried forward uncritically into abstracting and indexing journals, leads to frustration and time wasting.

The literature archive also contains misleading and often contradictory information, and therefore it is important that information should not be used uncritically. Factual data are continually being updated, with the introduction of new instruments and techniques, and old theories are superseded. Some false information may be included as a result of imperfections in the refereeing system. Conversely, however, over-zealous refereeing may result in useful information being suppressed.

More than one-third of the world's engineering literature is in languages other than English and this proportion is growing as more countries develop their technologies. The dangers of neglecting the Russian and Eastern European literature were recognised in the late 1950s and ambitious and expensive translation programmes were initiated by western governments. The share of less familiar languages from Asian countries is increasing rapidly and is likely to pose an even more difficult problem in the near future as China improves contacts with the western world. It is increasingly likely, therefore, that, having retrieved a document, the extra step in the information transfer process of acquiring a translation has to be overcome before it can be understood. Foreign-language literature and the availability of translations are discussed in Chapter 4.

Valuable information may not reach the engineer because he is ignorant of relevant information sources and how to use them effectively. The importance of this barrier has been recognised and proposals that lectures on the literature should be introduced into the curricula of engineering courses at all levels are slowly being put into effect (Wood, 1969). It is hoped that this book, in addition to providing a basis for such courses, will be used by practising engineers to alleviate their information problems. Others guides to the literature of engineering and related subjects are listed in Chapter 10.

LIBRARIES AND INFORMATION CENTRES

The printed word remains the main medium for the storage of information and inevitably the engineer will acquire a large collection of documents in the course of his work. Guidance on the organisation of personal document collections is given in Chapter 13.

The growth in the volume of information has been paralleled by an increase in the size and number of libraries to which the engineer may have direct or indirect access. The general aims of a library are to acquire the records of human knowledge in whatever form they may appear, and to store and organise them in a way which facilitates the retrieval of individual documents and the information they contain. The stock and services of an individual library are determined by the needs of its clientele. Owing to the increasing cost of providing comprehensive collections, it is no longer possible for a library to be self-sufficient. However, any library can act as an access point to the total library resources of the country, through inter-library co-operation and lending, backed up by the services of a comprehensive national library. It is possible, therefore, for an individual library to rationalise its acquisition and withdrawal policies, concentrating on the most relevant and up-to-date literature and relying on national resources for the remainder.

The services provided in libraries also vary widely according to the demands made on them by their users. The most fundamental service is to provide access to the material they contain, and in the majority of libraries the shelves are normally open to all users. However, in some libraries, which are required only to supply particular documents on demand, or for reasons of security, access to the shelves may be denied to all but the library staff. In most open-access libraries the material is arranged in a logical manner to facilitate its use. Journals may be shelved in alphabetical order of title, and then by date of publication, or some subject division may be made. Books are normally arranged according to a standard subject classification system, so that books on similar subjects are placed near to each other on the shelves (Maltby, 1972). A subject index to the classification scheme is provided. There may be separate sequences for different types of material, e.g. reports, standards, patents or microforms. The catalogue provides a key to the holdings of the library, usually allowing individual items to be located by means of an author, title or subject approach. Catalogues may be found in a wide variety of forms, the most common being a file of 5 × 3 in cards. However, microfilm catalogues are likely to become more common. The catalogues of large libraries are important bibliographic tools in their own right and may be reproduced and published for use in other libraries, e.g. *The Classed Subject Catalog of the Engineering Societies' Library* in New York has been published in 13 volumes (G. K. Hall, 1963). Supplementary volumes are published periodically. Libraries also provide facilities for the use of the material they contain in the form of reading rooms, microfilm readers and listening booths. Documents may be borrowed from the library for use elsewhere or, where allowed by the copyright law, reproductions of material may be supplied. Because of the variations in the ways in which different libraries achieve the same general aims, most libraries issue guides to their own particular systems. Several general guides to the effective use of libraries have been published. One example is *Using Libraries*, by K. Whittaker (3rd edn, Deutsch, 1972). Classification and indexing with special reference to engineering is discussed in Chapter 2.

Most libraries which cater for the needs of the engineer extend their services beyond the provision of documents to include the information contained in them. In recent years librarians with subject qualifications, who can understand the language and the problems of the specialist, have become more common and added to the efficacy of library services. Literature searches may be performed by the library staff to prepare bibliographies or to provide specific information on request. Abstracts and indexes to the literature may be produced and current awareness lists circulated to users. The library staff may liaise between the engineer and computer-based information services. A translations service may also be offered. Further, 'specialist information centres', where the service is extended to include the critical evaluation of information and the preparation of authoritative state-of-the-art reviews in specific areas, have been developed on an international basis (Mountstephens, 1971).

At the apex of the system of libraries and information centres in the United Kingdom is the British Library, which was established by Act of Parliament in 1972. The new library system was constituted from the existing resources of four major institutions—the British Museum Library (which included the National Reference Library of Science and Invention), the National Central Library, the National Lending Library for Science and Technology and the British National Bibliography. The declared aims of the British Library, published in the White Paper *The British Library* (Cmnd 4572) in 1971, are 'to preserve and make available for reference at least one copy of every book and periodical of domestic origin and of as many overseas publications as possible', and 'to provide an efficient central lending and photocopy service in support of the other libraries and information systems of the country'.

The sections of the British Library of most interest to the engineer are the Lending Division (BLLD) and the Holborn Division of the Science Reference Library (BSRL). The BLLD is situated at Boston Spa in Yorkshire, close to the geographical centre of the United Kingdom and major road and rail routes. It is, therefore, in an excellent position to offer rapid and efficient loans and photocopy services from its vast collections in science and technology. The loans service is available only via other libraries and not to individuals. However, material can be consulted on demand in the public reading room, which also houses a comprehensive collection of abstracting and indexing journals and reference books. The Holborn Division of the BSRL is based on the former Patent Office Library in Chancery Lane, London WC2, supplemented by appropriate material from the British Museum Library. The Library is particularly strong in the mathematical and physical sciences, engineering and technology, and includes comprehensive sets of British and foreign patents. The collections, most of which are on open access, are available for reference use only. However, a photocopy service is offered either at the Library or by post.

Other libraries serving the needs of engineers may be found in a variety of situations, including industrial organisations, universities, polytechnics or technical colleges, government departments or research organisations, research associations and trade development associations, learned and professional societies, and public libraries. Most of the libraries are members of Aslib, which acts as a clearing house to sources of information and provides an information service to its members. Access to other libraries and information sources can also be achieved via local and regional co-operative schemes organised mainly by public libraries, but including all other types of library. Details of the stock and services of individual libraries and co-operative schemes can be found in *Aslib Directory. Volume 1: Information Sources in Science, Technology and Commerce*, edited by B. J. Wilson (Aslib, 1968). The history and development of special library services are surveyed in *Special Library and Information Services in the United Kingdom*, by J. Burkett (2 vols, 3rd edn, Library Association, 1972–74). Volume 1 is subtitled *Industrial and Related Library and Information Services* and volume 2 is called *Government and Related Library and Information Services*.

A similar pattern is reflected in the library systems of other developed countries. Further details of particular libraries and information centres may be found in the following directories.

American Library Directory (Bowker, biennial)
Directory of Information Resources in the United States (National Referral Center for Science and Technology, 1971)
Directory of Special Libraries and Information Centers, by A. Kruzas (2nd edn, Gale, 1968)
European Library Directory: a Geographical and Bibliographical Guide, by R. C. Lewanski (Leo S. Olschki, 1968)
Guide to European Sources of Technical Information, by C. H. Williams (3rd edn, Hodgson, 1970)
International Library Directory: A World Directory of Libraries (2nd edn, Gale, 1968)
Subject Collections: A Guide to Special Book Collections in Libraries, by L. Ash (4th edn, Bowker, 1973)
World of Learning: Directory of the World's Universities, Colleges, Learned Societies, Libraries, Museums, Art Galleries and Research Institutes (Europa, annual)

REFERENCES

Allen, T. J. (1970). 'Roles in Technical Communication Networks', in *Communication among Scientists and Engineers*, by C. E. Nelson and D. K. Pollock, 191–208 (Heath)
Basu, R. N. (1972). 'Barriers to Effective Communication in the Scientific World', *IEEE Transactions on Professional Communication* PC-15 (2), 30–33
Bottle, R. T. (1973). 'Scientists, Information Transfer and Literature Characteristics', *Journal of Documentation*, 29 (3), 281–294
Brearley, N. (1973). 'The Role of Technical Reports in Scientific and Technical Communication', *IEEE Transactions on Professional Communication*, PC-16 (3), 117–119
Garvey, W. D. (1970). Johns Hopkins University. Center for Research in Scientific and Technical Communication. *Role of the National Meeting in Scientific and Technical Communication*, PB 202 367, CFSTI
Lufkin, J. M. (1966). 'The Reading Habits of Engineers', *IEEE Transactions on Education*, E-9 (4), 179–182
Maltby, A. (1972). *Classification in the 1970s* (Bingley)

Meadows, A. J. (1974). *Communication in Science* (Butterworths)

Mountstephens, B. (1971). *Quantitative Data in Science and Technology* (Aslib)

Passman, S. (1969). *Scientific and Technological Communication* (Pergamon)

SATCOM (1969). United States National Academy of Sciences – National Academy of Engineering. Committee on Scientific and Technical Communication. *Scientific and Technical Communication*

Waldhart, T. J. (1974). 'Utility of Scientific Research; the Engineer's Use of the Products of Science', *IEEE Transactions on Professional Communication*, **PC-17** (2), 33–35

Wolek, F. W. (1969). 'The Engineer: His Work and Needs for Information', *Proceedings. 32nd Annual Meeting of the American Society for Information Science*, **6**, 471–476

Wood, D. N. (1969). 'Library Education for Scientists and Engineers', *Bulletin of Mechanical Engineering Education*, **8**(1), 1–8

2

Classification and indexing in engineering

A. Gomersall

To most engineers the methods of classifying and indexing used in the special or academic libraries which aim to meet their needs offer at best a limited facility for retrieving the ever-increasing quantity of information they require in their work. Many now find that the proliferation of retrieval systems, both manual and mechanised, only serves to confuse rather than guide them directly to the information they require. The book stock in the library is usually in some form of classified order; the periodical collection in alphabetical arrangement, or, worse, in a different classified order from that of the books; company reports are arranged by number; microform reports and other literature in separate files; and probably all these different forms of information have separate indexes for retrieval. Additionally, many of the larger libraries now offer on-line terminals into computer data banks which hold references to stock stored by the library already in the various arrangements mentioned above.

The various procedures of information retrieval are all involved with manipulation of groups of documents arranged either alphabetically by author or subject, or by running number. These groups are assigned headings of some kind usually known as index terms or descriptors extracted from the index language used by the particular library. The engineer may then retrieve information on his specified topic by examining this language, selecting the index terms which most closely describe his problem, and then searching some form of file or index incorporating references to the groups of documents. Naturally, to obtain an adequate retrieval performance it is essential that in the indexing operation some level of consistency must have been applied, and so the use of a controlled vocabulary has become an accepted part of the indexing and retrieval processes. This controlled vocabulary exists essentially to narrow the gap between the indexer's vocabulary and that of the user. It provides a control over synonyms and near-synonyms, jargon, etc., to prevent different indexers from using different terms to express the same subject matter, and to guide the user to the correct search terms. Since the early 1960s many controlled vocabularies or languages have been presented in the form of thesauri which display the broader and narrower relationships between descriptors, and also present the more tenuous relationships between a concept such as equipment and its components, processes and properties, materials and properties, etc. The alternative form is the more traditional classification scheme represented by the Universal Decimal Classification, or specially constructed faceted classification schemes such as the new edition of the Bliss Classification.

This controlled vocabulary can be applied in two different ways to index documents and retrieve from the collection. First, it may be used pre-coordinately to create at the indexing stage a single theme, however complex, to uniquely define the whole subject of a document. For example, if the theme *'steam turbine blade vibration'* is created, the three separate classes *'steam turbines', 'blades' and 'vibration'* are being pre-coordinated. This form of pre-coordinate system is applied when the classification schemes mentioned above are used, many of which are in current use in engineering libraries. The theme is usually represented in the search files by a notation or class number—for example, 621.165-25:534 or Db Je Tj both representing *'steam turbine blade vibration'*. The second method is to apply the post-coordinate form, when the indexer does not combine concepts into themes, but lists descriptors consisting of single words or multi-word concepts as discrete items. Thus, the topic *'steam turbine blade vibration'* is indexed by assigning the separate index terms *'steam turbines', 'blades', 'vibration'*. The searching stage, in such a post-coordinate system, is the only time at which these descriptors are brought together in a theme, often by the use of edge-notched or optical coincidence cards, or possibly through on-line or off-line computer retrieval systems (see *Figure 2.1*). Many modern storage and retrieval systems combine the ideas of pre- and post-coordination, so that concepts which frequently occur together may be combined at the indexing stage in a pre-coordinated term—for example, *'axial flow compressors'*—and then coordinated with others at the searching stage.

As technology has become more complex and new ideas and developments appear more frequently, so the terminology of technology has had to follow suit. As a result, many of the traditional

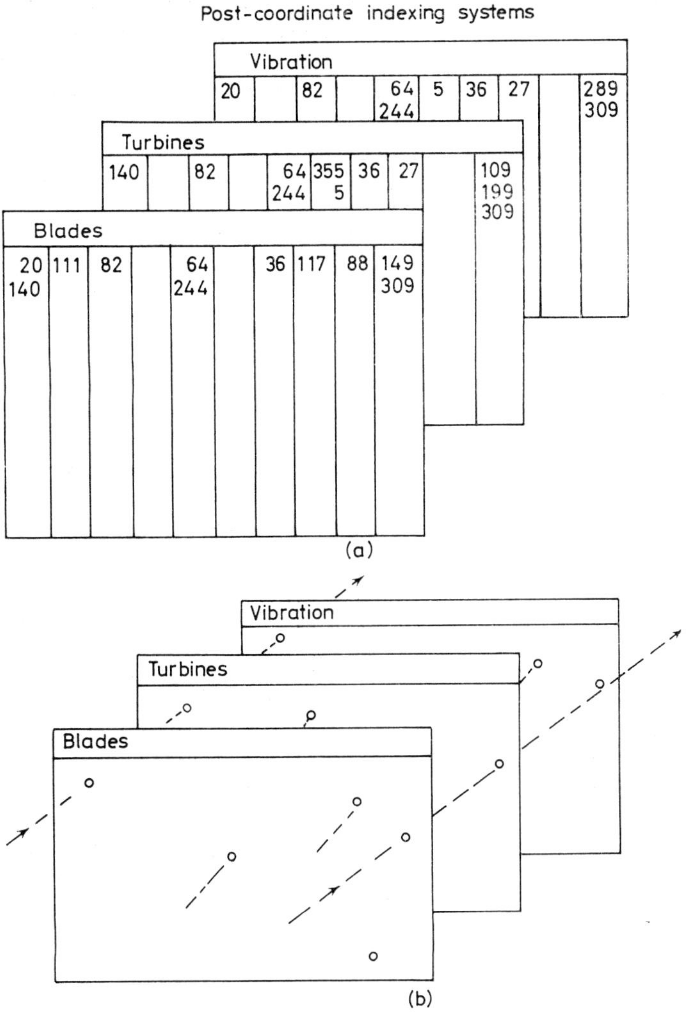

Post-coordinate indexing systems

Vibration

| 20 | | 82 | | 64 244 | 5 | 36 | 27 | | 289 309 |

Turbines

| 140 | | 82 | | 64 244 | 355 5 | 36 | 27 | 109 199 309 |

Blades

| 20 140 | 111 | 82 | | 64 244 | 36 | 117 | 88 | 149 309 |

(a)

Vibration

Turbines

Blades

(b)

Figure 2.1 (a) Uniterm cards; (b) optical coincidence cards

classification systems which have provided strictly enumerative lists of index terms have become increasingly inadequate, in that there is no facility whereby the classifier (indexer) can create new concepts by combining the listed index terms. Consequently, many of the recently developed schemes incorporate rules to enable combination of existing index terms in various ways to form the required more specific concepts. Such vocabularies are synthetic, and the most used classification scheme to be based on such principles is the Universal Decimal Classification (UDC), in which the predominant synthetic device is the colon. The latest edition of the Bliss Classification, which appeared in 1972, also offers similar facilities for synthesis and facet analysis.

PRE-COORDINATE SYSTEMS

UDC covers the whole field of knowledge, but the two sections of most interest to engineers are 5 and 6: 'Pure Sciences' and 'Applied Sciences'. The British Standards Institution is the official British editorial body and is currently preparing the full English edition. This was begun in 1943 and completion is expected in the near future, so users are left with some schedules already 30 years old and obviously badly in need of revision. Fortunately for the engineering industry, many of the recently completed schedules cover specific technologies such as electrical engineering, metallurgy, etc. In the absence of the whole of the full edition, *BS 1000*, the abridged edition, *BS 1000A*, produced in 1961, is used in many libraries. Within sections 5 and 6 considerable subdivision is possible—for example:

6	Applied sciences
62	Engineering science
621	Mechanical and electrical engineering
621.3	Electrical engineering
621.35	Applied electrochemistry
621.357	Industrial applications of electrochemistry
621.357.6	Electroforming

Such subdivision enables very specific information to be given a unique place in the classified catalogue, and to be retrieved directly by reference to an alphabetical index to the classified file.

However, using the schedules only provides the user with a strictly enumerative scheme presenting the problem mentioned previously of trying to accommodate new concepts without a place in the schedules—for example, fluidics, thyristors, explosive forming, etc. To provide UDC with the facility of synthesis, common and special auxiliaries have been introduced to act as common facets and facet indicators. The colon is the most widely used of these synthetic devices enabling the classifier to link two or more UDC numbers denoting related concepts of approximately equal value, e.g.

669.1 : 546.22	Sulphur in iron and steel
697.38 : 691.11	Effect of hot air central heating on wood
	(building material)

The colon, however, is so generally used nowadays that it lacks precision and, particularly if UDC is used in a mechanised retrieval system, the colon will have to be replaced by several more precise devices. Other auxiliaries frequently used are:

Common auxiliaries of place (1/9), e.g. coal mining in Wales 622.33(429)
'Plus' or 'and' sign (+), e.g. Mining and Metallurgy 622+669
Common auxiliaries of point of view (.00) e.g. 622 Mining: industry, engineering, etc. 622.001 Research, development
Special (auxiliary) sub-divisions (— . . . , 0 . . . ,   ). In engineering the — is the indicator for parts and shapes, and in electrical engineering the 0 introduces various facets: 621–578 Clutches for machinery control and regulation. 621.3.04 Machine and transformer components

As can be seen from the schedule details and examples of auxiliaries given, the drive for more specificity and flexibility in UDC has tended to produce extremely long class numbers, particularly when the colon is used to link subjects within a basic class. This notational problem coupled with the separation of science and technology, the many out-of-date concepts within the schedules and the lack of a full edition with alphabetical index has resulted in the development of many special schemes by industrial libraries. By careful facet analysis, better grouping of related concepts can be achieved, and shorter simpler notations can be allocated.

An analytico-synthetic classification scheme can provide enormous facilities for detailed description of the subject of a document. By grouping related concepts within facets representing particular aspects of a subject, the ability is provided to combine the concepts at the classifying stage to represent extremely complex subjects, e.g.

Db Jb Ns.h Pk Xl Fatigue testing of low alloy steel blades for steam turbines: Db representing *'steam turbines'*; Jb, *'blades'*; Ns.h, *'low alloy steel'*; Pk, *'fatigue'*; and Xl, *'testing'*.

Unfortunately, the ability to retrieve information directly on *'fatigue of blades'* in this example is lost in such a pre-coordinate system, and only by additional combinations of class numbers, alphabetical chain indexing or selective permutation of the individual parts of the class number can retrieval be improved. Thus, the flexibility of the post-coordinate system becomes even more attractive. This attraction was emphasised in 1969 when the English Electric Company in revising its *Faceted Classification for Engineering* provided the facility for using the revision either as a traditional pre-coordinate classification scheme or as a post-coordinate system. *Thesaurofacet*, as the scheme is called, will be described later.

POST-COORDINATE INDEXING SYSTEMS

These retrieval systems, often used for indexing reports and other special collections in industrial libraries, are based on one or other of two principles, term entry or item entry, with the former easily the most popular in the majority of libraries. The item entry principle is based on the use of a unique record for each document, microfiche, etc., with all information relating to that document contained on the record. Term entry, on the other hand, relies upon individual records for each piece of information (descriptor, author, etc.) and this record carries the identities of all documents to which the descriptor has been allocated. In retrieval using item entry systems, each record in the file has to be examined when a search is made, usually by 'needling' edge-notched punched cards (see *Figure 2.2*). Searching

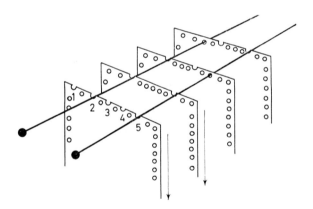

Figure 2.2   Edge-notched punch cards

is laborious even when the file comprises only a few hundred records, and physical limitations exist in the allocation of notches for descriptors on the cards. Complex coding is often needed to offset these limitations. Term entry systems at their most fundamental level are typified by Uniterm cards or optical coincidence cards with one card used for each descriptor in the controlling language. Searching here consists of selecting the terms required for retrieval, withdrawing the cards from the file and matching common numbers on the

Uniterm cards (*Figure 2.1a*) or matching coincident holes in the optical cards (*Figure 2.1b*). The main advantages offered by term entry over item entry retrieval systems are that the whole file does not have to be searched, the system is compact and easily operated by clerical staff, and coding is not required for indexing. A disadvantage exists in that a separate record of the file of documents has to be maintained to correlate with the numbers on the Uniterm cards or optical coincidence cards, with the latter having a maximum capacity of between 10 000 and 15 000 documents.

Most of these systems, both term and item entry, operate using a controlled vocabulary or thesaurus for purposes already described earlier in this chapter, and much development work has been carried out on these over the last few years. The Engineers' Joint Council *Thesaurus of Engineering and Scientific Terms* (1967) is possibly the most significant controlled language to come out of America, and indeed the structure of most thesauri produced since that date has been based on *TEST*. In the thesaurus generic terms are indicated by BT (Broader Terms) with their specific terms NT (Narrower Terms); and RT (Related Terms) producing linkages for a variety of more tenuous relationships, e.g. part–whole relationships such as piston–diesel engine, and relationships between processes and their end-products. An additional feature in *TEST* emphasises the close relationship between thesauri and conventional classification schemes. This is a *Subject Category Index*, which is in effect a broad classification with 22 major subject fields each subdivided into groups. Since 1967 many other thesauri have been constructed, often in specific fields within technology and usually based on EJC rules for construction, although one or two have a diagrammatic form of presentation. A list of the more important thesauri is given at the end of the chapter.

THESAUROFACET

As the successor to the third edition of English Electric Company's *Faceted Subject Classification for Engineering*, *Thesaurofacet* integrates a faceted classification with a thesaurus, each incomplete without the other, and clearly shows the refinements and developments on thesaurus construction which have been evolving since the mid-1960s. Indeed, since the publication of *Thesaurofacet* in 1969, the Construction Industry Research and Information Association has produced a similar scheme for its own subject areas, and Croghan (1970) has compiled a thesaurus-classification for non-book materials.

Thesaurofacet consists of two sections—the schedules and the thesaurus—the terms appearing twice, once in each section, the link between the two locations being the class number. If the user has a specific term in mind, he enters the thesaurus at that term. For example, in looking for information on *'television camera tubes'* (see *Figure 2.3*) he will be directed via the class number MCE to the schedule *'electron tubes'*. Here the usual visual hierarchal display gives the generic term *'cathode ray tubes'* and other broader generic terms above, while the narrower term is *'television colour camera tubes'*. Related terms displayed more satisfactorily in the schedules are *'image converter tubes'* and others. To obtain additional information about his query, the user now returns to the thesaurus, where the entry for *'television camera tubes'* gives another broader term and three other related terms.

Within each area of technology the *Thesaurofacet* schedules are displayed according to facet principles. Within a specific subject field facet analysis is used to group the terms into fundamental categories, and within these categories grouping into sub-facets is carried out (see *Figure 2.4*).

The thesaurus serves as the index to the schedules, but also controls synonyms and word forms in the manner of a conventional thesaurus. Because the main BT/NT relationships are shown in the classification schedules, these are not shown in the thesaurus, but additional or auxiliary hierarchies are displayed using the symbols BT(A) and NT(A). For example, in *Figure 2.4,*

Hydraulic servomotors   WEF
BT(A)   Hydraulic motors

The first entry above indicates that the term is located in the *'servocomponents'* schedules WE subordinate to *'servomotors'*. It also indicates that the term in addition is a species of *'hydraulic motors'*. *Thesaurofacet* is being used in engineering libraries in the United Kingdom for conventional pre-coordinated classified catalogues and the shelf arrangement and filing of documents. It is also being used as the source of terms for alphabetical printed indexes. Although a final citation order is not laid down by the compilers of *Thesaurofacet*, in practical situations the classifier is advised to select a preferred order of concepts for the main entry of the class number prior to permutation for indexing purposes, particularly when the notation is used for document arrangement. The scheme may also be used with post-coordinate systems, whether for term or item entry coordinate indexing or for computerised information retrieval systems. Additionally, *Thesaurofacet* is also a useful tool at

Television Camera Tubes   MCE *Class number*

Information not in classification schedules	UF	Camera tubes (television)
		Emitrons
		Iconoscopes
		Image iconoscopes
		Image orthicons
		Orthicons
		Pick-up tubes (television)
		Vidicons
	RT	Photomultipliers
		Phototubes
		Television cameras
	BT(A)	Television apparatus

(II) CLASSIFICATION SCHEDULES

Information not in Thesaurus	M	*ELECTRONIC ENGINEERING*
Related terms (RT):	MA	**ELECTRON TUBES** (Cont'd) *BT*
(i) *Image converter tubes*	MBT	Electron beam deflection tubes *BT*
(ii) *Image intensifiers*	MBV	Indicator tubes (tuning)
(iii) *Storage tubes*	MBW	Trochotons
(iv) *Television picture tubes*	MC	Cathode ray tubes *BT*
(v) *Television colour picture tubes*	MC2	RT { Image converter tubes
	MC4	{ Image intensifiers
	MC6	Storage tubes
Narrower terms (NT):	**MCE**	**Television camera tubes**
Television colour camera tubes	MCI	Television colour camera tubes *NT*
Broader terms (BT):	MCL	RT { Television picture tubes
(i) *Cathode ray tubes*	MCO	{ Television colour picture tubes
(ii) *Electron beam deflection tubes*		
(iii) *Electron tubes*	MCO	X-ray tubes

CONVENTIONAL THESAURUS ENTRY

		Television Camera Tubes
	UF	Camera tubes (television)
		Emitrons
		Iconoscopes
		Image iconoscopes
		Image orthicons
		Orthicons
		Pick-up tubes (television)
		Vidicons
	RT	Image converter tubes
In the Thesaurofacet underlined		Image intensifiers
items are shown in the classification		Photomultipliers
schedules and not in the thesaurus		Phototubes
		Storage tubes
		Television cameras
		Television colour picture tubes
		Television picture tubes
	NT	Television colour camera tubes
	BT	Cathode ray tubes
		Electron beam deflection tubes
		Electron tubes
		Television apparatus

Figure 2.3 The complementary parts of the *Thesaurofacet*

	SCHEDULES		THESAURUS

SCHEDULES

PD   FLUID POWER DEVICES
(UF common attributes 'fluid powered'.)
    * Fluidic devices   PR2
    * Fluidic valves   PL

PD2   **Hydraulic equipment**
PD3   Hydraulic fluids
PD4   Hydraulic accumulators
PD5   Diaphragm hydraulic accumulators
PD6   Piston hydraulic accumulators
PDB   Boosters (hydraulic)
PDC   Differential pistons
PDH   Hydraulic reservoirs
PDJ   Hydraulic cylinders

PE   **Hydraulic operated devices**
(UF common attribute 'Hydraulic'.) See thesaurus entry for individual devices.
PE2   Hydraulic motors
    * Hydraulic servomotors   WEF
    * Hydraulic starters   WEQ
PE4   Positive displacement hydraulic motors
PE6   Rotary hydraulic motors
PE7   Gear rotary hydraulic motors

WE   SERVOCOMPONENTS
    * Electric control equipment   JV
    * Fluidic devices   PR2
    * Fluid valves   PL
    * Flying controls   RS2
    * Guidance components   WIT
    * Servoamplifiers   K4
    * Switches   K4
    * Temperature control instruments   XQS
    * Transducers   X X

WE2   **Controllers**
    * Electric controllers   JV2
    * Temperature control instruments   XQS
WE3   Mechanical controllers
WE4   Hydraulic controllers
WE5   Pneumatic controllers
Combine with notation from appropriate schedule for variables controlled. for example:
WE2/WC5   Pressure controllers

WEB   **Servomotors**
    * Electric servomotors   JLK
    * Hall effect synchros   KWP
WED   Mechanical servomotors
WEF   Hydraulic servomotors
WEH   Pneumatic servomotors

THESAURUS

Hydraulic Power Transmission Systems   Q5A
   RT   Fluid power engineering
   NT(A)   Hydraulic brakes
       Hydraulic clutches
       Hydraulic couplings
   BT(A)   Hydraulic operated devices

Hydraulic Presses   TAM/PE
Synth
   RT   Hydraulic accumulators
       Hydraulic cylinders
       Hydraulic rams
S BT   Presses
S BT(A)   Hydraulic operated devices

Hydraulic Rams   QIF
   RT   Hydraulic presses
   BT(A)   Hydraulic operated devices

Hydraulic Reservoirs   PDH
   UF   Reservoirs (hydraulic)

Hydraulic Seals   QMV
   BT(A)   Hydraulic operated devices

Hydraulic Servomotors   WEF
   RT   Hydraulic control systems
   BT(A)   Hydraulic motors
       Hydraulic operated devices

Hydraulic Shapers   TRH/PE
Synth
S BT   Shapers
S BT(A)   Hydraulic operated devices

Hydraulic Starters   WEQ
   UF   Starter motors (hydraulic)
   RT   Starting
   BT(A)   Hydraulic motors
       Hydraulic operated devices

Hydraulic Steering Gear   QG9/PE
Synth
S BT(A)   Hydraulic operated devices
S BT   Steering gear

Hydraulic Test Tunnels *use* Water Tunnels

Hydraulic to Fluidic Transducers   PU6

Hydraulic Turbines *use* Water Turbines

Hydraulic Valves   PLE
   NT(A)   Hydraulic gate valves
   BT(A)   Hydraulic operated devices

Hydrazine Compounds   HKC

Hydrazine Nitrate   HKC/HK6
Synth
S BT   Hydrazine compounds
S BT(A)   Nitrates

Hydrazoic Acid   HK8

Hydrazines   HKB

Figure 2.4 *Thesaurofacet*: hydraulic servomotors schedules and thesaurus

the search stage in natural-language and free-text systems as an aid in search strategy planning. It scores over such schemes as UDC or the previous English Electric Company faceted scheme for this purpose because it is better able to display synonymous terms, and to display the interrelationships between pre-combined concepts and subject fields. It is possible that the ideas utilised in the construction of *Thesaurofacet* will become the standard procedures for the construction of controlled languages for information retrieval in the next few years.

MECHANISATION

It has already been mentioned above that a specially constructed scheme such as *Thesaurofacet* is suitable for use in mechanised retrieval, but several attempts have been made to use conventional classification schemes in such systems. UDC is probably the only conventional classification scheme which meets most of the demands mechanisation makes upon a classification. With the current revisions in the schedules, improvements in auxiliaries and notation, and its development as a switching language, UDC is the only international scheme covering all sciences and technologies which has the necessary specificity and the simplest possible structure needed in computerised retrieval. A decade of demonstration and tests has shown that UDC can be mechanised for schedule control, selective dissemination of information (SDI), retrospective retrieval, and schedule listing. The British Steel Corporation has used it for SDI; AWRE has produced an alphabetical subject index to it; and projects sponsored by the American Institute of Physics and the American Meteorological Association have shown that, at least in physics, meteorology and the geological sciences, UDC is a feasible proposition for mechanised indexing of abstract journals and on-line computer retrieval. The projected use of UDC in mechanised retrieval may receive new impetus if the current project for a world Scientific Information System—UNISIST, sponsored jointly by UNESCO and ICSU—develops further. UDC is already being widely recommended as an international switching language for use in this system. At present UDC has schedules in many languages and uses an internationally familiar notation of arabic numerals. Admittedly, much work needs to be done on the use of UDC as an interdisciplinary reference tool for the many thesauri already available, but the basis is there and the administrative body responsible for UDC, the International Federation for Documentation, is already setting up pilot projects.

INFORMATION SERVICES

Many engineers will already be aware of the various computer-based information services which have appeared on the market recently. Prominent amongst these in engineering are INSPEC and COMPENDEX.

INSPEC (International Information Services for the Physics and Engineering Communities), offered by the International Information Service Division of the Institution of Electrical Engineers, developed in the late 1960s as an integrated computer-based system, from which are provided the three *Science Abstracts* journals (*Physics Abstracts, Electrical and Electronics Abstracts, Computer and Control Abstracts*), the corresponding current awareness publications *Current Papers*, SDI services, and, recently, on-line retrospective searching facilities (RETROSPEC).

COMPENDEX (Computerised Engineering Index), first in operation in 1969, is a computerised version of the well-known abstracts journal *Engineering Index.*

Both systems as they have developed have been faced with the usual problems of classification and thesaurus control. INSPEC has developed a thesaurus and unified classification scheme both covering the fields of physics, electrotechnology, computers and control engineering. The thesaurus may be used, as with *Thesaurofacet*, either as a controlled-language system for coordinate indexing and manual or mechanised searching, or as a natural-language aid at the searching stage for suggesting new terms and expanding the concepts in a natural-language search profile for the SDI services and on-line retrospective searching. The thesaurus format follows the pattern set by *TEST* in 1967, presenting an alphabetical display, a hierarchal display and a classified listing of thesaurus terms. Additional relationships offered are TT for top term of hierarchy and CC for unified classification code comprising one or more 6-letter codes. In its controlled-language mode this thesaurus forms the basis of the headings used in the six-monthly subject indexes of the three abstracting journals.

In its uncontrolled or free-language mode the thesaurus may be used as an aid in searching the free-indexing which was chosen by INSPEC after extensive experiment to form the basis of its indexing for machine retrieval and manipulation. This free-indexing comprises terms selected, from the title, abstract or text of the documents or the mind of the indexer, to delineate the subject content of the document. Its use and application in selective dissemination is described in the INSPEC *SDI User Manual.*

The unified classification is used in the abstracting journals and

Current Papers series for arranging the items in a framework which is acceptable to the majority of subscribers. It is also used in the SDI and retrospective search services, either by itself or in association with the thesaurus, for restricting the area of file to be searched and improving precision.

COMPENDEX is also based mainly on so-called free-text searching, which means that by formulating a search profile the user has a free choice regarding what terms he will use. It is interesting that although Engineering Index Incorporated has produced a list of controlled *Subject Headings for Engineering (SHE)*, the terms are added only as auxiliaries to the COMPENDEX entries as separate access points. The operators of COMPENDEX obviously feel that greater efficiency can be obtained by using natural language for profiling, although the user is then faced with the problem of employing in his profile all possible terms and local variations of those terms to produce efficient retrieval (fibre and fiber; acoustooptic, acousto-optic, acousto optic, etc.).

INSPEC and COMPENDEX are probably the two best-known SDI and retrieval systems both catering for engineers in all fields, but many individual companies, academic institutions, and research establishments offer their employees similar services on more restricted scales.

BIBLIOGRAPHY
Aitchison, J. *et al.* (1969). *Thesaurofacet: a Thesaurus and Faceted Classification for Engineering and Related Subjects* (English Electric Company)
Aitchison, T. M., Hall, A. M., Lavalle, K. H. and Tracy, J. M. (1970). *Comparative Evaluation of Index Languages. Part II; Results. Report R70/2* (INSPEC, Institution of Electrical Engineers)
Aitchison, T. M. and Tracy, J. M. (1969). *Comparative Evaluation of Index Languages. Part 1; Design. Report R70/1* (INSPEC, Institution of Electrical Engineers)
Atomic Weapons Research Establishment, Aldermaston (1972). *Alphabetical Subject Index to the UDC*
Binns, J. and Bagley, D. E. (1961). *A Faceted Subject Classification for Engineering,* 3rd edn (English Electric Company)
British Standards Institution (1963). *BS 1000c. Guide to the Universal Decimal Classification (UDC)*
British Steel Corporation—Edinburgh (1965). *Colville's Computerisation Project*
Croghan. A. (1970). *A Thesaurus-Classification for Physical Forms of Non-Book Media* (the author)
Engineers' Joint Council (1967). *Thesaurus of Scientific and Engineering Terms*
Freeman, R. R., Atherton, P. *et al.* (1965–69). *Reports AIP/UDC 1–9* (American Institute of Physics)
Lloyd, G. A. 'The Universal Decimal Classification as an International Switching Language', *Subject Retrieval in the Seventies. Proceedings of an International Symposium, University of Maryland, May 14–15,* 116–123

Rigby, M. (1971). 'The UDC in Mechanised Subject Information Retrieval', ibid., 126–142
Roberts, M. *et al*. (1970). *Thesaurus for the Construction Industry* (Construction Industry Research and Information Association; 1st Draft)
Unesco (1970). *Joint Unesco/ISCU Study on the Feasibility of a World Science Information System (UNISIST)*. Final Report

Information on INSPEC is available from: INSPEC, Institution of Electrical Engineers, Station House, Nightingale Road, Hitchin, Herts. Information on *Engineering Index* and COMPENDEX is available from: Thompson Henry Ltd., The Bear House, London Road, Windlesham, Surrey, GU20 6LQ.

SUPPLEMENTARY LIST OF SELECTED THESAURI IN THE ENGINEERING SCIENCES
American Institute of Chemical Engineers (1961). *Chemical Engineering Thesaurus*
American Petroleum Institute (1971). *Information Retrieval System, Subject Authority List*, 8th edn
American Society for Metals (1968). *Thesaurus of Metallurgical Terms*
Communauté Européenne de l'Energie Atomique (1966–1967). *Euratom Thesaurus* (2 vols.)
Copper Development Association (1971). *Thesaurus of Terms on Copper Technology*, 5th edn.
International Road Research Documentation (1972). *IRRD Thesaurus* (Transport and Road Research Laboratory)
Institute of Textile Technology (1966). *Textile Technology Terms—an Information Retrieval Thesaurus* (Textile Information Center)
National Aeronautics and Space Administration (1967). *NASA Thesaurus*
Netherlands Armed Forces Technical Documentation and Information Centre (1963). *TDCK Circular Thesaurus System*
The City University of New York (1967). *Urbandoc Thesaurus*
Welding Institute (1969). *A Thesaurus of Welding and Allied Processes*

3

Journals, conferences and theses

D. R. King

Three forms of the primary literature are dealt with in this chapter. The relative importance of this material to the engineer, in the electrical field at least, has been shown by Coile (1969), with his analysis of citations to articles in 45 journals of the Institute of Electrical and Electronics Engineers (IEEE).

Table 3.1 TYPES OF INFORMATION SOURCES

Type	Number of citations	%
Periodicals, journals	13 763	61·9
Reports, monographs, memoranda	2 438	10·9
Meetings, conferences, symposia	1 969	8·9
Books	3 504	15·8
Dissertations, theses	295	1·3
Other	266	1·2
	22 235	100·0

Coile further analysed these data to provide details of the useful lifetime of published or presented papers. He found that 50% of the references to meetings were to meetings held within the last 2 years, and 75% to meetings held within the last 4 years. For journal articles he found the useful lifetime to be slightly longer, with 50% of the citations to material just under 3 years old, and 75% just under 7 years old. He also showed an increase in the rate with which this obsolescence occurs in this sample of the periodical literature, in a comparison of the above figures with those obtained in earlier surveys. This is attributable to the increase in number of journals referred to between the surveys, and the corresponding increase in the number of citations.

JOURNALS

In this section the terms 'journal' and 'periodical' are taken to be synonymous, and they are taken to describe publications containing material from a number of sources, which are usually, but not necessarily, produced at regular intervals in a consecutively numbered sequence with no foreseen end. Their main value lies in the fact that the information they contain is usually more up-to-date than that which is available in other publications. Also, the technical quality and originality of the articles they publish is in most cases guaranteed by the system of refereeing used by most journal editorial boards, whereby a submitted article is passed to an expert in the field for judgement before the paper is accepted for publication. This method, even with its faults, such as increasing the time between submission and publication, and possible lack of objectivity on the part of the referee, plays a large part in sustaining a reasonable standard of quality of journal articles.

The journal has been in existence for over 300 years since the publication of the *Journal des Sçavans* and the *Philosophical Transactions of the Royal Society* in 1665, and its growth in numbers since then has been impressive, to perhaps 50 000 current titles. Exact figures are difficult to obtain, but at the end of 1972 the then NLLST (National Lending Library for Science and Technology) was receiving 40 192 current titles with a further 4484 on order (Urquhart, 1973). These figures are for journals covering all subjects, and no recent estimates for the number of engineering journals appear to be available. The number of scientific and engineering journals is still increasing, however, and at an exponential rate, according to Price (1965), with a doubling period of 15 years.

Types of journal

It has been affirmed by Herschmann (1970) that there are three main functions of the primary journal: as a means of recording knowledge, for disseminating information, and for conveying prestige and recognition. Of the various types of journal available today, none fulfils one of these functions to the exclusion of the others, but bias towards one of them is often readily discernible. The various categories of journal commonly encountered are most easily distinguished by their source. Those emanating from learned and professional societies are very often of two kinds: the authoritative fundamental research type, usually, if not wholly, incorporating the Society's *Proceedings* or *Transactions*; and the more informative 'newsy' type of journal. The field of engineering is liberally endowed with both, most of which will be discussed in later subject chapters. However, some examples can usefully be given here. Since 1873, albeit irregularly, the ASCE (American Society of Civil Engineers) has published its *Proceedings*, from, at the most recent count, its 13 divisions. A useful provision in each issue of the *Proceedings*, incidentally, is the inclusion of reference cards to each article contained, giving abstract, keywords and reference. The companion ASCE informative, rather than archival, publication is *Civil Engineering*, containing a typical selection of items, ASCE news, meetings calendar, book reviews, new product information, classified advertisements and topical articles. The IEE (Institution of Electrical Engineers) with its even more venerable *Proceedings* (1871–) also produces a 'newsy' journal entitled *Electronics and Power* (the *Journal of the IEE*) which has a special section for Institution news. The American Society of Mechanical Engineers, the Institute of Electrical and Electronics Engineers and the Institution of Civil Engineers all produce a similar range of publications.

Many commercial organisations produce journals for prestige purposes. Often they are glossy exercises in public relations, combined with slick advertising, but some can be useful for their technical content, and, as all the material they contain emanates from the one source, apart from giving an insight into the research and development being carried out in the organisation, they can give a great deal of information on the current trends within a specialised field. The *Bell System Technical Journal*, for example, is devoted to the scientific and engineering aspects of electrical communication, and the *Brown Boveri Review* and Lawrence, Scott and Electromotors Ltd's *LSE Engineering Bulletin* to power engineering. The last of these, incidentally, is produced for the company's technical staff, but is available to all those interested.

More commercially orientated, but still with some technical content, are the ICI (Imperial Chemical Industries) *Fluon Engineering News* and the CIBA-GEIGY *Technical Notes*, which both provide a forum for their companies' products, be they present successes or new developments.

Commercial journals are produced for profit by commercial publishers and usually attempt to plug the gaps in the primary journal literature which are left by the society publications. They form the growth area in the literature of almost every subject, and their proliferation is a contributory factor towards the problems faced by the engineer trying to combat today's flood of information. Many are interdisciplinary, linking together information from a combination of related topics, or produced to fulfil a demand in a new or rapidly growing field of study, thus relieving the pressure for publication on the more established journals. There does seem to be an unfortunate trend to jump on to the latest profit-making bandwagon with an almost unprofessional rapidity in some cases—for example, the recent marked increase in the number of journals dealing with conservation, pollution and the environment. This is not to condemn the practice of providing a forum for the exchange of information and ideas on important new issues, but some control in this type of situation could only be an improvement. Another drawback with commercial journals is that very few provide any form of cumulative index over and above the annual author and subject indexes.

Perhaps the most widely read commercial journal is *New Scientist*, published by New Science Publications, which, with its popular approach to all matters scientific and technical, may make many an engineer's literature search on a Friday afternoon more bearable. Pergamon is a major contributor of engineering journals, numbering among its publications *Solid-State Electronics*, the *International Journal of Heat and Mass Transfer* and *Materials Research Bulletin*, the latter being somewhat unconventional with its 'rapid handling procedure', whereby a paper may be submitted directly to an associate editor with a request for it to be communicated to the *Bulletin* with neither delay nor further scrutiny, although for those authors who prefer it, conventional refereeing is offered. Other typical commercial journals are the *International Journal of Electronics*, published by Taylor and Francis Ltd, the *Journal of Materials Science* (Chapman and Hall) and the *International Journal for Numerical Methods in Engineering* (Wiley). Two other well-known publishers in this field are Academic Press and Elsevier, and in fact both of these provide good examples, along with Pergamon, of the latest trend of producing 'international' journals, often containing material in two or more languages, usually with at least an English abstract.

Certain hybrids occur where journals are published by commercial publishers on behalf of societies or institutes. Pergamon, for example, publish the *Journal of Terramechanics* for the International Society for Terrain Vehicle Systems, and American Elsevier publish *Combustion and Flame* for the Combustion Institute. The format of this type of journal is very similar to that of the commercial journal, with added notes and meetings information concerning the society involved.

The final group of journals to be dealt with by publisher are those produced by various institutions, such as universities, research establishments and governmental agencies. The Department of Systems Engineering, the University of Lancaster, publishes the *Journal of Systems Engineering*, and the IMechE (Institution of Mechanical Engineers) publishes for UMIST (the University of Manchester Institute of Science and Technology) the *International Journal of Mechanical Engineering Education*. Both of these journals deal with the development of techniques within their stated fields rather than fundamental research, which is, however, dealt with in the National Bureau of Standards' *Journal of Research. Section C, Engineering and Instrumentation*. Details of British governmental research being carried out by its various research establishments is included in such publications as the Building Research Establishment's *BRE News*, published by the Department of the Environment. Finally, the Battelle Memorial Institute publishes various journals, including the *Battelle Research Outlook*, from its Columbus Laboratories in the USA.

Like commercial journals, trade journals are produced for profit, but with a different approach, although the dividing line can be very thin. They usually have a judicious blend of advertising, industrial news, product information, details of forthcoming events, topical articles, statistics, book reviews, etc., are an invaluable source of information to the engineer, and owing to their advertising content are usually inexpensive. Many trade journals, such as *Heating/Piping/Air Conditioning*, published by the Reinhold Publishing Company and *Machinery and Production Engineering* (Machinery Publishing Company), include an index to advertisers, and a postage-free enquiry service for fuller details of products advertised within the journal. Fulfilling a similar role to that of the trade journal are the technical newspapers such as *Building Design* and *Electronics Weekly*, the major differences being the obvious one of form and more frequent publication.

Another type of trade journal, with similar content, is the controlled circulation journal, which is usually distributed free of charge to suitably qualified personnel, e.g. *Electronic Equipment News*, published by Mercury House Business Publications.

Problems with journals

BS2509:1970. Specification for the Presentation of Serial Publications, Including Periodicals gives in great detail the preferable form for the final publication, including recommendations for layout of title leaves, the form the title should take, the numbering and dating of issues, volumes, etc., pagination, the provision of contents tables and indexes, and the presentation of articles and other contributions. If every journal were to be produced to this standard, many of the frustrations met during a literature search would be avoided. At the moment, however, serious problems still exist, mainly associated with the style or format of article and journal titles. With the former the lack of solid technical information in an eyecatching title can lead to the article being read by only a few of those for whom it was intended. Problems with journal titles usually arise from confusion with journals of similar name, and faulty or ambiguous contractions of title. The former can only be solved by co-operation between publishers to avoid similar titles, such as *Engineering* (London), and *Engineering* (Toronto), and the three journals entitled *Elektronik* originating in Kobenhavn, Munich and Stockholm, and greater care when giving and noting journal citations. The problem with title contractions is aggravated by the varying systems used by different publishers, although many do recommend use of abbreviations from the *World List of Scientific Periodicals* (4th edn, 3 vols, Butterworths, 1963–65), which have been revised according to *BS4148:1967. Specification for the Abbreviation of Titles of Periodicals, Part 1, Principles.* There is now a more recent version of this, issued in 1970, and this is to be supplemented by *Part 2: Word Abbreviation List.* An interesting article on serials standardisation has been written by Paul (1970), with particular reference to the *United States Standard for Periodicals: Format and Arrangement* (1967). A variant of periodical title contractions more usually found in storage and retrieval systems is *CODEN.* This provides a unique four-letter code for titles, and now includes an arbitrary fifth letter to expand the coverage possible. The key to this system is the American Society for Testing and Materials' *CODEN for Periodical Titles* (2 vols., 1967), plus supplements from 1968 (to be published by the American Chemical Society, 1975–   ).

Apart from the problems associated with individual journal articles, titles, etc., there are those problems produced by the enormous number of journals now published. The state has long been reached where an individual, assuming that he can trace all items of interest to him, will experience difficulty in locating all these items, and find it impossible to read and assimilate them all. Libraries also find it impossible to buy all the material that would be useful to them (although this is admittedly no new situation), and in many cases could not possibly store it all anyway. One parameter reflecting this situation is the growth in the use of inter-library loan services; requests supplied by the NLL (now the BLLD) have increased from

779 000 in 1968 to 1 411 500 in 1972 (Urquhart, 1973). How far periodicals and journals can continue in their present form, and what, if anything, could effectively, and economically, replace or complement them, has been the subject of much debate, and many journal articles (thus compounding the problem). One series of articles (Reynolds *et al.*, 1970), comprising papers presented before the American Chemical Society, deals with the problems of the primary journal in general, and its future development. Microforms have been on the scene for a number of years, and can provide a space-saving medium for the archival publications held by a library—for example, the back runs of journals. User resistance to this medium is being reduced as familiarity with it increases, but problems still exist. There is a lack of standardisation between the manufacturers of the readers, cassettes not being interchangeable, etc., but, more important, the straight substitution of microfilm or microfiche for hard copy of journals only reduces the space problem for the storage of materials, and does not actively assist the engineer to keep up with the flood of information in his field. The increase in the number of the pre-liminary communications type of journal containing digests of work, followed up in more detail at leisure, has been seen as one partial solution to the individual's problem, and also as an aid to the publisher to reduce his backlog of papers awaiting publication. The AIAA (American Institute of Aeronautics and Astronautics) has experimented with three different approaches to the problems of speed and effectiveness of communication in its journals (Dugger, Bryans and Morris, 1973). Their 'miniprint' experiment involved the use of a type-size intermediate between that of full-size hard copy and microfiche, such that a magnifying glass was required to read it. The SDD (Selective Dissemination of Documents) system was an attempt at individually tailored distribution of, initially, hard copy and, secondly, microcards containing abstract and request technique, of preprints and articles, based on interest profiles. Also, 'synoptics' were utilised. These are extended abstracts of typically two pages length. The original papers are available on request from AIAA if a meetings paper, or otherwise from NTIS (National Technical Information Service) at cost. Of these alternatives, only 'synoptics' have been kept as a regular feature of AIAA publications. 'Miniprints' proved less acceptable than author-prepared copy, the AIAA publications output was too small for the data base for an SDD system, and not enough of the Institute's members were interested in it at practical cost levels. The two systems of extended abstract plus complete article availability from a central repository, and separate article distribution associated with interest profiles periodically matched against a large data base, have both been suggested as improvements

on the present primary journal system. Maxwell (1973) and Staiger (1973) have put forward arguments for the method each favours, but until such systems can be proved economically viable on a large scale, the journal as we know it cannot be challenged.

Indexes

Indexing services, such as *Engineering Index* and *British Technology Index*, endeavouring to cover a wide range of journals within their stated fields, are dealt with in Chapter 11, but those supplied with individual journal runs can also be most useful. The majority of journals nowadays are supplied with some form of annual index to facilitate the search for articles dealing with a certain topic, or written by a specific author. Author indexes are usually reasonably straightforward, but subject indexes can be of variable quality and comprehensiveness, and with some a very real danger exists of missing articles relevant to the search being performed. Another pitfall when using the indexes to a journal is that when the number of volumes, and therefore indexes, exceeds perhaps ten in number, the strain on the user can increase rapidly and cause careless errors. To improve this situation, and also increase the speed of the search, some publishers (especially good in this respect are the professional societies) provide cumulative indexes covering a number of volumes. ASME (American Society of Mechanical Engineers) have produced a 77-year index covering 1880–1956 in two sections, and one for 1957–70. ICE (Institution of Civil Engineers) have four cumulative indexes for their publications from 1917 to 1969. Especially useful with most of the society cumulative indexes is the inclusion of entries, not only for their main *Proceedings* or *Transactions,* but also for abstracts, reviews, letters, etc., and any other special publications or monographs sponsored by the society.

Guides and lists

For the identification of periodicals within a certain subject field, the verification of publication details, or the determination of the location of journals, various bibliographical tools have evolved. Both M. J. Fowler's *Guides to Scientific Periodicals* (Library Association, 1966) and volume 1 of A. J. Walford's *Guide to Reference Material* (3rd edn, Library Association, 1973) give details of useful aids, but the more important sources will be discussed here. The most valuable subject guide to periodicals is *Ulrich's International Periodicals*

Directory, published by Bowker. This is now in its 15th edition, covering the period 1973–74, and contains approximately 55 000 entries for in-print periodicals published throughout the world. New editions are produced in alternate years, with an annual updating supplement. Entries are arranged alphabetically within broad subject divisions, 'Engineering' being split into 'General', 'Chemical', 'Civil', 'Hydraulic' and 'Mechanical', with a separate section for 'Electricity and Electrical Engineering'. Quite full information is given, comprising title, sponsoring organisation, languages used in the text, date of first issue, frequency, subscription rate, editor, and publisher. Any special features of the journal, such as the inclusion of reviews, advertisements, bibliographies, abstracts or statistics, are also noted, as are details of indexes available, and those indexing/abstracting services which regularly include the journal. Also included in the entries for the first time in this edition are the ISSN (International Standard Serial Number), a unique code assigned to a serial for identification purposes, and the Dewey Decimal Classification number for each journal. A title and subject index is included, and also two other useful sections: one listing journal cessations since the previous edition, the other noting additions. A companion volume *Irregular Serials and Annuals: an International Directory* (2nd edn, Bowker, 1971) is similarly arranged.

Other general guides available are *Willing's Press Guide,* annually published by Thomas Skinner Directories; D. Woodworth's *Guide to Current British Journals* (2nd edn, Library Association, 1973), now including the previously separate publication, the *Directory of Publishers of British Journals,* as volume two; and, very useful for subscription information, the *EBSCO Librarians Handbook, a Guide to Periodicals/Serials* (annual).

One selected list of special interest to engineers is *Technical Journals for Industry: United Kingdom.* This is one of a series produced for the International Federation for Documentation (FID No. 415, 1970) by Loughborough School of Librarianship, and is aimed specifically at industrial information requirements including, as it does, only currently published journals. A very useful listing for the UK is *Current Serials Received by BLL, March 1974,* which gives the holdings as of that date of what is now the British Library, Lending Division. Three alphabetical listings, albeit with minimal information, are given: current non-Cyrillic titles, Cyrillic titles, and cover-to-cover translations of Cyrillic titles. With the rapid increase in the numbers of journals, a new edition of this would be most useful.

The major tool for locating scientific or technical periodicals in libraries in the UK is the *World List of Scientific Periodicals* (4th edn, 3 vols, Butterworths, 1963–65). This edition, 'the last that can appear in this format', contains just under 60 000 entries arranged alphabetically. A system of location symbols is used for the holdings of the nearly 300 libraries covered, and details of title changes are given. Updating supplements are now combined with the *British Union-Catalogue of Periodicals* (Butterworths), which first appeared in the late 1950s, and which, with its first supplement, provided coverage of serials in British libraries to 1960. The present title of the merged supplements is the *British Union-Catalogue of Periodicals incorporating World List of Scientific Periodicals. New Periodical Titles*. This first appeared in 1964 and is issued quarterly with two annual cumulations, one including all titles in the quarterly issues, the other having scientific, technical and medical entries only.

CONFERENCES

Conferences and their proceedings can be of great value to the engineer for both the contacts and the information that they produce. The documentation of conferences falls naturally, therefore, into two parts: that which gives notification of meetings, and that which gives details of the publications arising from them.

Details of forthcoming meetings are often given in the news sections of journals, especially in journals produced by the organisation or society sponsoring the meeting. Sometimes these include titles of papers or even abstracts. The Aslib publication *Forthcoming International Scientific and Technical Conferences* lists conferences, including British national meetings, in chronological order, each entry giving date, title, location and an enquiry address for the conference. Subject, location and sponsoring organisation indexes are included. The main issue is published in February each year with a cumulative supplement every 3 months. The World Meetings Information Center produces two aids, *World Meetings, United States and Canada*, and *World Meetings Outside the United States and Canada*, both quarterly. More information is given than in *Forthcoming International Scientific and Technical Conferences*, most useful being the details of publications arising from the meeting, and indexes of date, location, sponsor, keyword and deadline for abstracts or papers. Another source giving arrangements for preprints and proceedings volumes of conferences to be held is the *World Calendar of Forthcoming Meetings: Metallurgical and Related Fields*, published quarterly by The Metals Society. Entries appear in chronological order, and access is by title, location or organising body. *Scientific Meetings*, published quarterly by the SLA (Special Libraries

Association), is another useful tool, with alphabetical listings of sponsoring organisations and meetings titles, a chronological listing, and subject and geographical indexes. Finally, the *International Congress Calendar* is published annually by the Union of International Associations and has monthly supplements *New International Meetings Announced* contained in *International Associations.*

The preprints available at conferences are often overlooked as sources of information, and the programme of a conference, usually available on request in advance of the meeting, can be valuable in tracing those likely to be of most use. A new journal, which should take the labour out of this process, is *Current Programs*, a monthly current awareness service from the World Meetings Information Center. It is organised in subject sections, 9 of the 27 of direct interest to the engineer, each meeting entry giving the titles of papers to be presented, plus the author's name and his address. Subject, author and programme indexes are to be available quarterly and annually.

A paper presented at a conference may eventually appear in the literature in a number of ways: in a proceedings volume, in book form, as a conference serial, as a special issue of a journal or, if no official proceedings are published, as an article in a journal. Whichever of these occurs, the time delay between the conference and final publication is apt to be long, and, according to Mills (1973), although publication of proceedings in engineering is as good as for physical science and better than for life science and social science, within 12 months of the meetings being held, only 62% of proceedings that were to be published had in fact appeared. It would seem that for current awareness, reviews of meetings containing synopses of papers, such as those to be found in *Mechanical Engineering* for ASME (American Society of Mechanical Engineers), could be more highly utilised. No figures appear to be available on the proportion of conference papers in engineering which are eventually published, but Liebesny (1958), on a small sample, showed it to be only 48·5% of presented papers. A problem commonly met with proceedings that are published is the lack of indexes. Hanson and Janes (1960) found this, and Mills (1973) reaffirmed it, showing that only 15% of published proceedings in engineering had a subject index and only 28% had an author index, a reprehensible situation in view of the relevant *BS4446:1969. Presentation of Conference Proceedings*, which recommends that all conference proceedings should have an author and a subject index.

If individual papers from a conference are published as articles in a journal, then the usual bibliographical tools can be expected to locate them, but abstracts and indexes are, on the whole, weak at finding complete proceedings, although the INSPEC abstracting journals do include a conference index. A variety of special aids has developed to deal with published proceedings, the one of most value being the *Index of Conference Proceedings Received* (BLLD, July 1973), formerly the *Index of Conference Proceedings Received by the NLL* (1964–June 1973). Published monthly, it is in two sections, the first being an accession list of conference proceedings with brief details, the second containing a simple keyword subject index. The two are linked by the BLLD conference number. A useful aid in the index to differentiate between the entries when checking a keyword is the inclusion of the date of the conference. This is especially helpful when using the annual cumulative index. A useful cumulation entitled *BLL Conference Index 1964–1973* (British Library, Lending Division, 1974) has enabled a retrospective search to be made in a much shorter time. The InterDoc Corporation produce the *Directory of Published Proceedings, Series SEMT* (1965–   ) in 10 issues per year plus an annual cumulation. Entries are arranged chronologically, and editor, location and combined keyword/title/sponsor indexes are included. The Union of International Associations has produced a number of guides to published proceedings. The *Yearbook of International Congress Proceedings* (2nd ed, 1970) covers meetings held in the years 1962–69. Entries are arranged chronologically, and indexes by organisation, author/editor and subject/keyword are provided. For earlier meetings the first edition covered the years 1960–67, and a *Bibliography of Proceedings of International Meetings* was produced for each of the years from 1957 to 1959. *Proceedings in Print* (1964–   , bi-monthly) is arranged in two sections, the current section for meetings in the preceding 2 years, and the retrospective section. An index is provided and in December the annual cumulative issue is published. The final source of information on complete proceedings worthy of mention is the now somewhat dated, but still useful, *World List of Scientific Periodicals*. This contains a supplement on *Periodic International Congresses*.

For information on individual conference papers the CCM Information Corporation produces the *Current Index to Conference Papers: Science and Technology,* which is an amalgamation of the previous CICP publications for engineering, the life sciences and chemistry. Entries are arranged by conference title under broad subject headings, with details of individual papers and their authors. Subject and author indexes are provided in each issue, and a meeting title and sponsor index in each annual cumulation.

Exhibitions and trade fairs can play a prominent part in the work of the engineer, and information about them can be obtained from special guides, as well as from the news sections in the popular journals. The *Program of Trade Promotions,* a quarterly supplement to *Trade and Industry,* contains four sections, the first listing overseas trade promotions in the next 2 years for which Department of Trade and Industry support is under consideration, the remaining three listing regular trade fairs in Europe, outside Europe and in the United Kingdom. The *Exhibition Bulletin* (monthly), published by the London Bureau for the Centre of Exhibition Information, gives chronological and location listings of forthcoming exhibitions and fairs in Great Britain and overseas. Finally, *World Convention Dates* (monthly), by Hendrickson Publishing Company, provides a calendar of forthcoming events, mainly American, arranged by location and date.

THESES

Theses and dissertations can be valuable sources of information and contain detailed bibliographies, and although parts of the most worthwhile theses subsequently appear as journal articles, it is often more instructive to refer to the original theses, as long as it is remembered that they are written to satisfy the demands of external examiners and have not been subjected to any refereeing system. Several British universities produce their own annual lists of theses—for example, *University of London Theses and Dissertations Accepted for Higher Degrees.* Typically, this is arranged in subject sections, includes entries for both master's degrees and doctorates, and has an author index.

The best source of information on British theses is the Aslib *Index to Theses Accepted for Higher Degrees in the Universities of Great Britain and Ireland.* This is an annual listing arranged under subject headings, with author and subject indexes, but is, unfortunately, slow in appearing. For current awareness the *BLLD Announcement Bulletin: a Guide to British Reports, Translations, and Theses* (1973–), previously *NLL Announcement Bulletin,* can be useful. Entries are arranged under broad subject headings, but no index is available.

American master's theses are catered for by two services. *Masters Theses in Pure and Applied Science* (annual) is published by the Thermophysical Properties Research Center, Purdue University. *Masters Abstracts* (1962–), published quarterly by University Microfilms (UM), includes the theses from, at the last count, 28 institutions only. A cumulative author and subject index is contained in the December issue of each volume. For American doctoral theses there is *Dissertation Abstracts International: B the Sciences and Engineering* (1969–), published monthly by University Microfilms. This was previously known as *Dissertation Abstracts* (1952–69) and *University Microfilms* (1938–51). As a reflection of its most recent title change, it now includes some dissertations from non-American universities. Two very good characteristics of this source are the full ordering information for each entry and the length of the abstracts, which are arranged in 43 subject groups. There are author and keyword indexes in each issue, annually cumulated. For some time now, using the University Microfilms DATRIX (Direct Access To Reference Information, a Xerox service) system, it has been possible to search for title-keywords back to 1938, through 275 000 dissertations, and to obtain copies of required items on microfiche or as hard copy. A further development is the *Comprehensive Dissertations Index 1861–1972* (UM, 1973). In 30 bound volumes, or on microfiche, this indexes circa 400 000 dissertations, by author/title and subject/keyword. The volumes are subject divided and an author index is also available, as are annual supplements. *American Doctoral Dissertations* (University Microfilms, annual) is a complete listing of all doctoral dissertations accepted by American and Canadian universities.

Various sources exist for tracing European theses; for example, for French works, the *Catalogue des Theses et Ecrits Academiques,* and supplement D of *Bibliographie de la France;* for German theses, the *Jahresverzeichnes der Deutschen Hochschulschriften* and *Deutsche National Bibliographie*. Obtaining foreign theses may prove difficult, but a publication which may avoid an overseas loan is *Foreign Theses in British Libraries*, by R. S. Johnson (SCONUL, 1971). This lists holdings of foreign theses arranged by university accepting the thesis, and gives any locations of the theses in Britain.

REFERENCES

Coile, R. C. (1969). 'Information Sources for Electrical and Electronics Engineers', *IEEE Transactions on Engineering Writing and Speech*, **EWS 12** (3), 71–78
Dugger, G. L., Bryans, R. F. and Morris, W. T. (1973), 'AIAA Experiments and Results on SDD, Synoptics, Miniprints and Related Topics', *IEEE Transactions on Professional Communication*, **PC16** (3), 100–106
Hanson, C. W. and Janes, M. (1960). 'Lack of Indexes in Reports of Conferences', *Journal of Documentation*, **16** (2), 65–70
Herschmann, A. (1970). 'The Primary Journal: Past, Present and Future', *Journal of Chemical Documentation*, **10** (1), 37–42
Liebesny, F. (1958). 'Lost Information: Unpublished Conference Papers', *Proceedings of the International Conference on Scientific Information*, 475–479

Maxwell, R. (1973). 'Survival Values in Technical Journals', *IEEE Transactions on Professional Communication*, **PC16** (3), 64–65

Mills, P. R. (1973). 'Characteristics of Published Conference Proceedings', *Journal of Documentation*, **29** (1), 36–50

Paul, H. (1970). 'Serials: Chaos and Standardisation', *Library Resources and Technical Services*, **14** (1), 19–30

Price, D. J. de S. (1965). *Little Science Big Science* (Columbia University Press)

Reynolds, H. L. *et al.* (1970). 'Introduction to Symposium on Primary Journals', *Journal of Chemical Documentation*, **10** (1), 26–49

Staiger, D. L. (1973). 'Separate Article Distribution as an Alternate to Journal Publication', *IEEE Transactions on Professional Communication*, **PC16** (3), 107–112

Urquhart, D. J. (1973). 'NLL Progress 1972', *NLL Review*, **2** (6), 173–176

4

Translations

D. N. Wood

THE LANGUAGE PROBLEM

Depending on what is classed as worthwhile and on how engineering is defined, there are something like 90 000–100 000 worthwhile engineering articles published each year. Of these approximately half are in languages other than English.

The proportion of English-language material varies considerably from one branch of engineering to another, as does the number of papers in other languages. To illustrate this an analysis, by language, of the literature indexed in the engineering sections of the 1972 volume of *Bulletin Signalétique* is presented in *Table 4.1*. A further

Table 4.1

Subject	E	R	G	F	S	I	J	O
Civil engineering	31·8	3·9	15·8	34·2	0·7	3·0	0·4	10·0
Transport engineering	35·8	9·9	28·3	13·5	0·1	4·0	2·5	4·8
Aeronautical and space engineering	65·0	11·4	10·3	12·4	–	0·7	–	0·2
Biomedical engineering	56·0	3·1	16·4	19·0	0·5	1·5	2·2	1·3
Mechanical engineering	29·2	21·6	23·3	19·6	–	2·2	1·4	2·5
Chemical engineering	50·4	16·8	12·4	11·3	0·4	0·5	3·2	5·0
Electrical and electronic engineering	40·5	26·8	16·4	6·0	0·6	0·9	0·3	8·5

E, English; R, Russian; G, German; F, French; S, Spanish; I, Italian; J, Japanese; O, Other.

insight into the linguistic spread of engineering literature is given by the data in *Table 4.2*. This is an analysis, by language, of the journals indexed, either completely or in part, in the American

Table 4.2

Language	Number of titles	%
English	1 627	72·8
German	195	8·7
French	111	5·0
Russian	81	3·6
Japanese	74	3·3
Scandinavian (mainly Swedish)	38	1·7
Italian	26	1·2
Spanish	20	0·9
Polish	20	0·9
Dutch	7	0·3
Portuguese ⎫ Rumanian ⎪ Hungarian ⎬ Serbo-Croat ⎪ Czech ⎭	20	0·9
Multilingual	15	0·7
TOTAL	2 234	100·0

abstracting journal *Engineering Index.* In the case of both *Bulletin Signalétique* and *Engineering Index,* the figures are only indicative of the spread. Neither tool is comprehensive in its coverage of the literature and, as is the case with all abstracting services, there is a tendency to stress homeland references at the expense of those in other languages.

Despite the volume of non-English-language material published and abstracted, the use made of it by engineers appears to be very low. In a study at Sheffield University in 1971 (Hutchins, Pargeter and Saunders, 1971) it was shown that out of 4292 items borrowed from the library by engineers 97·5% were in English and a further 1·5% were journals written partly or mainly in English. In this respect, however, engineers are not notably worse than pure scientists (in whose case the comparable figure was 93%) or indeed workers in the social sciences (98·6%). Neither do the Sheffield engineers appear to be exceptional when compared with workers elsewhere in the country, for the over-all use of the science and technology literature at the BLLD shows a similar heavy bias towards English-language publications.

A further insight into the use made of non-English-language material is afforded by citation studies. In the Sheffield investigation, for instance, it was found that in the publications and theses of students and staff in the engineering faculty 94% of the citations were to English-language publications. The figure compares with 89% for pure scientists. A similar situation was revealed in an investigation carried out by Syracuse University. This included a citation analysis of the *Proceedings of Institute of Radio Engineers* (1960–62) and *Industrial and Engineering Chemistry* (1962–64) and showed that nearly 94% and 90%, respectively, of the citations were to publications in English (Syracuse University Research Unit, 1965). In all cases the most frequently borrowed and cited non-English items were German, with French and Russian the two other commonly used languages.

As to the most heavily used foreign-language titles, there is little published information. However, a survey carried out in 1969 at the NLL showed that during a 3-month period the following titles were most in demand:

	No. of times borrowed
Industrie Chimique Belge	81
Chemie-Ingenieur-Technik	80
Plaste und Kautschuk	44
Neue Hütte	40
Archiv für Eisenhüttenwessen	35
Chimica e l'Industria	35
Chimie et Industrie	35
Revue Générale de Thermique	34
Metall	33
Metallurgia Italiana	31
Chemische Technik – VEB Deutsche Verlag für Grundstoffindustrie	30
Praktische Metallographia	30
Materialprüfung	28
Fizika Metallov i Metallovedenie	26
Stahl und Eisen	26
Zeitschrift für Metallkunde	25
Adhäsion	24
Metallurgia	24
Starke	23
Zeitschrift für die Gesamte Textilindustrie	22
Jernkontorets Annaler	21
Izvestiya AN SSSR Metally	20

The low use of foreign-language material can be attributed to one or more of the following factors: the general inability of engineers to cope with the languages concerned; the fact that some workers regard anything not published in English as worthless; and a lack of knowledge concerning the availability of information in foreign languages.

Concerning language ability, the results of a study conducted from the NLL in 1965 (Wood, 1967) show that UK engineers are inadequately equipped to cope with anything but French and even in that case only 46·6% can manage without referring frequently to a dictionary. Comparable figures for German and Russian are 16·5% and 1·5%. These results, which place engineers at the bottom of the language ability league table, were supported by the Sheffield research referred to above. This showed that, making only occasional use of a dictionary, 39% could cope with French and 11% with German.

Notwithstanding the fact that some engineers would claim that anything of importance will be published sooner or later in English, both the NLL and Sheffield studies showed that most workers appreciate the value of foreign information and are seriously hampered by the language barrier. The NLL survey revealed that 74% of almost 600 engineers had recently encountered a paper in a foreign language which they would like to have read, but could not because of the language problem. Of these, 62% had encountered the problem within the month prior to the survey. In 46% of the cases the language causing difficulty was German, and the figures for Russian, Japanese and French were 32%, 7% and 5%, respectively. In the Sheffield study engineers were asked to indicate for each of a number of countries whether they considered them important sources of information. In descending order of indicated importance the countries were USA, 86%; UK,73%; USSR, 39%; Germany, 37%; Japan, 28%; France, 26%; and Scandinavia and the Netherlands, 16%. Furthermore, 43% of the Sheffield survey population felt they had probably missed foreign work of importance.

The language problem, however, is not the only reason why much relevant information in foreign languages goes unused. Several studies have shown that the most frequently used method of locating references is the scanning of bibliographies and reference lists at the end of other articles. It follows that if very few foreign-language papers are cited, very few will be read. However, the increasing amount of guidance which students are receiving on the structure and use of scientific and technical literature should bring about more sophisticated searching techniques and this in its turn can be expected to generate a greater awareness of foreign-language publications. This development of course will increase the number of language problems.

TRANSLATIONS INDEXES

Since it appears desirable that engineers and scientists should have access to information in languages other than English, the question arises as to what can be done to overcome the barrier that exists. Long-term solutions might be: the teaching of more languages at school level; the development and promotion of special language courses for engineers, e.g. the intensive course in Japanese which has recently been developed at Sheffield University; the provision within research establishments and universities of services which could enable workers to pick the bones out of an article without going to the trouble and expense of obtaining full length trans- lations, i.e. a service similar to that being operated by the British Library, Science Reference Library; and the use of computers to produce 'instant' translations. Although developments are taking going to the trouble and expense of obtaining full-length trans- lations that, for the time being, most engineers will have to turn for help.

Up to the immediate post-war years translations were produced on very much an *ad hoc* basis with organisations and individuals producing and collecting only those translations of interest to them. In the 1950s, however, it became obvious that in order to avoid duplication of cost and effort, national and international co-opera- tive schemes were necessary. During the last 15–20 years a number of developments have taken place which have led to the creation of national and international centres which collect translations and publicise their availability.

In the UK one of the first such co-operative schemes to be launched was the Aslib Commonwealth Index of Unpublished Translations. It was started in 1951 and is a location index of translations (com- pleted and in progress) from all languages into English. It is in three sections: journal articles; patents and standards; and books, theses and reports. Currently containing over 450 000 entries, it is being added to at a rate exceeding 10 000 per year. Requests for locations will be accepted from any individual or organisation in the UK, and approximately 20 000 requests are handled annually. Copies of the Commonwealth Index are also available in Australia, Canada and New Zealand.

In the late 1950s the DSIR Lending Library Unit, later to become the National Lending Library (NLL) and now known as the British Library, Lending Division (BLLD), began collecting translations. Initially the emphasis was placed on translations from the difficult languages into English. In 1967, however, the coverage was expanded to include translations from all languages into English. The BLLD

collection currently includes nearly 250 000 individual article translations, complete files of over 250 journals which are translated
regularly into English (see p. 55) and several thousand translated
books. The books and journals are obtained through commercial
channels, but individual articles are usually acquired direct from the
originators, e.g. government departments and agencies, industrial
firms, universities and research associations. They include about
3000 per year from British organisations and about 15 000 per year
from American sources, including the National Translations Center
(NTC), National Technical Information Service (NTIS) and the Joint
Publications Research Service (JPRS) (see below). All translations
received at the BLLD are indexed in card catalogues. Article translations are indexed under original journal title, and books under
author. These indexes enable the BLLD to provide an information
service on the availability of translations, and the Library is currently
dealing with around 40 000 enquiries annually. In the absence of
information at Boston Spa the BLLD checks with Aslib before
notifying the requestor.

In addition to collecting and indexing translations the BLLD
publicises its holdings through announcement bulletins. From
1959 to 1970 details of mainly UK translations were published
in *NLL Translations Bulletin* and its predecessor, *LLU Translations Bulletin*. Since 1970 solely British translations have been
announced together with other semi-published material in *BLLD
Announcement Bulletin* and its predecessor, *NLL Announcement
Bulletin*.

The only other major multi-disciplinary collection of translations
in the UK is maintained at the British Library, Science Reference
Library (formerly NRLSI). Over 30 000 individual article translations are held and the library subscribes to most of the commercially
available journal translations. The latter are listed in their 'Holdings
of Translated Journals'. A card index is available and where translations are held the original is also marked to refer readers to the
location of the translation.

In the United States, as in the UK, most of the developments
concerned with the production, collection and bibliographical control
of translations have taken place since 1945 and in particular in the
last 15 years. In 1947 the Office of Technical Services was formed to
coordinate the work of collecting, translating and evaluating information on scientific and technical developments in Germany and
Japan. By 1953 the increasing importance of Russian scientific and
technical information had been recognised and with financial support
from the National Science Foundation a Scientific Translations
Center was established at the Library of Congress, and at the same

time the Special Libraries Association established a collection of translations at the John Crerar Library in Chicago. Between 1953 and 1956 the Library of Congress issued *Bibliography of Translations from Russian Scientific and Technical Literature.*

By 1959 the importance of collecting translations and making them available was widely recognised and international co-operation schemes began to be developed between North America and Europe. At the centre of these was OTS, and a publication known as *Technical Translations* became the main vehicle for announcing translations. It was published twice monthly between 1959 and 1967 by the Clearinghouse for Federal Scientific and Technical Information (CFSTI) and contained translations into English produced by the US Government and its Agencies and any others of which it had been notified by organisations in both North America and Europe (including the NLL). Translations were listed in subject groups and cumulative indexes were produced semi-annually.

In 1967 responsibility for publicising translations was divided. CFSTI, renamed the National Technical Information Service (NTIS), was given the responsibility for announcing US Government-sponsored translations only. This was done in *US Government Research and Development Reports,* a twice-monthly publication started in 1958 which also listed other semi-published material originating in the government sector. Translations can be distinguished by the 'report' series to which they have been assigned, e.g. TT, PB-T, JPRS. Cumulative indexes were produced quarterly and annually. In 1972 the title of the publication changed to *Government Reports Announcements (GRA).*

Some of the translations listed in *GRA* are also listed in a separate publication. These are the translations issued by the Joint Publications Research Service (JPRS). They include translations on all aspects of science and technology, social science and the humanities in the Soviet Union, Eastern Europe and the developing countries. About 30 000 articles are issued each year in 3000 documents and details are given in *Transdex – Bibliography and Index to the United States Joint Publications Research Service (JPRS) Translations.* This commenced in July 1970, having previously been issued in four separate parts concerned, respectively, with East Europe, Soviet Union, China and Asia and International Developments. Between 1970 and 1972 the scientific and technical material was also announced separately in a publication called *Sci/Tech Quarterly Index.* Some of the most useful documents made available through the JPRS programme are abstracting publications. These are issued in two series, *USSR and East European Scientific Abstracts* and *Communist Chinese Scientific Abstracts.* Each is issued regularly in subject sections, and those most likely to be of use to engineers are: 'Cybernetics Computers and Automation Technology'; 'Electronics and Electrical Engineering'; and 'Engineering and Equipment'.

The function of collecting and publicising translations in the non-government sector in the US was given to the SLA Translations Center (renamed the National Translations Center) at the John Crerar Library, Chicago. This centre was also charged, with the help of an NSF grant, with the production of an index to all translations. The published version of this index is entitled *Translations Register —Index (TR-I)*. The 'Register Section' of this twice-monthly guide contains details of NTC's new accessions, while the 'Index Section' includes details of the same translations together with information on translations listed by NTIS in *Government Reports Announcements* and translations collected by BLLD in the UK. In the 'Register Section' translations are arranged in subject order, but in the 'Index Section' they are listed in alphabetical order of the journals, etc., from which the translations are made. Under each journal title the entries are in chronological order, and a code indicates the source from which the translation can be obtained (see *Figure 4.1*). *Translations Register — Index* cumulates quarterly and annually.

Organised on exactly similar lines to *TR-I* is *Consolidated Index of Translations into English*. This provides in a single volume details of all those translations listed previously in *SLA List of Translations* (1954), *Bibliography of Translations from Russian Scientific and Technical Literature, Translations Monthly* (SLA, 1955–58) and *Technical Translations*. Over 160 000 translations are listed.

In view of the similar terms of reference of the BLLD and NTC, i.e. to collect translations from all languages into English, a close working relationship has been established between the two organisations. An exchange of translations and index cards takes place and the bibliographical tools produced by NTC function effectively as guides to the BLLD's holdings.

In Europe national translation centres have been established in many countries, their functions being to collect translations mainly into their own languages. In France the Centre National de la Recherche Scientifique is responsible for this activity and publishes a monthly index entitled *Bulletin des Traductions — CNRS*. Translations into German are collected in the DDR by the Deutsche Akademie der Wissenschaften — Berlin, and in West Germany by the Technische Informationsbibliothek der Technischen Universitat in Hanover. The latter publishes *Nachweise von Uebersetzungen*. As well as supporting national collections, some European countries, and indeed some countries outside Europe, support the operation of the European Translations Centre (ETC), which is housed in the

```
ELE              * * * * *  JOURNAL CITATION INDEX  * * * * *                    ENE

ELEKTROTECHNISCHE ZEITSCHRIFT            ELTEKNIK
  1937 N16 P417-420   3774.5 (1725) <NLL*>    1969 N2 P6-9        9022.81 (2789) <NLL*>
  1941 V62 N1 P3-16   72-20900 <*>
                                         EMBERIZA
ELEKTROTECHNISCHE ZEITSCHRIFT, AUSGABE A   1970 V1 N2 P49-60   C-10771 <NRC>
  1962 V83 N1 P8-12     9022.09 (4652) <VLL*>
  1962 V83 N23 P776-780 72-21118 <*>     ENERGETICHESKOE STROITELSTVO
  1966 V87 N1 P9-16     9022.81 (2575) <NLL*>  1970 N1 P46-48   72-12251-09C <*>
  1966 V87 N20 P713-717 3774.5 (1836) <NLL*>
  1967 V88 P645-647     9022.09 (4807) <VLL*> ENERGETIK
  1967 V88 P647-652     9022.09 (4845) <NLL*>   1958 V6 N5 P3-6      9022.09 (4584) <NLL*>
  1967 V88 N1 P21-26    9022.09 (4687) <NLL*>   1966 V14 N11 P20-22  9022.03 (4160) <NLL*>
  1967 V88 N3 P74-79    9022.09 (4598) <NLL*>   1966 V14 N12 P19-23  9022.09 (4645) <NLL*>
  1967 V88 N9 P213-217  9022.09 (4724) <NLL*>   1967 V15 N3 P38      9022.03 (4198) <NLL*>
  1967 V88A P534-537    INC-67/363 <NLL*>       1968 V16 N2 P11-12   9022.09 (5005) <NLL*>
  1968 V89 P183-189     9022.09 (4952) <NLL*>   1968 V16 N4 P25-26   9022.03 (4287) <NLL*>
  1968 V89 P197-203     9022.09 (4964) <NLL*>   1968 V16 N4 P36-37   9022.09 (4292) <NLL*>
  1968 V89 P213-218     9022.09 (4967) <NLL*>   1968 V16 N5 P5-8     9022.03 (4290) <NLL*>
  1968 V89 N6 P121-126  3774.5 (1789) <NLL*>    1968 V16 N10 P27-30  9022.31 (4319) <NLL*>
  1968 V89 N6 P131-135  3774.5 (1791) <NLL*>    1969 V18 N12 P33-34  9022.31 (4474) <NLL*>
  1968 V89 N7 P146-450  9022.09 (4891) <NLL*>
  1968 V89 N11 P263-265 3774.5 (1818) <NLL*>  ENERGETIKA, PRAGUE
  1968 V89 N15 P356-361 3774.5 (1838) <NLL*>    1959 V9 N6 P289-292  72-20415 <*>
  1969 V90 P69-92       9022.09 (5216) <NLL*>   1966 V16 N10 P513-515 9022.09 (4727) <NLL*>
  1969 V90 N25 P656-662 9022.823 (582) <NLL*>   1968 V18 N4 P151-154 9022.09 (4988) <NLL*>
  1969 V90 N25 P675-678 9022.823 (572) <NLL*>
  1970 V91 N3 P149-152  3810.97 (467) <NLL*>  ENERGETYKA, WARSZAWA
  1970 V91 N3 P170-172  9022.09 (5493) <NLL*>   1964 V18 N12 P365-370 9022.09 (4593) <NLL*>
  1971 V92 N3 P125-130  3810.97 (463) <NLL*>
                                            ENERGIA ES ATOMTECHNIKA
ELEKTROTECHNISCHE ZEITSCHRIFT, AUSGABE B.      1965 V18 N12 P557-561 9022.09 (4625) <NLL*>
  1959 V11 N10 P348-404 72-20631 <*>           1970 V23 P215-222   5828.4F (20929) <*NLL> 0
  1963 V15 N10 P286-289 72-20146 <*>
                                            ENERGIA ELETTRICA
  1965 V17 N13 P347-395 3774.5 (1890) <NLL*>   1966 V43 P412-423   5828.4F (7392) <NLL*>
  1965 V17 N15 P445-499 9022.81 (2592) <NLL*>  1966 V43 N4 P215-220 9022.09 (5076) <NLL*>
  1966 V18 P989-991     72-10292-13K <*>       1967 V44 N2 P73-78   9022.09 (4870) <NLL*>

  1968 V20 N6 P133-137  6196.3 (632) <NLL*>  ENERGIA NUCLEARE
  1968 V20 N16 P441-447 9022.09 (5031) <NLL*>   1971 V18 N1 P45-54  WH 318F <NLL*>
  1968 V20 N23 P669-673 9022.81 (2586) <NLL*>
                                            ENERGIAGAZDALKODAS
ELEKTROTEKHNIKA                                1969 V10 N3 P136-138 5828.4F (20858) <NLL*>
  1966 N12 P19-22      3774.5 (1847) <NLL*>
  1966 V37 N6 P17-19   9022.09 (4810) <NLL*> ENERGIE, MUENCHEN
  1967 N1 P49-51       3774.5 (1643) <NLL*>    1958 V10 N5 P175-185 3774.5 (1941) <NLL*>
  1967 N8 P20-24       8313.4 (5558) <NLL*>    1965 V17 N10 P401-405 9022.09 (4386) <NLL*> P
  1967 N8 P25-27       3774.5 (1728) <NLL*>    1966 V18 N5 P177-180 9022.09 (4604) <NLL*>
  1968 N1 P23-25       3774.5 (1845) <NLL*>    1966 V18 N7 P311-316 9022.09 (4591) <NLL*>
  1968 N6 P39-41       INC 746-310 <NLL*>      1967 V19 N5 P163-165 9022.09 (4432) <NLL*>
  1968 N10 P1-6        3774.5 (1886) <NLL*>    1967 V19 N12 P393-398 9022.09 (4831) <NLL*>
  1968 N10 P6-8        3774.5 (1885) <NLL*>    1968 V20 N1 P11-15   72-10585-21R <*> 0
  1968 N10 P16-18      3774.5 (1887) <NLL*>    1969 V21 P202-206    9022.09 (5662) <NLL*>
  1968 V39 N1 P50-51   INC-68/276 <NLL*>       1969 V21 N6 P18-27   9022.09 (5213) <NLL*>
  1968 V39 N4 P9-12    9022.09 (5007) <NLL*>
  1968 V39 N6 P46-51   9022.310 (4334) <NLL*> ENERGIE NUCLEAIRE
  1968 V39 N10 P1-6    9022.31 (4323) <NLL*>    1965 V7 N1 P3-14    RISLEY 2217F <NLL*>
  1968 V39 N10 P6-8    9022.310 (4324) <NLL*>
                                               1968 V10 P9-15      9022.09 (5156) <NLL*>
ELEKTROTEKNISK TIDSSKRIFT                      1968 V10 P181-186   WINDSCALE 474 <NLL*>
  1966 N12 P1-5        72-12368-20A <*>         1969 V11 P345-355   72-10103-11F <*>
                                               1969 V11 N2 P84-92  LTR/TR-352 <=>
ELEKTROWAERME INTERNATIONAL (TITLE VARIES)     1970 V12 N6 P514-520 PISLEY 2157F <NLL*>
  1968 V26 N6 P220-227 6196.3 (1152) <NLL*>
  1969 V27 N3 P102-109 6196.3 (1121) <NLL*>  ENERGIE UND TECHNIK
  1970 V28 N4 P213-219 6196.3 (1099) <NLL*>    1967 V19 N9 P312-318 9022.09 (4847) <NLL*>
  1970 V28 N6 P332-342 6196.3 (1112) <NLL*>    1968 V20 P284-289    9022.09 (5174) <NLL*>

ELETTROTECNICA                              ENERGIETECHNIK
  1947 V34 P74-79      6196.3 (1136) <NLL*>    1965 V15 P439-445    9022.09 (5522) <NLL*>
  1966 V53 N7 P482-485 9022.09 (4535) <VLL*>   1966 V16 N12 P543-530 9022.09 (4654) <NLL*>
  1967 V54 N4 P296-302 3774.5 (1726) <NLL*>    1967 V17 P114-117    9022.09 (4660) <NLL*>
  1967 V54 N8 P571-580 3774.5 (1747) <NLL*>    1967 V17 N5 P107-200 9022.09 (4784) <NLL*>
  1967 V54 N11 P832-836 9022.09 (5012) <VLL*>  1967 V17 N5 P205-206 9022.09 (4780) <NLL*>
  1968 V55 N3 P237-243 9022.09 (5008) <VLL*>   1968 V18 P31-33      9022.09 (5015) <NLL*>
  1969 V56 N9 P534-544 6196.3 (1075) <NLL*>    1968 V18 P247-250    9022.09 (5149) <NLL*>
                                               1968 V18 N2 P72-75   9022.09 (5018) <NLL*>
ELEVAGE ET INSEMINATION, PARIS
  1968 P4-7            5828.4F (7699) <NLL*> ENERGIEWIRTSCHAFTLICHE TAGESFRAGEN, FRANKFURT

                              PAGE  26
```

Figure 4.1   A sample page from *Translations Register – Index*

University of Delft in The Netherlands. Established in 1960, ETC is charged with collecting, indexing and publicising translations from the difficult languages (Russian, Japanese, Chinese, Romanian, Bulgaran, Czech, Polish, Serbo-Croat, etc.) into any of the languages of Western Europe. Since 1967 ETC has published a *World Index of Scientific Translations.* Originally published quarterly, it is now issued monthly, with every third issue being a quarterly cumulation. The last issue per volume is a comprehensive cumulation for the whole year. Currently the publication comprises two sections. The first gives details of the translations acquired by ETC and these are listed in broad subject groups, e.g. 'Electronics and Electrical Engineering' and 'Mechanical Industrial, Civil and Marine Engineering'. The second is a citation index with articles, patents, standards, etc., arranged according to the title of the original publication. A 5-year cumulation (1967–71) of the index section is available.

All the above centres will deal with enquiries and can supply information about, or copies of, translations recorded in their indexes. English-speaking engineers are advised to make use of BLLD, or NTC, depending on whether they reside in the UK or the USA. For English-language material ETC can offer no better service, but may be useful to contact where translations into German and/or French would be acceptable.

Each of the organisations so far mentioned concerns itself with translations in all subject fields, although science and technology tends to dominate. A few organisations exist, however, which prepare, collect and announce translations in particular subject areas. One of the largest of these is Euratom. In co-operation with the UK Atomic Energy Authority and the US Atomic Energy Commission, Euratom maintains a collection of, and indexes, translations in the nuclear energy field. These are listed monthly in *Transatom Bulletin,* which cumulates annually, and has large cumulations covering the periods 1960–65 and 1966–70. Other special collections and indexes are maintained by such organisations as the Machine Tool Industry Research Association, the Rubber and Plastics Research Association and the British Iron and Steel Institute. However, most of the material in these collections is available sooner or later through the national centres.

A point to note about most of the translations available from or through the channels referred to above is that they are in semi-published form. Many are available as Xerox or off-set copies made from typed manuscripts of variable quality. The vast majority, however, are distributed in the form of 35 mm microfilm or, in the case of the more recent ones, as microfiche. Enlarged copies of these filmed versions can usually be provided by the libraries or translations centres which hold them, but the cost can be quite considerable in the case of a long translation.

COVER-TO-COVER AND SELECTIVE TRANSLATIONS

Brief mention has already been made of the existence of cover-to-cover translations. These are regularly produced translations of complete issues of certain foreign-language journals. This activity started in the 1950s in an attempt to make scientists and engineers more aware of scientific and technological developments in countries such as the Soviet Union. Particularly in the early years most cover-to-cover translations were produced by, or with the financial support of, the US and UK Governments. In the US the National Science Foundation was the main funding body. In the UK the NLL, now part of the British Library, has controlled most of the government-supported activity. There are currently over 250 scientific and technical periodicals being translated. Most are translations from the Russian, but a few Polish, Japanese and Chinese and at least one German journal are also translated from cover to cover. They constitute a valuable means of keeping abreast of overseas developments. Unfortunately, because of inevitable delays in acquiring the original journal, translating and editing it and publishing the translation, most cover-to-covers appear several months behind the original. In some cases the time-lag is measured in years.

In recent years there has been a growing realisation that much of the material appearing in cover-to-cover translations is of little value to readers and that in certain fields at least a more discriminating programme of translations was called for. As a result there has been a gradual increase in the number of selective translation journals. Some of these, such as *Russian Engineering Journal,* are based on one original journal with, say, only 75% of the contents being translated. Others, such as *Heat Transfer — Soviet Research,* consist of selected articles from various Russian journals.

In the case of Japanese material much selective translation work is undertaken in Japan itself; presumably with the object of promoting Japanese science and technology in western countries. Unfortunately, a large proportion of the translations are not subject to adequate bibliographical control, and many articles are published in English without any reference to the fact that they are translations of Japanese articles from other journals.

A list of those cover-to-cover and selective translation journals most likely to be of interest to engineers is given in the appendices to this chapter. Further details of these and other translated journals can be obtained from *A Guide to Scientific and Technical Journals in Translation,* compiled by C. J. Himmelsbach and G. E. Brociner (2nd edn, Special Libraries Association, 1972) and *Translations Journals 1973,* compiled by M. M. A. Knul (ETC, 1973).

BOOK TRANSLATIONS

Many thousands of scientific and technical books are published each year throughout the world. Most of these are, like English-language books, in the nature of review publications and many have got English-language equivalents, a fact which obviates the need for translations. However, in some countries — for example, the Soviet Union — a good deal of primary information is published in book form and translations of these works are frequently produced.

The BLLD attempts to acquire all worthwhile books translated into English and currently possesses around 15 000.

Most book translations are produced commercially and are announced in the national bibliographies of the countries in which they are published, but there are also a number of special guides. Several of those mentioned previously index translated monographs to some extent, but the most comprehensive list is *Index Translationum*. This has been published annually since 1949 by UNESCO, although it is somewhat belated in its appearance. The latest issue covers translations published in 1972. A cumulated index to those translations published in Australia, Canada, USA, Ireland, New Zealand, South Africa and the UK and listed in *Index Translationum* between 1948 and 1968 has recently been published by G. K. Hall under the title *Cumulative Index to English Translations*. The work is in two volumes and lists some 44 000 translations in a single alphabetical sequence according to the name of the author.

GETTING A TRANSLATION PREPARED

Despite the existence of many thousands of article and book translations, the bulk of the world's literature remains in its original language and the chances that a translation of any particular item will be available are only about 1 in 6. If the information is thought to be important, the next step should be to decide whether a translation is worth commissioning.

If one only has the title to go on, it may be worth obtaining a copy of the original paper. Many journals publish English abstracts of the papers they carry and some, particularly Japanese journals, contain translated summaries and/or conclusions, and carry English captions to the figures and tables. In most cases it will then become obvious whether the paper is really going to be useful. It may be that the tables, figures and conclusions are all that are required anyway.

Should the article itself not offer any help to the English-speaking reader, it might be worthwhile searching for an abstract in an English language bibliography such as *Engineering Index* or *Chemical Abstracts*. Some of the more informative abstracts published by abstracting services contain enough information to obviate the need once again for a full translation.

Another step which might usefully be taken, before commissioning a translation, is to sit down with a translator and obtain a brief verbal explanation of what the paper is about. This kind of service is currently provided by the British Library, Science Reference Library, and could possibly be instituted with profit by more special and university libraries.

Should it be decided that a full translation is absolutely necessary, it may be possible to obtain this through a service provided by the firm or university. In the UK, for instance, 40% of the industrial organisations employing more than 100 people provide a translation service for German and French and 30% do so for Russian (Wood, 1967). More and more universities are following the lead of industry and creating information departments through which such services can be obtained.

For engineers in the UK help in obtaining translations of any but French, German, Italian and Spanish material is obtainable from the British Library, Lending Division at Boston Spa. The library will arrange for the translation of an article or, on occasion, a book, subject to the following conditions: the item must not already be available in translation or scheduled for translation; the translation must be required for research or private study; and the requestor must agree to edit the translation from a technical point of view. A charge is made for the service and requestors are notified of the cost before the translations are undertaken. Translations produced under this scheme are announced in the *BLLD Announcement Bulletin* and made available to the public as Xerox copies of a typed manuscript. Books are issued by the Library as paperbacks or occasionally published through commercial channels.

Where this scheme is not appropriate, engineers may consult various registers to obtain the names and addresses of translators. Some public libraries maintain lists of translators resident in their localities and university libraries are beginning to compile similar lists — particularly of staff members who are willing to undertake translation work.

Aslib maintains a Register of Specialist Translators containing details of over 200 translators with both subject and linguistic qualifications. However, this service is available to members only.

The Institute of Linguists publishes an *Index of Members of the Translators Guild* which provides information on the Institute's Fellows and Members.

There are a number of published directories which may be of help in locating suitable translators. They include: *Directory of Technical and Scientific Translators and Services,* by P. Millard (Crosby Lockwood, 1968); *International Directory of Translators and Interpreters,* by B. Pond (Pond Press, 1967); *Translators and Translations; Services and Sources in Science and Technology,* by F. E. Kaiser (2nd edn, Special Libraries Association, 1965); and *Who's Who in Translating and Interpreting,* by A. Flegon (Flegon Press, 1967).

The cost of having translations prepared varies considerably according to the language and technical difficulty. For 'technical' material individual translators in the UK are likely to charge currently (1975) between £6.00 and £8.00 per thousand words from French and German, £8.00–£15.00 for Russian and the equivalent of between £20.00 and £25.00 per thousand words for Japanese. If translations are requested through a translation bureau, prices might well be higher.

REFERENCES

Hutchins, W. J., Pargeter, L. J. and Saunders, W. L. (1971). *The Language Barrier: A Study in Depth of the Place of Foreign Language Materials in the Research Activity of an Academic Community* (Postgraduate School of Librarianship and Information Science, University of Sheffield)
Syracuse University Research Unit (1965). *A Study of the Frequency with which Russian, French and German Scientific Articles are Cited in Selected American Journals* (University of Syracuse)
Wood, D. N. (1967). 'The Foreign Language Problem Facing Scientists in the United Kingdom — Report of a Recent Survey', *Journal of Documentation,* **23** (2), 117–130

Appendix 1 COVER-TO-COVER TRANSLATIONS OF POTENTIAL INTEREST TO ENGINEERS

Title of original journal	Years available in translation	Title of translation	Availability
Avtomaticheskaya Svarka	1959–	*Automatic Welding*	Welding Institute
Avtomatika i Telemekhanika	1956–	*Automation and Remote Control*	Plenum (CB)
Avtomatika i Vychislitelnaya Tekhnika	1967–1969	*Automatic Control*	Faraday Press
	1969–1971	*Automatic Control*	Allerton Press
	1972–	*Automatic Control and Computer Sciences*	Allerton Press
Avtomatika Tele- mekhanika i Svyaz	1967–	*Railway Automation Telemechanics Translation Series Section F*	Railroad Engineering Index Institute

Title of original journal	Years available in translation	Title of translation	Availability
Avtomatika	1968–	*Soviet Automatic Control*	Scripta Publishing Corporation (SPC)
Avtometriya	1966–1969	*Autometry*	Plenum (CB)
	1970	*Automatic Monitoring and Measuring*	Plenum (CB)
	1971–	*Automatic Monitoring and Measuring*	Scientific Information Consultants (SIC)
Avtomobilnaya Promyshlennost	1957–	*Automobile Industry*	NTIS
Defektoskopiya	1965–1967	*Defectoscopy*	Plenum (CB)
	1968	*Soviet Journal of Nondestructive Testing*	Plenum (CB)
Denki Tsushin Gakkai Zasshi	1963–1967	*Electronics and Communications in Japan*	IEEE/Scripta Electronica
Denshi Tsushin Gakkai Ronbunshi	1968–1969	*Electronics and Communications in Japan*	IEEE and SPC
	1970–	(1) *Electronics and Communications in Japan*	
		(2) *Systems, Computers, Control*	SPC
Doklady AN SSSR	1956–1962	*Proceedings of the Academy of Sciences of the USSR Section Chemical Technology*	Plenum (CB)
	1963–	*Doklady Chemical Technology*	Plenum (CB)
Elektricheskie Stantsii	1955–1959	*Power Station*	SIC
	1972–	*Soviet Power Engineering*	Ralph McElroy Co.
Elektrosvyaz i Radiotekhnika	1954–1959	*Telecommunications*	SIC
	1957–1962	*Telecommunications*	IEEE and SPC
	1963	*Telecommunications and Radio Engineering*	IEEE and SPC
Elektrotekhnika	1965–1966	*Soviet Electrical Engineering*	Faraday Press
	1966–		RCS Journals
Energomashinostroenie	1957–1959	*Power Generation Engineering*	SIC
Fiziko Teknicheskie Problemy Razrabotki Poleznykh Iskopaemykh	1965–	*Soviet Mining Science*	Plenum

Title of original journal	Years available in translation	Title of translation	Availability
Gidrotekhnicheskoe Stroitelstvo	1967–	Hydrotechnical Construction	American Society of Civil Engineers
Inzhenero-Fizicheskii Zhurnal	1965–	Journal of Engineering Physics	Plenum (CB)
Izmeritelnaya Tekhnika	1958–	Measurement Techniques	Plenum (CB)
Izvestiya AN SSSR Mekhanika i Mashinostroenie	1959 only	Bulletin of the Academy of Sciences of the USSR Mechanics and Mechanical Engineering	SIC
Izvestiya AN SSSR Mekhanika	1963–1965	Engineering Journal	NTIS
Tverdogo Tela (1969–) formerly	1965	Soviet Engineering Journal	Faraday Press
Inzhenernyi Zhurnal	1966–1967	Mechanics of Solids	Faraday Press
Mekhanika Tverdogo Tela (1966–1968) formerly Inzhenernyi Zhurnal (1961–1965)	1967–	Mechanics of Solids	Allerton Press
Izvestiya AN SSSR Mekhanika Zhidkosti i Gaza	1966–1969	Soviet Fluid Mechanics	Plenum (CB)
	1970–	Fluid Mechanics	Plenum (CB)
Izvestiya AN SSSR Metallurgiya i Gornoe Delo	1963–1964	Russian Metallurgy and Mining	SIC
Izvestiya AN SSSR, Metally	1965–	Russian Metallurgy	SIC
Izvestiya AN SSSR Otdel, Tekhnicheskikh Nauk Metallurgiya i Topiivo	1960,1962	Russian Metallurgy and Fuels	SIC
Izvestiya AN SSSR Otdel, Tekhnicheskikh Nauk Energetika i Avtomatika	1959	Bulletin of the Academy of Sciences of the USSR Dept. of Power Engineering and Automation	SIC
Izvestiya AN SSSR Tekhnicheskaya Kibernetika	1963–1967	Technial Cybernetics	NTIS
	1963–	Engineering Cybernetics	IEEE and SPC
Izvestiya Vysshikh Uchebnykh Zavedenii Aviatsionnaya Tekhnika	1958–1959	Aviation Engineering	SIC
	1963	Aviation Engineering	NTIS
	1966–1967	Soviet Aeronautics	Faraday Press
	1967–	Soviet Aeronautics	Allerton Press

Title of original journal	Years available in translation	Title of translation	Availability
Izvestiya Vysshikh Uchebnykh Zavedenii Elektromekhanika	1958–1959	Electrical Engineering	SIC
Izvestiya Vysshikh Uchebnykh Zavedenii Priborostroenie	1962–	Title varies	NTIS
Izvestiya Vysshikh Uchebnykh Zavedenii Radiotekhnika	1965–1966	Soviet Radio Engineering	NTIS
	1967–	Radioelectronics and Communications Systems	NTIS
Izvestiya Vysshikh Uchebnykh Zavedenii Radioelecktroika	1958–1959	Radio Engineering Series	SIC
	1959–1963	News of Higher Educational Institutions Radio Engineering	NTIS
Izvestiya Vysshikh Uchebnykh Zavedenii Technologiya Tekstilnoi Promyshlennosti	1960–1972	Technology of the Textile Industry USSR	Textile Institute
Izvestiya Vysshikh Uchebnykh Zavedenii Tsvetnaya Metallurgiya	1958–1959	Nonferrous Metallurgy	SIC
Khimicheskaya Promyshlennost	1969–	Soviet Chemical Industry	Ralph McElroy Co.
Khimicheskoe i Neftyanoe Mashinostroenie	1965–	Chemical and Petroleum Engineering	Plenum (CB)
Khimicheskoe Mashinostroenie	1964–	Chemical Engineering	Plenum (CB)
Kibernetika	1965–	Cybernetics	Plenum (CB)
	1965–	Cybernetics	NTIS
Koks i Khimiya	1958–1959	Coke and Chemistry	SIC
	1959–	Coke and Chemistry	British Carbonization Research Association
Liteinoe Proizvodstvo	1969–	Russian Castings Production	British Cast Iron Research Association
Meditsinskaya Tekhnika	1967–	Biomedical Engineering	Plenum (CB)
Metallovedenie i Termicheskaya Obrabotka Metallov	1954–1959	Metallurgy and Metal Treatment	Brutcher
	1959–	Metal Science and Heat Treatment	Plenum (CB)
Metallurg	1958–1959	Metallurgist	SIC
	1957–	Metallurgist	Plenum (CB)

Title of original journal	Years available in translation	Title of translation	Availability
Optiko-Mekhanicheskaya Promyshlennost	1966–	Soviet Journal of Optical Technology	American Institute of Physics
Osnovaniya Fundamenty i Mekhanika Gruntov	1964–	Soil Mechanics and Foundation Engineering	Plenum (CB)
Priborostroenie	1959–1966	Instrument Construction	BLLD
Pribory i Systemy Upravleniya	1967–1970	Soviet Journal of Instrumentation and Control	Transcripta Journals
	1971–	Soviet Instrumentation and Control Journal	Transcripta Journals
Pribory i Tekhnika Eksperimenta	1958–	Instruments and Experimental Techniques	Plenum (CB)
Prikladnaya Mekhanika	1966–1967	Soviet Applied Mechanics	Faraday Press
	1967–	Soviet Applied Mechanics	Plenum (CB)
Problemy Prochnosti	1969–	Strength of Materials	Plenum (CB)
Put i Putevoe Khozyaistvo	1967–	Railway Research and Engineering News—Section E	Railway Engineering Index Institute
Radiotekhnika i Elektronika (see also	1957–1961	Radio Engineering and Electronics	IEEE and SPC
Elektrosvyaz i Radiotekhnika)	1961–	Radio Engineering and Electronic Physics	
Stal	1957–1959	Steel	SIC
	1959–1970	Stal in English	Metals Society
Stanki i Instrument	1959–	Machines and Tooling	Production Engineering Research Association (PERA)
Svarochnoe Proizvodstvo	1959–	Welding Production	Welding Institute
Teoreticheskie Osnovy Khimicheskoi Technologii	1967–	Theoretical Foundations of Chemical Engineering	Plenum (CB)
Teploenergetika	1954–1959	Thermal Power	SIC
	1964–	Thermal Engineering	Pergamon
Tsvetnye Metally	1960–	The Soviet Journal of Non-ferrous Metals	Primary Sources
	1960–	Nonferrous Metals	SIC
Vestnik Mashinostroeniya	1957–1959	Engineering Journal	SIC
	1959–	Russian Engineering Journal	PERA

Title of original journal	Years available in translation	Title of translation	Availability
Vsesoyuznyi Nauchno-Issledovatelskii Inst.Zheleznodorozknogo Transporta Vestnik	1965–	*Railway Research and Engineering News – Section D*	Railway Engineering Index Institute

Appendix 2 JOURNALS CONSISTING OF TRANSLATIONS OF ARTICLES SELECTED FROM VARIOUS FOREIGN JOURNALS

Title of journal	Originating date	Availability
Bulletin of the Japan Society of Mechanical Engineers (selected articles from *Journal of the Japan Society of Mechanical Engineers, Transactions of the Japan Society of Mechanical Engineers*)	1958	Nihon Kikaku Kyokai
Denshi Gakkai Zasshi (*Electronics and Communications in Japan*) (selected articles from Sections A and B of *Electronics and Communications in Japan*	1970	Institute of Electrical and Electronic Engineers/Scripta Publishing Corporation (SPC)
East European Metals Review (selected articles from East European journals)	1969	Metal Information Services
Electrical Engineering in Japan (selected articles from *Denki Gakkai Zasshi*	1963	(IEEE/SPC)
Electric Technology USSR (selected articles from *Electrichestvo*	1957	Plenum
Heat Transfer – Japanese Research (selected articles from Japanese journals)	1972	SPC
Heat Transfer – Soviet Research (selected articles from Russian journals)	1969	SPC
International Chemical Engineering (selected articles from Eastern European and Asiatic sources)	1961	International Chemical Engineering
Journal of Cybernetics. Transactions of the American Society for Cybernetics (selected articles from Russian and Japanese journals)	1971	SPC
Journal of the Textile Machinery Society of Japan (selected articles from *Sen'i Kikai Gakkai Ronbunshu* and *Sen'i Kogaku*)	1955	Textile Machinery Society of Japan
Review of the Electrical Communication Laboratory (selected articles from *Denki Tsushin Kenkyusho Jitsuyoka Hokoku* and other journals)	1953	Electrical Communication Laboratory

Title of journal	Originating date	Availability
Selected papers from the Journal of the Society of Naval Architects of Japan (selected articles from *Nihon Zosan Gakkai Ronbunshu*)	1957	Society of Naval Architects of Japan
Steel in the USSR (selected articles from *Stal* and *Izvestiya Vysshikh Uchebnykh Zavedenii Chernaya Metallurgiya*)	1971	Metals Society
Tetsu to Hagane Overseas (*Journal of the Iron and Steel Institute of Japan*, English version)	1961–1965	Iron and Steel Institute of Japan
Transactions of the Japan Society of Civil Engineers (translations or abstracts of all articles in *Doboku Gakkai Ronbun Hokokushu*)	1969	Japan Society of Civil Engineers
Transactions of the National Research Institute for Metals (selected articles from various Japanese journals)	1959	National Research Institute for Metals
Transactions, Iron and Steel Institute of Japan (continued from *Tetsu to Hagane Overseas*) (translations from *Tetsu to Hagane*, other Japanese journals, and also original articles)	1966	Iron and Steel Institute of Japan
Translations on Soviet Transportation (selected articles on transportation in the Soviet Union)		NTIS
Translations on USSR Electric Power (selected articles from Russian journals)		NTIS
Translations on USSR Industrial Affairs (this includes material formerly published in the JPRS series *USSR Industrial Development, Automotive Industry, Electronics and Precision Equipment and Metalworking Equipment*)		NTIS
Translations on USSR Resources (this includes material formerly published in *USSR Industrial Development Electric Power, Fuels and Related Equipment, Translations on USSR Labor* and *USSR Industrial Development, Metallurgy and Minerals*)	1969	NTIS

5

Reports*

V. J. Benning

WHY REPORTS ARE OF VALUE TO ENGINEERS

The technical report is an important means used by engineers to communicate the results of their research and development activities. Articles in the scientific and technical press were probably the subject of earlier technical reports. The converse, however, is not always so, and some important research and development remains only in report form. Such reports are our main concern in this chapter.

Initially, reports are available from the originating organisation or its agents. A high charge may be made for reports if the organisation is attempting to retrieve some of the cost of the work described in them. Reports deposited in national report collections are announced in the report announcement bulletins and the availability of the reports is indicated in the citations. Some reports, which are little more than extracts from laboratory note-books, are communicated as memoranda to management or to other engineers having an interest in the subject. The distribution of these reports can be strictly controlled and an indication can be given in a report of any restrictions it is desired to place on its use.

Table 5.1 summarises the availability of the various main types of report.

An advantage of a technical report is that the information it contains can be made available with the minimum of delay. Reports, even those which are intentionally bulky to include all the detail

---

*A companion volume, *Use of Reports Literature*, edited by C. P. Auger, covers reports literature more fully than is possible in a single chapter.

Table 5.1

Type of report	Availability
Private communication	Sent only to selected individuals
Reports with a restricted distribution	Not generally available in the interests of national defence or owing to commercial limitations
Reports containing valuable 'know-how'	Available at a price, sometimes considerably above that of the cost of reproduction
'Unlimited' reports	Sometimes freely available from the originator, frequently sent on initial distribution or as the result of an exchange agreement. Held by the national centres, and single copies available at the price of reproduction

essential to a thorough understanding of the work described, can be produced quickly and cheaply in small quantities.

Many factors influence the age of the reports of current interest to an engineer. A major factor is the age of the reports which are cited in the sources he is using to become aware of the existence of reports. Reports more than 2 years old are consulted when patent action is contemplated; and when an engineer is working in a branch of his subject with which he is not too familiar, this may generate a demand for some old reports. Review articles and bibliographies frequently include references to reports covering many years.

To get some quantitative measure of the age of reports requested from the Department of Industry's Technology Reports Centre (TRC), a month's requests were examined (*Figure 5.1*). Not all the requests handled by TRC during June 1973 are included in the histogram. Four main categories of document are omitted—translations; conference papers; reports which required clearance before general release, their dates, therefore, bearing no relation to the date when they were first available to the public; and reports available from TRC in large numbers. Included in this latter category are hard-copy reports produced by the Ministry of Defence and the Department of Industry. These are sold by TRC at a price less than the cost charged for reproduction of a report when only one original copy is available.

It is interesting to note that very nearly half the requests received were for reports less than 9 months old. Less than 6% of the requests were for reports more than 18 months old. This demonstrates the value of the technical report to engineers requiring recent information, as publication inevitably introduces a delay for editing and refereeing. The result could also indicate that TRC's pattern of report supply is conditioned by the announcement services. The availability of more accessible retrospective search facilities in the future could give the engineer more information on older material available for problem solving.

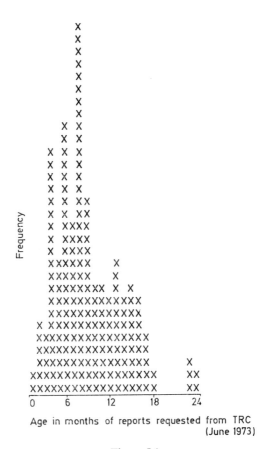

Figure 5.1

A variety of house styles must be expected in reports, and many reports contain extensive appendices, tables and figures. To encourage some degree of standardisation in report writing *BS 4811:1972 Presentation of Research and Development Reports*, was produced.

The sheer volume of reports available to the engineer should persuade him that they are worth his consideration when he is confronted with a problem, but more of this later.

HOW THE ENGINEER LEARNS ABOUT TECHNICAL REPORTS OF INTEREST

There are many reasons why reports arrive unrequested on the engineer's desk. These include his membership of a research and development committee, his organisation's membership of an industrial research association, or the fact that someone who knows his interests is keeping him informed.

The Department of Industry's Technology Reports Centre (TRC) is the only information centre in the UK dedicated to technical report literature. Through TRC the engineer has access to the formal report announcement bulletins.

Four formal methods of becoming aware of the existence of reports are:

The Technology Reports Centre's *R & D Abstracts*, a journal of abstracts of science and technology reports (semi-monthly)

The British Library, Lending Division's *BLLD Announcement Bulletin* (monthly)

The National Technical Information Service's *Government Reports Announcements (GRA)* (semi-monthly)

National Aeronautics and Space Administration's *Scientific and Technical Aerospace Reports (STAR)* (semi-monthly)

These are the main sources of information concerning openly available reports of interest to engineers. Some report citations appear in the major abstract journals, e.g. *Chemical Abstracts* and *Engineering Index*. In the field of atomic and nuclear physics the Atomic Energy Commission's *Nuclear Science Abstracts (NSA)* includes report series. The *Monthly Catalog of United States Government Publications*, issued by the Superintendent of Documents, US Government Printing Office, Washington, DC 20402, contains US reports listed under the US Government Department for whom they were printed. No abstracts are given, but many of these government reports appear abstracted in *GRA* or *ERIC*'s (Educational Resources Information Center) *Research in Education*. The British reports appear abstracted in *GRA* or *ERIC*'s (Educational Re- are abstracted in *R & D Abstracts*. Data bases containing the contents of *R & D Abstracts*, *Government Reports Announcements*

(GRA), *Scientific and Technical Aerospace Reports (STAR)*, and *Nuclear Science Abstracts (NSA)* are updated periodically. The data bases are used to provide current awareness publications, selective dissemination of information (SDI) and retrospective searching by TRC.

Less formal methods of becoming aware of the existence of reports usually follow the awareness that a research project is in progress. The results of government-sponsored research forms a large part of the report literature. Although the terms of government contracts in the UK ask only that final reports should be deposited at an information centre, interim period reports can be made available with the agreement of the contractor. In the United States the Smithsonian Institution provides *Notice of Research Project (NRP)*. Information on over 100 000 projects is added each year. Some organisations disseminate lists of the reports they produce, e.g. the National Engineering Laboratory, the United Kingdom Atomic Energy Authority and the British Steel Corporation. The engineer can write to an organisation producing reports in his own field of interest to ascertain how he can be placed on their mailing list. In the case of industrial research associations, some form of membership may be necessary. Atomic energy report series are described in Roland Smith's *Guide to UKAEA Documents* (5th edn, HMSO, 1973). Reports of interest to engineers are produced by the Reactor Group and the Research Group. The establishments in the Reactor Group are situated at Risley, Winfrith, Dounreay, Windscale and Springfields. The two establishments in the Research Group are the Atomic Energy Research Establishment (AERE) Harwell and Culham Laboratory. The Atomic Weapons Research Establishment (AWRE) was transferred to the Ministry of Defence (Spring 1973), but many AWRE reports, library bibliographies and translations were made available to the public. Two Science Research Council establishments produce reports of interest to nuclear engineers: Daresbury Laboratory, Warrington, Lancs.; and Rutherford Laboratory, Didcot, Berks. Another Science Research Council establishment is the Appleton Laboratory, Ditton Park, Slough, Bucks. (formerly the Radio and Space Research Station).

The engineer should not only be aware of the vast store of reports available to him in his own subject field or that in which he has to work, but should also take the necessary steps to ensure that his own reports are added to this store.

As an indication of the date of the reports requested from TRC has been given, it may be of interest to indicate the date of the reports included in two issues (one month) of *R & D Abstracts*. Once again considerable judgement was exercised in deciding what constituted

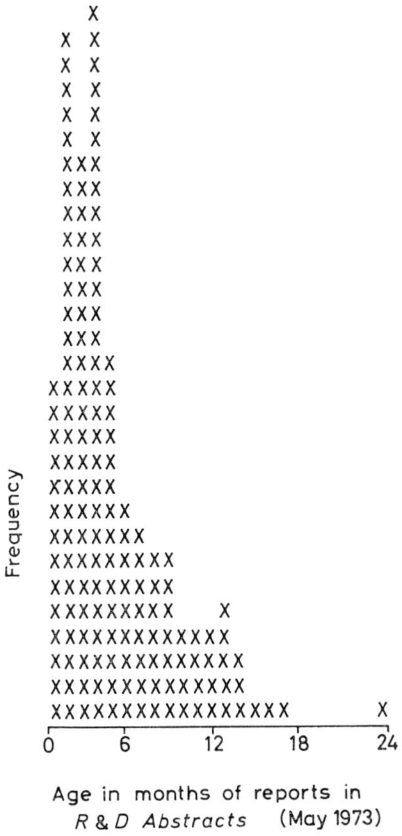

```
                    X
                  X X
                  X X
                  X X
                  X X
                  X X
                  X X X
                  X X X
                  X X X
                  X X X
                  X X X
                  X X X
                  X X X
                  X X X X
                  X X X X X
                  X X X X X
                  X X X X X
                  X X X X X
                  X X X X X
                  X X X X X X
                  X X X X X X X
                  X X X X X X X X X
                  X X X X X X X X X
                  X X X X X X X X X      X
                  X X X X X X X X X X X X
                  X X X X X X X X X X X X X
                  X X X X X X X X X X X X X
                  X X X X X X X X X X X X X X X      X
                  ┌──────┬──────┬──────┬──────
                  0      6      12     18     24
```

Frequency

Age in months of reports in
R & D Abstracts   (May 1973)

Figure 5.2

a report in order to produce the histogram shown in *Figure 5.2*.
Nearly 80% of the reports were less than 9 months old, and less than
5% were more than 18 months old.

THE SOLUTION OF SPECIAL PROBLEMS
THE ENGINEER ENCOUNTERS,
ASSISTED BY TECHNICAL REPORTS

In a report it is possible to include detail which is essential information to the engineer wishing to avoid a repetition of the work.
Examples are wind tunnel results, computer programs and exact
procedures in all branches of engineering. Wind tunnels are expensive

to run, and if a repetition of tests can be avoided, the savings are great. During the last few years, the number of reports including tabulated computer programs has increased. Sometimes a magnetic tape or a pack of punched cards is provided with a report. Care must be taken to ensure that the configuration and conditions in the report are those required. All reported activities should be examined critically, bearing in mind that reports have not been submitted to the scrutiny required for publication. Sometimes two reports will be found of tests made apparently under similar conditions but giving different results. If such conflicting results are found after a thorough search of the literature, or if the results are not in keeping with the engineer's own experience, then it is necessary to repeat the work.

In some organisations it is mandatory that the reports should receive certain signatures before they are issued. These signatures give the report some degree of authority. In fields where the advance of technology is rapid (for example, methods of welding) reports more than a few years old can be misleading, and it is advisable in such cases to approach the organisation producing the reports to obtain the latest 'state of the art'.

Frequently, when faced with a problem, one feels that someone must have had just this problem. How did they solve it? Here a thorough search of the report literature can provide the answer. A series of period reports or progress reports may be discovered describing how someone else tackled the problem. If the work had been government-sponsored, then a contract number may assist in the retrieval of the reports.

R & D Abstracts and *Government Reports Announcements* are available as subject sub-sets of the bulletins. These are a convenient method of receiving parts of the bulletins for current awareness and also for retrospective searching under broad subject headings. Four of the Technology Reports Centre's *R & D Report Announcements* (issued semi-monthly) are:

Selected Report Announcements—Aerospace Technology edition
Selected Report Announcements—Electronics and
                                    Communications edition
Selected Report Announcements—Materials, Design and
                                    Manufacturing Methods edition
    Selected Report Announcements—Physics edition

The magnetic tape from which *Government Reports Index (GRI)* and *Government Reports Announcements (GRA)* are prepared is also used to prepare the *Weekly Government Abstracts (WGA)*.

Using these sub-sets of the bulletins, it is possible to make a comprehensive search, in a specific field, from some 60 000 research reports each year. Rather more easily, and not expensively, full subject searches in many fields of technology can be obtained through TRC's *RECON* information retrieval service. This has data files going back, for example, to 1962 for *STAR* and 1970 for *GRA*.

In addition to the report announcement bulletins, many services are available to assist the engineer with special problems in his own field. Many organisations, realising that some of their reports are of interest only to a few engineers, distribute summaries of their reports. These summaries are of particular use to management. The Construction Industry Research and Information Association (CIRIA) distribute *Report Profiles*. For example, CIRIA *Report Profile 43* is on *Curing Concrete* and also includes a list of other CIRIA reports on the use of concrete in construction. CIRIA reports are available on sale to CIRIA members and also on sale to non-members at a higher price. Review reports, containing a bibliography, can be particularly useful if the engineer has to work in a subject field not very familiar to him. Such review reports contain references to other reports, which is uncommon in journal articles and technical books. The Defense Documentation Center, Alexandria, Va., USA, produce bibliographies from their large reports data base which can give the engineer an extensive survey of the reports available on certain subjects. For example:

AD-752400   DDC-TAS-72-64
Packaged circuits, Report Bibliography
June 1954–May 1972

AD-752700   DDC-TAS-72-66
Pyrotechnics, Report Bibliography
November 1956–February 1972

HOW THE ENGINEER CAN OBTAIN REPORTS

The easiest way to request a report from one of the large national depositories of reports is to use a report accession number. These accession numbers are given to a series of holdings. At the Technology Reports Centre (TRC) the reports are given a T accession number. The Transport and Road Research Laboratory *Report LR-561* on *The Calculation of the Distribution of Temperature in Bridges* has been given the TRC accession number T73-06350. At TRC there are large holdings of reports with AD and N accession numbers which are given to two report series in the USA. If the AD or the N number given to a report in the USA is known, then when that

report is received in TRC, the AD or N accession number is used
and the report is not given a T accession number, e.g. Cranfield
Institute of Technology *Report E & C-3* on *A Preliminary Investigation into a Machine Tool Spindle Drive System* has the accession
number N73-15505. There is inevitably a degree of overlap between
the series and where more than one accession number for a report
is cited, they should all be submitted in a request for that report.
This is important when requesting microfiche, as a report is rarely
held in fiche form under more than one accession number.

Table 5.2 NUMBER OF ACCESSIONS TO FOUR REPORT SERIES
DURING 1970, 1971, 1972 AND 1973

Accessions	1970	1971	1972	1973
R & D Abstracts	6 325	6 510	9 030	8 924
AD-700000 Series (*GRA*)	17 744	18 252	18 656	18 081
PB Numbers (*GRA*)	7 990	8 844	8 656	7 920
STAR N Series	33 200	28 799	23 984	23 931

The accession numbers are useful for filing reports in microfiche
form, but there is a case for filing hard-copy reports under well-
recognised report numbers, e.g. in a series such as that for Cambridge
University Engineering Department (CUED).

The British Library, Lending Division (BLLD) file the British
reports they acquire under the report number series. Each series is
then given a unique shelf number, e.g. the BLLD shelf location for
CUED-C-MAT-TR-5 on *The Theory For Fatigue Failure under
Multiaxial Stress-Strain Conditions* is 9106.17. One disadvantage of
this system is evident when the report does not have a well-recognised
report number.

Some report series are treated as irregular publications and appear,
as such, on the open library shelves. This is the case at the British
Library, Science Reference Library. In order to use these facilities, it
is frequently necessary to identify a report series code. Attempts
have been made to list the many report series produced by thousands
of originators all over the world. One notable example is that edited
by H. F. Redman and L. E. Godfrey entitled *Dictionary of Report
Series Codes* (2nd edn, Special Libraries Association, 1973).

The first line of bibliographical information of an item in the
TRC *R & D Abstracts* bulletin has: the TRC accession number, the
corporate author's reference number, and the monitoring agency
reference number. Any one of these three numbers identifies the
report. Report number indexes are available which cross-refer these
numbers. A report on *An Ionospheric Electric-Field Instrument*

appears in *R & D Abstracts* with the first line of bibliographic information as follows:

T73-02745   GCA-TR-72-3-A   AFCRL-72-0621

Reference to 'Redman and Godfrey' gives the information that a report number of the form GCA-TR-(Year)-(No.)-(LTR) belongs to Geophysics Corporation of America, Bedford, Mass., USA. There are other report series commencing with the letters GCA, e.g. Graham, Cowley and Associates Inc. and Geophysics Corporation of America, Boston, but only the series for GCA Bedford, Mass., has the particular form, viz. TR-73-3-A. AFCRL can only mean Air Force Cambridge Research Laboratories, Mass., as indicated in the list of report series codes. Time spent studying the way in which entries are made in the various report accession bulletins, and familiarising himself with the associated indexes, will help the engineer to use them.

Requests for reports should be made on a recognised request form. These may be obtained from the library or centre the engineer intends to use. It is important that any instructions given on these request forms be observed.

WHO THE ENGINEER CAN APPROACH IN CASE OF PROBLEMS

If the engineer is served by a technical information service, then he will be assisted in obtaining and learning about reports of interest, but the problems are the same as if he were unaided.

When the problem is who to approach if the existence and the bibliographical details of a report, but no accession number or location reference, are known, the answer is to contact the Technology Reports Centre in the UK. The staff at TRC are prepared to search the report abstract bulletins and, if necessary, obtain the report from overseas. This is a service that TRC has provided for many years to UK government and industry. Sometimes, when the bibliographical details of a report are incomplete, the experience of the staff at TRC will enable them to identify the report.

If an application is made to a report centre or a library for a report, it is advisable to quote at least one other piece of bibliographical information in addition to an accession number, e.g. an author. When the accession number is not known, then as much information as is known should be quoted in the request. It is of particular interest in the case of difficulties to know where the reference was quoted.

Reports not openly available may present problems. If the report is known to have some restrictive marking, then the requestor must prove his entitlement to the information. Some difficulty may be experienced before contact is made with the right organisation. Here TRC can frequently suggest who should be approached.

In the past, booksellers have not handled reports, but now that more reports are being priced, some booksellers are prepared to handle requests for reports.

Many government reports are published by Her Majesty's Stationery Office (HMSO), which publishes *Sectional Lists* of government publications for each department. *Sectional List No. 3* covers publications produced under the auspices of the Departments of Trade, Industry, Energy, Prices and Consumer Protection.

Estimates made on *Nuclear Science Abstracts* are that one-quarter of the citations are to reports; 40% of these are of non-US origin. The Atomic Energy Research Establishment (AERE) at Harwell can give assistance on US Atomic Energy Commission and, of course, UK Atomic Energy reports. The International Atomic Energy Agency (IAEA), Vienna, have a data base for nuclear science (peaceful applications) and the International Nuclear Information System's *INIS Atomindex* is produced semi-monthly with twice-yearly cumulative indexes.

HANDLING SPECIAL COLLECTIONS OF REPORTS

A large organisation may have a reports library and a staff whose main responsibility is reports. The library will be a registered borrower of the British Library, Lending Division in the UK, and engineers should channel their requests for reports through their library who will be responsible for returning the report. In a smaller organisation the reports will be handled along with other types of document. If an engineer has easy access to a library, then he will master the filing and shelving system used in that library so that he can retrieve the reports he requires.

Many reports are now available in microfiche (fiche) form and a personal library of reports in this form is easy to collect and store. Fiche, now the 98-page 24× format (4× 6in transparency) used by technical report centres, may be purchased very cheaply from BLLD and TRC in the UK.

Examining an issue of *STAR* for 6 June 1973, it is found that more than 80% of the reports have less than 96 pages. This is the number of pages which can be accommodated on one fiche. (Less than 13% of the reports require two fiche and less than 4% require three or more fiche.) From size considerations alone, the present fiche are thus a suitable method for storing reports. The fiche are always stored in paper wallets, so that two or more fiche in a wallet, separated by sheets of paper, are no problem.

The engineer must decide how he is to file his personal collection of reports. He may decide to use date, subject or name order. If he uses accession number order, then an index to the collection will be required. There is no reason why the index in his organisation's main library should not be used to give access to his personal library. In fact, the engineer should attempt to avoid keeping indexes which are easily accessible in the library. Hard copy can be stored in racks, cupboards or drawers, care being taken that each document is clearly identifiable. Difficulty arises with plastics spines, which can come adrift, and sticky tape binding the spine, which can cause two documents to stick together.

The engineer will be receiving an increasing number of computer print-outs in the form of either SDIs or retrospective bibliographies. There is no reason why these should not be filed as reports in his personal collection. Since they will be subject-orientated, they can be enhanced by marking the relevancy alongside the citations. A single tick can indicate highly relevant items and slashed ticks can be used to indicate lesser degrees of relevance. It is quite easy to tear computer print-outs neatly so that irrelevant citations can be destroyed. Suitable cardboard folders are available which hold computer print-outs and these can be indexed and treated as hard-copy reports.

No attempt has been made to disguise the fact that there are reports of interest to an engineer which are not openly available. He has only to consider the reports produced by his own organisation to see that this is true. Open availability is sometimes withheld from a report although the author is able to make it available to individuals. For this reason the engineer needs to give attention to resolutely seeking out technical reports relevant to his interests.

Appendix 1   USEFUL ADDRESSES

Aslib,
3 Belgrave Square,
London SW1X 8PL

Atomic Energy Research
Establishment, Harwell,
Didcot, Berks.

British Library, Lending Division,
Boston Spa,
Wetherby,
Yorks. LS23 7BQ

British Steel Corporation,
Information Services,
Corporate Development Library,
Hoyle Street,
Sheffield S3 7EY

Her Majesty's Stationery Office,
49 High Holborn,
London WC1V 6HB (callers only)
PO Box 569, SE1 9NH (trade and
London area mail orders)

British Library,
Science Reference Library,
25 Southampton Buildings,
London WC2 1AW

Technology Reports Centre,
Orpington,
Kent BR5 3RF

Appendix 2   RESEARCH ASSOCIATIONS PRODUCING REPORT
SERIES OF INTEREST TO ENGINEERS

Name	Address	Telephone	Report Series
British Carbonization Research Association	Wingerworth, Chesterfield, Derbyshire S42 6JS	0246–76821	Research reports
British Cast Iron Research Association	Bordesley Hall, Alvechurch, Birmingham B48 7QB	073–92–66414	Reports
British Ceramic Research Association	Queens Road, Penkhull, Stoke-on-Trent ST4 7LQ	0782–45431	Research papers, Special reports, Technical notes
Construction Industry Research and Information Association	6 Storey's Gate, London SW1P 3AU	01–839–6881	Reports, Technical notes

(CIRIA has no laboratory of its own but places contracts with other research associations)

Electrical Research Association	Cleeve Road, Leatherhead, Surrey	037–23–74151	Technical reports
Furniture Industry Research Association	Maxwell Road, Stevenage, Herts SG1 2EW	0438–3433	Research notes, Research reports, Information reports
British Glass Industry Research Association	Northumberland Road, Sheffield S10 2UA	0742–686201	Research reports, Technical notes, Information circulars
Building Services Research and Information Association	Old Bracknell Lane, Bracknell, Berks. RG12 4AH	0344–25071	Laboratory reports, Technical notes, Information circulars
British Hydromechanics Research Association	Cranfield, Bedford MK43 0AJ	023–045–422	Research reports, Technical notes
Machine Tool Industry Research Association	Hulley Road, Hurdsfield, Macclesfield, Cheshire SK10 2NE	0625–25421	Research reports, Notes for designers, Special reports
Motor Industry Research Association	Watling Street, Nuneaton, Warwicks. CV10 0TU	0682–68541	Research reports, Foreign vehicle analysis reports

Name	Address	Telephone	Report Series
BNF Metals Technology Centre	Grove Laboratories, Denchworth Road, Wantage, Berks. OX12 9BJ	023–572992	Research reports, Development reports
Production Engineering Research Association	Melton Mowbray, Leics. LE13 0PB	0664–4133	Research reports
Rubber and Plastics Research Association of Great Britain	Shawbury, Shrewsbury, Shropshire S74 4NR	093–944–383	Reports
Sira Institute	South Hill, Chislehurst, Kent BR7 5EH	01–467–2636	Research reports
British Ship Research Association	Wallsend Research Station, Wallsend, Northumberland, NE28 6UY	0632–625242	Research reports, Technical memoranda
Spring Research Association	Doncaster Street, Sheffield S3 7BB	0742–77451	Research reports, Production briefs
Steel Castings Research & Trade Association	East Bank Road, Sheffield S2 3PT	0742–28647	
Welding Institute	Abington Hall, Abington, Cambridge CB1 6AL	0223–891–162	Members' reports

Appendix 3 GOVERNMENT RESEARCH (excluding Defence)

Outside the atomic energy field, reports produced by the industrial research establishments of the Department of the Environment (DOE), the Department of Energy (DE) and the Department of Industry (DI) will be the main series of interest to engineers:

Name	Address	Telephone
Building Research Establishment (DOE)	Garston Watford WD2 7JR	092–73–74040
National Engineering Laboratory (DI)	East Kilbride, Glasgow G75 0QU	035–52–20222
National Physical Laboratory (DI)	Teddington, Middlesex TW11 0LW	01–977–3222
Safety in Mines Research Establishment	Sheffield S3 7HQ	0742–78141
Transport & Road Research Laboratory (DOE)	Crowthorne, Berks.	034–46–3131
Warren Spring Laboratory (DI)	Stevenage, Herts.	0438–3388
Water Research Centre, Medmenham Laboratory (DOE)	Marlow, Bucks. SL7 2HD	049–166–531
Water Research Centre, Stevenage Laboratory (DOE)	Stevenage, Herts.	0438–2444

6

Patents as a source of information for the engineer*

P. Meinhardt

BRITAIN

British patent specifications

A very substantial portion of engineering designs, methods and processes, used in the past and today, are described in patent specifications; this is true, in particular, regarding significant modern developments. Therefore, patent specifications are an important source of information for the engineer engaged in research and development.

The Patents Act 1949 provides that a complete specification shall particularly describe the invention and the method in which it is to be performed and shall disclose the best method for performing the invention which is known to the applicant. A patent is a kind of bargain between the inventor and the Crown: the Crown grants to the inventor a monopoly for his invention for a limited number of years. The inventor in turn discloses the invention with all its details so that after the expiry of the monopoly the public has full knowledge of the invention and can use it freely. If the inventor does not fulfil his part of the bargain and does not disclose his invention sufficiently and fairly and does not disclose the best method of performing it with all essential details, the patent will be invalid.

These legal provisions ensure that British patent specifications constitute a useful and reliable source of information for the engineer. Some foreign patent laws are less strict.

*Another volume in this series, *Mainly on Patents*, edited by F. Liebesny (1st edn, 1972), covers this topic more fully than is possible in a single chapter.

The first thing an engineer receiving a patent specification, which may have a bearing on his problem, should do is to look at the drawings; in the rare case where a patent has no drawing—for example, some metallurgical patents—he should look at claim one. The drawing will usually enable the engineer to see at a glance whether the specification is useful for his purpose. If on the strength of the drawing or the claim the patent appears of interest, the engineer should read the complete specification and all claims. The solution given in the complete specification may be the one the engineer has been looking for or it may guide him in the right direction. Frequently the specification will indicate the prior art; such a general description of the prior art—while not protected by the patent—will often lead the engineer towards his goal. Often an old patent may have lapsed for the reason that it was not possible to put the invention into practice because at the time the right materials, manufacturing techniques and control apparatus were not available. Some of these old inventions may now be a practical proposition. Thus, the development engineer can often start his own programme from lapsed patents.

By studying a series of patent specifications—lapsed or in force—in the names of a specific company (available in the name index published by the Patent Office) it is possible to follow that company's pattern of the development on a given subject.

Infringement in Britain

If an engineer comes to the conclusion that a patent specification is relevant for his problem he must look at it from two angles: (1) pure information and research, and (2) putting the invention into practical commercial operation.

In case a research engineer merely wants to find out what is known about his problem, or if he merely wants to make experiments in his laboratory, or if he simply wants to write a book or an article, he can freely use all the information he finds in a patent specification owned by a third party. As far as the patent law is concerned, the engineer can make use of the specification, the drawings and the claims for the purposes of pure information and research only. There is, however, copyright in patent specifications; quotations from such specifications are thus subject to the usual limitations for copyright material. In case the engineer invents an improvement to the invention covered by the patent of the third party, he can freely apply for a patent protecting that improvement.

On the other hand, in case an engineer wants to put an invention covered by a patent belonging to a third party into practical commercial operation he must be much more careful and consider whether there is a danger of patent infringement. The engineer should first look at the dates on top of the patent specification. The normal term of a British patent is 16 years from the date of the complete specification. In special cases there may, however, be an extension of up to 10 years. Moreover, there are proposals to extend the life of a British patent to 20 years. If the product is to be exported or manufactured abroad, there may be corresponding foreign patents with a term longer than that of the British patent. If the date of the British complete specification is older than 16 years, it is most likely that the patent has expired. If the patent is more than 26 years old, there is no danger whatsoever of infringement.

If any of the dates on a patent belonging to a third party fall within the last 26 years, a patent agent should be consulted and asked to ascertain whether the British patent is still in force. If export or manufacture abroad is contemplated, the existence of corresponding foreign patents should also be investigated. If the patent agent reports that the British patent has lapsed because of expiry of the normal term of 16 years or earlier because of non-payment of renewal fees, the engineer can go ahead freely with manufacture, sale or use in Britain. On the other hand, if the agent reports that the British patent is still in force, the engineer and the patent agent must study the claims. If there is any possibility that the design or method which the engineer wants to put into practice is covered by any of the claims, a thorough investigation concerning infringement and validity must be undertaken (see Meinhardt, *Inventions, Patents and Trade Marks* (1971), Chapters 29 and 30). In complicated cases Counsel should be consulted. If, according to the advice of the patent agent or Counsel, the patent is not infringed, the engineer can again go ahead freely. If the patent would be infringed by designs or methods planned by the engineer, he must either change his plans or try to obtain a licence from the patentee or wait until the patent expires. If the engineer succeeds in designing around the claims of the patent of the third party, he may consider patenting his alternative design.

Search for a British patent

A research and development engineer must investigate all British patent specifications relevant to his problem. Otherwise he may be wasting time by trying to find solutions which others already have found, or worse by arriving at a solution which he and his firm cannot put into practice because it would infringe a patent belonging to a third party. A development engineer of a large company with its own patent department will usually know the British patent specifications relevant to his problem because they have been circulated to him; otherwise he should acquaint the patent department with any new problem he is tackling and ask for copies of the relevant patent specifications. The same applies to engineers in laboratories or experimental stations of research institutes established by the government, by universities or by sectors of industry.

An engineer employed by a smaller firm or working in a smaller laboratory or an inventor working independently will not have the facilities offered by an internal patent department. Difficulties in finding relevant patent specifications may also arise if an engineer of a large company wants patent specifications outside the normal field of operation of the company. In all these cases a search must be instituted.

Searches for British patent specifications are best undertaken at the British Library, Science Reference Library, located in the building of the Patent Office at Southampton Buildings, off Chancery Lane, London WC2, commonly known as the Patent Office Library.

Many patent specifications and other British Patent Office publications can also be consulted in the Science Museum Library in South Kensington, and a number of public libraries in large provincial towns. In smaller towns the public librarian will be able to indicate the nearest centre where patent specifications are available.

In companies with an internal patent department the search will be undertaken by that department. In companies or institutions utilising the services of an independent patent agent he should be asked to organise the search. If an individual inventor or other member of the public requiring a search does not know a patent agent, he can obtain names and addresses of patent agents in his neighbourhood from the Chartered Institute of Patent Agents, Staple Inn Buildings, London WC1. Names of patent agents will also be found in the yellow pages of the telephone directories.

A company patent department and large firms of patent agents may have searchers on their staff. Frequently, however, they utilise the services of independent professional searchers. An individual inventor can approach a professional searcher direct; he will find names and addresses in the yellow pages under 'patent searchers'. Usually it is better for an individual inventor or a small company to instruct a patent agent to organise the search because the patent agent will at the same time be able to advise his client about the danger of infringing the patent of a third party and about filing a patent application for any invention devised by the inventor.

An engineer, or other member of the public, can undertake the search for patent specifications in which he is interested himself. Such a personal search will save fees and will be useful to give the engineer a general idea of the state of the art in his field; it may bring forward one or more patent specifications relevant to the problem on which the engineer is engaged. Searching, however, is a skilled job requiring knowledge and experience. An engineer not familiar with the classification system would waste much time if he endeavoured to undertake a search himself. Moreover, there is a real danger that a person without such knowledge and experience will miss vital specifications. Indeed, the mass of patent specifications is so great today that even an experienced searcher may occasionally miss a relevant specification.

There are over one million published British patent specifications (for further statistics, see page 87). Finding the specification or specifications relevant to a particular problem is therefore no mean task. Fortunately, it is not quite like finding a needle in a haystack, because the Patent Office has devised an extensive system of classification which will guide the searcher along the paths leading to the specification he is looking for.

The Patent Office has published several pamphlets and booklets for searchers (see Bibliography). Of these, the most useful booklet is entitled *About Patents—Patents as a Source of Technical Information* (1971). The booklet (hereinafter referred to as GB1 gives a practical example of how to carry out a search. The following quotation is taken from the booklet (pp. 6–7) with permission of HMSO:

> The classification system is set out in the Classification Key and is divided into eight very broad sections; these broad sections are divided into 40 divisions and these divisions into over 400 headings under which can be found about 65 000 sub-classifying or indexing terms identifying individual features or subjects. Each heading and each index term is identified by a code composed of letters and numbers. As technology advances and expands the coded terms need to be revised at intervals of 12 to 18 months.
>
> In order to locate a subject in the Classification Key it is first necessary to know under which heading to look. To help with this a reference index to the Key is available. The first section of the index consists of an alphabetical list of catchwords, each catchword being followed by the code and title of the heading

under which the subject is indexed in the Classification Key, or more usually a number of such references. Part Two of the reference index (Structure of the Classification Key) sets out the divisions and headings of the Classification Key. In Part Three the scope of each heading is clarified by an extract from the Key giving lists of those subjects covered by the heading. Also listed are those subjects which, having some connection, might be thought to be covered by the heading but which are in fact to be found elsewhere.

Other examples of searches are given by Newby (1967, p. 125) and by Lees (1965, p. 85).

By means of the classification key the searcher can thus find the heading, the division and the code under which patent specifications relevant to his problem are classified. Once he has found the correct code, his search will have narrowed down considerably. Frequently, specifications relevant to a problem confronting an engineer will be classified under two or more codes because they relate to more than one subject of technology.

Every British patent specification shows the code(s) under which it is classified near the top of the first page under the legend 'Index at Acceptance'. This legend is chosen because the classification key may be changed subsequently.

The next step is for the searcher to order from the Patent Office 'file lists' for the relevant codes. Booklet GB1 (pp. 11 and 17–18) gives full particulars about the 'file lists' available and methods of narrowing down the quantity of 'file lists' to be ordered. Each 'file list' is prepared by the Patent Office to meet a client's specific order. A study of the 'file lists' will show the searcher the identification numbers of the specifications which may be relevant to his problem.

If only a few specifications appear important, the searcher can look at them in the library or order them from the Sales Branch of the Patent Office (see p. 89). If the 'file lists' contain a large number of potentially relevant specifications and the searcher wants to restrict the number of specifications to be ordered, he should study the appropriate volume of the *Abridgments of Patent Specifications* (see below); this will enable him to find in a short time the important specifications warranting a detailed study.

The staff of the British Library, Science Reference Library in the Patent Office Building are very helpful in case searchers are confronted with difficulties. Assistance on classification can also be obtained by writing to the Patent Office Classification Section T.3, Southampton Buildings, London WC2A 1AV. The searching duties of the Patent Office, however, are confined to examining the novelty of applications for patents, and the Patent Office will not undertake searches for members of the public.

Abridgments and other British Patent Office publications

The Patent Office publishes illustrated abridgments of all British patent specifications. The abridgment contains a summary of the essential features of the invention, one or more key drawings (if any) and the text of the widest claim. The abridgments are prepared by the Patent Office examiners. The abridgments are grouped in 25 volumes covering one or more divisions of the classification key; the volumes are divided into sub-volumes comprising about 25 000 specifications. An engineer engaged in research and development on his own or as employee of smaller companies should purchase at least the current volume of abridgments in his field, and use this as a start for his searches. Large firms and patent departments will possess the complete series of current and back volumes of abridgments in their field of operation.

Engineers who want to be kept informed currently about the latest developments in their field should subscribe to the weekly edition of the abridgments in the groups which concern them. When an engineer perusing the abridgments comes across a patent of particular interest to him, he should procure the complete specification.

The abridgments of British patent specifications which go back to 1855 are an extremely useful condensed source of information for the engineer.

Abridgments of patent specifications are published currently in the journals of the Patent Offices of the USA, Canada and Germany. The systematic arrangement of abridgments in groups of bound volumes for every field of technology is, however, a unique facility offered to engineers by the British Patent Office.

Large companies and research-orientated small companies which find that abridgments alone are not sufficient for their purpose can place a standing order on the Patent Office for copies of all newly published patent specifications under a particular classification code.

Companies and engineers may want to know what their competitors are doing as regards inventing and patenting. This information can be found in the name index of patentees published weekly in the *Official Journal* of the Patent Office. Bound volumes for each series of 25 000 patent specifications arranged by names of patentees with cross-references to assignees of the patent rights are also available.

If an engineer has found a patent specification belonging to another party relevant to his problem, he or his patent agent can ask the Patent Office for a result of the official search relating to the patent application which led to the grant of the patent. The Patent Office will then supply a list of those earlier British and perhaps US patent specifications which the examiner considered relevant to the novelty of the invention. The engineer can then study the abridgments or the complete specifications of the patents listed.

For selected headings of the classification 80-column punched-card systems of patent specifications with appropriate term lists are available from the Patent Office for those fields of technology where the Patent Office has instituted mechanised retrieval systems. Details of the relevant headings are given in booklet GB1 (p. 19). Large companies possessing their own sorting and retrieval equipment can buy the punch-card systems and carry out the searches themselves.

USA

In view of the extensive research and development work undertaken in the USA, US patent specifications are a very important source of information for the engineer.

The US Patent Office operates its own system of classification, which is different from the British system. In the US system there are about 352 main classes, subdivided into more than 66 000 subclasses. Booklets issued by the US Patent Office useful for searches are listed below (p. 90).

Engineers interested in recent developments in a particular field can place an order for the current supply of US patent specifications in the appropriate sub-class as and when issued. Abstracts of new US patents showing the principal claim and principal drawing are published each week in the US Patent Office *Official Gazette,* but they do not contain a précis similar to that in the British abridgments. The official abstracts are not available in classified volumes, but there are several publications by private publishers containing classified abstracts of US patents in fields such as chemistry and electricity; for details see Newby (1967, p. 81).

Microfilms of US patent specifications dating from 1966 as well as microfilm sub-class lists for older patents are available.

Searches into US patent specifications can be undertaken at the US Patent Office in Arlington, Va., just outside Washington, or at certain libraries throughout the USA. Copies of most US patent specifications are available at the Patent Office library in London. If export to the USA or manufacture in the USA is contemplated, it is advisable to arrange a search into US patent specifications through a British patent agent or direct through a US patent attorney. For a directory of US patent attorneys, see p. 90.

Apart from US patent specifications, the scientific library of the US Patent Office contains over 8 million foreign patent specifications in bound volumes, the official journals of most foreign patent offices, 120 000 scientific books and about 90 000 volumes of periodicals devoted to science and technology.

US patent specifications frequently list older patents and literature found during examination of the application. These references will give a development engineer valuable leads to information on his problem.

CANADA

The majority of Canadian patent specifications are in English; some, depending on the choice of the applicant, are in French. Copies can be obtained from the Canadian or from the British Patent Office (see pp. 89–90).

The Canadian Patent Office uses its own system of classification, which is similar to the US system of classification, but has been modified to meet Canadian requirements.

There are about 400 classes divided into over 25 000 sub-classes. The headings of the main classes are shown each week in the *Patent Office Record*. A list of publications by the Canadian Patent Office is given on p. 91. Standing orders can be placed for the weekly supply of new Canadian patent specifications in any sub-class.

Since 1969 the *Patent Office Record* contains an illustrated abstract of every patent issued during the week. The abstract is similar to the British abridgment, but is prepared by the applicant and not by the examiner.

The majority of patents granted in Canada will correspond to patents granted in the USA or Britain for the same invention. Therefore, by and large, an engineer will find Canadian inventions by searching US and British patents. He can, however, study the original Canadian patent specifications at the library of the Patent Office in Ottawa or place orders by post (see p. 90). Photocopies of Canadian patent specifications are available from the British Patent Office.

If export to Canada or manufacture in Canada is contemplated, it is advisable to undertake a search in Canada in order to avoid infringement. This can be done by instructing a Canadian patent agent direct or through a British patent agent. A list of Canadian patent agents is obtainable from the Patent and Trade Mark Institute of Canada (see p. 91). The Canadian Patent Office will not undertake a search for members of the public, but the classification examiners will assist searchers to find the appropriate fields of technology.

GERMANY

After British and US specifications, the next most important source of information for the engineer is patent specifications of the Federal Republic of West Germany.

The German Patent Office operates its own classification system. The present system was introduced in 1955 and last revised in 1968. It comprises 89 classes, about 500 sub-classes, and roughly 3000 subgroups. For new patents the German system will gradually be changed over to the international classification (see p. 86), but with a more detailed subdivision.

Copies of German patent specifications, namely Auslegeschriften (published examined specifications) and Offenlegungsschriften (published unexamined specifications) and Gebrauchsmuster (petty patents), can be inspected at the German Patent Office in Munich and with limitations at the British Patent Office in London. Patent specifications published since 1968 are available as microfilm aperture cards. Standing orders can be placed for copies of specifications for particular sub-divisions as and when issued. (About orders, see p. 91.)

The official journal of the German Patent Office is the *Patentblatt*; this contains illustrated abridgments of patent specifications published each week.

Lists of patent specifications and some sub-groups of patent specifications are available as microfilm aperture cards. Searches are facilitated by a loose-leaf book containing the group sub-divisions of the patent classes (Gruppeneinteilung der Patentklassen) (see p. 91) and an alphabetical catchword register of 1955: the latter is out of print, but can be inspected at the Patent Office.

In the search room of the German Patent Office a searcher can look at folders with copies of all German patent specifications, in a particular German sub-class. Such a service is available also for British, US and French patent specifications corresponding to a German sub-class; this facility, which is not available in the British Patent Office, is particularly useful for an engineer interested in foreign patent specifications relating to a given field.

A list of professional searchers (Patentberichterstatter) is available in the search room of the German Patent Office. On a list of German Patent Agents, see p. 92).

A German patent application is published unexamined 18 months after its priority date, while a British patent application is published only after examination and acceptance, which may be about $3\frac{1}{2}$ years after the priority date. Therefore, if a British engineer wants to find out at the earliest possible date what his competitors are patenting, he should place an order (see p. 91) for unexamined German patent applications in the corresponding German sub-class. If the competitor has filed a German application, the British engineer will receive a copy up to 2 years earlier than if he confines his search to British specifications.

INTERNATIONAL

In order to avoid the infringement of foreign patents, knowledge of foreign patent specifications is important for engineers working for firms engaged in export or manufacturing abroad. In such cases a search for relevant foreign patents should be arranged through a British patent agent or direct instruction to a local agent abroad. A preliminary search can also be undertaken by looking at copies of foreign specifications in the British Patent Office Library, but this is not sufficient, because vital foreign patents may be missed.

Knowledge of foreign patent specifications will also be important for an engineer who wants to have a really comprehensive survey of patent literature on his particular problem. In most cases a search into British, US and West German patents will be sufficient, because patentees in other countries will in all probability have filed patents in at least one of these countries. However, in special cases a patent search in one or more of the following countries might be undertaken: Australia, Austria, Belgium, Canada, France, German Democratic Republic, Italy, Japan, Netherlands, South Africa, Sweden, Switzerland, USSR. Multi-country searches are best done through the International Patent Institute (see p. 87). No international patents have been granted at this date. Under the convention on the Grant of European Patents signed in 1973 European patents will, however, be granted by the European Patent Office situated in Munich within a few years.

An international classification for patents was established in accordance with a convention of 1954 and revised in 1972. The international classification comprises 103 classes, 594 sub-classes and nearly 30 000 sub-groups. Member countries, including Britain and others, show the international classification symbol on their patent specifications in addition to the national symbol. The international classification with a catchword reference index is available in English, German and French (see p. 92).

Searches into patent specifications and other technical literature are undertaken by the International Patent Institute, PO Box 5021, The Hague, Holland. For British engineers a search by this Institute is particularly useful if information is required on patent specifications in several of the minor countries mentioned on p. 86. Members of the staff of the Institute have a sound knowledge of the languages of those countries (reading knowledge only of Japanese); in addition, they master Arabic, Bulgarian, Czech, Danish, Hebrew, Malay, Norwegian, Polish, Portuguese, Serbo-Croat and Spanish, and have a reading knowledge of Chinese. The Institute has a collection of more than 8 million patent specifications from different countries increasing at a rate of over 200 000 p.a., as well as over 30 000 textbooks and journals; a booklet on the services of the Institute is available from the above address.

An International Patent Documentation Centre was established in Vienna in 1972; the address is A.1010 Vienna, Fleischmarkt 3. The Centre is organised under the auspices of the World Intellectual Property Organisation (WIPO) of Geneva (see p. 92). At present it is too early to say how useful the Centre will be for engineers requiring information on patent specifications.

There is a Committee for International Co-operation on Information Retrieval among Patent Offices (ICIREPAT) with headquarters at WIPO, Geneva. Its aim is to develop 'shared use' indexing systems for mechanical and computer retrieval of information in patent specifications. The progress of ICIREPAT has been slow, and it may take many years before engineers get benefit from its work.

File lists and punch-card systems of certain foreign patent specifications in a few specialised fields are obtainable from the British Patent Office; for details, see p. 19 of booklet GB1.

Comprehensive information on foreign patents is published by Derwent (see p. 92): e.g. *German Patents Abstracts* (weekly), *Soviet Inventions Illustrated* (weekly), etc.

See also: *Foreign Patents: A Guide to Official Patent Literature*, by F. J. Kase (Oceana Publications, 1972).

SOME STATISTICS

Britain

Well over 1 million patent specifications were published by the British Patent Office from 1916 to 1970 (booklet GB1). The rate of increase is about 50 000 p.a. Taking pre-1916 and post-1970 patents, there are at a rough guess more than $1\frac{1}{4}$ million British patent specifications.

USA

The first US patent was issued in 1790, and since 1836 the US Patent Office has published nearly 4 million patent specifications. The rate of increase is about 90 000 p.a.

Canada

Since 1791 over 90 000 patent specifications have been published by the Canadian Patent Office. The rate of increase is approximately 30 000 p.a.

Germany (West)

Since 1877 more than 2 million German patent specifications have been published.

World

According to a paper read on 29 May 1974 to the Chartered Institute of Patent Agents by Lord Eccles, Chairman of the British Library, 709 510 new British and foreign patent specifications were received by the library in 1973. According to figures given at a meeting of the German Documentation Society held in 1971 (see p. 92), about $9\frac{1}{2}$ million patent specifications had been issued by that date by the seven most important patent offices, and about $14\frac{1}{4}$ million by all patent offices of the world. The growth rate is frightening; according to the present trend some 21 million patent specifications will have been issued by 1980. The rate of increase may, however, diminish when European and EEC patents are granted.

It must be borne in mind that patents for the same invention are often granted in different countries. The ratio between inventions and patent specifications in the same 'patent family' is about 1:3·5. Even at this ratio there are about 4 million inventions to be found in patent specifications.

BIBLIOGRAPHY

Britain

Publications by the British Patent Office available by post from the Sales Branch, Orpington, Kent BR5 3RD, or personally at 25 Southampton Buildings, off Chancery Lane, London WC2:

About Patents—Patents as a Source of Technical Information (1972. Booklet. Referred to as GB1 in the text)
Patent Searches (1972). Pamphlet
Searching British Patent Literature (1971). Booklet
Classification Key. In 25 group units. Various dates
Structure of the Classification Key (1971). Booklet
Reference Index to the Classification Key. Several volumes at various dates
Classification Key—Forward Concordance (1963). Booklet
Classification Manuals. For certain products. Various dates
Subject Matter File Lists. For British patents. To order
Subject Matter File Lists. For certain sub-classes of foreign patents. To order
Index to Name of Applicants. Various dates
Abridgments, Classified, Illustrated. Bound volumes of various dates or current weekly
British Patent Specifications. Starting in 1617 and complete since 1853. Individual copies
Foreign Patent Specifications. Selected individual photocopies
Official Journal (Patents). Weekly
Classified Selected Patent Specification Service. On standing order
Classified Punched Card Systems Service. For certain sub-classes on standing order

Further details, prices, methods of payment and particulars on other Patent Office publications, outside the scope of the present book, are given in booklet GB1 and in a list available from the Sales Branch.

Books

How to Find Out About Patents, by F. Newby (Pergamon, 1967)
Inventions, Patents and Trade Marks, by P. Meinhardt (Gower Press, 1971)
Mainly on Patents, edited by F. Liebesny (Butterworths, 1972)

Patent Protection, by C. Lees (Business Books, 1965)

Register of Patent Agents. Annual publication by the Chartered Institute of Patent Agents, Staple Inn Buildings, London WC1V 7PZ

USA

Publications available from the US Department of Commerce, Patent Office, Washington DC 20231:

US Patent Specifications. Individual copies
Foreign Patent Specifications. Selected individual photocopies

Publications available from the US Government Printing Office, Washington DC 202402:

General Information Concerning Patents (1972). Booklet
Manual of Classification. Loose-leaf book. Various dates
Development and Use of Patent Classification Systems (1966)
Index of Patents (by subject matter). Annual
Index of Patents by Names of Patentees. Annual
Classification Definitions. Various dates
Weekly Class Sheets. On standing order
US Patent Office Official Gazette. Weekly
Directory of Registered Patent Attorneys and Agents
Patent Specifications by Sub-Class. Individual or standing orders

Publications available from the National Technical Information Service, US Department of Commerce, Springfield, Va. 22151:

Microfilm Copies of US Patent Specifications. Since 1966 arranged by number or by category; on subscription by category only
Microfilm Sub-Class Lists of Patent Specifications (up to 1968)

Publications available from private publishers:
Classified indexes of US patents in fields such as chemistry, electricity, fertilisers, resins, polymers and petroleum; for details, see Newby, p. 81.

Canada

Publications by the Canadian Patent Office, available from Information Canada, Ottawa K1A 089 or personally at the Patent Office, Place du Portage, Hull, near Ottawa:

Manual of Classification
Class and Sub-class Definitions
Index to the Classification
Canadian Patent Specifications. Individual copies
Foreign Patent Specifications. Selected individual photocopies
Lists of Patents for Specific Subject Matter
Classified Canadian Patent Specification. In any sub-class on standing
order
The Patent Office Record (Weekly)

The following publication is available from the Canadian Patent and Trade Mark Institute, P.O. Box 553, Station B, Ottawa K1P 5T4:

List of Canadian Patent and Trade Mark Agents

Germany

Publications by the German Patent Office available by post from Deutsches Patentamt, Dienststelle Berlin, Schriftenvertrieb, 1000 Berlin, Gitschinerstr. 97, and for some publications (personally only) at the Patent Office, Munich, Zweibrückenstr. 12:

German Patent Specifications. Individual copies
Classified German Patent Specifications. Upon issue on standing
order
German Petty Patents (Gebrauchsmuster). Individual microfilm
aperture cards
German Patent Specifications Since 1968. Individual microfilm
aperture cards
Classified Lists of Patent Specifications Since 1968 (and some older)

Publications by Carl Heymanns Verlag, 5 Köln 1, Gereonstr. 18:
Patentblatt. Weekly. This is the Official Journal of the German
Patent Office
Index to Names of Applicants. Quarterly
*Group Sub-divisions of German Patent Classes (Gruppeneinteilung
der Patentklassen)*

Publications by WILA Verlag, 8 München 21, Landsbergerstr. 19a:
Abridgments of German Patent Specifications. Illustrated (Auszüge
aus den Offenlegungsschriften unexamined and Auslegeschriften
examined and Gebrauchsmuster). Monthly

German Patent Specifications. Individual copies
German Petty Patents (Gebrauchsmuster). Individual copies

A list of German Patent Agents who will arrange searches and advise on the result of searches is available from the Patentanwaltskammer, München 5, Morassisstr. 2.

International and statistics

Publications by Derwent Publications Ltd., Rochdale House, Theobalds Road, London WC1X 8RP:

British Patents Abstracts
German Patents Abstracts
Soviet Inventions Illustrated
World Patents Index

Publications by Morgan Grampian Books Ltd., West Wickham, Kent:

International Patent Classification (2nd edn, 1974) (with catchword index). In English, German and French
Concordance List. International Classification and German Classification. The German version is available also from Carl Heymanns Verlag (see above)

Publication by Deutsche Gesellschaft für Dokumentation EV, 6 Frankfurt/Main, Schubertstr. 1:

Committee for Patent Documentation (Ausschuss für Patent Dokumentation). Reports of annual meetings containing reports of papers read on many aspects of patent documentation in Germany and internationally

International Patent Institute, P.O. Box 5021, The Hague, Netherlands:
Booklet on the Services of the Institute

WIPO (World International Intellectual Property Organisation) in conjunction with BIRPI (United International Bureaux for the Protection of Intellectual Property), P.O. Box 18, 32 Chemin des Columbettes, 1211 Geneva, Switzerland:

Industrial Property. Monthly journal in English, French and German. The December issue of each year contains international statistics on patent specifications

7

Standards

J. Brown

A general definition of standardisation is the establishment, by authority, by custom or by general consent, of rules, disciplines, techniques and other defined conditions which have to be followed to enable a society or particular sections of it to function smoothly and efficiently. Thus, this definition applies to statutory regulations, industrial practices, trading conditions, business behaviour, social customs, educational qualifications, units of measurement, and so on.

INDUSTRIAL STANDARDS

For the main purpose of trade and commerce, industrial standards are used. An industrial standard is a precise statement defining the requirements that can normally be met, with a minimum of variety, in a reproducible and economic manner as the basis of the best current techniques. Such standards may be based on company, industry, national, regional or international practices.

Industrial standards fall into the following classifications:

Dimensional: to secure uniformity, interchangeability and simplification of the types and sizes of one product.

Qualitative: to assess fitness for purpose.

Methods of test: to provide a uniform, efficient and economic basis of comparison between products.

Methods of use (Code of Practice): to define the correct application of methods, materials and appliances.

Definitions: to secure precision in descriptions.

Glossaries: to secure uniformity in the use of terms.

The primary condition for industrial standardisation is the maximum economic production, i.e. for the provision of goods more cheaply and more quickly and with speedier servicing facilities. In addition, it gives the user assurances of quality, performance, and accurate replacement.

PHYSICAL STANDARDS

A distinction needs to be made between physical standards and industrial standards. The former apply to natural phenomena which are accurately determined and are not subject to change with advancing knowledge, e.g. time based on the sidereal day; temperature.

CONSUMER STANDARDS

Greater emphasis has been given in recent years to the preparation of standards for consumer goods which are primarily intended for the personal use of the ultimate buyer. The development of such standards has not kept pace with industrial standardisation because of the general absence of adequate testing methods, the high cost of laboratory testing and the unorganised position of the ultimate consumer. Greater co-operation between manufacturing, distributing and consumer organisations has developed to permit of the formulation of standards for consumer goods.

NATIONAL STANDARDS

Standardisation began in private firms and then developed to become the concern of an entire industry. Experience, however, has shown abundantly that for such work to reach its highest efficiency and to proffer the greatest benefits to the whole community it should be carried out nationally under some central co-ordinating body. The British Standards Institution (BSI) fulfils this function in the United Kingdom. As the national standardisation centre, it is concerned essentially with the avoidance of overlapping, duplication and possible conflict, and economises effort. BSI has thus demonstrated to the engineering industry and other industries the immense advantages of national co-operation through a central co-ordinating body.

The initiation of a national standard may come from several

sources: from a particular section of industry; trade association; government; learned society; academic body; local authority; consumer; or any other responsible source. In some cases the need for a standard stems from international trade and the need to increase export business. National standards are prepared by BSI committees on which all interested parties are represented, and these standards are adopted by common consent. New scientific techniques are being applied through standardisation in design, production, purchasing and distribution.

INTERNATIONAL STANDARDS

The two main international organisations concerned with standardisation are: The International Organization for Standardization (ISO), and The International Electrotechnical Commission (IEC). BSI takes an active part in this important international work.

Countries in certain regions of the world also consult together to achieve harmonisation of standards for trading in those particular areas. For example, in Europe CEN (European Committee for Standardization) and CENELEC (European Electrotechnical Committee for Standardization) exist for the purpose of achieving regional standards for the free flow of trade in the European Economic Community and, in some cases, EFTA countries. Likewise, a pan-American organisation, COPANT (The Pan-American Standards Commission), exists to create a common market between North and South American countries. In Eastern Europe an organisation (COMECON) has been formed to facilitate trading between the eastern bloc countries. An African Regional Standards Organization is being formed.

ENGINEERING DESIGN

Standardisation needs to be introduced at the design stage to achieve the full economic benefits which it provides. While each company has to evolve its own standards policy, this should be determined within the broad framework of national and international standards. Considerable time and effort of designers is saved by eliminating the need for special drawings by calling up national standards, thus leaving them free to concentrate on creative rather than routine work. As far back as 1923, *BS 308: Engineering Drawing Practice* was introduced for the purpose of standardising engineering draughtsmen's conventions to remove the diversity of practice which existed in different firms and industries, because at that time the methods used by draughtsmen led to an unnecessary amount of work in manufacture and inspection owing to variations in procedures specified in drawings.

The current edition covers all the essential features for the harmonisation of design drawings throughout industry and is aligned in all essential particulars with the international agreements of the International Organization for Standardization (ISO). Thus, considerable economies in time and effort of designers and draughtsmen can be achieved by adopting the conventions, symbols and recommendations in *BS 308;* drawings prepared in accordance with the Standard are also readily understood and avoid misunderstanding by artisans, purchasers, customers and the like in this and other countries.

Associated standards, e.g. *BS 2517: 1954. Definition of Terms Used in Mechanical Engineering*, also contribute towards easing the work of designers and draughtsmen. In addition, the standardisation of graphical symbols and illustrations for use in various kinds of diagrams has been accomplished, e.g. for diagrams of hydraulic and pneumatic systems, electrical circuits and process plants, in order to eliminate unnecessary work in drawing offices.

The saving to industry arising from the introduction of the use of standards in drawing offices is phenomenal and runs into several million pounds each year. Standards contribute towards the reduction of designers' and draughtsmen's time and effort, and the avoidance of misunderstandings and delays. The introduction of standard drawing paper sizes, particularly in changing to metric practice with the accompanying easement in the storage of drawings, also provides economies, especially to those firms who need to look at the wider international scene.

It has been recognised that in several important industrial fields the laying down of basic requirements of good design is of paramount importance in order to contribute to the economic use of materials, processes, suitable maintenance and proper operation of equipment. This has led to the preparation of *British Standards* for important industrial equipment such as cranes, boilers, air receivers, gas cylinders, pressure vessels, materials-handling equipment, to mention just a few. Such properly thought-out standards are of great help to the designer, allowing him to concentrate on the solution of vital problems and the development of new ideas.

These design standards embody the best principles arising from the accumulated experience over many years of the wide sections of industry, including manufacturing and operating experience in different environments and under varying conditions of service, together with advances in knowledge arising from research, development in the use of materials and methods of manufacture. A more sophisticated approach to design is now necessary because of the need to utilise to the full the physical properties of materials and, hence, economise in their use. The standardisation of test methods for determining the chemical, physical and mechanical properties of materials and for providing a reliable basis of comparison of material behaviour also assists the designer in making an economic assessment of material for particular duties. In addition, considerable reliable test data of materials are now available in *British Standards* to enable designers to make a full assessment of materials for their particular needs and to avoid the 'over-designing' of products. In the past there has been a tendency for designers to apply large factors of safety because of their lack of knowledge of materials under certain conditions of application and because of their lack of confidence in some manufacturing techniques. The reliable information provided in *British Standards* enables designers to apply lower and more realistic factors of safety and, hence, use the materials economically according to their mechanical properties and the service conditions envisaged.

Industrial management is realising that standardisation applied in the early stages of design is a potentially powerful aid towards more efficient and profitable production.

PRODUCTION

Dimensional unification is probably the most elementary feature of standardisation to provide economic production. It involves the specification of size according to the strength of the material used and the determination of the magnitude of tolerance suitable for the product. It would be uneconomic to choose unnecessarily close limits of tolerance for parts where the tolerance was unimportant and, hence, increase production costs; on the other hand, it could be disastrous to use coarse tolerance where a degree of precision was essential. Careful thought must, therefore, be given to limits and fits to obtain the right balance of size for economic production. The importance of this subject cannot be over-emphasised; it has in fact, been the subject of international discussion for some considerable time, and international agreement on limits and fits has been reached.

On the shop floor, standards are in constant use as a production aid. Practical information is provided on manufacturing equipment such as lathes, drills, reamers, cutting tools, gauges, measuring instruments, hand tools, and the like. Uniformity in these manufacturing tools ensures their quick and easy replacement, thus avoiding holdups in production runs, and providing the reliability assurance for mass production. Standards provide for interchangeability of equipment and machines, thus allowing of easier servicing and maintenance. A good example is the *British Standard* for fractional horsepower motors which fixes the dimensions of mounting bases, spindle lengths, shaft diameter and centre-line height from base, so that electric motors of any make are interchangeable for all normal uses.

A paramount feature afforded by standardisation in production is variety reduction — not so much for the end product, but for the components that go into it. Why, for example, use a number of different sizes of fasteners on a particular project when one size would do? Sensible assessments of this kind have resulted in the saving of thousands of pounds to companies which have applied a standards approach in order to obtain the utmost economy from their available resources.

PURCHASING AND DISTRIBUTION

The use of standards relieves the purchaser of the need to draw up his own specifications and all he need do is to quote the relevant *British Standard* and, where appropriate, choose from the standard the requisite grade for his application. The use of standards ensures that the purchaser gets exactly what he wants and usually when he wants it, and at an economic price, through concentration of the manufacturing effort on fewer varieties, ease of ordering from different sources to a single standard — in many cases receiving supplies from stock, and providing a sound basis for comparison of products made at different places to the same general pattern.

Purchasers generally recognise the need for standardisation of the various components which they have to purchase for assembling in the products of their firms. Such components must be accurate not only for speedy assembly, but also for the subsequent replacement of worn or defective parts. There must be a close relationship between purchasing and standards.

Standardisation plays a crucial part in the economic distribution of goods. It provides a means for the quick and easy handling of commerce from the common carton for the storage of small individual items, to freight containers for the bulk handling of goods. The essential equipment to assist the speedy and safe handling of goods has been standardised. Such equipment is in evidence in the docks. Here can be seen all kinds of standardised equipment such as conveyors, cranes, fork-lift trucks, pallets, containers, grain suction elevators, slings, hooks, etc., essential to the handling of the multifarious kinds of goods being loaded on and unloaded from ships. Similarly, standardised equipment is being increasingly used for the handling of air freight.

It is useful to record that the *British Standard Packaging Code* embodies in a form suitable for industry all the wealth of experience built up over many years in the shipment of all kinds of goods to all destinations of the world. This *Standard Code* provides a sound guidance enabling the exporter to package his product economically and to be reasonably sure that it will arrive at its destination in a condition fit for use.

INTERNATIONAL TRADE

In international trade standards have become a vital factor and, as new countries become industrialised, the influence of standardisation will grow. It is essential, therefore, that the UK take a prominent part in any international agreement involving standardisation, to ensure that all overseas markets remain open to Britain. The tempo and scope of international standardisation work is increasing, and the number of international agreements being reached on standards is growing apace. Britain, or indeed any country, or any particular section of an industry in any country that holds aloof from these international activities, does so at its economic peril.

Countries participate in international standardisation in order to secure the unification of basic requirements in national standards, and, wherever possible, to help in the harmonisation of Statutory Rules and Regulations so as to reduce the disparities which could become increasingly damaging to international trade.

Standardisation is a potent factor in the economic life of an industrial country such as Britain, and when given the right impetus it contributes to increased productivity. There is a growing recognition in most countries that economic production, purchasing and distribution must be approached from the point of view that there is in the background of all these activities a national interest best served by wise measures of standardisation in order to triumph in the international markets of the world.

Constructive standardisation on an international plane is imperative both to remove barriers to trade and to promote economic intercourse.

METRICATION

The urge for greater exports has led British industry to the decision that this country should adopt the metric system as the primary system of weights and measures. In accordance with this recommendation, the President of the Board of Trade made a statement to Parliament on 24 May 1965 alerting industry to the need to change to the metric system and proposed that this change should be substantially completed within a period of 10 years. Standardisation is crucial to the change, and *British Standards* in the metric system have been prepared covering basic materials, tools, components, screw threads, gauges, measuring instruments, etc., to assist industry in implementing the change to metric. The essential basic series of standards required for making the change principally in mechanical engineering and the complementary standards was completed by the end of 1972. Fortunately, the task of providing national standards in metric units has been helped considerably by the agreements which have been reached in the international discussions that have taken place since World War II.

The simplicity and the ease of application of the metric system have led to the general adoption of this system virtually throughout the world. The metric system is used in countries with approximately 85% of the world's population and approximately 70% of the world's national product. Between the years 1875 and 1960, 82 countries adopted the metric system. Since 1960 many countries, including the UK and the Commonwealth, have moved steadily towards the adoption of the metric system, and it is predicted that by the end of the twentieth century the metric system will have been adopted throughout the world.

Useful references on the metric system or the Système International d'Unités (SI) are:

BS 3763:1970. The International System of Units (SI)
PD 5686:1972. The Use of SI Units
Changing to the Metric System (4th edn, HMSO, 1972)
SI: The International System of Units (HMSO, 1970)

STANDARDS ORGANISATIONS

Towards the end of the nineteenth century, several industrial nations became acutely aware of the need for the introduction in their countries of systematic national standardisation and took action to set up national standardisation organisations. Brief details of some of these organisations are given below.

United Kingdom

British Standards Institution, 2 Park Street, London W1A 2BS

Origin: In 1901 the Engineering Standards Committee was formed and was succeeded by the British Engineering Standards Association (BESA), which was incorporated in 1918 and granted a Royal Charter in 1929. In 1931 its scope of work was enlarged to include standardisation and simplification for all industries and it assumed the title 'British Standards Institution' (BSI).

Publications: British Standards, Codes of Practice, Drafts for Development. In addition, BSI issues regularly: *British Standards Yearbook* (issued annually), listing the complete series of *British Standards*; *Annual Report*, containing a broad review of BSI activities; *BSI News*, a monthly journal containing details of new and revised standards and items of interest in the field of standardisation; and various other publications and pamphlets on general matters of standardisation.

United States of America

There are government, trade, learned and other institutions in the USA dealing with matters of standardisation; information on three of these American bodies is given below.

American National Standards Institute (ANSI), 1430 Broadway, New York, NY 10018

Origin: In 1918 the American Engineering Standards Committee was formed to co-ordinate the development of voluntary national standards in the USA; in 1928 was reorganised and renamed 'American Standards Association' (ASA); and in 1969 changed its name to 'American National Standards Institute' (ANSI).

Publications: American National Standards with the prefix ANSI, although some publications still have the prefix ASA. In addition, ANSI issues regularly: *The Magazine of Standards*, a quarterly journal; *The ANSI Reporter*, a bi-weekly news letter; and *Standards Action*, a bi-monthly journal, soliciting views on proposed new and revised *ANSI Standards*.

American Society for Testing and Materials (ASTM), 1916 Race Street, Philadelphia, Pa 19103

Origin: In 1882 a meeting was held in Europe to discuss the subject of unifying methods of testing materials, and this led a few years later to the formation of the International Association for Testing Materials. An American section of this

Association was formed in 1898, and in 1902 the Society was incorporated as the American Society for Testing Materials. In 1961, in view of the broad scope of this Society, particularly its attention to research, the word 'and' was inserted in the title, which now reads 'American Society for Testing and Materials'. The ASTM is an international Society with members in over 70 countries.

Publications: ASTM Standards are published every year in the *Book of ASTM Standards* in 47 parts plus index. In addition, ASTM issue regularly: *ASTM Standardization News*, issued monthly (formerly *Materials Research and Standards*); and *Journal of Testing and Evaluation*, issued six times a year (formerly *Journal of Materials*). A list of publications of the Society is also published from time to time containing research papers in most of the technical fields served by the Society. *Special Technical Publications* (STPs) are issued from time to time, about 35 each year, containing papers presented at research symposia. *ASTD/Data Sheets* are also issued and these are closely allied to STPs. *ASTM Proceedings* are issued annually.

National Bureau of Standards (NBS), Washington DC 20234
Origin: The NBS is an American Society principally concerned with research and development in the physical sciences and engineering. It was established in 1901 to formulate standards for electricity, temperature and sub-divisions of mass and length. Since then the NBS functions have increased enormously, and it now deals with all phases of development work in physics, mathematics, chemistry and engineering to improve standards and measurement methods.

Publications: Dimensions/NBS, issued monthly; *Journal of Research: Section A. Physics and Chemistry,* issued bi-monthly. *Section B. Mathematics and Mathematical Physics,* issued quarterly. *Section C. Engineering and Instrumentation,* issued quarterly.

Germany

Deutsches Institut für Normung e.V. (DIN), 4–7 Burggrafenstrasse 1 Berlin 30, and Kamekestrasse 2–8 5 Köln 1
Origin: In the early part of 1917 the Verein Deutscher Ingenieure (VDI—Society of German Engineers) formed a Standards Committee for General Mechanical Engineering, and later in that year it was succeeded by Normenausschuss der Deutschen Industrie (Standards Committee of the German Industry). A fur-

ther change took place in 1926 to form the Deutscher Normen-ausschuss (DNA) because its work had grown beyond the industrial field. In 1975 its title was again changed to Deutsches Institut für Normung e.V. (DIN). *Publications:* All standards have the prefix DIN. In addition, DIN issues regularly: *DIN Mitteilungen*, a bi-monthly journal; *Catalogue of DIN Standards*, a yearly publication; *DK-Mitteilungen*, a bi-monthly publication on UDC matters; and *Elektronorm*, a monthly bulletin on electrotechnical matters of standardisation; special lists of DIN standards translated into English, French and Spanish and *Normungs-Literatur* (literature on standards), listing all work on standards, published from time to time.

France

Association Française de Normalisation (AFNOR), Tour Europe, Cedex 7 92080 Paris-La Défense
Origin: In 1918 a Commission Permanente de Standardisation (Permanent Standardization Commission) was created by decree, but ceased operating a few years later. The Association Française de Normalisation (AFNOR) was founded in 1926 to fill the gap left by the Commission Permanente.
Publications: French Standards with the prefix NF. In addition, AFNOR issues regularly: *Courrier de la Normalisation*, a bi-monthly journal; *Bulletin Mensuel de la Normalisation Française*, a monthly bulletin; *Catalogues of French Standards*; and *Guides de l' Acheteur* (Buyers' Guides).

USSR

Gosudarstvennyj Komitet Standartov Soveta Ministrov SSSR (GOST), Leninsky Prospekt 9b Moskva 117049
Origin: After 1918 standardisation was carried out by the Bureau of Industrial Standardization under the main economic department of the Supreme Soviet for National Economy (SSNE), but systematic work on state standardisation began in 1925 when the Council of People's Commissars of the USSR issued a decree setting up the first central standardisation body, attached to the Council of Work and Defence. The Committee on Standards, Measurements and Measuring Instruments attached to the Council of Ministers of the USSR is the State organ responsible for standardisation and metrology throughout the Soviet Union.

Publications: USSR State Standards have the prefix GOST; *Industry Standards* (OST); *Republic Standards* (RST); *Enterprise Standards* (STP).

State Standards are mandatory for all enterprises, organisations and institutions of the Union or Republics and for the local authorities in all sectors of the national economy of the USSR and the federated Republics. *State Standards* are adopted by the Committee on Standards, Measurements and Measuring Instruments attached to the Council of Ministers of the USSR, except for those of fundamental importance, which are adopted by the Council of Ministers itself.

Industry Standards are mandatory for all enterprises and organisations of the relevant sector of the national economy, and are adopted by the Ministry or Department that is the largest producer of such goods.

The standards of federated Republics are mandatory for all enterprises and organisations under each Republic, regardless of the sector of the national economy to which they belong. They are adopted either by the Councils of Ministers of the federated Republics or by the Planning Departments of those Republics.

Enterprise Standards are compulsory only within the enterprise in question and are adopted by the directorate of the enterprise.

In addition, the USSR issues the following publications: the journal *Standards and Quality* with the supplement *Reliability and Control of Quality*, published monthly; the journal *Measurement Techniques* with the supplement *Metrology Matters*, published monthly; *Express-Standards*, published 5 times a month; *Catalogue of State Standards*, published annually; *Catalogue of Information on Foreign Standards*, published monthly; *Catalogue of COMECON Recommendations*, published annually; and *Catalogue of ISO Recommendations*, published annually. General information is provided in *Standardization, Metrology and Quality in the USSR* (thematic studies, 5–6 publications a year) and *Standardization, Metrology and Quality Abroad* (a series of thematic studies, 9–10 publications a year). Also, standard reference data and other works on standardisation are issued by the USSR.

Brief details of other national standards organisations are listed in *Table 7.1*. The establishment of national standards organisations by the newly developing countries is constantly under consideration.

Table 7.1

Country	Title of Organisation	Abbreviation	Prefix to standards	Publications
Australia	Standards Association of Australia	SAA	AS	Annual List of Publications / Annual Report / Monthly Information Sheet
Austria	Österreichisches Normungsinstitut	ÖN	ÖNORM	Ö Norm (monthly bulletin) / Annual Catalogue of Standards
Belgium	Institut Belge de Normalisation	IBN	NBN	Annual Report / Annual Catalogue of IBN Standards / Revue IBN (monthly journal)
Brazil	Associação Brasileira de Normas Técnicas	ABNT	ABNT	Boletin (bi-monthly journal) / ABNT Standards (in volume form)
Bulgaria	Comité de la Qualité, de la Normalisation et de la Métrologie	KKCM	BDS	Monthly Bulletin listing State Standards, approved Amendments to Standards, etc.
Canada	Standards Council of Canada	CSA	CSA	Standards/Canada (bi-monthly journal) / CSA Quarterly Review
Chile	Instituto Nacional de Investigaciones Tecnologicas y Normalización	INDITECNOR	INDITECNOR	Revista Chilena de Racionalización
Colombia	Instituto Colombiano de Normas Técnicas	ICONTEC	ICONTEC	Monthly News Bulletin
Cuba	Dirección de Normas y Metrologia Ministerio de Industrias	NC	UNC	NC Bulletin (issued bi-monthly)
Czechoslovakia	Úřad pro normalizaci a měřeni	CSN	CSN	Normalizace (a monthly periodical) / Měrová technika (a monthly periodical)

Country	Title of Organisation	Abbreviation	Prefix to standards	Publications
Denmark	Dansk Standardiseringsraad	DS	DS	*Danish Technical Review* (issued monthly) / *Standard NYT* (monthly) / *Annual Report* / *List of Danish Standards* (issued bi-annually)
Finland	Suomen Standardisoimislitto r.y.	SFS	SFS	*SFS Tiedotus* (a bulletin issued monthly) / *SFS Standards* (issued annually)
Greece	Ministry of National Economy	NHS	ENO	An announcement on the publication of each Standard appears in the national press
Hungary	Magyar Szabványügyi Hivatal	MSZH	MSZ	*Lists of Standards* (issued from time to time)
India	Indian Standards Institution	ISI	IS	*ISI Bulletin* (issued monthly) / *Standards – Monthly Additions* / *Annual Report*
Indonesia	Jajassan 'Dana Normalisasi Indonesia'	DNI	NI	*Berita DNI* (bi-monthly periodical)
Iran	Institute of Standards and Industrial Research of Iran	ISIRI	ISIRI	*ISIRI Bulletin* (issued from time to time)
Iraq	Iraqi Organization for Standards Planning Board	IOS	IOS	*Bulletin* (issued quarterly)
Ireland	Institute for Industrial Research and Standards	IIRS	IS	*Monthly Journal* / *Irish Standards Yearbook* (issued periodically) / *Annual Report*
Israel	Standards Institution of Israel	SII	SI	*Quarterly Journal* / *Catalogue of Israel Standards* (issued annually)

Country	Organization			Publications
Italy	Ente Nazionale Italiano di Unificazione	UNI	UNI	*Unificazione* (issued quarterly) *List of UNI Standards* (issued periodically) —
Japan	Japanese Industrial Standards Committee Ministry of International Trade and Industry	JISC	JIS	*Standardization Bulletin* (issued bi-monthly) *Standards Announcements* (issued bi-monthly)
Democratic Republic of Korea	Committee for Standardization of the Democratic People's Republic of Korea	CSK	—	*Quality Control* (monthly bulletin) —
Republic of Korea	Korean Bureau of Standards	KBS	KS	*SIM Standard Yearbook*
Lebanon	Lebanese Standards Institution	LIBNOR	LS	*Annual Report*
Malaysia	Standards Institution of Malaysia	SIM	MS	*SIM News* (to be issued at regular intervals)
Mexico	Dirección General de Normas	DGN	—	*Catalogue of Standards* *Annual Report*
Netherlands	Nederlands Normalisatie-Instituut	NNI	NEN	*Normalisatie* (issued monthly) *Catalogue of Standards*
New Zealand	Standards Association of New Zealand	SANZ	NZSS	*Monthly Newsletter* *NZ Quarterly Bulletin* *Index of NZ Standards* (issued annually)
Nigeria	Nigerian Standards Organization	NSO	NIS	*List of Industrial Standards* (issued periodically)
Norway	Norges Standardiseringsforbund	NSF	NS	*Annual Report* *Catalogue of Norwegian Standards* (issued annually) *Standardisering* (quarterly bulletin)

Country	Title of Organisation	Abbreviation	Prefix to standards	Publications
Pakistan	Pakistan Standards Institution	PSI	PS	*PSI Standards Bulletin* (issued quarterly) *Annual Report*
Peru	Instituto de Investigación Tecnológica	ITINTEC	INANTIC	——
Philippines	Bureau of Standards of the Philippines	KP	PTS	——
Poland	Polski Komitet Normalizacji i Miar	PKNIM	PN	*Normalizacja* (issued monthly) *Biuletyn PKN* (official bulletin of the *PKNIM* (issued monthly) *Catalogue of Polish Standards (PN)* *Catalogue of Industrial Standards (BN)* (issued from time to time)
Portugal	Repartição de Normalização	IGPAI	NP	*Standardization Bulletin* *Catalogue of Portuguese Standards* Index cards on definitive *Portuguese Standards*
Romania	Institutul Român de Standardizare	IRS	STAS	*Standardizarea* (issued monthly)
Singapore	Singapore Institute of Standards and Industrial Research	SISIR		——
South Africa	South African Bureau of Standards	SABS	SABS	*Standards Bulletin* (issued monthly) *Annual Report* *Yearbook of SABS Standards and Codes of Practice*
Spain	Instituto Nacional de Racionalización y Normalización	IRANOR	UNE	*Racionalización* (issued bi-monthly)

Country	Organisation			Publications
Sri Lanka	Bureau of Ceylon Standards	BCS	CS	
Sweden	Sveriges Standardiserings-kommission	SIS	SIS	*Catalogue of Swedish Standards,* with titles in Swedish and English (issued annually) *SIS-nytt (SIS News)* (issued monthly)
Switzerland	Association Suisse de Normalisation	SNV	SNV	*VSM/SNV Standards Bulletin* (issued monthly)
Thailand	Centre for Thai National Standard Specifications	CTNSS	THAI	*CTNSS Annual Report*
Turkey	Türk Standardiari Enstitüsü	TSE	TS	*Standard* (magazine issued monthly) *Catalogue of Turkish Standards* (issued from time to time)
United Arab Republic	Egyptian Organization for Standardization	EOS	ES	*UAR Standards Bulletin* (issued bi-monthly in Arabic and English) *Standards Yearbook in Arabic and English* (with yearly supplements)
Venezuela	Comisión Venezolana de Normas Industriales	COVENIN	NORVEN	
Yugoslavia	Jugoslovenski zavod za Standardizaciju	JZS	JUS	*Standardization* (issued monthly) *Catalogue of Yugoslav Standards* (issued annually)

Standards organisations have also been formed in the following countries:

Cameroons	Direction de l'Industrie (Service de normalisation)
Ecuador	Instituto Ecuatoriaro de Normalización
Ivory Coast	Bureau Ivoirien de Normalisation
Jamaica	The Bureau of Standards
Jordan	Directorate of Standards, Ministry of National Economy
Kenya	Kenya Bureau of Standards
Malawi	Malawi Bureau of Standards
Saudi Arabia	Saudi Arabian Standards Organization

International

The International Organization for Standardization (ISO) and the International Electrotechnical Commission (IEC) are the main international organisations concerned with standardisation.

International Organization for Standardization (ISO), 1 rue de Varembé, 1211 Geneva 20
Origin: In 1926 the International Federation of the National Standardizing Associations (ISA) was formed and 20 countries joined this Association. It functioned until World War II and the ISA officially ceased work in 1942. In 1944 ISA was succeeded by the United Nations Standards Co-ordinating Committee (UNSCC), consisting of the 18 Allied countries, but in 1946 discussion took place in London, when recommendations were made to establish a more permanent international organisation for standardisation. These recommendations were adopted by 25 nations, and a new organisation entitled ISO was established in 1947.

The ISO is an international non-governmental organisation and has consultative status with United Nations and many of its agencies. It also maintains liaison with many other international organisations. The membership of ISO now consists of 60 member countries.

Publications: The ISO publishes *International Standards* and *Technical Reports* on all subjects except those concerned with the electrical and electrotechnical industries (see IEC, below). In addition, ISO issues regularly *ISO Memento,* issued annually; *ISO Catalogue of International Standards, Recommendations and Draft International Standards,* issued annually and *The ISO Bulletin,* issued monthly.

International Electrotechnical Commission (IEC), 1 rue de Varembé, 1211 Geneva 20
Origin: Towards the end of the nineteenth century electrical engineers in many countries became acutely aware of the need for international standardisation in the electrical field, and several congresses were held during this period. In 1904 at the St Louis (USA) International Electrotechnical Congress a resolution put forward by R. E. Crompton (UK) was approved, and the International Electrotechnical Commission (IEC) came into being in 1906 and has functioned continuously since that time. The first President was Lord Kelvin, and the IEC held its first meeting in London in 1908. In 1947 the IEC became

affiliated with ISO as its Electrical Division, but preserving its technical and financial autonomy. The members of the IEC are the national committees, one from each country, with representatives from the interested sections of the electrical industry in each country. The membership of the IEC consists of 44 countries.

Publications: The IEC publishes *IEC Recommendations* and these are intended to serve as the basis for national standards. In addition, IEC publishes regularly *Catalogue of IEC Publications*, issued annually; *IEC Bulletin*, issued quarterly; *IEC Report of Activities*, issued annually; and *The IEC Handbook*, issued annually.

Other international organisations concerned with standards include the following.

Non-governmental

European Committee for Standardization (CEN). Founded 1960. Membership consists of national standards bodies of EEC and EFTA countries. The *European Standards* prepared under CEN are intended to be adopted without deviation by member countries.

European Electrotechnical Committee for Standardization (CENELEC). Founded in 1960 as CENEL, renamed in 1972. Counterpart of CEN. Members consist of EEC and EFTA countries with Finland as associate member. The *European Electrotechnical Standards* prepared by CENELEC are intended to be adopted without deviation by member countries.

International Commission on Rules for the Approval of Electrical Equipment (CEE). Founded 1946. Comprises national electrotechnical committees of 19 European countries with Australia, Canada, Iceland, Japan, South Africa and the USA as observer members.

International Special Committee on Radio Interference (CISPR). Set up under the aegis of IEC, bringing together IEC member committees, and other international organisations in the electrical, broadcasting and transport fields.

Pan American Standards Commission (COPANT). Founded 1961. Comprises national standards bodies of USA and 11 Latin American countries. A co-ordinating organisation concerned with the regional implementation of *ISO International Standards* and *IEC Recommendations*.

Asian Standards Advisory Committee (ASAC). Set up in 1966 under the Economic Commission for Asia and the Far East (ECAFE) auspices to promote co-ordination between existing national standards bodies in the region and to assist the establishment of new ones.

Intergovernmental organisations

European Economic Community (EEC). The 'Common Market' founded by the Treaty of Rome 1957.

European Coal and Steel Community (ECSC). Founded by the Paris Treaty 1951.

European Atomic Energy Community (EURATOM). Founded by a second Rome Treaty 1957.

Commission of the European Communities (CCE). Thirteen senior officials appointed to head the administrative staff in Brussels and Luxembourg.

European Free Trade Association (EFTA). Austria, Norway, Portugal, Sweden, Switzerland, plus Finland and Iceland.

Organisation for Economic Co-operation and Development (OECD). Founded 1961. Comprises Western Europe, plus USA, Canada, Japan.

International Organization for Legal Metrology (OIML). Set up in 1955 to resolve the technical and administrative problems of legal metrology, raised by the construction, use and checking of instruments of measurement, and to facilitate co-operation between states in this field.

General Conference of Weights and Measures (CGPM). Membership drawn from those nations who are signatories to the Metre Convention.

United Nations agencies

Economic Commission for Europe (ECE). Acts to facilitate trade in Europe and notably prepared regulations associated with 'E' mark certification scheme (now operating for motor vehicle accessories).

Economic Commission for Asia and the Far East (ECAFE)
Economic Commission for Latin America (ECLA)
Economic Commission for Africa (ECA)
Economic and Social Committee (ECOSOC)
Food and Agriculture Organization (FAO)
General Agreement on Tariffs and Trade (GATT)
International Atomic Energy Agency (IAEA)
International Labour Organization (ILO)
Intergovernmental Maritime Consultative Organization (IMCO)
International Telecommunications Union (ITU)
UN Conference on Trade and Development (UNCTAD)
UN Educational, Scientific and Cultural Organization (UNESCO)
UN Industrial Development Organization (UNIDO)
World Health Organization (WHO)
World Meteorological Organization (WMO)
Codex Alimentarius Commission (CODEX). Created to implement joint FAO/WHO Food Standards Programmes

Other organisations

Organization for Liaison between the European Electrical and Mechanical Engineering Industries (ORGALIME), representing national trade organisations of the EEC/EFTA countries

Union of the Industrial Federations (UNICE) (parallel to CBI) of the EEC countries

Council of European Industrial Federations (CEIF) parallel to UNICE, outside the EEC

European Organization for Quality Control (EOQC)

International Federation for Documentation (FID)

International Organization of Consumers Unions (IOCU)

BIBLIOGRAPHY

British Standards Institution (1951). *Fifty Years of British Standards 1901–1951*
MacNiece, E. H. (1953). *Industrial Specifications* (Wiley)
Reck, D. (ed.) (1956). *National Standards in a Modern Economy* (Harper)

Sanders, T. R. B. (ed.) (1972). *Aims and Principles of Standardisation* (International Organization for Standardization)

Woodward, C. D. (1972). *BSI: The Story of Standards* (British Standards Institution)

Woodward, C. D. (ed.) (1965). *Standards for Industry* (Heinemann)

8

Product information

D. P. Easton

Every year vast sums of money are lost in engineering research and development departments because of the failure to control information about the nature and supply of bought-out components and materials.

If an engineer does not know every firm which can supply a product, he cannot select with confidence the ideal component for the design. If, lacking knowledge of new developments, he uses an out-of-date component in a design, the final product will itself be out-of-date before it is made.

If his information sources are inaccurate, incomplete or out-of-date, he will be faced with an unacceptable search time finding each new component. The writer once heard how in a large UK manufacturing group one Division spent a great deal of effort designing to prototype stage a new piece of test equipment only then to discover that another Division within the group had already designed, and for 3 years had been selling, an identical piece of equipment.

Our society just cannot afford either the waste of manpower or the loss of other discoveries that might have been made had it not been for this failure of communication. The facts and information are known by someone somewhere. The failure is our inability to marshal the facts and render them available to everyone, anywhere, at the right time.

Many engineering companies have not learned to recognise the serious consequences of half ignoring the problem. Good draughtsmen and engineers are aware of it because it is their time that is wasted on non-engineering, mainly clerical, search time. Bad engineers are perhaps only too happy to take the easy route offered by the first component they find and only too anxious to avoid the test of their engineering skills that is demanded by a more thorough search of available items. Other engineers who have experienced the frustration of inadequate information tend to believe that they have provided a solution by employing a technical librarian. The fact is that however skilled the technical librarian, he is just not provided with a large enough budget to handle the masses of product data required by a broadly based research and development organisation.

How technical librarians fail to meet the problem is demonstrated by the results of a survey conducted in a company that was spending substantial sums attempting to provide a service. Why they fail is revealed when one examines the sources of information available to them.

A SURVEY OF ENGINEERING TIME LOSSES

Some years ago Technical Indexes Ltd. conducted a survey among engineers and draughtsmen working in the research and development departments of a leading UK company. The summarised replies from just under 2000 men were as follows:

Estimate the number of hours spent each week searching for information on materials and components for use in the product you design/manufacture.   3.36 hours

Do you require to see the supplier's catalogues before finally specifying his product?

Always: 49%   Sometimes: 48%   Never 3%

Do you limit your selection of components to those upon which you already have information?

Always: 11%   Sometimes: 74%   Never: 15%

Does the 'Company Information Centre' hold a satisfactory catalogue collection?

Yes: 34%   No: 66%

Is it complete?   No: 87%
Is it up to date?   No: 69%
Is it product indexed in real depth?   No: 74%

Do you keep a personal catalogue file?

Yes: 60%   No: 40%

Is your file satisfactory?

       Yes: 18%   No: 82%

In other words:

At least 10% of the working week is being spent looking for product information.

The suppliers' catalogues are an extremely important source of that information.

A substantial number of engineers admit that if they have information on a suitable component, they do not bother to look any further for something better or cheaper.

The company's information centre is failing to provide a satisfactory catalogue collection.

Because of the inadequacy of the company's collection, engineers make their own private collections.

Private collections are of little use, not only because they are too small, but also because colleagues who might need that information have no way of knowing that it is there in someone else's desk.

THE INADEQUACY OF INFORMATION SOURCES

It is tempting to complain that if the suppliers want customers to purchase their products, the onus is firmly on the supplier to ensure that they are kept informed about their products. The sources of product data are almost entirely advertising material. Marketing people spend enormous sums of money trying to inform their current and potential customers: directly mailed catalogues, visits from sales representatives, press releases resulting in journal editorial, exhibitions, journal advertising, etc., etc. However, to whom was attributed the statement that he knew half of his advertising costs were wasted but that he only wished he knew which half?

The problem is one of size and of timing. There are too many engineers to inform; there are too many products for the engineer to know about. Each supplier plans his advertising efforts very carefully to announce a new product, or to remind the market about an old one, but from the individual engineer's point of view there is a hopelessly unplanned avalanche of bits and pieces of valuable information each arriving when it is not needed.

Perhaps the one thing for which suppliers can be blamed is not presenting their information in a standard format. The literature comes in all shapes and sizes in spite of *BS 1311:1955. Sizes of Manufacturers Trade and Technical Literature (Including Recommendations for Contents of Catalogues)*.

The problem of course is that the catalogue is designed above all to persuade and sell, and no one can persuade two or three competitors that their products should be presented in an identical fashion when there might be a sales advantage in presenting them in a unique manner. The individual can complain about aspects of the suppliers' efforts to keep him informed, but he cannot really blame them for a problem that is fundamentally his own.

The engineer may ask his colleagues if they can remember where to obtain the product, but they cannot remember everything and so he turns each month to his technical journal, but is met by the familiar problem of timing and memory. The journals carry advertisements from suppliers, and skilfully report important new developments. From time to time they review 'the state of the art', but most of the time they are concerned with the expert interpretation and presentation of the 'news' of the industry. And news must never be confused with knowledge. News is current awareness. News is what happened last month, what is happening now, and is concerned with what, in the light of current developments, is likely to happen next month and next year. Product knowledge is a retrievable store of valid current information: the summation of old news with the rumour, speculation and deaths removed. So the journal is a valuable and necessary source of product information, but its function is to provide news rather than knowledge.

The engineer may attend an exhibition where he does have the opportunity of carrying out a thorough survey of available products . . . once a year . . . or every other year. A valuable information source, but once again it does not today answer *today*'s query. Exhibitions of course are excellent places to collect catalogues that can be stored in the private collection, or, better, handed to the technical librarian for the corporate catalogue library.

The engineer may look up his problem item in a buyer's guide or a directory that is kept in the technical library. The value of the directory is that it can tell you 'who makes what' and, therefore, where to ask for the catalogues and data sheets. Thus, an answer to the query can be obtained within the 10–14 days that it will take for the catalogues to arrive. There is obvious good sense in leaving to someone else the tiresome task of collating the 'who makes what' information. But how accurate are most directories? After all there is little point in waiting 10–14 days only to discover that many of the addresses were inaccurate and many of the companies listed do not supply the item sought, and then 6 months later to realise that many of the suppliers of this item were not in fact listed in the directory.

The reason for the inadequacy of most directories is to be found in the price. They are too cheap. Collecting, sorting and disseminating information is costly, and most directories cannot (for their price) employ sufficient resources to ensure the accuracy and comprehensiveness that the engineer requires of his information sources. Furthermore, the buyer's guide only contributes to the first stage of an engineer's enquiry—who makes a fairly broad group of product types? It cannot tell him who can supply precisely what he wants. It does not provide detailed information upon the product itself.

Product information sources are bad because they come in what is to the engineer a disorganised flood from thousands of different suppliers in differing formats, at times not necessarily ideal for the engineer. Therefore, bad decisions and delayed decisions occur. Product design is poorer than it need be. Design time is longer than it should be. Product price will be uncompetitive or profit margins too low for continued engineer job security. The solution is that product information sources must be rationalised by the engineer and not the supplier, who is in no position to do much more than he does at present.

COLLECTION OF CATALOGUES

Catalogues may be collected internally or externally. If an internal system is chosen, the most likely candidate for the job is the company librarian, who is the expert on gathering, classifying and retrieving information. Catalogues come to the company free of charge and may be channelled through him for filing. However, the size of the problem can be assessed from figures showing that suppliers to the electronic engineering industry use 200 000 pages of catalogues to describe their products. A similar volume of material is produced for the chemical, electromechanical and general engineering industries. The updating and classification of this material would obviously be an onerous task for the librarian, and not even the largest unit could support the budget required, although it could justify a high expenditure from the savings in engineering time alone. But not even the largest units need to spend so much if the job is being done by a specialist who recovers his costs from a large number of users.

This leads to the alternative external method. If external specialists are to collect, classify and disseminate the information, then the ideal system has to be found. The engineer is the end-user, so he must dictate its form. Engineers are not information specialists, so they must be able to ask questions of the data bank in straightforward engineering terms and not have to put their question into a special form just for the sake of the system. In many information systems this problem is overcome by interposing a librarian between the engineer and the data bank. The librarian translates the engineer's question into terms understood by the data bank. Such an interposition, although unavoidable in some circumstances, is clearly undesirable. A conversation through an interpreter is bound to be less fluent than a direct one.

Answers may be required to questions relating to the supplier of a type of product, the supplier of a precise product with precise parameters, the performance characteristics or dimensions of a specific product produced by a specific supplier, the type of product, or combination of products, which can perform a particular function, etc. The questions vary from the specific to the general: some require a short alphanumeric answer; some require a graphic reply; and some call for the opportunity to browse through a general mass of information.

The variety of question and answer clearly calls for a flexible system employing an extremely versatile storage and display capability.

If there is only one possible answer to the question, the engineer will not welcome the statement, 'Don't know'. If there are 20 possible answers, a reply providing only five answers will not enable him to make the optimum selection. Thus, the data bank needs to be as comprehensive as possible. If there are two realistic answers, he does not want to see 150 possible replies, because this will leave him with the irksome task of sorting the two possibles out of the 150 replies.

Thus, the classification system must be fine enough to provide only the answers he asks for with a minimum of excess information, but not so fine as to restrict the browsing facility that is so necessary in the design process.

Enquires are usually urgent. The answer is wanted today, this morning, this minute—not in a week's time. The system will therefore have to provide direct on-line access to the data bank.

Information sources that the information specialist requires are mostly contained in the catalogues and data sheets published by suppliers. Even these do not always provide complete information, and so, if he wants a perfect system, the specialist will have to approach each supplier for missing information to include in the data bank.

The first costs are those attributable to collecting the catalogues and additional information, keeping it all up to date, recording, and classifying. These are heavy costs, requiring considerable numbers of expert staff, but not astronomic when considered in relation to a large number of users. The sting is in the cost of storing, retrieving and disseminating. Quite a number of computer-based systems have been devised. They all envisage a network of terminals on-line to a powerful central computer. When consideration is given to the quantity of information to be stored, the range of questions and the variety of answer format required, it is not too difficult to see why none of them has so far materialised as an economic possibility for the great majority of engineers. The cost of not rationalising information sources will continue as a heavy and not entirely necessary burden on engineering companies; and although a perfect solution may not be available for a long while, economic compromise is possible.

COMMERCIALLY AVAILABLE SERVICES

In the USA Information Handling Services Inc. have for many years been providing a vendor catalogue service called VSMF. Catalogues of suppliers to certain industries are microfilmed and classified. The engineering user has a microfilm reader or reader/printer, a supply of cassettes of microfilmed catalogues, and an index book as an access to the information. The files are updated at regular intervals. They are reasonably comprehensive and reasonably indexed, and answers to questions are immediately available without the need for a systems interpreter. This is not as perfect as the ideal solution, but is a workable economic solution that saves a great deal of engineering time.

In the UK Technical Indexes Ltd. offers similar services for a range of engineering disciplines, and Barbour Index Ltd. deals with architects and the construction industry. None of these systems is expensive, because the costs are spread over a large number of users. These services are of course tailored to an industry and not to a specific company, so the individual technical librarian still has the minor problem of providing an ancillary service.

The great advantages of commercially provided services are:

The company's technical librarian is freed from an irksome major task that in any case he does not, and cannot, adequately perform himself.

To use the systems requires no special training other than familiarisation with very simple equipment.

The specialised, expensive and frequently boring work of creating an information bank is carried out consistently by one central body and only a small proportion of the cost is borne by each user.

MICROFILM SYSTEMS

If systems employing computer terminals have been rejected, at least for the present, because of an unacceptable cost level, the best alternative appears to be microfilm. And microfilm is not perfect. Few would deny that it is preferable to have the original printed literature on the desk. Once the correct pieces of literature have been brought to him, the individual finds them more convenient to use, and more pleasant to read, than a microfilmed copy projected on to a screen or than an abstract from a computer-held data bank also projected on to a screen. The compromise has to be accepted in exchange for the facility of getting the right information on the desk at the right moment.

Microfilm comes in a variety of formats, and it is important to appreciate that the format best suited to one task is not ideal for a different job.

Aperture cards—most engineers are familiar with these punched-card size cards having an aperture large enough to take a single 35 mm frame upon which has been filmed a single engineering drawing or a small number of associated documents. They are perfect for storing bulky engineering drawings, for sorting them, for issuing copies and for mailing copies very cheaply.

Microfiche—a sheet of film (about 6 in $\times$ 4 in) containing up to 98 A4 pages. These are ideal for reports, articles and papers that are likely to be requested as individual items, and where relatively small (in publishing terms) quantities of each item are called for. The equipment for reading fiche is often portable, usually fairly cheap, and, because there are few moving parts, generally very reliable. The suppliers of vendor catalogue systems have almost unanimously rejected fiche in favour of 16 mm roll film cassettes, even though catalogues might appear at first sight to fall within the ideal category for fiche. The paramount reason for rejection is 'file integrity'. A file containing 2000 or 3000 separate fiche calls for disciplined use if it is to remain complete and orderly. The commercial vendor catalogue systems aim to be 'self-servicing'. That is, they design the systems for direct access by the inquirer without the need for a librarian to be interposed between inquirer and data bank. The aim is to provide more rapid access to the information and to save librarian-time just as much as it saves engineer-time. A file of several thousand fiche is rapidly disordered when there is free and frequent access to it, and the result is that the system ceases to be fully self-servicing.

Roll film in cassettes—16 mm film in lengths of 30 or 60 m contained in protective cassettes. The readers are more expensive than fiche readers, but most casual users seem to find them easier to use. Each single unit in the file, namely a cassette, can hold as many as 20 000 pages, none of which can be disordered or rearranged. File integrity is almost perfect and a self-servicing system is achievable.

FURTHER READING

Barber, R. (1970). 'A Retrieval System for Product Data', *Information Scientist*, **4** (1), 3–10
Ford, M. (1972). 'The Technical Indexes System for the Control of Trade Literature', *Aslib Proceedings*, **24** (5), 284–292
Kelbrick, N. (1971). 'Trade Literature as a Library Material', *Library Association Record*, **73** (4), 65–67
Kennington, D. (1969). 'Product Information Services—Some Comparisons', *Aslib Proceedings*, **21** (8), 312–316
Short, M. P. (1971). 'Micropublishing of Packaged Information Systems', *Journal of Micrographics*, **5** (2), 75–80
Wall, R. A. (1968). 'Trade Literature Problems', *Engineer*, **225** (5851), 453–454; **225** (5852), 489–491

9

The publications of governments and international organisations

D. F. Francis

GREAT BRITAIN

Her Majesty's Stationery Office (HMSO) is a prolific publisher. There are a number of works devoted, *inter alia*, to the elucidation of parliamentary and non-parliamentary publications, green papers, blue books and the like, e.g. P. Ford and G. Ford's *A Guide to Parliamentary Papers* (Blackwell, 1956), J. G. Ollé's *An Introduction to British Government Publications* (2nd edn, Association of Assistant Librarians, 1973) and J. E. Pemberton's *British Official Publications* (2nd edn, Pergamon, 1973).

All users of HMSO publications should be aware of the various catalogues describing them. The *Daily List of Government Publications* and the monthly *Government Publications Issued during* [e.g.] *March 1973* enable a current check to be kept, but more generally useful is the annual catalogue *Government Publications* [e.g.] *1972*. As with the daily and monthly lists, the annual catalogues detail legislative material first, but some three-quarters of each volume is occupied by lists of publications classified by ministry, department or body responsible for them, arranged alphabetically. Thus, knowing the departments likely to publish material in his field, the researcher can see at once a list of such publications issued for the year.

The monthly and annual catalogues are provided with author, title and subject indexes, but a caveat must be given that they are not very detailed, and that some types of relevant publications are omitted, e.g. United Kingdom Atomic Energy Authority reports, and Boiler Explosion enquiries. Most useful for subject work are the *Sectional Lists*. These list in-print material covering either the publications of a particular ministry or department, or items from several departments on a particular subject. Their frequency varies, and the lists do change from time to time as departments or ministries are reorganised or change their function. A selection of those most likely to interest engineers follows, with numbers and titles as at Summer 1975.

No. 3. Departments of Trade, Industry, Energy, Prices and Consumer Protection
No. 5. Department of the Environment
No. 8. Aeronautical Research Council
No. 22. Department of the Environment (Transport)
No. 45. Institute of Geological Sciences
No. 61. Building
No. 67. Ministry of Defence

The Department of the Environment (DOE), which came into being at the end of 1970 with the unification of the former Ministries of Transport, Housing and Public Works, is particularly noteworthy. A number of important publications are issued under its aegis: in particular, those of the Transport and Road Research Laboratory, with its annual report *Road Research* and *Road Notes* series, and *Technical Papers*; the Building Research Station, with its monthly *Building Science Abstracts* (1928–); the Fire Research Station; the Hydraulics Research Station; and the Water Research Centre with its *WRC Information* (formerly *Water Pollution Abstracts*, 1927–74) (1975–). Railways, water supply, sewerage and air pollution are other topics dealt with by DOE. Early in 1973 a catalogue was issued of its 1971 publications, classified by subject and provided with a detailed index. This proposed annual will be a welcome addition to existing bibliographical tools, where the general HMSO catalogues and *Sectional Lists Nos. 5, 22* and *61* must be consulted.

The Department of Trade and Industry (DTI) was also formed in 1970, by combining the Board of Trade with the Ministry of Technology. The latter had merged with the Ministry of Power, so by 1970 had responsibility for electricity, gas, coal, oil, mines, petroleum, atomic energy, and iron and steel. In 1974, these various functions were diversified again into separate Departments of Trade, Industry, Energy, and Prices and Consumer Protection, but, at the time of writing, the pattern of publishing remains similar. *Sectional List No. 3* records the most important publications in the appropriate subject areas. Of the many activities covered, those of the National Engineering and National Physical Laboratories are among the most important.

The National Engineering Laboratory (NEL), East Kilbride, Glasgow, was set up in 1949. It has a research programme of its own, and also undertakes commissions for industry, particularly in fluids, machinery and materials. Its *Annual Report* and (also annual) *Heat Bibliography* (1948–) are particularly worthy of note. An annual *Index of NEL Publications* is available from the above address, which includes short summaries of the reports listed.

The National Physical Laboratory (NPL) was founded in 1900, and has its headquarters in Teddington, Middlesex. Its work is divided into three groups: that of the Engineering Sciences Group concerns ships and other marine vehicles, computer usage, fluid dynamics, acoustics and information technology; that of the Materials Group 'maintaining the national standards of measurement . . ., providing definitive data on the engineering properties of substances, and . . . advising on meaningful ways of specification of materials'; and that of the Measurement Group 'the establishment of internationally acceptable basic standards of measurement, the development of new techniques . . ., and the furtherance of their use . . .'.

The NPL issued an *Annual Report* up to 1968/9, whence each group in turn has issued its own report. Monographs are also published, recent ones including F. Wormwell's *Corrosion of Metals Research* (1973) and D. E. J. Walshe's *Wind-Excited Oscillations of Structures* (1972). A complete catalogue of publications is available from the NPL.

The joint departments issue several reference tools and bulletins. *Technical Services for Industry* (1963–) is a detailed directory of information and services available from government organisations and government-sponsored bodies, appearing every 3 years or so, while *New Technology*, a monthly supplement to *Trade and Industry* (1970–) (previously *Board of Trade Journal*), mentions all new DTI publications. The Technology Reports Centre issues 24 times a year *R & D Abstracts* (1969–) and there are various other newsletters and journals.

Section List No. 3 also records the publications of, for example, the national and regional gas and electricity authorities; the National Coal Board; the Warren Spring Research Laboratory, with its work on pollution; and the Safety in Mines Research Establishment: also of various bodies dealing with metrication, mines and quarries, some aspects of patents and inventions, shipping and aviation.

There are several other areas of government involvement in engineering and technology. The Ministry of Defence groups its publications into sections for the Navy, Army and Air Force departments. Particular attention is drawn to the military engineering side of the Army department's publications, concerning such topics as concrete, bridges and applied geology. The Aeronautical Research Council now comes under the Ministry of Defence. It issues two series, *Reports and Memoranda . . .* and *Current Papers. . . .* The former are considered to have permanent value, and are printed by a photosetting process; while the latter are usually of more ephemeral interest, and are reproduced by photocopying or stencilling.

The United Kingdom Atomic Energy Authority has an express policy of publicising its non-classified information as widely as possible, and to that end issues free a monthly *UKAEA List of Publications Available to the Public.* Also useful is J. R. Smith's *Guide to UKAEA Documents* (5th edn, HMSO, 1973).

Finally worth noting is *Sectional List No. 61,* which collects together the material of various government departments concerned with building. In addition to those of the DOE, appropriate publications of the Departments of Health and Social Security, Education and Science, Employment and others are listed.

The monthly *BLLD Announcement Bulletin* (which first appeared in January 1971 as the *NLL Announcement Bulletin,* superseding *British Research and Development Reports*) lists reports, translations and some theses. The entries are grouped into broad subject areas, e.g. aeronautics; electronics and electrical engineering; mechanical, industrial, civil and marine engineering and naval architecture; nuclear science and technology, etc. A good many of the reports stem directly or indirectly from government departments, so that the *BLLD Announcement Bulletin* is another important means of subject access to government documents.

While legislation on a particular topic *can* be traced through the lists of statutes in the HMSO catalogues, a simpler way is to use *Halsbury's Statutes of England* (3rd edn, Butterworths, 1968–). All the statutes in force are presented classified into subject groups. Thus, for example *Electricity* is found in volume 11; *Mines, Minerals and Quarries* in volume 22; and *Patents and Designs* in volume 24. A current statutes service keeps the series up to date.

USA

The United States Government Printing Office claims to be the world's most prolific publisher, and covers a wide subject spectrum. There are, by corollary, a large number of catalogues and guides to its vast output. Only a few of the most apposite will be considered here. Further information can be obtained from, *inter alia,* L. F. Schmeckebier and R. B. Eastin's *Government Publications and their*

128   *Publications of governments and international organisations*

Use (2nd edn, Brookings Institution, 1969), especially Chapters 2, 3 and 17. Some indication of the range of literature in engineering subjects can be obtained from S. Wynkoop's *Subject Guide to Government Reference Books* (2nd edn, Libraries Unlimited Inc., 1972). This annotated bibliography of over 1000 items lists some 50 items in its *Engineering* sub-section of *Science and Technology*. Other sub-headings include *Earth Sciences* and *Environmental Sciences*. It is intended that this work be updated every 2 years.

A more specific, and very useful, source is J. Bereny's *Guide to Locating US Government Technical Information, Technology and Patents* (TTA Information Services Company, 1972). This identifies most of the major technical information sources of the US government agencies; explains and provides practical identification of major publications and the various announcement services; describes NASA (National Aeronautics and Space Administration) and AEC (Atomic Energy Commission) technology utilisation programmes; and lists patents of the former available for foreign and domestic licensing. It also tells how to order the documentation. Its four sections describe the major catalogues of government publications; the technical information available from the individual agencies; government information data services; and sources of information on government patents. It is in loose-leaf format to accommodate updating supplements, and all its information is most easy to assimilate.

The Official *Monthly Catalog. United States Government Publications* lists publications by title, grouped under what it calls 'Government authors', i.e. the various bureaux and other publishing offices. Although not an absolutely complete record of US government publishing, it is indispensable to those wishing to keep abreast of current output. Each issue has an index including subjects and some titles (more detailed than the indexes in the HMSO catalogues), and an annual index follows the December number. There are several dozen entries under *Engineering* and *Technology* in the 1974 annual index, for example, as well as many more under more specific headings. Decennial indexes are produced from time to time. Departments producing regular material on these topics include the Atomic Energy Commission, Civil Aeronautics Board, Commerce Department (especially its National Bureau of Standards and National Technical Information Service), Defense Department, National Aeronautics and Space Administration (NASA) and the National Science Foundation. Particular attention is drawn to the February issue of the *Monthly Catalog. . .*, which includes an appendix *Directory of United States Government Periodicals and Subscription Services*. This is a list of some 1000 titles, arranged alphabetically, and giving details of price, frequency, and so on. A few examples of the many dealing with technical matters are: *Air Pollution Abstracts* (1968–) (monthly); *USSR and East European Scientific Abstracts* (having sections entitled *Engineering and Equipment* and *Materials Science and Metallurgy*) (1973–); the quarterly *Journal of Research of the National Bureau of Standards* (1959–), especially series C, *Engineering and Instrumentation*; *NASA Technical Briefs* (1958–) (irregular); the semi-monthly *Nuclear Science Abstracts* (1947–); and *Scientific and Technical Aerospace Reports* (1963–), a semi-monthly abstracting journal.

The *Public Documents Department Price List Series* is the nearest equivalent to the HMSO *Sectional Lists*. The various lists detail publications available for purchase, so while not being a complete subject guide, they give a fair indication of recent material in specific subject areas. The lists are revised from time to time, and some do go out of print, but the following selection all had revised editions published in or about 1972:

No. 15. Geology
No. 25. Transportation, Highways, Roads . . .
No. 53. Maps, Engineering, Surveying
No. 64. Scientific Tests, Standards, Mathematics, Physics
No. 79A. Space Missiles . . . NASA and Satellites . . .
    Research and Technology
No. 84. Atomic Energy . . .

Lists have also been issued on, for example, *Irrigation, Drainage and Water Power* (No. 42), *Soils . . . Soil Survey . . .* (No. 46), *Mines . . .* (No. 58) and *Radio and Electricity, Electronics, Radar and Communications* (No. 82).

Use of the *Monthly Catalog . . .* and the *Price List Series* will provide information about recent publications. Also valuable are the US Government of Commerce, National Technical Information Service's *Government Reports Announcements, Government Reports Index* and *Government Reports Topical Announcements,* described fully by Bereny (see p. 128).

OTHER COUNTRIES

For Australia there is a monthly catalogue, cumulating in the annual *Australian Government Publications* issued by the National Library of Australia. It lists the official publications of the Commonwealth, the six states and the territories. A topic index is provided. All

Australian government publications also appear in the weekly *Australian National Bibliography.*

Now published by Information Canada, Canada's official publishing agency, a comprehensive list of publications is provided by the bilingual *Canadian Government Publications: Monthly Catalogue,* with an annual cumulation. The internal arrangement is similar to the HMSO catalogues, with a list of parliamentary publications, then those of the departments, and finally a list of periodicals. Detailed indexes are provided.

The fullest list of the official publications of France is to be found in the *Bibliographie de la France: Biblio.* This comes out weekly, but government publications appear in a special supplement 'F', which is issued only a few times each year. (Many scientific and technical government publications *do* get into the regular weekly lists, however.) There is an annual index. That for 1974 appeared in the fourth issue of the *Bibliographie . . .* in 1975.

There is no formal catalogue of the official publications of the German Federal Republic, but they are all recorded in the weekly *Deutsche Bibliografie Wochentliches Verzeichnis* with its half-yearly and 5-yearly cumulations. The same publishers issue a biennial *Verzeichnis Amtlicher Druckschriften,* supported by the Federal Government. Its value is somewhat limited by the time-lag between the dates of the publications listed and that of the catalogues themselves, that for 1961/2 not appearing until 1969.

For Japan there is a monthly bulletin of government publications, *Seifu Kankōbutsu Geppō,* providing full bibliographical details. Although there is an official printer, the ministries and departments publish and distribute their documentation themselves. The Japanese national current bibliography, *Zen Nihon Shuppa-Butsu So-Moku-roku,* also lists official publications.

As printing and publishing in the USSR is entirely state-controlled, all publications might be regarded as official. A number of bibliographies exist, but none exclusively for government publications as such. The annual *Ezhegodnik Knigi USSR* does list them. Otherwise trade catalogues and those of important institutions, e.g. the Academy of Sciences Library, can be used.

INTERNATIONAL ORGANISATIONS

There are a vast number of organisations throughout the world producing literature in appropriate subject areas ranging from international, government-supported bodies such as the United Nations, down to local institutions and societies. Reference to the under-mentioned directories will draw attention to the major ones. A few, which by the amount and/or importance of their publications are most worthy of attention, will be discussed individually. A useful starting point is A. P. Harvey's *Directory of Scientific Directories: a World Guide . . . including . . . Engineering, Manufacturing and Industrial Directories* (2nd edn, Hodgson, 1972).

The *Yearbook of International Organisations* (Union of International Associations, annual) details over 4000 organisations, including international non-governmental associations, inter-government associations, EEC and EFTA committees, international foundations and international relations institutes. Indexes by English and French name or organisation, by country and by acronym are provided, as well as a list classified by subject (again in English and French). Over 150 bodies are listed in the 1974 issue under the heading *Technology*.

Other useful publications of the Union of International Associations include its *Directory of Periodicals Issued by International Organisations* (2nd edn, 1959) and the monthly journal *International Associations* (1949–), which includes material to update the *Yearbook*. . . .

The *Directory of European Associations* (CBD Research) is planned to comprise at least three parts. The first (published in 1971) covers trade, industrial and professional organisations. The second volume, dealing with learned, research and scientific societies, was published in 1975, and is a useful additional source. The entries in the existing volumes are classified by subject, and include mention of the organisations' publications, where appropriate. A subject index in English, French and German is provided, together with an organisation index.

Another specialised source is the *Guide to World Science* (Hodgson, 1968–1970). This is in 20 volumes, with the general arrangement of a directory of scientific institutions in particular countries. However, volume 20 deals specifically with international organisations.

A few other directories will receive mention in noting the publications of some of the major international organisations. Finally here, attention is drawn to United States Library of Congress, General Reference and Bibliography Division's *International Scientific Organisations: a Guide to their Library, Documentation and Information Services . . .* (Library of Congress, 1962). This gives details of nearly 700 organisations, arranged alphabetically by title. Although this work is now somewhat dated, its lists of the organisations' publications are extensive, and the index copious.

The United Nations and its related bodies publish a vast number of reports, reference works, periodicals and other literature. While their primary interests are not normally in the field of engineering or technology, by the sheer volume of output these subjects receive not inconsiderable attention.

Of the various catalogues and bibliographies issued under the auspices of the United Nations, *the United Nations Document Index* 1950–1973 (often known as UNDI) was the official list. It appeared 11 times a year, listing and (separately) subject indexing all the current publications of the UN (with a few minor exceptions). An annual cumulated catalogue and subject index followed the eleventh issue. A cumulated index to volumes 1–13 (1950–1962), published by Kraus, was due in mid-1974, and a further index to volumes 14– 24 (1963–1973) is planned.

Beginning with the first issue for 1974, the bibliography appeared in a new format with the title *UNDEX: United Nations Documents Index,* in three separate series: A. *Subject Index,* B. *Country Index* and C. *List of Documents issued,* appearing 10 times a year.

United Nations Publications; Checklist 1945–71: Books in Print (1971) provides a partial means of retrospective searching and B. Brimmer's *Guide to the Use of United Nations Documents* (Oceana, 1962) is, despite its date, still very useful. A more recent guide is H. N. M. Winton's *Publications of the United Nations System: a Reference Guide* (Bowker, 1972), and the quarterly *International Bibliography, Information and Documentation* (1973–) aims to provide a current service of bibliographical information on the publications of the United Nations and its specialised agencies. The arrangement is under 43 subject headings, including Energy and Fuel, Nuclear Sciences and Technology, Space Sciences . . ., and Technology.

The International Atomic Energy Agency (IAEA), having its headquarters at Kärntnerring 11, A-1011, Vienna, was set up in 1957 'to accelerate and enlarge the contribution of atomic energy to peace, health and prosperity throughout the world and to ensure . . . it is not used to . . . military purpose'. The coverage of its publications includes nuclear technology (research and power reactors, etc.). A general catalogue of publications is issued, available in sections which include *Industrial Applications* and *Reactors and Nuclear Power*. Particularly important publications include the multi-volume *Directory of Nuclear Power Reactors* (1959), the quarterly *Atomic Energy Review* (1963–), *INIS Atomindex* (1970–) (a monthly subject bibliography with semi-annual cumulations), the *IAEA Bulletin* (1959–), appearing six times a year, and the thrice-yearly *List of Bibilographies on Nuclear Energy* (1960–).

The International Civil Aviation Authority (ICAO), based at the

International Aviation building, 1080 University Street, Montreal, came into being in 1947. It has among its terms of reference the development of air navigation, aircraft design, and the use of new technical methods and equipment, and also concerns itself with such related fields as communications and accident reports. The *Catalogue of ICAO Publications* covers material currently available, while the *Weekly List of Publications* records new works. Annual cumulations have appeared. Two periodicals to note are the annual *Aircraft Accident Digest* (1950–) and the *ICAO Technical Publications*, the quarterly supplement to the monthly *ICAO Bulletin* (1946–).

The International Telecommunication Union (ITU), with head-quarters at Place des Nations, CH-1211, Geneva 20, includes among its aims the promotion and development of technical facilities, and the collection and publication of information on telecommunication matters, covering the telegraph, telephone, radio and television. A general catalogue of its publications is issued. Particularly important are the annual *Report on the Activities of the International Tele-communications Union in 19–* and the monthly *Telecommunication Journal* (1934–).

The United Nations Educational, Scientific and Cultural Organi-sation (UNESCO), with headquarters at 7–9 Place de Fontenoy, Paris 7e, has a range of activities and publications as broad as is implied by its title. An annual *Catalogue of UNESCO Publications* (which is also available in sections, including one on science) listing in-print publications is updated by the bi-monthly *List of UNESCO Documents and Publications,* while the *General Catalogue of UNESCO Publications and UNESCO Sponsored Publications 1946–59* (1962), with supplements in 1964 and 1969 covering, respectively, the years 1960–1963, and 1964–1967, is a complete bibliography. Noteworthy UNESCO publications include the *World Directory of National Science Policy-Making Bodies* (4 vols., 1966–), *World Guide to Technical Information and Documentation Services* (1969), *Bibilography of Interlingual Scientific and Technical Dictionaries* (5th edn, 1969) and the *Bibliography of Monolingual Scientific and Technical Dictionaries,* by E. Wüster (2 vols., 1955 and 1959).

OTHER INTERGOVERNMENTAL ORGANISATIONS

The Organisation for Economic Co-operation and Development (OECD), 2 rue André Pascal, 75-Paris 16, was set up in 1961. It comprises all the states of Western Europe, as well as Australia, Canada, Japan, Turkey and the USA. While its main activities concern economies and trade, it also embodies the European Nuclear

Energy Agency and has committees for technical co-operation, science policy, industry, oil and energy. The OECD is responsible for several apposite publications, including *International Scientific Organisations: Catalogue* . . . (1965, and supplement 1966) giving information on about 70 institutions; the *Country Reports on the Organisation of Scientific Research* (1963–1966) in 18 volumes; and the somewhat dated, but still useful, *Liste des Associations Internationales Interéssant les Industries Mécaniques et Éléctriques 1961* (1962), with information on 89 associations. *Guide to European Sources of Technical Information* is a former publication of the OECD now issued by Hodgson. Its third edition in 1970 was edited by C. H. Williams.

The Commonwealth and its various committees, although having some interest in science and technology, do not contribute greatly to the sum of publications, their activities being largely of an advisory or hortatory nature. The Commonwealth Engineering Conference does issue a *Booklet* (revised in 1971), and the Commonwealth Scientific Committee publishes conference reports and directories of research, while the Council of Commonwealth Mining and Metallurgical Institutions issues reports of its congresses, held every 4 or 5 years. Other Commonwealth bodies active in technological subjects include the Commonwealth Advisory Aeronautical Research Council, the Commonwealth Committee on Mineral Resources and Geology, the Commonwealth Consultative Space Research Committee and the Commonwealth Telecommunications Organisation.

Sometimes overlooked is the fact that Britain's entry into Europe entails membership not only of the European Economic Community (EEC), but also of the European Coal and Steel Community (ECSC) and EURATOM, the European Community for Nuclear Energy. Apart from the various reports and documents issued as part of its information service, the ECSC published, through its Bureau de Terminologie, a five-volume, five-language glossary *Termes Techniques* (1964) covering all aspects of steel utilisation, including construction, prefabrication, bridges, etc.

EURATOM also publishes reports, both of its own activities and deriving from its information service, a quarterly journal *Eurospectra* (1962–) and *Abstracts and Indexing Journals in the Nuclear and Borderline Fields* (1964) (a list of some 500 such journals in 35 subject areas); and is responsible for a monthly *Transatom Bulletin* (1960–) detailing translations of East European literature of nuclear interest. Five-yearly cumulations have appeared for 1960–1965, and 1966–1970.

Two other European inter-governmental bodies involved with technical subjects are the European Space Research Organisation

(ESRO) and the European Launch Development Organisation (ELDO), both founded in 1962. ESRO publishes an annual report and various notes and memoranda, while ELDO has its annual *ELDO Technical Review* and various technical memoranda. In addition, they publish jointly the *ESRO/ELDO Bulletin* six times a year.

NON-GOVERNMENTAL INTERNATIONAL ORGANISATIONS

There are, of course, too many non-governmental international organisations to enumerate separately, and reference to the aforementioned directories will bring most of them to light. Mention may be made, however, of the Union of International Engineering Organisations (UIEO) and its member institutions. The UIEO, often known by its French name and acronym Union des Associations Techniques Internationales (UATI), was set up in 1951 at the instigation of UNESCO, and receives some of its funds therefrom. Its headquarters are at 62, rue de Courcelles, 75-Paris 8e. The main function of UIEO is promoting co-ordination and co-operation between member organisations by arranging conferences and so on, although in its early days there was collaboration with UNESCO in the publication of technical dictionaries.

The member organisations, with date of foundation, address and brief notes on their publications, are as follows:

International Association for Bridge and Structural Engineering (IABSE), founded in 1929, c/o Ecole Polytechnique Fédérale, 8006 Zurich, publishes an annual *Bulletin*, twice-yearly volumes of *Publications* . . . (of original scientific work), congress papers and reports of working commissions.

International Association for Hydraulic Research (IAHR) (1935), Raam 61, P.O. Box 177, Delft, publishes the quarterly *Journal of Hydraulic Research* (1963–), a *Directory of Hydraulic Research Institutions and Laboratories* (1971) and congress proceedings.

International Commission on Irrigation and Drainage (ICID) (1950), 48 Nyaya Marg, Chanakyapuri, New Delhi 21, publishes a bi-annual *Bulletin*, an annual *Bibliography on Irrigation, Drainage, Flood Control and River Training, Congress Transactions* and various monographs, including *Multilingual Technical Dictionary on Irrigation and Drainage* (1967). This has over 12 000 English and French terms grouped into 16 chapters.

International Commission on Large Dams (ICOLD) (1928), 22 et 30 avenue de Wagram, 75-Paris 8e, publishes a *World Register of Dams* (1964, and supplement 1966); a multilingual *Technical Dictionary on Dams* (2nd edn, 1962), defining some 4500 terms in ten languages; *Dam Terminology: a Glossary of Words and Phrases related to Dams* (1970), with a previous Norwegian edition *Ordbok for Dambygging* (1968); and various reports, bulletins and conference proceedings.

International Conference on Large High Tension Electrical Systems (usually known by its French acronym CIGRE) (1921), 112 boulevard Haussman, 75-Paris 8e, publishes the bi-monthly *Electra* (1931–   ), *Congress Proceedings* and other reports.

International Federation of Automatic Control (IFAC) (1957), Graf-Recke Strasse 84, POB 1139, 4 Düsseldorf 1, issues the *IFAC Information Bulletin* (irregular) and the bi-monthly *Automatica* (1963–   ) (actually published by Pergamon).

International Federation of Surveyors (usually known by its French title Fédération Internationale des Géometres, and acronym FIG) (1926), Kiedrichstrasse 6, 62 Wiesbaden, has among its publications the *FIG Bulletin* and congress reports, while Argus published for it the *Dictionnaire Multilingue de la Fédération . . .* (1963). This is a French–German–English dictionary of about 5500 terms.

International Gas Union (IGU) (1931), c/o Institution of Gas Engineers, 17, Grosvenor Crescent, London, SW1 7ES, has as its principal publications the *International Safety Code for International Transmission of Fuel Gas by Pipeline* (1965) and *Safety Code for Compressor Regulating and/or Measurement Stations in Gas Transmission Systems* (1972), while also issuing congress reports.

International Institute of Welding (IIW) (1948), 54 Princes Gate, Exhibition Road, London, SW7 2PG, issues two quarterlies, *Welding in the World* (1963–   ) and *Bibliographical Bulletin for Welding and Allied Processes* (1949–   ), the latter actually published by La Soudure Autogène; the Institute is also responsible for the *Multilingual Collection of Terms for Welding and Allied Processes* (Institute of Welding, 1955–1965) in five parts.

International Society for Soil Mechanics and Foundation Engineering (ISSMFE) (1936), c/o Institution of Civil Engineers, Great George Street, London, SW1, has published *Technical Terms,*

Symbols and Definitions in English, French, German, Swedish, Portuguese, Spanish, Italian and Russian, used in Soil Mechanics and Foundation Engineering (3rd edn, 1967), which defines over 1600 terms and explains more than 70 symbols. Conference proceedings are also published.

International Union of Public Transport (known by its French initials UITP) (1885), 19 avenue de l'Uruguay, 1050 Brussels, publishes the *UITP Biblio-Index: Transport-Verkehr* (1962–), a quarterly with references on cards; the *UITP Review* (1952–) (also quarterly); congress proceedings; and various technical reports, statistics and bibliographies.

Permanent International Association of Road Congresses (PIARC) (1909), 43 avenue du President Wilson, 75-Paris 16e, publishes a quarterly *Bulletin*; various congress and committee reports; and, in association with the World Touring and Automobile Organisation, the *International Traffic Engineering Vocabulary* (1957), which defines about 300 terms in six languages.

World Energy Conference (WEC), formerly the World Power Conference (1924), 5 Bury Street, St. James's, London, SW1Y 6AB, publishes transactions of all its conferences, various statistical yearbooks, and other surveys and reports.

10

Standard reference sources

K. W. Mildren

The aim of this chapter is to draw attention to some of the various sources available to the engineer in his search for factual information. Types of material discussed are guides to the literature, encyclopaedias, dictionaries, handbooks and directories.

GUIDES TO THE LITERATURE

A necessary prerequisite to the effective retrieval of information is knowledge of the relevant sources which should be consulted. Anyone unfamiliar with reference material in a particular subject may find it useful, initially at least, to study the guides to the literature available in that field. There are several of these compilations for engineering and related disciplines.

Critical annotations are given in volume 1 of *Guide to Reference Material,* by A. J. Walford (3rd edn, Library Association, 1973). This volume is concerned with science and technology and the entries are arranged in UDC order. Although biased towards works published in Britain, it is also international in scope. In the United States C. M. Winchell's *Guide to Reference Books* (8th edn, American Library Association, 1967, supplements 1968 and 1970) is more popular. *Science Reference Sources,* by F. B. Jenkins (5th edn, MIT Press, 1969), gives a selected list of sources arranged by subject, while E. J. Lasworth's *Reference Sources in Science and Technology* (Scarecrow, 1972) is more comprehensive, containing fairly long listings of items. Although D. J. Grogan's *Science and Technology: an Introduction to the Literature* (2nd edn, Bingley, 1973) is a sampler aimed primarily at students of librarianship and information science, it is still a useful aid to the practitioner. S. Herner's *A Brief Guide to Sources of Scientific and Technical Information* (Information Resources Press, 1970) is substantially the content of a short course given in 1967 under the sponsorship of the Panel on Education and Training of the Committee on Scientific and Technical Information (COSATI). Coverage of Russian material in science, technology and medicine can be obtained from volume 5 of *Guide to Russian Reference Books,* by K. Maichel (Hoover Institution, Stanford University, 1967).

Electrical and electronic engineering sources are easily traced by consulting the following books: *Electronics: a Bibliographical Guide,* by C. K. Moore and K. J. Spencer (2 vols., Macdonald, 1961 and 1965); *Electronics—a Bibliographical Guide—3,* by L. Corbett (Macdonald, 1973); *Handlist of Basic Reference Material in Electrical and Electronic Engineering,* by E. M. Codlin (6th edn, Aslib Electronics Group, 1973); *Electronics Industry: a Guide to Information Sources,* by G. R. Randle (Gale Research Company, 1968); and the now dated *How to Find Out in Electrical Engineering,* by J. Burkett and P. Plumb (Pergamon, 1967).

A useful guide to mechanical engineering literature is *Mechanical Engineering: the Sources of Information,* by B. Houghton (Bingley, 1970). Coverage of metallurgy is supplied in *Guide to Metallurgical Information,* by E. B. Gibson and E. W. Tapia (2nd edn, Special Libraries Association, 1965). More specific in the metals field is *How to Find Out in Iron and Steel,* by D. White (Pergamon, 1970). *Non-Ferrous Metals—a Bibliographical Guide,* by K. Boodson (Macdonald, 1972), is a substantial work containing 4335 items numbered sequentially through 59 sections. Another publication by E. M. Codlin is *Cryogenics and Refrigeration: a Bibliographical Guide* (2 vols., Macdonald, 1968 and 1970).

Other guides of interest to engineers are: *Physics Literature: a Reference Manual,* by R. H. Whitford (2nd edn, Scarecrow, 1968); *How to Find Out About Physics,* by B. Yates (Pergamon, 1965); *How to Find Out About Mathematics,* by J. E. Pemberton (2nd edn, Pergamon, 1969); *Guide to the Literature of Mathematics and Physics,* by N. G. Parke (2nd edn, Dover, 1958); *Guide to the Literature of Engineering, Mathematics and the Physical Sciences,* by S. Goldman (2nd edn, Johns Hopkins University, 1964); and *A Guide to Computer Literature,* by A. Pritchard (2nd edn, Bingley, 1972).

The American Society for Engineering Education produces a series of short guides to the literature in the following subjects: electrical and electronic engineering, mechanical engineering, chemical engineering, metals and metallurgical engineering, and industrial engineering.

ENCYCLOPAEDIAS

Multi-volume encyclopaedias are sometimes the most useful readily accessible sources of information. Three examples of such works are *Encyclopaedia Britannica, Chambers Encyclopaedia* and *Encyclopedia Americana.* In 1974 *Encyclopaedia Britannica* changed its format and title, becoming *Britannica 3,* consisting of the single-volume *Propaedia,* the 10-volume *Micropaedia,* and the 19-volume *Macropaedia. Americana*'s 1971 edition has 30 volumes, while *Chambers* (1973) restricts its entries to 15 volumes. The best-known multi-volume work devoted to science and engineering is the *McGraw-Hill Encyclopaedia of Science and Technology* (3rd edn, 1971), which is supplemented with an annual yearbook.

A single-volume compilation is *Engineering Encyclopedia,* by F. D. Jones and P. B. Schubert (3rd edn, Industrial Press, 1963), containing 4500 concise entries. *Engineering Eponyms,* by C. P. Auger (2nd edn, Library Association, 1975), is an annotated bibliography of some named elements, principles and machines in mechanical engineering and is arranged alphabetically. Examples of non-alphabetical encyclopaedias are the *Universal Encyclopedia of Machines: or How Things Work* (Allen and Unwin, 1967) and *Materials and Technology: a Systematic Encyclopaedia . . .* (8 vols., Longman-de Bussy, 1968–).

There are several specialist encyclopaedias such as *Newnes Concise Encyclopedia of Electrical Engineering,* by M. G. Say (Newnes, 1962), which covers all branches of power production and utilisation. *Encyclopaedia of Engineering Materials,* by H. R. Clauser (Reinhold, 1963), contains 300 signed articles. A more recent work is *Encyclopedia of Instrumentation and Control,* by D. M. Considine (McGraw-Hill, 1971), a very good encyclopaedia covering this comparatively new technology. An encyclopaedia aimed at purchasing managers is *Materials Handbook,* by G. S. Brady (10th edn, McGraw-Hill, 1971). Some compilations with short entries may often nearly resemble a dictionary, e.g. *Encyclopaedia of Hydraulics, Soil and Foundation Engineering,* by E. Vollmer (Elsevier, 1967). Another outstanding work, although now in need of updating, is *Encyclopedic Dictionary of Electronics and Nuclear Engineering,* by R. I. Sarbacher (Pitman, 1960). Nine volumes, updated by supplementary issues (1966–1971), make up the *Encyclopaedic Dictionary of Physics,* by J. Thewlis (Pergamon, 1961–1964). This is a vital acquisition for any scientific or engineering collection, since it covers a wide subject field and is written to about graduate level.

DICTIONARIES

Engineering, like other disciplines, relies heavily on dictionaries as sources of explanation, and as a consequence many specialist defining, translating and other dictionaries are available for consultation.

Defining dictionaries

Perhaps the best-known general work is the *Dictionary of Science and Technology,* edited by T. C. Collocott and A. B. Dobson (Chambers, 1974), which replaces *Chambers's Technical Dictionary* (3rd edn, 1958). Other recent compilations are the *Dictionary of Scientific and Technical Terms* (McGraw-Hill, 1974); *Glossary of Contemporary Engineering,* by J. D. Beadle (Macmillan, 1972); and *Engineering and its Language,* by B. Scharf (Muller, 1971). More specific defining dictionaries are especially prevalent in electronics and electrical engineering. A really excellent authoritative publication is the *IEEE Standard Dictionary of Electrical and Electronic Terms* (Institute of Electrical and Electronics Engineers, 1972). Another essential source of reference is the *Dictionary of Electronics and Nucleonics,* by L. E. C. Hughes (Chambers, 1969). The *Arlington Dictionary of Electronics* (Arlington Books, 1971) contains circuit diagrams and is aimed at a broad audience, not only specialists in the subject. Compiled by A. K. M. Green, the *Glossary of Coined Terms in Electronics* (4th edn, Aslib Electronics Group, 1969) is a handy reference to about 350 common terms. Other dictionaries worthy of note are R. F. Graf's *Modern Dictionary of Electronics* (4th edn, Foulsham, 1972); J. Markus's *Electronics and Nucleonics Dictionary* (3rd edn, McGraw-Hill, 1966); and the revised impression of K. G. Jackson's *Dictionary of Electrical Engineering* (Butterworths, 1973). Even narrower subject coverage appears in R. A. Bones's *Dictionary of Telecommunications* (Newnes–Butterworths, 1970), containing about 3000 concise definitions.

In mechanical engineering the two compilations of interest are *A Dictionary of Mechanical Engineering Terms,* by J. G. Horner (9th edn, Technical Press, 1967), and the *Dictionary of Mechanical Engineering* by J. L. Nayler and G. H. F. Nayler (Newnes, 1967). In the metals field reasonable coverage is given in the following works: *A Dictionary of Metallurgy,* by A. D. Merriman (Macdonald and Evans, 1958); *Dictionary of Metallurgy,* by D. Birchon (Newnes, 1965); *The Iron and Steel Industry: a Dictionary of Terms,* by W. K. V. Gale (David and Charles, 1971); and the *Dictionary of Ferrous Metals,* by E. N. Simons (Muller, 1970).

Civil engineering lacks a really extensive dictionary on the subject, although adequate definitions appear in J. S. Scott's *Dictionary of Civil Engineering* (2nd edn, Penguin, 1965) and A. Nelson and K. D. Nelson's *Dictionary of Applied Geology, Mining and Civil Engineering* (Newnes, 1967). The last-named also compiled the *Dictionary of Water and Water Engineering* (Newnes–Butterworths, 1973).

Translating dictionaries

Much technical information is published in Russian and German, and so it is essential for any scientific or technical reference collection to contain a number of translating dictionaries in these languages. The main Russian source is the *Russian–English Scientific and Technical Dictionary*, by M. H. T. Alford and V. L. Alford (Pergamon, 1970), in two volumes containing over 100 000 entries from 94 disciplines. A shorter dictionary is the *Concise Russian–English Scientific Dictionary*, by A. Blum (Pergamon, 1965), which has appendices of grammatical forms and abbreviations. A somewhat different slant is given in M. E. Zimmerman's *Russian–English Translators Dictionary: a Guide to Scientific and Technical Usage* (Plenum, 1967), which contains a series of typical examples from scientific and technical sources and is not a dictionary of terms.

German translation may be aided by consulting the following general dictionaries: *Dictionary of Modern Engineering, English–German, German–English,* by A. Oppermann (3rd edn, Bowker, 1973); *English–German, German–English Technical and Engineering Dictionary,* by L. De Vries and T. M. Herrmann (2 vols., 2nd and 3rd edn, McGraw-Hill, 1967 and 1970); *Dictionary of Science and Technology, English–German, German–English,* by A. F. Dorian (Elsevier, 1967 and 1970); and *Polytechnical Dictionary, English–German, German–English,* edited by R. Walther (Pergamon, 1973 and 1967). More specialist German subject dictionaries have also been published, e.g. *Technical Dictionary of Production Engineering,* edited by R. Walther (2 vols., Pergamon, 1972), and *Dictionary of Civil Engineering and Construction Machinery and Equipment,* by H. Bucksch (2 vols., Bauverlag GMBH, 1968). Electrical and electronic engineering is again well represented, although some items are in need of new editions. The *Dictionary of Electronics, Communications and Electrical Engineering, English–German, German–English,* by H. Wernicke (2nd and 1st edn, Rohde and Schwarz, 1968 and 1964), contains 71 000 terms and abbreviations in volume 1 and 66 000 entries in volume 2. Other shorter compilations are the *Dictionary of Electrotechnology: German–*

English, by E. Hohn (Chapman and Hall, 1966), and the *Dictionary of Semiconductor Physics and Electronics, English–German, German–English,* by W. Bindmann (2 vols., Pergamon, 1966).

Multi-lingual dictionaries

There are numerous multi-lingual dictionaries in various engineering subjects and it is not profitable to list them all here, Many are published by the Elsevier Publishing Company and reference to its catalogues is recommended.

A handy source in the electrical field is the *Dictionary of Electrical Engineering, Telecommunications and Electronics,* by W. Goedecke (Pitman, 1965–1967), appearing in three volumes and treating English–German–French first in rotation in the different volumes. A multi-lingual glossary has been produced by the OECD entitled *Glossary of Terms and Definitions in the Fields of Friction, Wear and Lubrication—Tribology* (1969) in which English terms and definitions are given followed by a six-language index and an alphabetical index for each listed language.

Other dictionaries may be traced via the publications listed below:

Foreign Language and English Dictionaries in the Physical Sciences and Engineering, by W. Marton (US Department of Commerce, National Bureau of Standards, 1964)
A Bibliography of Scientific, Technical and Specialised Dictionaries: Polyglot, Bilingual, Unilingual, by C. W. Rechenbach and E. R. Garnett (Catholic University of America Press, 1969)
Scientific and Technical Dictionaries: an Annotated Bibliography, by W. R. Turnbull (Bibliothek Press, 1966)
Bibliography of Interlingual Scientific and Technical Dictionaries (5th edn, Unesco, 1969)

Dictionaries of symbols and abbreviations

About 5000 symbols are listed in *Shepherd's Glossary of Graphic Signs and Symbols* (Dent, 1971), while a comprehensive US compendium is *Standard Graphical Symbols: a Comprehensive Guide for Use in Industry, Engineering and Science,* by A. Arnell (McGraw-Hill, 1963). In addition, *BS 1991: 1961–1967,* in six parts, is devoted to letter symbols, signs and abbreviations.

Several dictionaries of abbreviations are available, the better ones being:

Abbreviations Dictionary, by R. De Sola (4th edn, Elsevier, 1974)
Acronyms and Initialisms Dictionary: a Guide to Alphabetic Designations, Contractions, Acronyms, Initialisms and Similar Condensed Appellations, edited by E. T. Crowley and R. C. Thomas (4th edn, Gale Research Company, 1973)
World Guide to Abbeviations of Organisations, by F. A. Buttress (5th edn, L. Hill, 1974)
A Dictionary of Acronyms and Abbreviations: Some Abbreviations in Management, Technology and Information Science, by E. A. Pugh (2nd edn, Bingley, 1970)
World Guide to Abbreviations of Associations and Institutions (2 vols., Bowker, 1972)
Dictionary of Russian Technical and Scientific Abbreviations, by H. Zalucki (Elsevier, 1968)
Dictionary of Abbreviations in English, German, Dutch and Scandinavian Languages, by W. Bluwstein (Soviet Encyclopaedia Publishing House, 1964)
Glossary of Russian Abbreviations and Acronyms (Library of Congress, 1967)
Dictionary of Electrical Abbreviations, Signs and Symbols, by R. Mehra (Simon and Schuster, 1971)
Anglo-American Abbreviations and Contractions Used in Electrical Engineering and Allied Fields, by P. Wennrich (VEB Verlag Technik, 1970)

HANDBOOKS, TABLES AND DATA SOURCES

Easy access to verified facts and data is as essential to an engineer as it is to a chemist or a physicist. Substantial personal collections of handbooks are not uncommon among engineers. Many engineers will possess a copy of *Kempe's Engineers Year Book* (Morgan Grampian) and no description of this incomparable two-volume work is necessary. Also in two volumes is the *Handbook of the Engineering Sciences,* by J. H. Potter (Van Nostrand, 1967 and 1968). *Engineering Manual,* by R. H. Perry (2nd edn, McGraw-Hill, 1967) is a practical reference book of data and methods in six engineering disciplines. Other noted though somewhat dated works are *Newnes' Engineers' Reference Book* (10th edn, 1965) and the *Handbook of Engineering Fundamentals,* by O. W. Esbach (2nd edn, Wiley, 1952). A much needed handbook of calculation procedures for solving problems in everyday engineering practice is *Standard Handbook of Engineering Calculations,* by T. G. Hicks (McGraw-Hill, 1972). The latest handbook dealing with mechanical engineering is the *Mechanical Engineer's Reference Book,* edited by A. Parrish

(11th edn, Newnes–Butterworths, 1973), while the classic reference book in the field is J. Baumeister's *Mark's Standard Handbook for Mechanical Engineers* (7th edn, McGraw-Hill, 1967). *Machinery's Handbook,* by E. Oberg and F. D. Jones (19th edn, Machinery Publishing Company, 1971), as its sub-title indicates, is a reference book for the mechanical engineer, draughtsman, toolmaker and machinist, and is necessarily comprehensive in scope, containing in fact about 10 sections. Another important specialist new edition to the literature is the *Tribology Handbook,* edited by M. J. Neale (Butterworths, 1973), containing a wealth of information supplied by over 100 contributors.

Four important handbooks for electrical and electronic engineers are *Electronics Engineers' Handbook,* edited by D. G. Fink and A. A. McKenzie (McGraw-Hill, 1975); *The Electrical Engineers' Reference Book,* edited by M. G. Say (13th edn, Newnes–Butterworths, 1973); *Standard Handbook for Electrical Engineers,* by D. G. Fink and J. M. Carroll (10th edn, McGraw-Hill, 1968); and *Electronic Engineer's Reference Book,* by L. E. C. Hughes and F. W. Holland (3rd edn, Heywood, 1967).

Appearing 13 years after the previous edition, the *Civil Engineer's Reference Book,* edited by L. S. Blake (3rd edn, Newnes–Butterworths, 1975), is a very welcome addition to the literature of the subject. The *Standard Handbook for Civil Engineers,* edited by F. S. Merritt (McGraw-Hill, 1968), is both practical and user-oriented and deals with construction methods and equipment as well as design techniques. An authoritative compilation in 26 sections, E. H. Gaylord and C. N. Gaylord's *Structural Engineering Handbook* (McGraw-Hill, 1968) treats the planning, design and construction of a variety of engineered structures. Design or analysis methods are described in F. S. Merritt's *Structural Steel Designer's Handbook* (McGraw-Hill, 1972), which also includes numerous tables, charts, formulae and data. Essential reading in the heavy construction field is J. A. Havers and F. W. Stubb's *Handbook of Heavy Construction* (2nd edn, McGraw-Hill, 1971).

A mathematical handbook specially written for scientists and engineers is the *Mathematical Handbook for Scientists and Engineers,* by G. A. Korn and T. M. Korn (2nd edn, McGraw-Hill, 1968). Intended to be used as part of a personal collection is the well-known *Handbook of Engineering Mechanics,* by W. Flugge (McGraw-Hill, 1962).

The 'rubber bible', i.e. R. C. Weast's *Handbook of Chemistry and Physics* (56th edn, CRC Press, 1975), is normally the first source one would consult for general chemical and physical data and is revised annually. A smaller but equally indispensable handbook is G. W. E.

Kaye and T. H. Laby's *Tables of Physical and Chemical Constants* (14th edn, Longmans, 1973). The *International Critical Tables* (McGraw-Hill, 1926–1933) in seven volumes were prepared under the auspices of the International Research Council and the National Academy of Sciences. The tables are now relatively incomplete, but the data included are critically evaluated and are compiled by experts in the various fields.

Another member of the Chemical Rubber Company series is R. E. Bolz and G. L. Tuve's *Handbook of Tables for Applied Engineering Science* (1970), intended as a 'first source of critically evaluated numerical data with condensed tables and with specific references to more extensive coverage'. Material at a more elementary level is contained in *Data and Formulae for Engineering Students,* by J. C. Anderson (Pergamon, 1967). Designed for practising engineers and students, *Basic Tables in Electrical Engineering,* by G. A. Korn (McGraw-Hill, 1965), includes tables of numerical functions, tables listing physical properties of materials, formulae and circuit data. *ASME Handbook: Engineering Tables* (American Society of Mechanical Engineers, 1956) is a collection of tables often wanted by engineers, but not commonly found in handbooks. A convenient source of engineering design data for various materials is *Materials Data Book for Engineers and Scientists,* by E. R. Parker (McGraw-Hill, 1967). An interesting loose-leaf publication is D. S. Lock's *The Engineers' Metric Data Manual and Buyers' Guide* (Sciences, Engineering, Medical and Business Data Ltd., 1973), comprising data sheets which list metric sizes together with details of materials, suppliers, availability and method of ordering.

Returning to mathematical tables again, no collection would be complete without a copy of R. C. Weast's *Handbook of Tables for Mathematics* (4th edn, Chemical Rubber Company, 1970). Mathematical functions are particularly important, and one of the better handbooks available is the *Handbook of Mathematical Functions,* by M. Abramowitz and I. A. Segun (Dover, 1965). A small booklet dealing with a very narrow topic is the handy *Tables of Laplace Transforms,* by W. D. Day (Iliffe, 1966). Mention must also be made of L. J. Comrie's *Chambers Six-Figure Mathematical Tables* (Chambers, 1970 and 1972) in two volumes, the first dealing with logarithmic values and the second listing natural values. Rapid consultation of conversion tables is sometimes needed and a well-used accurate source is S. Naft and R. De Sola's *International Conversion Tables* (Cassell, 1965), revised and enlarged by P. H. Bigg. Another booklet in common use is the National Physical Laboratory publication *Changing to the Metric System: Conversion Factors, Symbols and Definitions* (4th edn, HMSO, 1972).

DATA CENTRES

Although conventional data sources are still the most frequently consulted, there is a growing number of data centres appearing on the scene. Here the individual plays what may be termed a supporting role, and the data centre undertakes the tasks of collection, organisation and dissemination of the data. Further development of such centres is indicated in the future, and international co-operation is fostered through CODATA (Committee on Data for Science and Technology), which was formed in 1966.

Details of the data centres in operation in Britain may be obtained from a publication of OSTI (Office for Scientific and Technical Information) entitled *Data Activities in Britain* (1969). CODATA's *International Compendium of Numerical Data Projects* (Springer-Verlag, 1969) surveys and analyses more than 150 data centres in 26 countries and is arranged in six broad property categories: nuclear; atomic and molecular, including spectroscopy; solid state (crystallographic, mineralogical, electrical, magnetic); thermodynamic, including transport and solution; chemical kinetics; and other properties, including gas chromatographic and optical.

Many data centres are now using automated techniques, and the report of the CODATA Task Group on Computer Use on 'Automated Information Handling in Data Centers' is contained in *CODATA Bulletin,* (4) (1971). Centres of particular interest to engineers in the USA include the Cryogenic Data Center at the NBS (National Bureau of Standards); High-Pressure Data Center at Brigham Young University, Provo, Utah; Alloy Data Center at the NBS; Superconductive Materials Data Center at the General Electric Company's Research and Development Center, Schenectady, New York; Microwave Spectral Data Center at the NBS; and the data banks on infra-red spectroscopy of the American Society for Testing and Materials.

In the UK eight data centres are described in *Quantitative Data in Science and Technology*, by B. Mountstephens (Aslib, 1971). The most important centre for engineers is the Engineering Sciences Data Unit Ltd., known as ESDU, a private company owned by the Royal Aeronautical Society. ESDU provides working design and physical data in the form of data sheets for use in aeronautical, mechanical, chemical and structural engineering. The work is sponsored by the appropriate professional bodies. Each of ESDU's 14 Technical Committees, Steering Groups or Panels, which guide and monitor the work of the ESDU staff, consists of a group of practising experts whose services are provided on a voluntary basis. There is an annual subject index, location schedule and list of titles of all items published by ESDU.

DIRECTORIES AND YEARBOOKS

The two most prominent aids to the tracing of directories are the *Directory of Scientific Directories*, compiled by A. P. Harvey (2nd edn, Hodgson, 1972), and *Current British Directories*, edited by I. G. Anderson (7th edn, CBD Research, 1973).

Details of professional bodies and associations are always in demand, and the following compilations provide such information on the national and international level: *Directory of British Associations*, by G. P. Henderson and S. P. A. Henderson (4th edn, CBD Research, 1974); *Trade Associations and Professional Bodies of the United Kingdom*, by P. Millard (5th edn, Pergamon, 1971); *Directory of European Associations: Part 1, National, Industrial, Trade and Professional Associations*, by I. G. Anderson (CBD Research, 1971); *Directory of National Trade and Professional Associations of the United States* (5 vols., Columbia Books, 1966–1970); *Directory of Engineering Societies and Related Organisations* (7th edn, Engineers Joint Council, 1974); *Encyclopedia of Associations*, by F. G. Ruffner (3 vols., 5th edn, Gale Research Company, 1968; and *Yearbook of International Organisations* (14th edn, Union of International Associations, 1972. *World of Learning* (Europa Publications, annual) lists, in its two volumes, '24 000 universities, colleges, libraries, learned societies, research institutes and the 150 000 people active in them'. The *Commonwealth Universities Yearbook* (Association of Commonwealth Universities, annual) fulfils a similar function within its obviously restricted limits. Services available from government departments and associated organisations are described in *Technical Services for Industry* (Department of Industry, 1975).

RESEARCH INFORMATION

Research information is often difficult to obtain, especially in the industrial sphere, for obvious reasons. Universities and colleges disclose information much more freely, and details of current activities in engineering may be found in volume 1 of *Scientific Research in British Universities and Colleges* (HMSO), compiled annually by the Department of Education and Science. One is, however, forced to wonder if all the claims contained therein are justified. The subject index could be a little more specific and perhaps a KWIC index might be suitable for such detailed entries.

Another guide which does cover the industrial scene to some extent, in a general sense, in addition to its description of academic activities is *Industrial Research in Britain* (7th edn, Hodgson, 1972). An indication of research in progress at the individual level may be obtained by reference to the *List of Research Grants Current on 1st October*, an annual publication of the Science Research Council. A more specialist listing is found in publications such as the *Register of Research in Machine Tools and Production Engineering* (HMSO, 1972), compiled by the Department of Trade and Industry. On the wider plane, details of European research activities is provided in *European Research Index*, edited by C. H. Williams (2 vols., 2nd edn, Hodgson, 1969). Hazan International, based in Paris, is a representative of the Smithsonian Science Information Exchange, which collects, every year, information on about 100 000 research projects currently in progress in the USA. Descriptions of specific research projects, together with the names of researchers and organisations, are available to individuals on payment of a fee.

BIOGRAPHICAL INFORMATION

In addition to the general series of *Who's Who* . . ., there are several biographical directories solely devoted to engineers, e.g. *Who's Who of British Engineers 1974–1975*, edited by R. A. Baynton (4th edn, Eurobooks, 1974); *Engineers of Distinction* (Engineers Joint Council, 1970); and *Electrical Who's Who and Yearbook 1973* (Northwood Publications, 1973). Also, many institutions publish lists of members which may be used to verify qualifications, etc., e.g. *Yearbook and List of Members 1974* (Institution of Electrical Engineers, 1975).

TRADE DIRECTORIES

The problems associated with product information are discussed in Chapter 8, and the deficiencies of trade directories are highlighted. Nevertheless, these catalogues are useful in many cases as a preliminary step or where detailed information is unnecessary or where the name and address of a firm is the sole requirement. The following types of lists appear in trade directories, though not all of them will be found in every catalogue: manufacturers' names, addresses, telex and telephone numbers; classified products and services; materials, plant and equipment; instruments and components; trade names; UK agents for foreign manufacturers; forthcoming exhibitions; lists of contractors, etc. A selection of available directories is given below:

Kelly's Manufacturers and Merchants Directory (2 vols., annual)
Kompass Register of British Industry and Commerce (2 vols., annual)
   (other Kompass directories are available for other countries)
Thomas's Register of American Manufacturers (11 vols., annual)

Engineer Buyers Guide (Morgan Grampian, annual)
Machinery's Buyers Guide (Machinery Publishing Company, annual)
Purchasing Directory for the Process Industries (Morgan Grampian, 1971)

British Instruments Directory (United Trade Press, 1971)
Electrical and Electronics Trade Directory (Peter Peregrinus, annual)
Electronics Buyers Guide (McGraw-Hill, annual)
Eurolec: Electronic Components, Materials and Sub-Systems; Companies and Contacts (David Rayner Associates, 1972)
Eurolec: Electronic Components, Materials and Sub-Systems; Locations (David Rayner Associates, 1970)
Instruments and Control Systems Buyers Guide (Chilton Company, annual)
Instruments, Electronics and Automation Directory (Morgan Grampian, annual)
Microelectronics Yearbook 1970–71 (Shaw Publishing Company, 1970)

British Plastics Yearbook (IPC Business Publications, annual)
Building and Contract Journal Directory (IPC Building and Contract Journals, annual)
Concrete Yearbook (Cement and Concrete Association, annual)
Construction Industries Buyers' Guide—Materials, Plant, Services (Biggar and Company, annual)
Sell's Building Index (annual)

UK Trade Names (4th edn, Kompass, 1972)

Other directories with price listings are also worthy of a place on the reference shelves, e.g.:

Laxton's Building Price Book (IPC Business Press Information Services, annual)
Spon's Architects and Builders Price Book (annual)
Spon's Mechanical and Electrical Services Price Book (annual)

Electrical and electronics components catalogues are always in demand, and many firms supply their prospective customers with details of their products in this way. A short list of components handbooks appears in the *Handlist of Basic Reference Material in Electrical and Electronic Engineering*, by E. M. Codlin (6th edn, Aslib Electronics Group, 48–52).

11

Abstracts and indexes, magnetic tape services, reviews and bibliographies

K. W. Mildren

ABSTRACTS AND ABSTRACTING JOURNALS

There must now be in existence about 2000 abstracting and indexing journals covering science and technology. The vast increase in the amount of published literature has meant that it is impossible for any individual to acquire, let alone read, all the material relevant to his subject. Wide scattering of information in the various forms of literature also makes retrieval difficult for the individual researcher. Abstracts help the reader to select articles to be read and in some cases substitute for the original paper, thereby saving valuable time. This is especially true for foreign material, since the technologist is usually only familiar with one or two languages in addition to his native tongue. An abstracting service in his own language enables him to select articles for translation with minimum effort and expenditure of time. Abstracts are used as both current awareness and retrospective retrieval tools. Individuals mainly use abstracting journals as an alerting service, with libraries and information centres making use of them for retrospective searching, although a researcher needing to undertake a literature search at the start of a project would also make use of their retrospective properties.

Technologists and engineers appear, on the evidence of several surveys, to make less use of abstracts than other groups. This implies that they are satisfied by other paths of communication or that they are unaware of the existence and potential use of abstracting services.

The next few pages are intended to provide information on the nature and scope of some of the services available to engineers.

Definition and lay-out

Various definitions of an abstract have appeared, but the most generally accepted is that noted by Ashworth (1967) and is to be found in the report of an International Conference on Science Abstracting organised by UNESCO in 1959—'an abstract is a summary of a publication or article accompanied by an adequate bibliographical description to enable the publication or article to be traced'.

An abstract normally starts with the title of the article (in English or the original language), followed by the name and initials of the author(s). Next is the bibliographical reference giving the name of the journal in standard form followed by the year, volume number and page numbers of the start and finish of the article. For patents and reports the associated source numbers are given. Finally, the abstract itself appears, sometimes followed by the name or initials of the abstractor.

Types of abstracts

Abstracts are informative, indicative or critical. Sometimes abstracts may also be described as mission-oriented, discipline-oriented or reader-oriented, the concepts being self-explanatory. The International Conference on Science Abstracting arrived at the following definitions: 'an informative abstract summarises the principal arguments and gives the principal data in the original publication or article' and 'an indicative abstract is a short abstract written with the intention of enabling the reader to decide whether he should refer to the original publication or article'.

Sometimes an informative abstract obviates the necessity of reading the article itself. The informative abstract will contain, if it is to be adequate, the following details: purpose (goals, objectives, aims of the research); method (experimental techniques); results (findings); and conclusions (interpretation or significance of results). Critical abstracts give the context of a work and evaluate the contribution it supplies to the advancement of knowledge. The abstractor needs to be an expert in the field who can judge the material in perspective.

Coverage and delay

Tests on abstracting journals have shown that there is a certain amount of overlap in the coverage of various abstracting journals and that there is a significant time-lag of several months between the original publication date and the appearance of the abstract in the abstracting journal.

Abstracting journals may cover one form of literature only, e.g. reports. Their scope may be national or international, and the abstracts themselves may appear in one or more languages. Also, the journal may provide coverage of a specific rather than a general subject field.

The interdisciplinary nature of some subjects makes it inevitable that duplication will occur in the citation of references in the abstracting journals. This overlap between services is not necessarily to be regarded as wasted effort. The more times a reference is given, the more likely people are to see it. The onus is on the user to find the services most useful to him and to use them efficiently.

Commercial services are sometimes supplemented by locally produced bulletins where they are found to be unsatisfactory for various reasons. The following advantages should be gained: speedier dissemination of information; better coverage (in both scope and form of literature); slanting of material to suit the individual needs of the organisation; and inclusion of non-published material.

Indexing of abstracting journals

Abstracting journals normally possess author and subject indexes, and some also contain report, patent number, conference, book or formula indexes. Author indexes are generally published with each issue and, in addition, there may be quarterly, semi-annual, annual or multi-annual subject and author indexes designed to improve retrospective searching. Subject indexes present various problems, not least those of terminology and vocabulary control. Lists of subject headings and thesauri have been produced as attempts to deal with consistency and ambiguities in terminology, and can be used to good effect by the individual searcher.

INDEXING JOURNALS

Indexing journals differ from abstracting journals in that they give no information on the material content of the articles cited, merely their bibliographical descriptions, which are usually listed under alphabetical subject headings or within classified subject fields. The intellectual effort involved is much less than that needed for an abstracting journal and, hence, indexing journals are able to avoid excessive time delay between the publication of an article and its appearance in the journal. Indexing journals are essentially current awareness devices, although they are also used to bring a retrospective search completely up to date.

SOME ABSTRACTING AND INDEXING JOURNALS

Engineering Index (EI). Annual 1885–   ; *Monthly* 1962–
Engineering Index is published monthly and annually by Engineering Index Inc. and is the major transdisciplinary index to the world's engineering literature. In 1974 over 85 000 abstracts and notations appeared in the index extracted from over 2000 journals, reports, conferences and symposia proceedings, books, standards, etc. This compares with a figure of 36 000 items indexed in 1960. All fields of engineering and technology are covered, and material is selected for inclusion on the basis of its engineering importance.

The abstracts and notations are arranged aphabetically by subject headings, these having been assigned from *SHE* (*S*ubject *H*eadings for *E*ngineering). *SHE* is an alphabetical list of terms used by *EI* as a controlled vocabulary and contains over 12 000 main headings and sub-headings. A copy of *SHE* (1972) may be obtained from Engineering Index Inc. or from their marketing representatives.

'See' and 'see also' references are given after main headings and sub-headings, and these must be searched, where relevant, if complete coverage is to be obtained. Each abstract or notation is sequentially numbered. There is an author index to each monthly issue, and the whole index, including the author index, cumulates annually. Since October 1972 the *EI Monthly* has been produced using a new computerised production system. Engineering Index Inc. estimate that this has reduced the time delay to one-third, i.e. from 4–6 months to 6–8 weeks.

A point to note when using *EI* is that, sometimes, conference and symposia papers are listed as one item under the general subject of the conference, with author entries for the individual papers given in the author index. This means that a search for a paper under its specific subject heading might prove fruitless unless all cross-references are followed up.

A microfilm edition of the *EI Annual* is also available. See p. 166 for *EI*'s magnetic tape version.

Applied Mechanics Reviews. 1948–

Published monthly by the American Society of Mechanical Engineers, *Applied Mechanics Reviews* is 'a critical review of the world literature in applied mechanics and related engineering science'. The signed abstracts are critical as well as informative and are arranged in six sections under 54 sub-headings. The main headings are rational mechanics and math methods, automatic control, mechanics of solids, mechanics of fluids, heat and combined fields. There is usually a feature article, with an extensive bibliography, in each issue. Publications indexed are journals (about 1000), technical reports, conference proceedings and books. An author index is included and this cumulates annually. In 1972 the annual subject index was replaced by a subject tabulation which is merely a guide to the broad subject headings used in the monthly issues. This is much less satisfactory than the previous subject index and is of little use to the retrospective searcher. A list of books reviewed appears each month and, although useful, some of the book reviews are rather long.

Nuclear Science Abstracts (NSA). 1948–

Nuclear Science Abstracts is published semi-monthly by the United States Atomic Energy Commission (USAEC) Technical Information Center and is an important comprehensive abstracting journal covering international nuclear science literature. The forms of literature included are reports (USAEC, etc.), books (above graduate level), conference proceedings, patents and journals. There are four indexes in each issue of *NSA*: subject, personal author, corporate author and report number. Each of these indexes is cumulated quarterly and annually. Multi-volume cumulations are also available. All documents are indexed by subject and personal author and, in addition, reports are indexed by corporate author and report number. The subject scope of *NSA* is defined in a booklet entitled *Subject Scope of Nuclear Science Abstracts* and is available from the USAEC Technical Information Center. Subjects covered include energy conversion, heat transfer and fluid flow, materials testing, vacuum engineering, environmental and earth sciences, instrumentation, materials, particle accelerators, physics (in 11 sections), reactor technology, etc.

Metals Abstracts. 1968–

Metals Abstracts is published monthly by The Metals Society and the American Society for Metals and is the successor to the *Review of Metal Literature,* formerly published by The American Society for Metals, and *Metallurgical Abstracts,* formerly published by The Institute of Metals.

The 25 000 or so abstracts produced per annum are a reflection of the wide international coverage of about 1000 journals, reports, monographs and conference proceedings. The abstracts are not too long (usually about 200 words) and are arranged in 33 sub-sections under seven main subject headings: constitution, metallography, mechanical properties, ores, foundry, engineering components and structures, and general and non-classified.

In addition to the author index in each issue of *Metals Abstracts*, there are separate author and subject indexes appearing monthly as *Metals Abstracts Index*, which is produced from a computer base and uses the *ASM Thesaurus of Metallurgical Terms* as its controlling vocabulary for the subject index. Users of *Metals Abstracts Index* are, therefore, recommended to use this thesaurus when drawing up a list of subject index entry terms. The computer program is designed to generate 'see' and 'see also' references for synonyms and narrower terms, but not for related terms. There are excellent guidelines for the use of the index in each issue. The abstract serial number in the index has four parts: year, issue number within that year, section number and number of the abstract within its subject section. Cumulations of author and subject indexes appear annually. The service is also available on magnetic tape for computer use and is known as *METADEX*.

International Aerospace Abstracts (IAA). 1961–
IAA is published semi-monthly by the Technical Information Service, American Institute of Aeronautics and Astronautics Inc. The accessions (more than 36 000 in 1974) cover world literature in the field of aeronautics and space science and technology. Journals (about 1200 per annum), books and conference proceedings are covered, but not reports, which are to be found in *Scientific and Technical Aerospace Reports (STAR)*.

The abstracts are informative and are arranged in 74 subject sections from aerodynamics to thermodynamics with a general category at the end. Each abstract has an accession number. There are five indexes: subject, author, contract number, meeting and paper and report number, and accession number. These cumulate quarterly, semi-annually and annually.

An unusual feature of the subject index is that a brief description of the content, and not title, of the item is given under the subject heading. Each item appears under several subject headings, thus giving multiple access to the subject content. In fact, two kinds of cross-references are used: 'use' references, which refer to alternative headings; and 'narrower term' references, leading to more specific subject headings. The indexing vocabulary is the *NASA Thesaurus* and should be consulted for a complete insight to the terminology and reference structure. Searching might be less arduous if a simpler numbering system were used, although there are explanatory notes at the beginning of each of the five well-produced indexes.

Electrical and Electronics Abstracts (EEA). 1898–
The continuation of *Science Abstracts Series B, EEA* is one of three abstracting journals published by the Institution of Electrical Engineers and the Institute of Electrical and Electronics Engineers Inc., in association with the Institution of Electronic and Radio Engineers and the International Federation of Automatic Control. It is produced monthly and is the foremost English-language abstracts journal in the field of electrotechnology, covering journals (about 1800), reports, books, dissertations, patents and conference papers on a worldwide basis, producing more than 40 000 entries per annum. Instead of subscribing to the complete service, it is possible to take one of the series of 24-page monthly *Key Abstracts*, which were introduced in 1975. Each of these periodicals is restricted to a specific subject discipline, e.g. solid state devices.

Each abstract is simply given a running number, and entries are arranged by means of a subject classification scheme which consists of main sections subdivided into subject groups, each of which is split into sub-sections.

In each issue there is a combined subject index of 2500 terms for *EEA* and *CCA (Computer and Control Abstracts)* which refers from subject headings to the sections in the classification scheme. A semi-annual subject index is generated which arranges entries under fairly broad subject divisions and gives their running abstract number, from which reference may be made to the abstract itself. Searching this subject index is a tedious task in places because of some rather non-specific indexing, and in the 4- and 5-year indexes the problem is correspondingly multiplied.

There are also author, report, bibliography, patent, book and conference indexes in each issue which cumulate every 6 months. Cumulative author and subject indexes are produced every 4 or 5 years (i.e. 1955–1959, 1960–1964, 1965–1968, 1969–1972) which aid retrospective searching.

Computer and Control Abstracts (CCA). 1966–
CCA is a replica of *EEA* for the fields of computers and control. It is the continuation of *Science Abstracts Series C.* More than 24 000 items a year are listed.

Both *EEA* and *CCA*, together with *Physics Abstracts*, are available on magnetic tape. All back runs of the three abstracting journals are available on microfilm in the following formats: 16 mm cartridge, 16 mm magazine, 16 mm reel and 32 mm reel.

ICE Abstracts. 1974–
This new publication from the Institution of Civil Engineers fills what has been a serious gap in the abstracting services for civil engineering. Civil engineers now have an abstracting journal devoted solely to their subject which appears ten times a year (monthly except August and December) and covers the publications of the Institution of Civil Engineers, the American Society of Civil Engineers, the American Concrete Institute, the Prestressed Concrete Institute and the Water Pollution Control Federation. In addition, articles from about 100 European journals are abstracted by qualified engineers. *ICE Abstracts* may be obtained on microfiche or on computer tapes, but is produced in A5 format for general use. It remains to be seen whether or not the present coverage is wide enough to satisfy the needs of the practising civil engineer in view of the surprising exclusion of journals published outside Europe, with the exception of those produced by the institutions in the USA previously mentioned.

Referativnyi Zhurnal. 1953–
Undoubtedly the largest and most comprehensive abstracting service for science and technology in the world, this Russian journal published by VINITI (All Union Institute of Scientific Information, Moscow) includes over a million abstracts per annum and covers about 21 000 journals, 6000 monographs and 100 000 patents and standards every year. It is selective and publishes material in at least 60 languages. Many of its 61 sections and numerous sub-sections are of value to the engineer. These include: aircraft and rocket engines, automation, telemechanics and computer technology, internal combustion engines, industrial economics, electronic engineering, electric communication, electrical and power engineering, physics, surveying, mining, space research, cybernetics, corrosion, engineering materials, machine elements, mathematics, mechanics, metallurgy, metrology, measuring instruments, pumps and compressors, refrigeration, industrial management, industrial transport, radio engineering, building and road machines, mechanical engineering, thermal power, turbine engineering, water transport, air transport, rail transport, nuclear reactors, etc.

Most of these sections appear monthly. The indexes vary, but each part has one or more indexes of some kind: author, subject, patent, formula. All items included are abstracts, having a Russian title, a bibliographical citation in the original language, a Russian abstract, an identification number, a letter to indicate the form of the original if it is not a journal article (e.g. patent) and a UDC (Universal Decimal Classification) number.

To use the journal the reader must at least be familiar with the Cyrillic alphabet. For further details on how to use the journal see *A Guide to Referativnyi Zhurnal*, by E. J. Copley (National Reference Library of Science and Invention (now the British Library, Science Reference Library), 1970).

Bulletin Signalétique. 1940–
This French abstracting service has recently been divided into 36 specific subject sections and eight interdisciplinary fields. Each section is available separately or as part of four group subscriptions covering physical science, earth science, biological and medical sciences and interdisciplinary sciences. Sections of interest to the engineer are physics, mechanics, electronics, nuclear technology, fuels and energy, metallurgy, welding and associated technologies, chemical engineering, public health engineering, mechanical engineering, civil engineering, transportation and space technology.

The abstracts are short (and in French), which makes them of limited value. About 7000 journals are covered plus reports, conference proceedings and dissertations. Most sections have subject indexes in addition to author indexes, and these cumulate annually. From 1972 the service has been available on microfiche.

Technisches Zentralblatt. 1951–
Published by the German Academy of Sciences, this service scans journals and patents in the fields of mechanical and electrical engineering, producing over 10 000 abstracts per annum. Recently, owing to policy changes, the abstracts have been reduced in length and are now single-sentence only, which greatly lessens their usefulness. Arrangement is by UDC in 29 sections, with author and subject indexes cumulating annually.

Japan Science Review—Mechanical and Electrical Engineering. 1954–
Edited and published by the Japan Documentation Society, this quarterly review provides a comprehensive bibliography, with abstracts in most cases, of Japanese scientific and technical contributions to the journal literature. The review covers about 260 journals plus reports and conference proceedings, and lists about 5000 abstracts a year. Fringe subjects such as applied physics, atomic energy, metallurgy, chemical technology, etc., are treated, but not comprehensively. Arrangement is by UDC. The author index cumulates

annually, and there is a guide to the subject classification, but no subject index.

USSR and East European Scientific Abstracts. 1973–
This English-language service is produced bi-monthly by the Joint Publication Research Service of the US National Technical Information Service. Only East European and Russian journals are scanned, and the informative abstracts are arranged in very broad subject groups. Fields covered of interest to engineers are: cybernetics, computers and automation technology, electronics and electrical engineering, engineering and equipment, materials, science and metallurgy and physics and mathematics. There are no indexes in the journal.

Abstracts of Romanian Scientific and Technical Literature. 1965–
This quarterly abstracting bulletin is published in either English, French or Russian by the National Institute for Scientific and Technical Information and Documentation, Bucharest. The informative abstracts are of articles from all Romanian scientific and technical periodicals, over 2000 appearing each year, arranged in 24 sections. There is a list of source journals in each issue in addition to author and subject indexes and a new books list (Romanian only).

Hungarian Technical Abstracts. 1949–
Only Hungarian journals, reports and dissertations are covered by this English-language service published quarterly by the Hungarian Central Technical Library and Documentation Centre, Budapest. The abstracts are listed under 14 broad subject headings. There is an author index in each issue which cumulates annually, and a subject index appears at the end of the year. Also included now is a table of content of articles featured in the publications of the Hungarian Academy of Sciences and of Hungarian Universities and in Hungarian technical journals and scientific publications.

Indian Science Abstracts. 1965–
Published monthly by the Indian National Scientific Documentation Centre, New Delhi, this journal includes work published in India and work done in India but published abroad. Coverage is of original articles, including short communications and review articles, appearing in about 700 periodicals plus conference and symposia proceedings, books, theses, patents and standards. Material of a popular nature is not included. Entries (13 000 p.a.) are numbered sequentially and classified by UDC. The author and subject indexes in each issue refer to the abstract number and cumulate annually.

The address of the author is given in the citation, the title of which is usually given in English irrespective of the language of the original.

Polish Technical and Economic Abstracts. 1951–
Arranged in 10 main subject groups, this is a selective, critical service covering contributions in Polish technical journals, books, pamphlets and other publications. The work is produced quarterly by the Center For Scientific, Technical and Economic Information. There appears to be a rather long time-lag, probably due to the selective and critical policy adopted.

CURRENT AWARENESS JOURNALS

Current awareness journals are produced for the sole purpose of helping individuals to keep in touch with current developments as quickly as possible. Primary consideration has to be given to the speed of compilation and distribution, sometimes to the detriment of other desirable properties. Hanson (1971) notes three kinds of current awareness journals:

Those reproducing title pages of journals.
Those giving lists of titles of articles appearing in periodicals together with their bibliographical citations. Normally, entries are arranged in subject sections.
Those more generally known as indexing journals. These are indexes giving title references to articles in periodicals and listed under alphabetical subject headings or by keywords, e.g. KWIC (Key-Word-In-Context).

Some examples are as follows.

Current Contents — Engineering, Technology & Applied Sciences.
1970–
Current Contents is produced weekly by the Institute for Scientific Information, and reproduces, in original format, the contents pages of more than 700 research journals in engineering and technology. Entries are arranged in subject groups, and each issue lists the journals carried in that issue and refers the reader to the relevant page. There is an author index, and subject index and address directory in each issue (first authors only). This is an excellent service and often includes material in advance of its receipt in England.

Pandex Current Index to Scientific and Technical Literature. 1967–
Pandex is an interdisciplinary index covering science, medicine and technology, published by CCM Information Corporation. It contains the contents pages of over 2300 journals listed alphabetically by title within 18 subject areas. Each article listed on the contents page has a sequential number assigned to it to which reference is made by the author and subject indexes. The subject index is a cross between a KWIC index and an alphabetical subject heading list. The publication, including indexes, cumulates quarterly and annually, and these cumulations are available on 4 × 6 microfiche and 16 mm microfilm. A magnetic tape subscription may also be taken out, for which the subscriber receives the total weekly input of bibliographical data plus complete programs for printout, retrospective searching and SDI (Selective Dissemination of Information).

Current Papers in Electrical and Electronic Engineering. 1964–
The scope, arrangement and coverage of the journal is the same as that of its sister abstracting journal (*EEA*), also published by INSPEC. Basically, the only difference between the two journals is that *Current Papers* has no abstracts, which implies that the notification of the existence of a particular article will be supplied more quickly by the current awareness journal. Readers may find it better to consult the abstracting journal, since the citations often appear in the same monthly issue of both journals.

Current Papers on Computers and Control. 1966–
As above, for the fields of computing and control.

ISMEC (Information Service in Mechanical Engineering) Bulletin. 1973–
Published twice monthly, the *ISMEC Bulletin* is the first wide-ranging current awareness journal to cover all aspects of mechanical and production engineering and management. Cumulative author and subject indexes are produced every 6 months. Also available is a magnetic tape service which includes all items contained in the *ISMEC Bulletin* and additional information, including subject indexing in depth.

VDE Schnellberichte. 1955–
This German current awareness journal is published by the German Association of Electrical Engineers. There are 24 issues a year listing about 10 000 titles from 300 native and foreign specialist journals, dissertations and research reports. Subject divisions used include nuclear engineering, power engineering, information theory,

measurement and control, automatic control, circuits, etc. Six-monthly author and keyword subject indexes are produced. A list of journals regularly scanned is enclosed in the first issue of the year.

British Technology Index (BTI). 1962–
A continuation of the *Subject Index to Periodicals,* this index, published by The Library Association, scans about 350 British technical journals. Entries are arranged under a series of alphabetical subject headings. Many 'see' references and related headings are used to lead the reader from his entry point to the subject heading under which a particular item is indexed. From 1972, an author index appears in each issue and leads the reader from the author's name to the first word of the appropriate subject heading in the index. The bibliographical citation is also given alongside each author entry in the author index. The entire index cumulates annually.

Applied Science and Technology Index (ASTI). 1958–
ASTI is a continuation of the *Industrial Arts Index* (1913–1957) and is published monthly (except July) by the H. W. Wilson Company. Cumulations appear quarterly and annually, but without an author index. Entries are arranged under alphabetical subject headings, with some articles being listed under more than one heading, thereby increasing the chance of retrieval. As only about 230 journals (mostly American) are scanned, it is not particularly effective (for its cost).

OTHER ABSTRACTING AND INDEXING JOURNALS

The journals mentioned in the previous pages are, for the most part, of a general nature. Many other services exist and may be traced through the following publications:

International Federation for Documentation (FID). *Abstracting Services. Vol. 1: Science and Technology* (2nd edn, 1969)
British Library, Lending Division (BLLD). *A KWIC Index to the English Language Abstracting and Indexing Publications Currently Being Received by the British Library, Lending Division* (4th edn, 1972)
National Federation of Scientific Abstracting and Indexing Services. *Guide to the World's Abstracting and Indexing Services in Science and Technology* (1963)
Owen, D. B. and Hanchey, M.M. *Abstracts and Indexes in Science and Technology: a Descriptive Guide* (Scarecrow Press, 1974)
Ulrich's International Periodicals Directory (15th edn, Bowker, 1973)

Abstracts also appear as a section in many specialist journals, e.g. *Production Engineer*. These and other services are discussed in the relevant chapters.

Science Citation Index (SCI)

Garfield (1964) defined a citation index as 'an ordered list of cited articles each of which is accompanied by a list of citing articles. The citing article is identified by a source citation, the cited article by a reference citation. The index is arranged by reference citations. Any source citation may subsequently become a reference citation.'

Shepard's Citations has been used by American lawyers since 1873, and the citation principle has been applied by Garfield to the scientific sphere, leading to the production of the *Science Citation Index*, which has been published regularly since 1964 by the Institute for Scientific Information (ISI).

Each year *SCI* indexes over 370 000 papers from 2400 selected journals. Analyses have shown that 1000 journals account for 90% of the significant literature, which implies that the 2400 journals covered by *SCI* will give coverage of most important material, provided that the journals are properly chosen. The publishers take great care to ensure that this is done by using an editorial board of 'experts' and by making large-scale citation analyses. By looking through the ISI list of source publications, the individual researcher can verify that the major journals in his field are covered. It will be seen from this list that engineering journals are covered to a greater extent than the title of the index would suggest.

SCI has three separate indexes: the *Citation Index* (containing a *Patent Citation Index*), the *Source Index* and the *Permuterm Subject Index*. They are published quarterly for the first three quarters of the year, the fourth issue being the annual cumulation. A cumulation for 1965–1969 is available.

Normally, the search of *SCI* starts from an article known to be relevant to the searcher's subject. The searcher looks up the author of this paper in the *Citation Index*. There he finds all items which have cited this particular article. Having noted down the author and journal reference of each citing item, he looks up each citing author in the *Source Index* and obtains the full title and co-authors of each citing article. At this point the searcher must select the articles of interest to him and obtain the full papers from the library. Each of these articles will usually have a bibliography, and each author cited in the bibliography can now be looked up in the *Citation Index*. Garfield called this process 'cycling' and defined it as 'examining the bibliographies of the papers you start with, and of source papers obtained, in order to locate additional relevant works. By looking up the latter in the *Citation Index* you find new citing sources.' The *Source Index* may also give citations of other relevant papers written by the source author which are not linked to the searcher's starting references. Even if the author of the entry paper is not listed in *SCI*, the paper will probably have references in its bibliography which may be used as starters in the *Citation Index*. If no specific author of interest is known, then a list of search terms must be compiled by the user. Taking a starting term, he now looks in the *Permuterm Subject Index (PSI)* and locates this as a primary and/or co-term. Under this term he finds all authors who have used this term in the title of their articles. The searcher can now enter the *Source Index* to obtain the title and bibliographical details of each article and can then obtain the full papers. He may then, if he wishes, generate a further search through the *Citation Index*.

SCI has certain advantages over conventional indexes. The terminology problem inherent in subject indexes is solved, since the citation is indexed, not the subject term(s) describing the content of the paper. The linkages between the papers are dependent on the author's citations and not directly on any subject descriptors, which gives rise to multidisciplinary coverage and the breaking down of rigid subject barriers. *SCI*'s most important property is that it brings the searcher forward in time from a known reference, irrespective of its date, provided it is cited once in a year. It is thus possible to ascertain whether a theory has been verified or a particular method improved, and so on. The negative search producing no relevant items may be desired and obtained.

One problem with *SCI* is the inaccurate citations given by authors. Unfortunately, it is impossible to check all references, and ISI are completely at the mercy of the writers in this respect.

SCI has been shown to be a useful retrieval device, especially in a rapidly increasing field, but it will not, and cannot be expected to, replace good abstracting services with adequate subject indexes used to their full potential. For further information on *SCI* see the works by Garfield (1964), Malin (1968), Martyn (1965, 1966) and Weinstock (1971), and also the various material published as trade literature by ISI, especially the *Science Citation Index Guide and Journal Lists* published annually.

MAGNETIC TAPE SERVICES AND SDI

Many of the computer-based abstracting and indexing journals are now commercially available on magnetic tape. This gives the opportunity for organisations with computer facilities to undertake their own SDI (Selective Dissemination of Information) and retrospective searching using one or more of these services (e.g. AWRE, Aldermaston). In addition, the publishers of these tape services also generate searches and/or provide SDI services on an individual basis for a given fee.

Briefly, SDI is a service which selects items containing information of interest to a receiver and automatically sends a notification of their existence to him, without a formal request from the receiver. The system may be manual or mechanised (the latter is becoming more common). It is essential to ascertain the user's interests in sufficient detail to allow the system to work at the desired level. Thus, the 'user profile' (the description of the user's interests) is probably the critical factor in an SDI system. Documents containing items of current information are collected and scanned and their contents described. These 'document profiles' are matched with the 'user profile', thereby providing the items of interest to the user. The construction of the 'user profile' is dependent on the specificity with which the user defines his subject needs. If this profile covers only his main needs, the user must also search other sources for his fringe areas of interest, thus using up more of his time. Perhaps he may write these fringe subjects into his profile and find that he receives too many notifications of relevant (and irrelevant) items. The right balance in the 'user profile' is therefore essential if the user is to avoid the receipt of too much material. An SDI system is probably best used to obtain notification of items in precise rather than general areas of interest, with other search methods being employed to cover wider fields.

Several tape services of interest to engineers are available. INSPEC offer a magnetic tape service, an SDI service and *Topics* (an SDI service based on standard profiles). For further details see the various advertising literature distributed by INSPEC, especially 'Finding the Correct Indexing Language', by T. M. Aitchison (a paper presented to the Association of Scientific Information Dissemination Centers at New York, 23 September 1970); *INSPEC and Selective Dissemination of Information*, by P. Clague; and the *SDI User Manual*.

PANDEX also offers a weekly magnetic tape service and an individualised search program. The latter is available on a weekly, bi-weekly or monthly basis. Further details on the nature, scope and cost may be obtained from CCM Information Corporation.

COMPENDEX is available on either a lease or licence basis from Engineering Index Inc. The tapes contain all the abstracts, full citations, subject headings and cross-references appearing in *Engineering Index Monthly*. In addition, access points are provided to give the user the choice of a complete or selective search of the items on the tape. These access points mean that it is possible to produce bibliographies and perform searches where the user's facilities do not permit of full text searching.

Card-A-Lert is the weekly card service by Engineering Index Inc. which brings direct to the engineer information, in the form of short abstracts, relevant to his own particular subject interests. Selection is made by the engineer from 38 groups and 171 divisions, and the fee paid is dependent on the number of divisions or groups to which the engineer subscribes.

Metals Abstracts Index is available on magnetic tape on a lease basis from 1968, and, in addition, tapes of the *Review of Metal Literature* for 1966 and 1967 have been produced. All the index terms and bibliographical citations used in *Metals Abstracts* are included on the tape. The abstracts themselves are not on the tape, reference to them being made via their serial number in the printed journal.

Nuclear Science Abstracts (1968–   ) is also available on tape.

At present the cost of these various tape leases seems excessive when compared with the price of their printed journal equivalents. For a discussion of the problems in using these external services see the paper by Corbett (1972).

For a list of available services see the following publications:

A Guide to Selected Computer Based Information Services, by R. Finer (Aslib, 1972)

Selected Federal Computer Based Information Systems, by S. Herner and M. J. Vellucci (Information Resources Press, 1972)

Survey of Commercially Available Computer-Readable Bibliographic Data Bases, by J. H. Schneider, M. Gechman and S. E. Furth (American Society for Information Science, 1973)

REVIEWS

Hysterical statements are frequently made about the information explosion and its ramifications. At the moment, a certain degree of control over the problem is exercised through the production of evaluated condensations of material appearing in the primary publications. This review literature, as it is called, is one of the most important aids to every engineer and scientist in his attempt to keep up with developments in his own and related fields.

Reviews, if they are to be of any use, must provide a brief outline of previous work, followed by a comprehensive critical evaluation of the new ideas, methods, results and conclusions put forward in the material under review, plus a good bibliography citing all items of any significant value. Goudsmit (1966) has succinctly described the most useful review as 'one that will be cited in the literature instead of the original articles it summarises'. Ideally, the reviews should be written by specialists in the field. However, the problem of persuading a research worker to devote his time and energy to such tasks is all too well known to editors of review publications.

Reviews appear in several forms:

As a review paper in a non-review journal. The Institution of Electrical Engineers, for instance, commissions review articles for inclusion in a special issue of the *IEE Proceedings*. Many other primary journals also contain reviews, with varying degrees of regularity.

As articles in review journals, e.g. *International Metallurgical Reviews*. These papers present the state of the art in a subject in the journal literature.

As longer review papers or surveys of progress collected in book form, and appearing as an annual or irregular serial. Within this category are those series containing papers presented at an annual conference at which the advances in a particular subject field are announced. A list of a selection of these series appears as an appendix to this chapter.

As a monograph which is a comprehensive review of a particular subject. There is usually a long delay in the publication of such works, which limits their value in the research field. However, they are very useful as teaching material on a wider plane.

Review articles from more than 2700 journals may be traced by using the *Index to Scientific Reviews,* published semi-annually and annually by the Institute for Scientific Information.

Thus, as all these review publications have been produced at an ever-increasing rate, so the time has long since been reached when it has become necessary to look for other forms of condensed literature in addition to reviews. Perhaps the terse literatures of aphorisms suggested by Bernier (1968, 1970, 1972) may be the answer. This we shall see in due course.

BIBLIOGRAPHIES

Any list of readings on a particular subject may be described as a bibliography. The bibliography may consist of a list of books, journal articles and/or any other form of literature. It may be current or retrospective, selective or comprehensive, and have limitations of time, language or form placed upon it.

There are two major general bibliographies of bibliographies: T. Besterman's *World Bibilography of Bibliographies* (5 vols., 4th edn, Societas Bibliographica, 1965–1966), arranged alphabetically by subject with an author index; and *Bibliographic Index,* published in April and August, with an annual cumulation, by the H. W. Wilson Company, being a subject list of bibliographies, each containing 40 or more citations. The *Classed Subject Catalog of the Engineering Societies' Library* (G. K. Hall, 1963), with annual supplements, gives a comprehensive list of bibliographies in its first volume.

Other useful aids in tracing books on particular subjects include the *British National Bibliography* (1950–), published weekly with quarterly, annual and multi-annual cumulations; *Subject Guide to Books in Print* (Bowker, 1957–), published annually; *Cumulative Book Index* (The H. W. Wilson Company, 1928–), published monthly with quarterly, annual and multi-annual cumulations; and *Technical Books in Print,* published by Whitaker *Books in English,* published bi-monthly by the British National Bibliography, is an author/title bibliography, although the 1970 and 1971 cumulations are arranged by Dewey Decimal Classification. The bibliography is compiled from the UK and US MARC data base.

Lists of recent scientific and technical books received for review appear in many journals, e.g. *Nature.* The journals of the various institutions contain useful lists of additions to their respective libraries. INSPEC's abstracting journals include a bibliography index, which is very useful as a means of tracing review articles and other papers containing an extensive bibliography.

Many research associations, industrial, government and other organisations produce subject bibliographies, and these are often on specialist topics of current interest.

Users should note that bibliographies available in their own library may be found by looking in the library catalogue under the appropriate heading, e.g. Bibliographies: Lasers. Reference to the call number will locate the bibliography.

BIBLIOGRAPHY

Adams, S. and Baker, D. B. (1968). 'Mission and Discipline Orientation in Scientific Abstracting and Indexing Services', *Library Trends,* **16** (3), 307–322
Ashworth, W. (1967). 'Abstracting', in *Handbook of Special Librarianship and Information Work,* 3rd edn, 453–481 (Aslib)
Bernier, C. L. (1968). 'Condensed Technical Literatures', *Journal of Chemical Documentation,* **8** (4), 195–197
Bernier, C. L. (1970). 'Terse Literatures: Terse Conclusions', *Journal of the American Society for Information Science,* **21** (5), 316–319

Bernier, C. L. (1972). 'Terse Literature Viewpoint of Wordage Problems—Amount, Languages and Access', *Journal of Chemical Documentation*, **12** (2), 81–83

Burkett, J. (1968). 'Published Indexing and Abstracting Services', *Trends in Special Librarianship*, 35–72 (Bingley)

Clague, P. (1964). 'The Coverage of Heavy Electrical Engineering Periodical Literature by Abstracts Journals', *Journal of Documentation*, **20** (2), 70–75

Collison, R. L. (1971). *Abstracts and Abstracting Services* (ABC–Clio)

Corbett, L. (1972). 'Problems in Using External Information Services—Attitudes of the Special Library and its Users', *Aslib Proceedings*, **24** (2), 96–110

Garfield, E. (1964). '*Science Citation Index*—a New Dimension in Indexing', *Science*, **144** (3619), 649–654

Gechman, M. C. (1970). 'Analysis of Technical Data Bases and Processing Services', *Proceedings of the American Society for Information Science*, **7**, 97–99

Goudsmit, S. A. (1966). 'Is the Literature Worth Retrieving?', *Physics Today*, **19** (9), 52–55

Hanson, C. W. (1971). *Introduction to Science Information Work*, 85 (Aslib)

Herling, J. P. (1961). 'Engineering Abstracting Services', *Special Libraries*, **52** (10), 560–565

Herner, S. (1954). 'Information Gathering Habits of Workers in Pure and Applied Science', *Industrial and Engineering Chemistry*, **46** (1), 228–236

Herring, C. (1968). 'Critical Reviews: the User's Point of View', *Journal of Chemical Documentation*, **8** (4), 232–236

Herring, C. (1968). 'Distil or Drown—the Need for Reviews', *Physics Today*, **21** (9), 27–33

Lancaster, F. W. (1971). 'The Evaluation of Published Indexes and Abstract Journals: Criteria and Possible Procedures', *Bulletin of the Medical Library Association*, **59** (3), 479–494

Malin, M. V. (1968). 'The *Science Citation Index*: a New Concept in Indexing', *Library Trends*, **16** (3), 374–387

Marquis, D. G. and Allen, T. J. (1966). 'Communication Patterns in Applied Technology', *American Psychologist*, **21** (11), 1052–1060

Martyn, J. (1964). 'Tests on Abstracts Journals', *Journal of Documentation*, **20** (4), 212–235

Martyn, J. (1965). 'An Examination of Citation Indexes', *Aslib Proceedings*, **17** (6), 184–196

Martyn, J. (1966). 'Citation Indexing', *The Indexer*, **5** (1), 5–15

Martyn, J. (1967). 'Tests on Abstracts Journals: Coverage, Overlap and Indexing', *Journal of Documentation*, **23** (1), 45–70

Weinstock, M. (1971). 'Citation Indexes', in *Encyclopedia of Library and Information Science*, Vol. 5, 16–40 (Dekker)

Wood, D. N. and Hamilton, D. R. L. (1967). *The Information Requirements of Mechanical Engineers* (Library Association)

Woodward, A. M. (1974). 'Review Literature: Characteristics, Sources and Output in 1972', *Aslib Proceedings*, **26** (9), 367–376

APPENDIX: REVIEWS

Advances in Aeronautical Sciences (American Institute of Aeronautics and Astronautics, 1958– , irregular)

Advances in Applied Mechanics (Academic Press, 1948– , irregular)

Advances in the Astronautical Sciences (American Astronautical Society, 1957– , irregular)

Advances in Automobile Engineering (Pergamon, 1963– , annual)

Advances in Bioengineering and Instrumentation (Plenum, 1966– , irregular)

Advances in Biomedical Engineering and Medical Physics (Wiley, 1968– , annual)

Advances in Chemical Engineering (Academic Press, 1956– irregular)

Advances in Communications Systems (Academic Press, 1965– , irregular)

Advances in Computers (Academic Press, 1960– , annual)

Advances in Control Systems (Academic Press, 1964– , annual), continued as *Control and Dynamic Systems: Advances in Theory and Applications* (1973–)

Advances in Corrosion Science and Technology (Plenum, 1970– , irregular)

Advances in Creep Design (Applied Science, 1971– , irregular)

Advances in Cryogenic Engineering (Plenum, 1960– , annual)

Advances in Electrochemistry and Electrochemical Engineering (Wiley, 1961– , irregular)

Advances in Electronic Circuit Packaging (Plenum, 1962– , irregular)

Advances in Electronics and Electron Physics (Academic Press, 1948– , annual)

Advances in Heat Transfer (Academic Press, 1964– , annual)

Advances in High Pressure Research (Academic Press, 1966– , irregular)

Advances in Hydroscience (Academic Press, 1964– , annual)

Advances in Machine Tool Design and Research (Pergamon, 1964– , annual), continued as the *Proceedings of the ( ) International Machine Tool Design and Research Conference* (Macmillan, 1972– , annual)

Advances in Materials Research (Wiley, 1967– , irregular)

Advances in Microwaves (Academic Press, 1966– , annual)

Advances in Nuclear Science and Technology (Academic Press, 1962– , irregular)

Advances in Optical and Electron Microscopy (Academic Press, 1966–, irregular)

Advances in Quantum Electronics (Academic Press, 1970– , irregular)

Advances in Radio Research (Academic Press, 1964– , irregular)

Advances in Semiconductor Science (Pergamon, 1959– , irregular)
Advances in Space Research (Pergamon, 1965– , irregular)
Advances in Space Science and Technology (Academic Press, 1959– , annual)
Advances in Water Pollution Research (Pergamon, 1964– , irregular)

Annual Review in Automatic Programming (Pergamon, 1960– , annual)
Annual Review of Biophysics and Bioengineering (Annual Reviews Inc., 1972– , annual)
Annual Review of Fluid Mechanics (Annual Reviews Inc., 1969– , annual)
Annual Review of Materials Science (Annual Reviews Inc., 1971– , annual)
Annual Review of Nuclear Science (Annual Reviews Inc., 1952– , annual)

Control and Dynamic Systems: Advances in Theory and Applications (Academic Press, 1973– , annual)
Critical Reviews in Bioengineering (CRC Press, 1971– , irregular)
Critical Reviews in Solid State Sciences (CRC Press, 1970– , quarterly)
Critical Reviews in Tribology (IPC Science and Technology Press, 1972– , annual)

Progress in Aerospace Sciences (Pergamon, 1961– , annual), incorporates *Progress in Aeronautical Sciences*
Progress in the Astronautical Sciences (Wiley, 1962– , irregular)
Progress in Astronautics and Aeronautics (Academic Press, 1960– , irregular)
Progress in Combustion Science and Technology (Pergamon, 1960– , irregular)
Progress in Construction Science and Technology (Medical and Technical, 1971– , annual)
Progress in Cryogenics (Academic Press, 1959– , irregular)
Progress in Dielectrics (Wiley, 1959– , irregular)
Progress in Heat and Mass Transfer (Pergamon, 1969– , irregular)
Progress in Infrared Spectroscopy (Plenum, 1962– , irregular)
Progress in Materials Science (Pergamon, 1949– , irregular)
Progress in Nuclear Energy (Pergamon, 1956– , irregular)
Progress in Nuclear Techniques and Instrumentation (North Holland, 1965– , irregular)

Progress in Operations Research (Wiley, 1961– , irregular)
Progress in Quantum Electronics (Pergamon, 1969– , irregular)
Progress in Water Technology (Pergamon, 1973– , irregular)

Recent Advances in Engineering Science (Gordon and Breach, 1967– , irregular)

See also *List of Annual Reviews of Progress in Science and Technology* (2nd edn, Unesco, 1969) and *Directory of Review Serials in Science and Technology*, by A. M. Woodward (Aslib, 1974).

12

Literature searching

K. W. Mildren

The retrieval of published information has become as onerous a task to the engineering profession as it has to any other group of individuals, and many of the problems encountered by an engineer in his search for information are very often the result of his own reluctance to come to terms with the situation as it is today. Licklider (1966) has said, when referring to the 'information explosion', that 'the best response to a warning of explosion is usually either to duck or to run, depending upon one's distance from the center of the scene'. Judging by the results of various surveys on information-gathering habits undertaken in recent years, it would appear that a considerable number of engineers have chosen the latter alternative. It may be of course that the 'need-to-know' is not as great as imagined by members of the information profession and, given the time at his disposal, the engineer may indeed be using this time to his best advantage. Most engineers would probably claim that this is the case.

Voigt (1961) has listed three approaches to information — current, everyday and exhaustive – and it is convenient to consider literature searching from these viewpoints.

CURRENT APPROACH

The current approach arises from the desire to be aware of what other people in the field are doing or have just completed. Generally speaking, most current information is obtained by direct contact with colleagues in the engineer's own organisation and outside it, especially with contacts made at conferences. In narrow subject areas correspondence plays a key role in the communication of new information.

Literature as such first appears on the engineer's desk in the form of reports or preprints or reprints of conference papers or journal articles, provided that the engineer has made the necessary contacts. If these do not reach him, the engineer must take a positive step and obtain and read current articles in the journals covering his particular field. Review articles are of course especially welcome, and even more so if they are critical or in a subject in which the engineer only requires a summary of the topic rather than a complete knowledge of the subject. Very few engineers do not read some primary journals on a regular basis, even if this is only limited to the publications of their own Institution.

Abstracting journals are a different matter and not all engineers are aware of their usefulness, even allowing for time delay, in locating references to current material of potential interest. Engineers, especially in academic institutions, need to be aware of new monographs and textbooks. Sources of information on such works, in addition to publishers' pre-publication literature and published catalogues and bibliographies, are the primary and abstracting journals.

Selective dissemination of information (SDI)

Throughout the discussion so far, the onus has been on the individual to locate the information himself from whatever source he can find. In certain subject areas and within certain organisations an SDI service is available to the engineer. This is a service which provides an individual with a list of references to a limited number of documents containing information of possible interest to him. Some SDI services are mentioned in the previous chapter. Such services are becoming increasingly important as a method of helping the engineer in his struggle to keep informed of current developments elsewhere with the minimum use of his own time. A survey by Hall, Clague and Aitchison (1972) indicates that scientists and technologists using the INSPEC service have been able to reduce the number of journals scanned and consequently they find it easier to keep up to date. However, the engineer still has to expend time in the construction and updating of his user profile, even if there is an information officer or librarian available to monitor the profile for him.

EVERYDAY APPROACH

The browsing nature of the 'current approach' is totally different and far removed from the so-called 'everyday approach', which may more clearly be defined as the 'factual approach'. When an engineer is faced with an information problem of a specific nature, which may be the need to know an established fact such as the viscosity of a particular fluid, or any physical property of a material, he will probably first consult the literature in his personal library. If this fails to provide the desired information, a possible resulting search pattern to find the answer from this point is shown on the flow chart in *Figure 12.1*. Given that the information concerned is likely to be

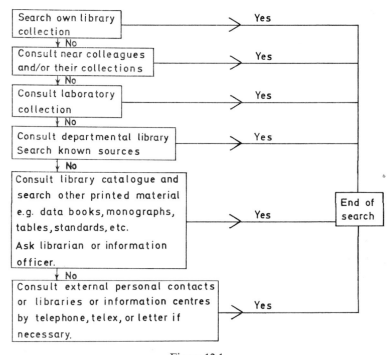

Figure 12.1

found in several sources on the majority of occasions, it is only likely that the final stage has to be used in a very few instances. The proviso is of course that there is a reasonable library or information service close at hand. Naturally, the shortest route to the information is taken, and this is normally via a source within the individual's own organisation. It is this kind of factual information which appears to be most in demand by engineers. In particular, information on materials, equipment and products contained in the literature distributed by industrial organisations is of prime importance.

EXHAUSTIVE APPROACH

An 'exhaustive' or 'retrospective' literature search may be carried out by two groups of individuals: those persons acting on their own behalf, i.e. engineers who are at the start of a project or research programme or perhaps are about to publish a paper on a particular topic and require all previous references on the subject; and those persons (librarians and information officers) who have been delegated the task by a requestor. The search strategy employed by these two groups of searchers should in theory be similar, although one expects in practice that the engineer may be less methodical than the professional searcher and consequently may take longer and be less efficient in the task, mainly owing to lack of familiarity with the material used in the searching process. Efficiency is not of paramount importance to the engineer who is making only one or two searches, but the professional searcher is unable to waste time and effort and must therefore employ effective search strategy and efficient procedures.

SEARCH PATTERN

The pattern of a comprehensive retrospective literature search is fairly orthodox. Carey (1966), Voress (1963), Hanson (1963) and Beltran (1971) have each provided various descriptions of the searching process. All writings on the subject must inevitably be very similar, since the basic technique is commonly accepted as the only way to produce a satisfactory result.

Simple flow charts are a useful way to demonstrate the steps followed in a particular procedure, and the chart contained in *Figure 12.2* indicates a possible route taken by a literature searcher.

The first two stages are very important in any search. It is essential to know the exact subject and scope of the search, and if the searcher is acting on behalf of a requestor, then he must interrogate this person until a satisfactory definition of his requirements has been obtained. The main points to be clarified are as follows: the subject of the search, any limitations of a technical nature, the level required within the subject in question, the number of references wanted (all or just a few major ones?), languages required or not acceptable,

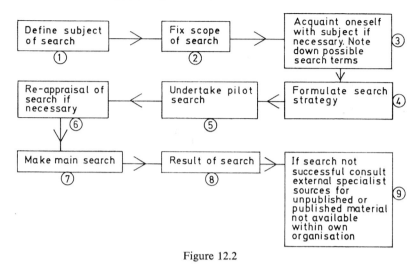

Figure 12.2

forms of literature acceptable/required (i.e. reports, journal articles, etc.), time limitation on search (how far back should the search be made?), form in which the results are expected (abstracts, bibliography?), and, of course, the completion date for the search. Stage 3 is unnecessary for the engineer acting on his own behalf, since he will draw up his list of search terms when he formulates his search strategy.

SEARCH STRATEGY

The search strategy, stage 4, is merely the planning of the quickest way to the relevant document references. The first step is to list the sources to be used and the order in which they will be searched. These retrieval systems may be manual or automated, and if any external services are used, then it is wise to send requests to these outside organisations as soon as possible, since there may be a delay involved.

Books are generally easily traced through the various national bibliographies, publishers' catalogues and local library catalogues. The various 'guides to the literature' will be a useful aid to the sources available in a subject if the searcher is unfamiliar with the literature. The major part of the search is usually concerned with scanning the abstracting and indexing journals which cover the journal, patent, report and conference literature. Which of these are used is dependent on the experience of the searcher, and it is here that the professional searcher has an advantage over the engineer who is not well acquainted with the secondary sources in his field. The other limitation on the sources used is their proximity to the searcher. Sources on or near the premises will be consulted first and will be the only ones employed if the search has to be finished in a short space of time. Assuming that the sources are available, then any searcher looking for information on sub-millimetre waves, for instance, will normally scan *Electrical and Electronics Abstracts* and possibly *Physics Abstracts* before consulting *Engineering Index*, simply because of subject coverage and ease of use. Thus, depending on time and sources available, an ordered list of bibliographical tools to be consulted is compiled.

A list of subject headings is now produced. These are only possible headings at this stage and may have to be modified as each source is consulted, since the chances are that the terminology and subject indexing employed in the secondary sources will be of a different nature and level. A more general subject entry term may be necessary because the subject headings used in the source are insufficiently specific. This will mean an addition to the subject heading list. The currency of the terminology used in the indexing may also be a factor involved. The use of a thesaurus or list of subject headings is particularly recommended at this stage, and is really essential when using abstracting and indexing journals which employ a particular one in their construction (examples of thesauri are *EJC Thesaurus, NASA Thesaurus, ASM Thesaurus INSPEC Thesaurus*; and *SHE* is a list of subject headings) (see p. 154).

PILOT SEARCH

A search through the last year or so of all sources to be used is the next advisable step, stage 5. This pilot search enables subject headings to be finalised and the documents retrieved can be analysed. From this analysis the scope of the search may be found to be too narrow or too broad, and any clarification and re-appraisal, stage 6, made at this point should save a considerable amount of time and effort in the main search.

MAIN SEARCH (STAGE 7)

If a bibliography on the subject can be located at the start of a search, this is most useful. It should provide the basic references and may limit the work needed to complete the search. These may be found by searching bibliographies of bibliographies, e.g. *Bibliographic Index*, or a more specialist bibliography of bibliographies such as H. Marienfeld's *Bibliographies 1960–1968: Mechanical Engineering, Machine Elements, Metals, Plastics, Mechanics, Thermodynamics, Combustion, Nuclear Energy* (Gesellschaft für Regelungstechnik und Simulationstechnik GMBH, 1969), or by scanning the national book bibliographies using title and subject approach. Bibliographies may also be traced by consulting the library catalogue, or even by using an abstracting journal with a bibliography index, e.g. *Electrical and Electronics Abstracts*.

As already stated, the greater part of the search is devoted to scanning the abstracting and indexing journals which the searcher has decided will yield the desired results and cover all forms of literature, i.e. journals, reports, patents, conferences, etc. These sources should be searched methodically backwards in time within the defined limits, if any, of the search. The key papers are located and obtained, and the references contained at the end of each paper can be followed up, producing the so-called 'mushroom' or 'snow-ball' effect. The first key reference is always the hardest to find. Review articles are most useful in providing extensive additional references.

Effective searching of abstracting journals is dependent on several factors. There is an art in effective searching which is often under-estimated. It requires experience, training, knowledge, exercise of judgement and perseverance on the part of the searcher. The use of the indexes to abstracting journals is a critical factor in any search. Cumulations of the indexes should be used where available. Cross-references may be relevant and ought to be checked. This can be a tedious process, especially when scanning a publication which has fairly broad indexing with many cross-references, e.g. *Engineering Index*. The thoroughness and quality of the indexes to abstracting journals also vary considerably, and the searcher has to be alert to changes in subject fields and indexing within a particular journal and must be prepared to adjust his search terms and retrace his steps as necessary.

The search having been completed, it may be noticed that certain prolific authors have been at work and it may be rewarding to under-take an author search through the sources used for the main search. Also, a particular periodical may continuously contain articles of interest, and the indexes to this journal should be scanned for other papers which may have been omitted from the abstracting journals for any reason, i.e. editorial policy, oversight. A search through the *Science Citation Index* should not be omitted from the search strategy, as this is the easiest way to update useful references written some time ago. A description of *SCI* is given on p. 164, and it is an extremely useful index for the retrospective searcher if the complete run of the index is available, including the quinquennial cumulation. Next, the search can be brought completely up to date by scanning the current awareness journals in the field and by reading the latest issues of the primary journals. Unpublished sources of information available to the searcher should also be covered, since these may yield important recent information.

Throughout the search it is vital to maintain good bibliographical records, usually on cards. It is essential to note down in full the source of any reference found, so that this may be consulted at a later date should the need arise. A card referring to an item subsequently deemed irrelevant should still be retained in the record file to indicate that the item has been evaluated.

RETROSPECTIVE TAPE SERVICES

The search described has been a manual search through conventional published sources, some of which have been available for retrospective use for a number of years in machine-readable form on magnetic tapes. However, these are rather expensive to operate and few organisations and institutions have been prepared to provide a service using tapes for sole use by their personnel. Problems associated with tape services are: several tapes would normally be needed to cover the literature adequately; they do not all have the same format and even if a standard format existed technical problems, usually cost problems, are still associated with them, although unlimited computer facilities would remove most of these if such services were available; problems of size and number of terms; problems of vocabulary control and indexing levels. However, some progress has been made in this area. For instance, the RECON (REmote CONsole) terminal link at the Technology Reports Centre (TRC) enables TRC to provide a retrospective search using the data bases employed in the ESRO/ELDO Space Documentation Service. These data bases are the INSPEC data base (1969–) *Scientific and Technical Aerospace Reports (STAR)* and *International Aerospace Abstracts (IAA)* (1962–), *Metals Abstracts* (1969–), *Engineering Index (EI)* (1970–), *Nuclear Science Abstracts (NSA)* (1968–) and *Government Reports Announcements (GRA)* (1970–). An interactive system such as RECON presents various problems at the user/computer interface, and these are discussed in a publication edited by D. E. Walker entitled *Interactive Bibliographic Search: The User/Computer Interface* (AFIPS Press, 1971).

The INSPEC services have also been extended to provide retrospective searches (RETROSPEC), but again covering a relatively short period backwards in time. In many cases the searcher must look for older material than that covered by the tape services, but nevertheless retrospective facilities such as RECON and RETROSPEC can be used by the searcher to cover part of his search and he can carry out the remainder of the search manually using the printed sources. From the point of view of the searcher, the key factors when deciding whether or not to use external retrospective tape services are the cost and coverage of the service, the relevance of the search product and the time taken to complete the search.

SEARCH RESULT (STAGE 8)

When the search is completed, the items discovered can be obtained either from the library within the organisation or on external loan from the British Library, Lending Division or other libraries. Some items will have already been obtained during the course of the search. If the searcher is acting for a requestor, he will present the search result to the enquirer in the desired form, i.e. on cards, as a bibliography, as an annotated bibliography or perhaps as a report. Follow-up searches may be necessary as a result of the information uncovered.

If the search does not produce the desired result, then the searcher should consult an external specialist source (stage 9), which may be traced via the relevant directories if not already known.

EXAMPLE SEARCH

An example of an extended search is described in H. Schenck's *Introduction to the Engineering Research Project* (McGraw-Hill, 1969, pp. 73–78). The topic considered is 'the design and construction of small transparent models of water-seepage systems', and the subject in question highlights the necessity for the searcher to be constantly aware of interdisciplinary sources when undertaking his search. Also demonstrated, by the retrieval of the final most useful reference in this particular search, is that relevant material is often to be found in the most unlikely places. Indeed it simply reinforces the uncertainty always facing the information seeker, in that the searcher can never be sure that all possible sources have been checked or indeed that the information sought exists at all and, if it does, that it has in fact been published in the open literature.

REFERENCES

Beltran, A. A. (1971). 'The Craft of Literature Searching', *Sci-Tech News*, **25** (4), 113–116

Carey, R. J. P. (1966). *Finding and Using Technical Information* (Arnold)

Hall, A. M., Clague, P. and Aitchison, T. M. (1972). *The Effect of the Use of an SDI Service on the Information Gathering Habits of Scientists and Technologists* (INSPEC)

Hanson, C. W. (1963). 'Subject Inquiries and Literature Searching', *Aslib Proceedings*, **15** (11), 315–322

Licklider, J. C. R. (1966). 'A Crux in Scientific and Technical Communication', *American Psychologist*, **21** (11), 1044

Voigt, M. J. (1961). *Scientists Approaches to Information*, 21 (American Library Association; ACRL Monograph 24)

Voress, H. E. (1963). 'Searching Techniques in the Literature of the Sciences', *College and Research Libraries*, **24** (3), 209–212

13

Personal indexes

A. C. Foskett

Most of us keep some kind of records of pieces of information that are of use to us. If we include mental records, then of course everybody does; but most of us find it worthwhile to make notes of items that are of particular significance. We may make a note of somebody's address; of a particular article that we found interesting or useful; of a piece of apparatus or equipment that we saw at an exhibition and thought would be useful at some time. To begin with, we can manage with a minimum of organisation of our notes, but as our collection grows, so will the need for us to impose some kind of formal record keeping; otherwise we shall lose items which we need when we most need them.

There are a variety of ways in which we may attempt to do this, and this chapter attempts to set out some of the practical details of how to manage a personal index of this kind. It does *not* set out to be a complete introduction to the theory of information retrieval, and indeed the amount of theory will be kept to the very minimum. For those who wish to pursue this particular aspect, there are now very many works which range through the topic at all levels.

The simplest way of keeping a personal index is what might be called the 'scrap of paper' technique. Names, addresses and telephone numbers are jotted down in our diary or address book; articles that we have found useful are jotted down, on a card perhaps; odd pieces of information that have come to hand are written down on the most handy piece of paper lying around, traditionally the back of an old envelope. It is fairly clear that this widely used method suffers from certain disadvantages. Old envelopes tend to be cleared up and thrown away; even when they are not, we often find that the information on them means nothing to us after a period of time, because we have forgotten the reason why we wrote it down. Diaries change from year to year, and we may find that last year's diary, with that vital telephone number in it, has been thrown away, or is at least not to hand. Even rather more official-looking cards will not serve very much purpose if we do not have somewhere to keep them, and some way of arranging them in order, so that we can find information when we need it.

CARD INDEXES

The next stage is thus to introduce an element of formality into the records that we keep of pieces of information, and the simplest way of doing this is to use standard size cards, such as the widely used library catalogue card, 12·7 × 7·5 cm, or rather larger cards, say 8 × 5 in, if we wish to record rather more details of any given piece of information. Each piece of information should be written on one of these standard cards, whichever size is chosen, and kept in a proper filing drawer.

While the collection of cards is small, the amount of organisation that is needed will also be small, and we can perhaps begin by separating out our addresses and telephone numbers into one alphabetical sequence. We may need to make more than one entry for a particular address; for example, we may have the name of a firm, but we may also have the name of a particular representative within the firm. In this kind of situation, it is useful to have more than one entry in the file. The advantage of this kind of file is that it is convenient to update; we can of course update an address book, by crossing out the old entries and putting in new ones, but this means that eventually our address book is cluttered up with masses of entries which have no further significance. Buried among these entries will of course be those that do still have some significance, and these we may overlook. There is also the point that the order in which we put entries into an address book is strictly a chronological one, within each initial letter, and if we have a lot of names under the same initial, we may have to hunt around for quite a while to pick out the one we are looking for in this chronological sequence. If we maintain the index on cards, we can remove old entries as they are superseded, and we can also keep the whole file in complete alphabetical order.

The next group of cards that we may wish to build up will be entries for documents, such as articles in periodicals or technical reports, patents and similar sources of information. It is quite simple to jot down the reference for these documents in any old fashion, but it is much better if some system of rules is chosen and all entries are made according to this. Inadequate references often cause problems later on when you try to obtain the document and find that you do not have enough information to do so. For a periodical article, the minimum amount of information includes the title, the author's name, the title of the periodical, details of the issue (i.e. volume number and full date) and page numbers. In addition to this information, it is often useful to list the author's affiliation, and if necessary add a brief abstract of the document or notes on its contents. If the information that we found useful consists of a few facts, it is probably worthwhile just jotting down the facts themselves and thus eliminating the need to refer to the original article. For a technical report, the corporate author—that is, the body responsible for the work—is normally more important than the individual author, and the report number or numbers takes the place of the volume number and part number of a periodical article. With patents, it is certainly necessary to list the number, the corporate author and the individual inventor, and it is often useful to list both the date of publication and the date of original application, which may be several years earlier; it frequently takes two years or so for a patent to be granted after application has been made, and this time lag can be an important one. There is a relevant standard, *BS 1629:1950. Bibliographical References,* and some scientific and engineering societies have also published guidelines; in addition, it is often useful to look at an abstracting journal in the appropriate subject field and follow its practice, since this means that your own index will be in the same kind of mode as the most commonly used published index. Whichever system is chosen, it must be followed as closely as possible; otherwise confusion and ambiguity is sure to result. Many librarians have been baffled by requests for the journal 'ibid.', which does not exist but simply refers to a previous reference in the same bibliography. Careful attention to the rules will avoid that kind of trap.

The next problem that arises is the order in which these references should be kept. Again, if there are not very many, it is no trouble to look through the pack of cards and pick up the one or two that are useful at the moment, and this 'sequential scan' can be quite useful in reminding us of items which we had forgotten about and which may be useful. In fact, it may be useful to do this kind of scan from time to time, just to keep in touch with the material put into the index. However, once the number of cards begins to grow it becomes tedious to look through them every time, and we want to try and arrange them in some kind of useful order. There are two ways in which this can be done very simply: one of these is to arrange by author's name; the other is to arrange by title. Which we choose will depend very largely on the kind of material we are including in our index. It should also be remembered that in many cases the author we shall be looking for will not be the individual, but the corporate author of the document. The third possibility is to arrange by journal title, but this is not highly recommended. It is true that we often remember the journal in which an article was published without remembering other details, but it is very easy to be mistaken in this respect. We would then be searching under the wrong journal title and wasting time. A chronological sequence suffers from the same kind of handicap; our memory as to the date at which we saw a particular article is very often indeed wildly inaccurate.

SUBJECT FILES

The author and title of a document are both 'manifest'; that is to say, they come with the package, and we do not have to make them up at all. It is fairly obvious that another useful way of arranging our cards would be by subject, but subject descriptions do not come with the package, and we may find ourselves getting involved in a certain amount of complicated effort in establishing subject terms for the articles in our collection.

When we start to consider subject entry, a further point arises — namely, that we are able to arrange a file of single entries in only one way. If we arrange our cards by title, then we cannot arrange by subject, and vice versa. Once we start thinking about subject approach, in fact, we commit ourselves to a rather more sophisticated approach than the one discussed so far.

We mentioned earlier a third group of entries that we might wish to consider making in our index: cards for information about pieces of equipment or odd data that we want to record. These items clearly lend themselves to the subject approach rather than the author and title approach, and can therefore be catered for in a subject file if we develop one.

Assuming that we decide that it is useful for us to adopt the subject approach, we are faced with a further decision as to whether we should use a classified approach or an alphabetical one. Most of us are familiar with the classified system used to arrange the books on public library shelves, the Dewey Decimal Classification. It would be possible for us to use this as a tool for our own personal indexes, but it is not very detailed, whereas the information we are likely to include in a personal index is probably going to be on quite specific topics. An alternative classification scheme is the Universal Decimal Classification. This is in some ways comparable with the Dewey Decimal Classification and indeed follows the same outline, but it is possible to build up more specific classification numbers by joining the numbers for different topics together by a colon. On the whole, however, the use of classification schemes in personal indexes is probably not a very good idea. Their use requires some skill if it is to be of value, and, in addition, it is necessary to compile an alphabetical index to the class numbers to be able to use the file properly. The amount of effort involved is therefore rather excessive in return for the advantages of using this kind of classification scheme. It is better within one's own file to stick to the alphabetical approach.

If we decide to do this, the next problem that arises is the one of selecting suitable terms to index our documents and information. Here again there are several existing lists. Some public libraries have 'dictionary' catalogues in which the subject entries are based on either *Sears List of Subject Headings* or the *Subject Headings Used in the Dictionary Catalogs of the Library of Congress*, usually shortened to *Library of Congress Subject Headings*. As an alternative to these lists, in recent years more specialised lists have been compiled, in particular the Engineers Joint Council's *Thesaurus of Engineering and Scientific Terms*, and *Thesaurofacet*, a very similar list published by the English Electric Company. There are also more specialised lists in particular areas of science and technology — for example, that published by the American Petroleum Institute. Use of one of these lists gives you a handy source of terms to use in your index, and also gives you considerable guidance on how to construct the index. However, most of these lists are relatively expensive, and few individuals will wish to purchase them for their own use.

Fortunately, quite a number of recent research projects have come to the conclusion that we will get quite reasonable results if we stick to the words that occur in the documents themselves, particularly in the kind of situation we are discussing, where the size of the index remains relatively small. A useful technique is to go through an article and underline the words which you feel are significant terms under which you would like to retrieve the document at some later date. Probably the best results will be obtained if you restrict your selection to those words which occur in the title and any abstracts of the article; to go through the whole article may mean that at a later date you retrieve a number of documents which treat only marginally the subject you are searching for, and this can be a little annoying.

When we have selected the terms that we wish to use to index a particular document, the next question that arises is precisely how we use these terms to compile an index. The way commonly used in libraries is to make several entries for each document, each one bearing complete information, and then file one card under each of the relevant terms. However, this is geared to the availability of multiple copies of cards, whereas in the personal index we will normally have made out only one card with all the details on, and do not wish to have to write out several. This suggests the idea of a file arranged by author or title, but with a supporting index file which leads from the subject indexing term to the author's name or the title. In this way, we only have to make out a card for each subject term and enter on it enough information to enable us to get from that term to the main file, where we will find the rest of the information. Unfortunately, once we start getting into any kind of indexing situation other than the most simple, this idea of two files becomes inevitable.

Using cards, then, we may finish up with three separate sequences. The first of these is our straightforward addresses file, which can contain quite a lot of useful information and will be called on quite frequently. The second sequence consists of a list of articles and other items of information, arranged either by title or by author. The third sequence will contain the subject index entries to enable us to refer to the appropriate articles in the second sequence. A card index of this kind can be maintained relatively easily and works quite effectively until our collection begins to contain several hundred documents, when we may consider that the card index is becoming a little too clumsy for us to continue maintaining it. The decision as to whether we should use a more sophisticated method thus really depends on the likely size of the index that we envisage maintaining. It is important to remember here that people have access to libraries which contain many indexing and abstracting services and similar reference works, and that in many situations it is better to rely on the library for any large-scale searching rather than on our own personal files. If we make a conscious decision to restrict the size of our own personal index, by weeding out at regular intervals those items which no longer are of use, and by maintaining fairly strict control over what we put in, it may well be possible for us to manage very satisfactorily with a simple index on cards of the kind described. If, however, we wish to go into business in a rather larger way, then almost inevitably it will be necessary to adopt more complex methods.

COORDINATE INDEXING

Once we start looking at subjects, we find that many of them, particularly those dealt with in periodical articles and similar items likely to be included in our personal file, are of the kind that may be described as composite. That is, they are not composed of a single concept but of a number of concepts acting together. For example, we have an article on aluminium alloys, which is of interest both from the point of view of containing information about aluminium and from that of alloys. We are therefore faced with the problem of indexing our collection and taking account of this kind of subject. The straightforward indexing described in the previous section can become rather complicated if we have very many of the composite subjects that I have mentioned; unfortunately, it seems that most scientific and technical literature nowadays is of this kind, and the situation is likely to grow worse rather than better. If we just use the single terms of our index, this will work, but it will mean that under a term such as 'aluminium' we shall have noted articles which deal with aluminium alloys, aluminium conductors, production of aluminium, and so on. If we are searching for a particular topic such as aluminium alloys, we have to look through a whole lot of entries dealing with aluminium in order to pick out the few dealing with alloys. Coordinate indexing is a means of getting round this particular difficulty. Instead of using ordinary catalogue cards of the kind already described, we use special cards which enable us to compare the entries under two or more headings conveniently, and thus select only those which relate to particular combinations of terms rather than all of those relating to the individual terms themselves.

The simplest form of coordinate index is the Uniterm Index. We construct our main file as before, but this time instead of arranging by author or title we arrange chronologically and give each document a number. Our subject index is now compiled on Uniterm cards, which consist of cards with a space for a heading at the top and ten columns ruled on the rest of the card. These ten columns are headed 0–9 (*Figure 13.1*), and to enter a document number, we simply put the number into the column appropriate to the final digit. For example, number 83 is entered in column 3, not column 8. In making out our card in the main accessions file, we add after the bibliographical details the terms under which we wish to index that particular article. We then make out the Uniterm card for each of these terms, if there is not already one in our subject index, and enter the document number on each of these cards. To take a concrete example, we may have a document on the casting of an aluminium alloy. In our index, we find that we already have cards for aluminium and alloys, but not one for casting. We therefore make out an additional card for casting, and enter the document number on all three cards. To search for information through the subject index, we think of the terms we are interested in, remove the matching cards from

Aluminium

0	1	2	3	4	5	6	7	8	9
10	21	72	33	34	25	66	17	78	9
90	131		133		95	86		88	59
140						116			119

Alloys

0	1	2	3	4	5	6	7	8	9
20		72		44	55	36	147		59
					95	86			

Figure 13.1 Uniterm cards

the index and compare them to see which document numbers appear on all of those we have selected. For example, to find any information that we have on the casting of aluminium, we would withdraw cards for casting and aluminium from the subject index and compare them, and in the particular instance cited we would, of course, find the number of the document already referred to on casting of aluminium alloys. We therefore finish our search of the subject index with a set of document numbers, and with this set of numbers we go back to our accessions file to find details of the documents themselves.

The system is a very cheap one to set up and to operate, and on the scale of the personal index it works very satisfactorily. We do not need to introduce very many controls over the terms we use to denote subjects, with the possible exception of synonyms and different forms of the same word. For example, it would probably be worth our while to merge cast and casting, rather than have two separate cards. We would also, of course, only have one card for aluminium and aluminum; this is an example of a synonym arising out of a different English and American usage, but there are many other kinds of synonyms, of course, and, in general, we will avoid some complications if we try to recognise these as they occur and select only one of the possible terms. Since our personal index will normally cover a subject field with which we are reasonably familiar, this should not present any problems, and in fact the terms we use in our subject index should be those which we would use in our everyday work.

The simplicity of the Uniterm Index is very attractive, but it should be fairly evident that in a large file searching the Uniterm cards can become rather tedious, and there is always the risk that we may miss a number which appears on the two cards we are searching. Because of this, we find that a rather more sophisticated approach is to use optical coincidence cards, otherwise known as peek-a-boo cards. In a peek-a-boo system the subject index cards are printed with a grid showing a number of positions (*Figure 13.2*),

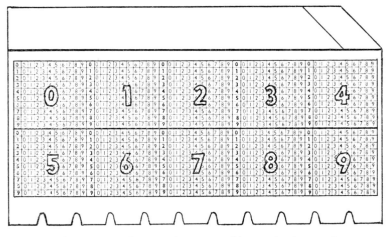

Figure 13.2   Diagram of a 1000 position peek-a-boo card

and instead of writing a document number on the card, we punch a hole at the position corresponding to the document number. Cards are available in a variety of sizes, the smallest having about 400 positions and the largest about 10 000.

There are a number of features of peek-a-boo cards which it is important to consider before launching out into this kind of index.

In the first place, the cards are very much more expensive than the simple cards required for a Uniterm index. In addition, we require a punch to make the hole, and this may cost anything from £5 upwards. There is also the point that each card only contains a limited number of holes. If we consider a 10 000 position card, for example, we find that on the day we add document 10 000, we have to start a new set of cards, because there is no position available for us to punch. Some libraries make use of this particular feature by starting a new index at regular intervals, often yearly. In this case, we have to estimate the number of documents that we shall want to include in the system in the course of a year. However, this idea may not lend itself so much to the personal index, where, in general, we tend to build up a collection over a considerable period of time. The essential thing is to decide how many documents we are likely to want to index in a reasonable span of years, and make sure that the cards we obtain are big enough to give positions for at least this number of documents. We shall also need some minor items of equipment, such as a sheet of hardboard with guides at two edges on which we can position the cards for punching; some kind of container for the cards will also be required, since some of them, particularly those with a large number of punching positions, can be quite sizeable. In a library, one would probably also find an illuminated box on which to place the cards when searching, but this is an unnecessary refinement for the individual user.

Procedure with peek-a-boo cards is very similar to that with Uniterm. We have to make out our card for the accessions file, listing on it the terms on which we wish to index the document; we then have to withdraw the appropriate cards from the subject file, or make out new ones if necessary, and enter the document numbers on to the card. This is done by punching out the appropriate position on the cards. The cards are then filed back in the alphabetical subject sequence. To search the file, we take the appropriate cards out, just as with Uniterm, but instead of having to search for matching pairs of numbers written on to the cards, we simply align the cards together and hold them up to the light. Those positions that are punched out on all of the cards we are looking at will of course show up as points of light, which can be translated into document numbers. The indexing method and the end product is just the same as with a Uniterm file; in effect, we pay for the increased facility that peek-a-boo cards give us for comparing the entry on different cards. The system is certainly well worthwhile if the index is to receive any heavy amount of use, but for personal purposes we may well decide that Uniterm wins out because of its cheapness.

EDGE-NOTCHED CARDS

Another way of mechanising the maintenance of a personal index is to use edge-notched cards (*Figure 13.3*). These consist of a small card, usually about 6 × 4 in or 8 × 5 in, with a row of holes punched round each of the edges. The middle of the card is blank, or we may

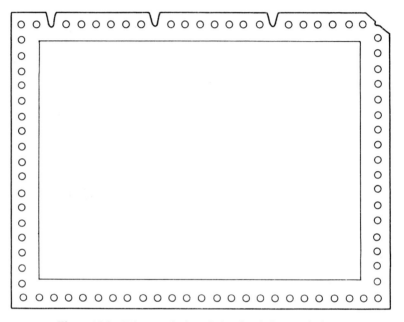

Figure 13.3   Edge-notched card showing holes notched out

find it printed with particular information; it is not too difficult to duplicate information for one's own purposes on a spirit duplicating machine. To use the cards, the appropriate holes are notched out so that they extend to the edge; if a needle is passed through that particular hole in the pack of cards, and the pack is shaken, any cards notched out at that point will drop and can be retrieved.

In use, edge-notched cards present a slightly different approach from the kind of index that we have been looking at so far. They are what is know as 'item cards', in that we make out one card, and only one card, for each item in our collection. We do not have to make out any extra cards for the subject or author indexing. In theory, we can index as many documents as we like by this method; in practice the number of cards that can be manipulated at any time is rather limited. The maximum number that can be needled at once is about 200; if we try to needle a larger pack than this, it becomes very difficult to get all the appropriate cards to drop—the cards tend to stick together. If we have a file of several hundred cards, therefore, we may well have to search it in sections, which can become tedious. The method therefore lends itself to relatively small collections which have to be exploited rather intensively. For example, edge-notched cards work well as a form of student record card in an educational establishment where the student numbers are relatively small. A system of this kind was set up at the College of Librarianship Wales, where the maximum number of students on any given course was about 150, and proved very suitable for the complex manipulation involved in compiling the College timetable. The choice of edge-notched cards is thus linked to the likely size of our collection; as most personal indexes tend to be fairly small, edge-notched cards are a practical proposition, but we should bear in mind that if our collection does grow beyond a few hundred documents, we may find that we have made the wrong choice.

The information that is written on the body of the card will be the information that we need in any index. However, we have the additional task of converting that information into notches round the edge of the card, by adopting some system of coding. Coding systems are basically of two kinds: direct and indirect.

Direct coding links each hole to a particular concept. For example, if the articles that we want to include in our index come very largely from half a dozen periodicals, we could allocate seven holes to coding for periodical titles. The major six would have one hole each; the seventh would be labelled 'others'. Direct coding is very easy to do, and of course searching the file is equally easy. It has, however, one major disadvantage, which is that we only have a limited number of holes round the edge of the card. Even on an 8 × 5 in card, we may well find that there are only 128 holes. If we have 128 holes, we can only cater for 128 concepts by direct coding. If we wish to index by subject, then it is certain that this number of possibilities will be quite inadequate unless we are content with a very broad coding indeed. The same is of course true of author's names. Direct coding is thus of value when we have a limited number of ideas to code, but breaks down when the number of concepts to be coded is indefinitely large.

In this situation, we are obliged to use indirect coding: the combination of two or more holes to represent a particular concept. There are various ways in which this can be done, and some edge-notched cards can be bought with indirect codes already printed on them.

One of the most common is the 7 – 4 – 2 – 1 code (*Figure 13.4*). This uses a field of four holes, the first being coded 7, the second 4, the third 2 and the fourth 1. Using these four holes or combinations of two of them, we can represent any number from 0 to 9. For example, 3 is represented by notching out 1 and 2; 8 is represented by notching out 7 and 1.

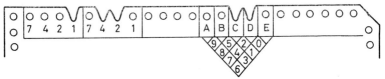

Figure 13.4   7-4-2-1 coding: first field 1; second field 6 (4+2). Pyramid coding: combination of holes C and D used for 2; A and D used for 7; B and D used for 4, etc.

Another method of coding is known as pyramid coding, and in this we use a five-hole field, and use pairs of holes to denote each number. This gives us the same opportunity to code from 0 to 9, but without the problem of false drops that we may find with 7 – 4 – 2 – 1 coding. With the latter, for example, when we search for '4' we will also find all the cards in which we have punched 5 and 6, because 4 forms part of those codes. In pyramid coding, this possibility is removed, at the cost of using one extra hole for each field.

By using, say, three fields, we can code numbers from 0 to 999. What we do is to use the first field for units, the second field for tens, and the third field for hundreds. We can of course extend the idea by using a fourth field for thousands and so on. We can in this way code not only numbers but also names, if we draw up a table of equivalents; some manufacturers of edge-notched cards supply such codes, and the present author has also prepared a set. Indirect coding of this kind is, however, restricted to one concept per field or set of fields. For example, if we have two authors for a particular document, we can only code one of them in our author field, because if we try to code more than one author in the same field, the number of false drops rises alarmingly. (One may consider what would happen in a 7 – 4 – 2 – 1 field if we punched out 7 and 4 and 1 and 2 . . .) As we will usually be able to manage with one author for a document, this may not be too serious; however, there are not many documents which would require only one subject entry, and we are therefore faced with the problem of how to code a number of entries of the same kind. One solution that has been suggested to this is known as 'superimposed random coding'. In this method, we allocate a large field, as large as possible, in fact; within this large field, we punch out pairs of holes, the number of which are chosen at random, to represent each concept. If we have n holes at our disposal, the number of possible pairs of codes is

$$n \frac{(n-1)}{2}$$

If, for example, we have 100 holes, then we have 4950 potential codes, and this will obviously be adequate for quite detailed subject indexing. In order to avoid false drops, which can occur when we punch codes for several concepts, it is recommended that we should not use more than about 45% of the available holes in any given card; in the example already quoted, with 100 holes in the random coding field, we should not punch out more than 45 holes for any given document, which implies a maximum of 22 concepts to be coded. Again, for the majority of purposes, this is ample for a personal index. The problem of false drops arises because the punching out of more than one pair of holes gives rise to spurious combinations; for example, if we punch out 20 and 32, 17 and 39, to represent two concepts, we have also entered the spurious combinations of 17 and 32, 20 and 39, 32 and 39, 17 and 20. By restricting the number of holes we punch out as shown, we can keep the number of false drops to an acceptable level. It is necessary to keep careful records of the codes that we adopt, and to ensure that we do not inadvertently use the same code for more than one concept. It may well be that the system is more trouble that it is worth for the usual personal index, and though it is a powerful method of entering a very large number of concepts into a very limited number of potential punching positions, its difficulties have to be carefully evaluated before it is adopted. *There is no point in constructing a personal index which we do not use because it is too complex.*

COMPUTERS

It is unlikely that an individual will have his own personal computer, even in these days of mini-computers, deck-top computers, and so on. However, a great many of us now have access to computers, and this may well provide a useful way of generating our own personal indexes. For example, ICI have a full set of computer programs to produce KWIC indexes, and these may be used to produce indexes of this type for individual members of staff. The system involves the usual accessions file, in which we make one entry per document giving the full details (*Figure 13.5*). The subject indexing is done through the title, or a title-like statement. Each entry consists of this title, manipulated by the computer to bring each significant word in turn to the filing position, which is approximately in the middle of

13927	JAPANESE MEDICAL	LIBRARIES*
13918	BRITISH MEDICAL	LIBRARIES 1953–1962*
13930	TITUTE OF ARCTIC AND ALPINE RESEARCH* THE	LIBRARIES AND LITERATURE OF C
13931	EARCH CENTER, UNIVERSITY OF WISCONSIN*THE	LIBRARIES AND LITERATURE OF C
13936	PUBLIC	LIBRARIES IN CZECHOSLOVAKIA*
13920	MEDICAL	LIBRARIES IN HOSPITALS*
13921	SCHOOLS*	LIBRARIES IN NURSE-TRAINING
13916	LITTLE BROTHER TO	LIBRAROCRAT*
13933	LANGUAGE AND THE LAW	LIBRARY*
13919	THE GENERAL HOSPITAL	LIBRARY*
13935	THE YOUNG STUDENT AND THE SCIENTIFIC	LIBRARY*
13932	SPECIAL LIBRARY SERVICES* THE JOHN CRERAR	LIBRARY, A COMPLEX OF SPECIAL
13934	RICAN MEDICAL ASSOCIATION* THE ARCHIVE	LIBRARY DEPARTMENT OF THE AME
13929	* DAPRATO	·LIBRARY OF ECCLESIASTICAL ART
13944	ING RURAL COMMUNITY*	LIBRARY SERVICE FOR THE CHANG
13926	TIONS*	A 'LIBRARY SURVEY OF 117 CORPORA

Figure 13.5 KWIC index. Number on left leads to accessions file for full
details of each entry. Part of some titles is lost if they exceed about 70 characters

the page. Each index entry is linked to the main file through the
accessions number which appears at one side of the page. It is
possible to introduce some elements of sophistication into the
straightforward KWIC programs, so that, for example, correspondence can be set out in a tabulated fashion, showing the date of a
letter and the correspondent involved, as well as the title of the
letter. All that the individual has to do is to supply the computer
centre with the necessary information and they will then go ahead
and produce the personal KWIC index. The system is cheap in
terms of computer time, and relieves the individual of the chore of
maintaining his own personal index. On the other hand, of course, it
also removes from him the ability to scribble comments on an entry
to remind him of additional points which may not have occurred to
him the first time round. This problem may be solved by the increasing use of computers on-line. With immediate access, we recover the
ability to browse which is lost in batch processing, and the computer
can give us all the flexibility of our own personal file, relieved of the
chore of organising and maintaining it. There are now an increasing
number of program packages available for this purpose. It seems
likely that the use of computers to maintain personal indexes, either
in this fashion or in some other, will become more widespread in the
future as we become more accustomed to having a computer at our
beck and call.

SUMMARY

A personal index can be valuable, by obliging us to formalise our own records and thus to introduce an element of organisation into what can be a very haphazard collection of memories, notes and scraps of paper. However, it loses its value if it becomes so complex that we do not use it. As far as personal indexes are concerned, simplest is best, and the Uniterm Index described here is probably one of the most effective kinds for this purpose. It should also be remembered that a personal index can only remind us of what we already know; it is very important that we should not overlook the resources available to us in libraries, which can extend far beyond the limits of our own personal information store.

BIBLIOGRAPHY AND NOTES

There are a number of works which deal with the theory of information retrieval, including Foskett, A. C. *The Subject Approach to Information* (2nd edn, Bingley, 1971). This contains a list of the major works on pp. 16–18. There are a few books specifically on personal indexes, including Foskett, A. C. *A Guide to Personal Indexes* (2nd edn, Bingley, 1970) and Jahoda, G. *Information Storage and Retrieval Systems for Individual Researchers* (Wiley, 1970).

Dewey Decimal Classification and Relative Index (18th edn, Forest Press, 1971). The individual user may find the 10th abridged edition, also published in 1971, more convenient to handle, but it is unlikely to be detailed enough for personal indexes.

Universal Decimal Classification (abridged English edition) (3rd edn, British Standards Institution, 1961). The full English edition is also available from the British Standards Institution in fascicules, but the individual is unlikely to find the 100 separate items convenient to handle. The abridged edition is satisfactory for many purposes if the synthetic devices such as the colon are used.

Sears List of Subject Headings, edited by B. M. Westby (10th edn, The H. W. Wilson Company, 1972)

Subject Headings Used in the Dictionary Catalogs of the Library of Congress (7th edn, Library of Congress, 1966)

Thesaurus of Engineering and Scientific Terms (Engineers Joint Council, 1967). The major American thesaurus.

Thesaurofacet (English Electric Company, 1970). A combination of a classification scheme and a list of subject headings covering engineering and related subjects.

Thesaurus (American Petroleum Institute, 1974). This is the 11th edition of the tool previously known as *Information Retrieval System Subject Authority List*.

Casey, R. S. *Punched Cards: Their Application to Science and Industry* (2nd edn, Reinhold, 1958). Although some sections of this are now outdated, the basic discussions of techniques for edge-notched cards have not been superseded. The *Guide to Personal Indexes* quoted above presents a simplified approach which is adequate for most purposes.

Chapter 21 in Casey, above, gives a detailed mathematical analysis of the superimposed random coding techniques. In a personal index, it is unlikely that many problems will arise even if the suggested limitations are not observed.

Matthews, F. and Shillingford, A. D. 'Variations on KWIC', *Aslib Proceedings*, **25** (4), 140–152 (1973)

Cook, C. E. *CAPRI—a Computer-Aided Personal Reference Index System for use by Individual Research Workers and Groups* (United Kingdom Atomic Energy Authority, 1971) (AWRE Report 063/71)

Burton, H. D. and Yerke, T. B. 'FAMULUS: a Computer-Based System for Augmenting Personal Documentation Efforts', in *Proceedings of the 32nd Annual Meeting of the American Society for Information Science (San Francisco, October 1969)*, 53–56 (Greenwood Publishing Corporation, 1969)

Both CAPRI and FAMULUS are geared to the use of large third-generation computers.

Lancaster, F. W. *Information Retrieval On-line* (Wiley; Becker and Hayes, 1973). Chapter 13 deals in some detail with the use of computers on-line for personal indexes.

14

Electrical power systems and machines

A. M. Parker

This chapter aims to cover the electrical aspects of the generation, transmission and distribution of electrical power, together with its utilisation in machines. It starts with a brief general survey of the field and then progresses along the path followed by the electrical energy from the power station, through the transmission and distribution networks to the user and his machine drives.

PRIMARY SOURCES AND ABSTRACTING SERVICES

The IEE in the United Kingdom and the IEEE in the United States both publish separate sections of their transactions dealing with the field covered by this chapter, and these together are widely regarded as the major English-language sources on the subject. The IEE's *Power Record,* published quarterly, and the *IEEE Transactions on Power Apparatus and Systems,* which appears bi-monthly, are devoted wholly to electrical power engineering, although in the former case an earlier view of papers can be gained from the monthly publication of the full *IEE Proceedings.*

In addition to its *Proceedings* forming an indispensable primary information source, the IEE's *Current Papers in Electrical and Electronics Engineering* provides an invaluable monthly guide covering a very wide range of periodicals and other publications. In particular, section 5 on 'Power Systems and Applications' should be regularly perused by researchers in this field. The *CEGB Digest,* published monthly by the Information Services of the Central

Electricity Generating Board, London, does not claim to be as exhaustive as the IEE publication, but it does offer a one-paragraph abstract of the papers listed and also a description of some of the translations undertaken by CEGB staff of foreign-language publications.

A further source of useful information on current papers in power is *Power Express* (International Physical Index Inc.), a monthly publication which lists relevant papers published in the USSR, providing for selected papers a fairly full abstract in English, together with salient diagrams and formulae.

POWER SYSTEMS

General

Because of the great importance of electricity supply in all countries of the world, and the way in which it impinges upon all aspects of life, the design and operation of power systems are subject to a number of constraints which are not wholly technical in nature. The siting of power stations, the type of fuel employed, the treatment of waste products, the routes of transmission lines and the transfer of power between separate companies or countries are examples of the types of problem in which economic, social, political and environmental factors have considerable weight in determining the shape of a system, and, hence, its technical performance. Some of these wider issues, and in particular a comparison of private and public ownership of the supply industry, can be found in *The Social Organisation of Electric Power Supply in Modern Societies,* by P. Sporn (MIT, 1971).

It is thus highly desirable that a researcher new to the power systems field have a good general idea of the constraints under which the system has to operate so that he will be able to evaluate the consequences which flow from his work in a realistic manner. For a good introduction to the field the reader is recommended to *Electric Power Systems,* by B. M. Weedy (2nd edn, Wiley, 1972), or *Electricity Supply,* by M. F. Buchan (Arnold, 1967). The approach is rather different in the two books, Weedy describing the components of a power system in his first chapter and then concentrating on the performance of the system as a whole, while Buchan devotes separate chapters to individual items of equipment.

In addition to these general textbooks, papers presented by the supply authorities themselves throw light upon the detailed problems encountered in the day-to-day operation of power systems, and it is possible to infer, from these papers and the discussions upon them, the future trends which may emerge in various parts of the world. Two international bodies provide a forum for this discussion and publish papers contributed by their member authorities. These are the Conference Internationale des Grands Réseaux Electriques (CIGRE), which publishes biennially the proceedings of its General Assembly and forms numerous study committees to investigate internationally detailed power system problems, and the International Union of Producers and Distributors of Electrical Energy (UNIPEDE), which publishes a monthly journal, originally in French but with an experimental English edition starting in mid-1973.

A good introduction to the economic factors in electricity supply can be gained from *Power System Economics,* edited by E. Openshaw Taylor and G. A. Boal (Arnold, 1969), while the book *New Approaches to the Design and Economics of EHV Transmission Plant,* by B. Jones (Pergamon, 1972), gives a very clear exposition of the way in which a transmission network is inherently shaped by its economics. More detailed facets of the economics of electricity supply have been dealt with in two IEE conferences for which the proceedings have been published:

The Economics of the Reliability of Supply (IEE Conference Publication No. 34, 1967)
Management of Transmission and Distribution Systems (IEE Conference Publication No. 76, 1971)

By their very nature the social, political and environmental factors, which have a considerable influence on decision-making at the highest level of power system management, are subject to the continued shifts and trends of public attitudes and opinions. It is very difficult, therefore, to offer a definitive published work in this area, but regular perusal of the periodicals *Electrical Review* (IPC), *Electrical Times* (IPC) and *Electrical World* (McGraw-Hill) is recommended. The first two are published weekly in the UK, and contain several pages of news affecting electrical power engineering as a whole, naturally concentrating on British conditions. They are particularly useful for reports of recent government decisions, public discussions and legislation affecting the industry. The third publication deals with similar topics from the US standpoint, and carries separate sections on the environment and the political scene. At the time of writing, two international conferences of relevance to this aspect of power supply were in the course of preparation, and could provide fruitful information. These were 'The Socially Undesirable Products of Power Generation in Engineering and Ways and Means of Tackling Them', sponsored by the Institution of Mechanical Engineers, London, in November 1973, and 'Energy, Europe and the 1980's' (IEE Conference Publication No. 112, 1974).

The generation of power

The present form of the power station is a result of a process of evolution occupying some 90 years in the case of thermal and hydro-electric types and 20 years for nuclear power stations. The accumulated experience and design knowledge, at least in British conditions, derived from this process has been collected in an eight-volume work: *Modern Power Station Practice* (2nd edn, Pergamon, for Central Electricity Generating Board, 1971). This contains a wealth of practical detail on the planning, layout and operation of power stations of various types, although the section dealing with hydro-electric schemes is necessarily rather limited. This topic, however, is the exclusive subject of the three-volume *Hydro-Electric Engineering Practice,* edited by J. G. Brown (Blackie, 1965–1966). Neither of the above texts claims an exhaustive analytical treatment of any aspect of power generation, but they are invaluable for enumerating the practical constraints under which each item of plant has to operate.

The generators themselves, although logically candidates for the general section on electrical machines later in this chapter, are a sufficiently important application to deserve some treatment in their own right at this point in the text. Forming the interface between mechanical and electrical power, they become the power sources for all electrical circuit studies on power systems, and so their properties are of fundamental importance in system studies.

Before the use of high-speed digital computers became common in power system analysis, the effect of generators on power systems was described in terms of a relatively few key parameters. These quantities, mainly reactances and time constants, were strategically chosen from results of a few standard tests, and were intended for use in hand calculations on stability margins. The methods used, which are still perfectly valid, but tend to give approximate answers on the conservative side, are admirably described in *Power System Stability: Synchronous Machines,* by E. W. Kimbark (Dover, 1968), and in the second volume of the comprehensive work *Circuit Analysis of A. C. Power Systems,* by E. Clarke (Wiley, 1943).

For a more accurate analysis of machine transient effects, especially for multi-machine studies, a more rigorous approach is required and is now feasible given access to a computer. The underlying theory to this approach can be found in *Synchronous Machines — Theory and Performance,* by C. Concordia (Wiley, 1951), or in *The General Theory of Alternating Current Machines,* by B. Adkins and R. G. Harley (Chapman and Hall, 1975). These introduce the differential circuit equations of a machine in terms of rotating frames of reference of which the classical source paper is 'Classification of the Reference Frames of a Synchronous Machine', by G. Kron (*Transactions of the AIEE,* **67,** 720–727, 1950).

Dynamic analysis of power systems using this approach is now widespread, and two typical examples in the literature are: 'A General-Purpose Turbo-Alternator Model', by G. Shackshaft (*Proceeding of the IEE,* **110,** 703–713, 1963), and 'Digital Computer Methods in Dynamic Response Analyses of Turbo-Generator Units', by W. D. Humpage and T. N. Saha (*Proceedings of the IEE,* **114,** 1115–1130, 1967). Shackshaft uses the basic equations to set up an analogue computer study, while the approach of Humpage and Saha is obviously digital.

The effect of a generator on its associated power system cannot be separated from the effects of its excitation and speed control systems. Suitable introductory texts in this connection are: *Introduction to the Dynamics of Automatic Regulating of Electrical Machines,* by M. V. Meerov (Butterworths, 1961), and the appropriate chapter of Kimbark above for automatic voltage regulators, while the effect of governors is described in 'Effect of Prime-Mover Speed Governor Characteristics on Power System Frequency Variations and Tie-Line Power Swings', by C. Concordia, S. B. Crary, E. E. Parker (*Transactions of the AIEE,* **60,** 559–567, 1941).

The incorporation of these controllers into dynamic system studies can be traced in 'Power System Stability — Digital Analysis Showing Effects of Generator Representation, Types of Voltage Regulators, and Speed Governor Systems', by H. E. Lokay and J. W. Skooglund (IEEE Power Industries Computer Application Conference, 1965), and in 'Effect of Turbine-Generator Representation in System Stability Studies', by H. E. Lokay and R. L. Bolger (*IEEE Transactions on Power Apparatus and Systems,* **84,** 933–942, 1965).

Direct conversion

While the rotating machine is the sole generator of electrical power in commercial quantities at present, there is a continual effort to achieve a breakthrough in the direct conversion of energy into electricity without using a mechanical conversion stage. Although nearly all the methods are still in the hands of the physicist rather than the engineer, useful reviews of possible techniques can be found in *Direct Energy Conversion,* by S. W. Angrist (2nd edn, Allyn and Bacon, 1971), and *Direct Energy Conversion,* by M. A. Kettani (Addison-Wesley, 1970), the latter text containing comprehensive references up to the date of publication.

A potentially valuable source of current information on the subject can be obtained from the papers presented to the yearly Intersociety Energy Conversion Engineering Conference in the United States. The sponsorship of the Conference rotates from year to year among the participating institutions, the sponsoring body being responsible for publishing the papers and discussion report. For example, the 1967 proceedings were published by the ASME, while those of 1971 were in the hands of the Society of Automotive Engineers.

Transmission and distribution plant

As the title of the section suggests, the author has found it convenient to consider separately the plant involved in the transmission and distribution of electrical energy, leaving the methods of predicting the performance of the interconnected equipment to the general analytical section which follows later.

For the transmission lines themselves the following references give a fairly complete review of the electrical, mechanical and thermal constraints involved, the detailed design features and operating characteristics: *Overhead Line Practice,* by J. McCombe and F. R. Haigh (3rd edn, Macdonald, 1966), and *Power Cables,* by C. C. Barnes (2nd edn, Chapman and Hall, 1966).

For EHV lines the IEE conference *Progress in Overhead Lines and Cables for 220 kV and above* (IEE Conference Publication No. 44, 1968) gives a useful guide to recent advances in the field, while for the lower voltage distribution networks *Electric Power Distribution,* edited by E. Openshaw Taylor and G. A. Boal (Arnold, 1966), provides practical information specific to those systems.

For the transformers used in power systems the comprehensive *The J and P Transformer Book,* by S. A. Stigant and A. C. Franklin (10th edn, Newnes–Butterworth, 1972), gives a wealth of practical design and operational details on transformers of various types. A similar field is covered in *Transformer Engineering,* by L. F. Blume *et al.* (2nd edn, Wiley, 1951), which is noted for its chapter on thermal performance.

In the field of switchgear *The J and P Switchgear Book,* by R. T. Lythall (7th edn, Newnes–Butterworth, 1972), has achieved a similar status to its transformer counterpart in providing a wide-ranging review of current practice. An interesting account of the development of switchgear to its present form can be found in *High Voltage Circuit-Breakers*, by V. Zajic (Artia, Prague, 1957). The protection of power systems has given rise to a large number of publications, of which *Analysis and Protection of Electrical Power Systems*, by D. Jones (Pitman, 1971), forms an excellent introduction to the subject, showing the interrelationship between a power system and its protective gear. In addition, the reader is recommended to two works, each of which sets out in detail the requirements for and practical realisation of system protection. *Protective Relays—Their Theory and Practice*, by A. R. Van C. Warrington (2nd edn, Chapman and Hall, 1968) is a two-volume text, in which the first volume establishes the requirements of protection systems and describes practical electromagnetic schemes, while the second deals with solid state developments. Both volumes carry a copious list of references. The second work is *Power System Protection*, by The Electricity Council (Macdonald, 1969), which is in three volumes. Based on a correspondence tuition course run by the electricity supply authorities in the UK, the three volumes cover, respectively, the background, the theory and the application of protection to a wide range of practical situations.

From a materials viewpoint, the feature which distinguishes power systems from most other electrical applications is the necessity for high-voltage insulation of nearly all items of equipment. The problems posed by high voltages and a description of insulating techniques in practical power systems can be found in *High Voltage Technology*, edited by L. L. Alston (Oxford University Press, 1968). The approach of *High Voltage Engineering*, by E. Kuffel and M. Abdullah (Pergamon, 1970), while outlining the same physical phenomena as the Alston volume, is directed at the laboratory production and measurement of high voltages. The specialised high-voltage testing procedures necesary for power system equipment are described in *Impulse Voltage Testing*, by W. G. Hawley (Chapman and Hall, 1959).

High-voltage direct current transmission

From the late 1950s considerable interest has been shown in the application of high-voltage direct current (HVDC) transmission links within predominantly a.c. systems, both for the feeding in of remote generation and also as a stabilising and fault-limiting agency in heavily interconnected systems. The original work in this field was *High Voltage Direct Current Power Transmission*, by C. Adamson and N. G. Hingorani (Garraway, 1960), which deals with the basic theory and describes the requirements for the equipment involved. It is interesting to compare their approach with that used in *Direct Current Transmission,* by E. W. Kimbark (Wiley/Interscience, 1971). Although both works cover essentially the same ground, a reading of both will prove fruitful, since the subject is covered from an academic viewpoint on the one hand, and by an established power systems engineer on the other. A second volume of the Kimbark is in the course of preparation, dealing inter alia with the impact of a d.c. link on its associated a.c. network.

The conference publication *High Voltage DC Transmission* (IEE Conference Publication No. 22, 1966) is also a useful source of operational details on practical HVDC schemes and indicates the lines of research in progress at that time. Another IEE conference publication in this field is *High Voltage DC/AC Power Transmission* (IEE Conference Publication No. 107, 1973).

Power system analysis

The size and complexity of modern power systems have required specialised techniques to be developed in dealing with the analysis of complete networks or sub-networks under operational conditions. In general, the problems encountered in the analysis fall into three main categories: (1) the solution of the network equations under given loading and generation conditions to give the complete load flow pattern; (2) a similar analysis under specified fault conditions, both types of analysis being carried out in the frequency domain under steady state conditions; (3) transient analysis, where the time domain is employed to obtain the response of the system to a specified change in network conditions, normally in the interests of evaluating stability margins. All three classes of problem are identified in *Computer Methods in Power System Analysis,* by G. W. Stagg and A. H. El-Abiad (McGraw-Hill, 1968), where a clear exposition of solution techniques in each category is given, together with numerous references.

The most significant adaptations of conventional circuit theory to deal specifically with the problems posed by power systems have been developed by Gabriel Kron. The basis of his method is to sub-divide the network into a number of smaller and more easily manageable circuits, with specially derived connection terms to compensate for the artificial subdivision. The process is described in *Diakoptics,* by G. Kron (Macdonald, 1963), and relies theoretically on formal tensor analysis, originally applied to the network field in *Tensor*

Analysis of Networks, by G. Kron (Macdonald, 1939). Although the theoretical basis of the method is rigorously developed in this work, a more convenient introduction to the subject can be gained from *Tensors for Circuits,* by G. Kron (2nd edn, Dover, 1959). A useful guide to the topological terminology and basic properties can be found in *Linear Graphs and Electrical Networks,* by S. Seshu and M. B. Reed (Addison-Wesley, 1961).

For the load-flow problem, the appropriate chapter in Stagg and El-Abiad is a good introduction to the methods available, and 'Numerical Techniques in Solution of Power System Load-Flow Problems', by M. A. Laughton and M. W. Humphrey-Davies (*Proceedings of the IEE,* **111**, 1575–1578, 1964), gives a comprehensive review of the literature up to the date of publication.

The basic method used for fault calculations is that of symmetrical components, a subject which receives a thorough treatment in the first volume of *Circuit Analysis of A.C. Power Systems,* by E. Clarke (Wiley, 1943), in which several alternative component systems are postulated. The extension of this theory into formalised analysis of large networks can be seen in 'Digital Calculation of Three-Phase Short Circuits by Matrix Methods', by H. E. Brown, C. E. Person, L. K. Kirchmayer and G. W. Stagg (*Transactions of the AIEE,* **79**, Part III, 1277–1282, 1960). Chapter 6 of Stagg and El-Abiad also indicates the structure of a practical digital computer program for short-circuit calculations as used by a major power company.

For transient stability studies the normal approach is to include dynamic representations for the rotating machines in the system, to provide instantaneous values of power infeed to the transmission network, whose equations are then solved in the frequency domain as for steady state studies. The solutions to these equations are then applied as terminal conditions to the machines. The machine representations have been mentioned before, and range from those used by Kimbark and by Clarke, in which the mechanical equations of motion are solved in a step-by-step fashion, to the inclusion of the effects of internal machine transients and of controllers, necessitating the use of state–space analysis and numerical integration techniques.

The degree of complexity of machine representation depends on the information required from a study. Examples of two widely different approaches are: 'Direct Calculation of Power System Stability Using the Impedance Matrix', by R. B. Shipley, N. Sato, D. W. Coleman and C. F. Watts (*IEEE Transactions on Power Apparatus and Systems,* **85**, 777–782, 1966), where the step-by-step approach is used to calculate stability limits for a large power system with minimal generator complexity, and 'Multinode-Power-System Dynamic Analysis', by W. D. Humpage, J. P. Bayne and K. E. Durrant (*Proceedings of the IEE*, **119**, 1167–1175, 1972), where the transient behaviour of the various generator circuits and controllers is taken into account.

An important feature of transient analysis is the representation of the dynamic performance of the system loads. This is the subject of an IEEE Committee Report 'System Load Dynamics—Simulation Effects and Determination of Load Constants' (*Transactions on Power Apparatus and Systems*, **92**, 1600–1609, 1973), which gives a clear account of the representations available and their effect on simulated system performance.

ELECTRICAL MACHINES

The provision of mechanical drives from electrical machines has undergone a gradual transition in recent years. Starting from a position where the particular torque/speed requirements of a load were met by design modifications to the machines themselves, there has been an increasing tendency to design a limited range of standard machines, and to shape their characteristics by adjusting electronically the magnitude of frequency of their supply, normally by using thyristors. As a recognition of this trend the subject will be tackled under three headings: the design of the machines themselves, the methods of solid-state control available and analysis of the resultant drive systems.

A good survey of the complete field can be obtained from *Automatic Control of Industrial Drives*, by C. H. Pike (Newnes–Butterworth, 1971), and the IEE Conference *Electrical Variable-Speed Drives* (IEE Conference Publication No. 93, 1972) covers the same area, with examples of several practical schemes.

Machine design and construction

For alternating current machines a convenient introductory work is *The Performance and Design of Alternating Current Machines*, by M. G. Say (3rd edn, Pitman, 1958), which enumerates the types of machine in general use, together with constructional details and characteristics. The synchronous machine has been mentioned in the power generation section, so that the references in this section deal mainly with the induction type, which is by far the most common machine in industrial use. *The Nature of Induction Machines*, by

P. L. Alger (Gordon and Breach, 1965), gives a thorough grounding in the theory of the machine and much detailed information on its design, performance and applications. *Induction Machines,* by B. V. Jayawant (McGraw-Hill, 1968), provides a review of recent developments, including pole-amplitude-modulation and linear machines, developing the theory from an electromagnetic field viewpoint rather than the more usual circuit approach. The work also includes an excellent bibliography.

Induction Machines for Special Purposes, by E. R. Laithwaite (Newnes, 1966), as well as describing variants such as the linear motor and brushless variable-speed motors, has an interesting treatise on the criterion of goodness in a machine.

In the field of direct current machines *The Performance and Design of Direct Current Machines,* by A. E. Clayton and N. N. Hancock (3rd edn, Pitman, 1959), has an equivalent status to the Say work mentioned above, forming a good starting point for studying the subject. Specialised applications of direct current machines can be found in *Direct Current Machines for Control Systems,* by A. Tustin (Spon, 1952).

Solid state machine controllers

The use of solid state devices for controlling machines has expanded rapidly in recent years, until virtually all types of machine can now be incorporated into an integrated and controlled drive system. A review of the types of controller available can be found in *Power Electronics,* by R. S. Ramshaw (Chapman and Hall, 1973), which deals with the use of the thyristor in conjunction with direct current, induction and synchronous machines, together with a brief introduction to the thyristor itself.

When used to control a.c. machines, the function of the thyristor is to produce a variable-frequency supply, using a d.c. link inverter (which inverts a d.c. supply obtained by rectifying a.c. mains) or directly by subdividing the mains frequency by suitable switching using a cycloconvertor. Both methods are described in *Thyristor Control of A.C. Motors,* by J. M. D. Murphy (Pergamon, 1973), which includes the effects of the resultant non-sinusoidal supplies on the machines. A comprehensive list of references is also given. For inverters 'Adjustable-Frequency Inverters and their Application to Variable-Speed Drives', by D. A. Bradley, C. D. Clarke, R. M. Davis and D. A. Jones (*Proceedings of the IEE,* **111**, 1833–1846, 1964), discusses a number of practical arrangements, while a systematic classification of circuit types can be found in *Principles of*

Inverter Circuits, by B. D. Bedford and R. G. Hoft (Wiley, 1964). The direct a.c.-to-a.c. conversion is described in *The Theory and Design of Cycloconvertors* by W. McMurray (MIT Press, 1972), which also gives an indication of the limitations and constraints placed upon such a method.

When a direct current machine is to be used, the thyristor controller serves to provide a variable-voltage armature supply by means of either a phase-controlled rectifier fed by a.c. mains or a chopper to reduce the mean value of an existing direct supply. Both types are dealt with in *Solid-State D.C. Motor Drives,* by A. Kusko (MIT Press, 1969).

Analysis of drive systems

As in electrical power systems, the dynamic analysis of complete motor/controller/load drive systems has become possible comparatively recently with the application of digital computers to the problem. Previously, most analytical work had concentrated on matching the steady state characteristics of motor and load or considering small excursions from the operating point. However, by developing a theory which enables the dynamic behaviour of the machine to be predicted, standard control theory and numerical analysis can be used to predict the behaviour of the drive as a whole. A work which goes a considerable distance towards bridging the gap between the 'classical' and the 'modern' approach is *Electric Machinery,* by A. E. Fitzgerald, C. Kingsley and A. Kusko (3rd edn, McGraw-Hill, 1971). Starting from the physical concepts of energy interchange and storage, the book progresses to the steady state performance of idealised and practical forms of machines, while stopping short of development of rigorous generalised equations. *Electromechanical Devices for Energy Conversion and Control Systems,* by V. Del Toro (Prentice-Hall, 1968), carries the analysis into the dynamic regime, using the transfer function approach to study the transient behaviour of a number of practical drives.

While *Electromechanical Energy Conversion,* by V. Gourishanker (International Textbook, 1965) starts from the same physical basis as Fitzgerald, Kingley and Kusko, it arrives at the formulation of the primitive machine idea, and thence to the generalised theory. This theory, which is a systematic method of writing down machine equations, is based on a coupled-circuit approach, and provides a means of predicting performance in the time domain. *Introduction to Generalised Electrical Machine Theory,* by D. O'Kelly and S. Simmons (McGraw-Hill, 1968), forms a good starting point for a study of the subject, and the formalised nature of the machine equations and their organisation is brought out to the full in *Matrix Analysis of Electrical Machinery,* by N. N. Hancock (Pergamon, 1964).

Having formulated the machine equations in this manner, and knowing the dynamic properties of the controller and the load, the problem of analysing the drive system is then reduced to one of applying well-proven control engineering concepts, such as state–space analysis, to their solution.

15

Electronics

D. Petersen

The subject matter of electronics embraces the study of the physical processes underlying the electrical properties of materials, the devices based on these materials and the circuits and systems incorporating such devices. It thus covers some branches of both physics and electrical engineering. However, the central feature peculiar to electronics is the presence of an active device in the circuit which behaves either intermittently as a switch (or gate) or continuously as an amplifier, both switch and amplifier operating under the control of an electrical signal. Practical active devices, of course, function only approximately as perfect switches or as linear amplifiers, and are invariably operating in an environment with spurious electrical signals which constitute 'noise' in the systems. Circuits which contain switches are termed switching (or digital) circuits; those that have amplifiers are termed linear (or analogue) circuits.

The original uses of electronics were in radio communication, but over the past 50 years applications have spread enormously into areas such as radar, control, instrumentation and computers, and these application titles, together with their sub-divisions, may form some of the library classifications of the subject.

Historically there have been three phases of development corresponding to the three major active device inventions: the thermionic tube (valve), the transistor and the integrated circuit. The valve was well-developed in many forms by about 1950, with hot-cathode high-vacuum tubes and cold-cathode gas-filled tubes performing a wide range of circuit functions. The year 1948 saw the invention of the transistor. This device has now developed from its original point-contact form to the state where the transistor and its associated passive devices, the resistor and capacitor, are all fabricated simultaneously on the same silicon chip by photolithographic and diffusion techniques to produce an integrated circuit. During the decade 1950–1960 the bipolar junction transistor (BJT) was developed with great commercial vigour and the resulting technological expertise enabled the unipolar or field effect transistor (FET) to be produced commercially in the early 1960s. Although the BJT continues to have superior high-frequency performance, the low cost and high packing density of the FET, especially in its metal-oxide-semiconductor (MOS) form, have made MOS technology the principal type used in the larger integrated circuits. The increasing complexity of integrated circuits has led to the use of computer-aided design (CAD) techniques to work out the interconnection patterns used in large-scale integration (LSI). The cost of an integrated circuit falls dramatically with mass production, but many users do not require the same circuit in vast quantities and for these users there are two viable alternatives at present: a standard 'off-the-shelf IC' which has discretionary wiring enabling the user to design his own interconnecting pattern, and thick-film techniques and multilayer wired boards with standard small-scale integrated circuits, giving so-called hybrid circuits.

At the extreme high-frequency end of the radio spectrum, the super-high-frequency (SHF) band, bipolar transistors are ineffective and components called transferred-electron devices have been developed. These are forms of solid state diodes which exhibit amplifying properties at certain frequencies. The high-power end of the market is covered by specially designed power diodes and thyristors (silicon controlled rectifiers). In order to handle high power, solid state devices must be comparatively large, and this leads to poor high-frequency performance. The thermionic valve is, therefore, still used in applications requiring very high output power at high frequency, such as radio transmitters. The thermionic valve in the form of a cathode ray tube remains the only practical form of display device for oscilloscopes, radar and television. For small displays on instruments the cold-cathode tube with its glowing cathode is very popular, although these may give way to liquid crystal displays. For the majority of electronic applications an integrated circuit is now available, and many electronic engineers are concerned with the design and development of systems in which the basic element is an integrated circuit. These systems fall into two broad categories, digital and linear, and for the various applications already mentioned both digital and linear integrated circuits are available.

Electronics may be conveniently divided into the broad classifications of physical electronics, which deals with the action of circuit elements and production technology and has sub-sections such as solid state physics, materials science and opto-electronics; and applied electronics, which covers the principles and techniques used in the circuits of computer, communication and control systems and electronic instruments. Applied electronics may be alternatively classified according to such features as the frequency spectrum used, the power range of the application or the mode of operation of the circuits, whether digital or linear.

TEXTBOOKS

During the past ten years there has been a vast increase in the number of textbooks dealing with electronics at all levels. United States companies have published many excellent books and these dominate the British market. Most of these textbooks either have been based on lecture courses or are highly specialised works in which each chapter is contributed by an industrial expert. The former usually offer exercises and are very suitable for private study, while the latter give the current state-of-the-art. Usually both types of books give good references to original sources. In the case of applied electronics, the textbooks have been kept up to date by frequent substantial revisions. Although technology frequently changes in electronics, the fundamentals of circuit analysis and design and the device physics have altered very little. Most basic textbooks, therefore, develop these fundamentals carefully and use current technology by way of illustration and example. In this way the books can have a reasonable life of about 5–10 years before too much revision is needed.

A particularly well written set of undergraduate textbooks are published in the McGraw-Hill Electrical and Electronic Engineering Series. Those covering basic physical and linear electronic principles are: *Electronic Circuits: Discrete and Integrated,* by D. L. Schilling and C. Belove (1968), and *Electronics: B.J.T's., F.E.T's. and Microcircuits,* by E. J. Angelo (1969). The principles of non-linear circuits are covered in *Pulse, Digital and Switching Waveforms,* by J. Millman and H. Taub (1965), and *Wave Generation and Shaping,* by L. Strauss (1970). The latest addition to this series, *Integrated Electronics: Analogue and Digital Circuits and Systems,* by J. Millman and C. C. Halkias (1972), covers both linear and non-linear electronic principles and is an excellent undergraduate text. All these books are published in cheaper form in the International Student Edition Series of McGraw-Hill, and all contain student exercises and chapter references.

The Microelectronic Series published by Van Nostrand are very well produced books and can be recommended, particularly *Electronic Integrated Circuits and Systems,* by F. C. Fitchen (1970); *Electronic Integrated Systems Design,* by H. R. Camenzind (1972); *MOS Integrated-Circuits: Theory, Fabrication, Design and Systems Applications of MOS/LSI,* edited by W. M. Penney and L. Lau (1972); *Analogue Integrated Circuit Design,* by A. B. Grebene (1972); and *Semiconductors for Engineers,* by D. F. Dunster (1969).

A very comprehensive series of teaching textbooks were published by John Wiley and Sons over the period 1964–1967, consisting of seven volumes, by the Semiconductor Electronics Education Committee (SEEC). These provide a good background to physical and circuital electronics under the following titles: *Introduction to Semiconductor Physics,* by R. B. Adler, A. C. Smith and R. L. Longini (1964); *Physical Electronics and Circuit Models of Transistors,* by P. E. Gray, D. DeWitt, A. R. Boothroyd and J. E. Gibbons (1964); *Elementary Circuit Properties of Transistors,* by C. L. Searle, A. R. Boothroyd, E. J. Angelo, P. E. Gray and D. O. Pederson (1964); *Characteristics and Limitations of Transistors,* by R. D. Thornton, D. DeWitt, P. E. Gray and E. R. Chenette (1966); *Multistage Transistor Circuits,* by R. D. Thornton, C. L. Searle, D. O. Pederson, R. B. Adler, E. J. Angelo and J. Willis (1965); *Digital Transistor Circuits,* by J. N. Harris, P. E. Gray and C. L. Searle (1966); and *Handbook of Basic Transistor Circuits and Measurements,* by R. D. Thornton, J. G. Linvill, E. R. Chenette, H. L. Albin, J. N. Harris, A. R. Boothroyd, J. Willis and C. L. Searle (1966). Some of the work covered by these texts has been brought up to date by two of the original team of authors in a single large volume, *Electronic Principles: Physics, Models and Circuits,* by P. E. Gray and C. L. Searle (Wiley, 1969), in which the trend towards integration and computer-aided design and analysis has been followed.

Prentice-Hall have published a useful series in four volumes, *Physical Design of Electronic Systems,* by the staff of the Bell Telephone Laboratories (1971). These contain a wealth of practical details on materials and production technology. In the Prentice-Hall Electrical Engineering Series there are several books which are well worth consideration, in particular *Linear Active Network Theory,* by L. DePian (1962), which gives a good account of electronic circuit principles which do not rely on topical devices, and *Solid State Physical Electronics,* by A. van der Ziel (1957). *Electronic Devices and Circuit Theory,* by R. Boylestad and L. Nashelsky (1972), is a very well illustrated book using basic modern electronics.

The Modern Electrical Studies Series published by Chapman and Hall contains several monographs of electronic interest. Particularly

218   *Electronics*

recommendable are *The Thyristor and its Applications,* by A. W. J. Griffin and R. S. Ramshaw (1965); *Electrical Noise,* by R. A. King (1966); *Transistor Physics,* by K. G. Nichols and E. V. Vernon (1966); and *An Introduction to Masers and Lasers,* by T. P. Melia (1967).

For more advanced specialist reading McGraw-Hill have published the Texas Instruments Electronic Series, which consists of a set of books written by the engineering staff of Texas Instruments Inc. which emphasise practical aspects of electronics but give the theoretical back-up with very full references to original sources. *Semiconductor Circuit Design,* edited by B. Norris (2 vols., Texas Instruments, 1972), gives excellent coverage of practical electronic design and is full of illustrations and circuit data. A recent addition to this series, which covers much new work that has not appeared elsewhere, is *MOS/LSI Design and Application,* by W. N. Carr and J. P. Mize (1972). A similar series is the Motorola Series in Solid-State Electronics, one of which, *Analysis and Design of Integrated Circuits,* by a team from Motorola Inc. (McGraw-Hill, 1968), gives a very full account of bipolar digital gates.

There are very many excellent individual textbooks covering all aspects of electronics at every level. *Electronics,* by A. van der Ziel (Allyn and Bacon, 1966), takes an unusual and stimulating view of the subject and has a good chapter on noise. A very full and authoritative treatment is given in *Noise: Sources, Characterisation and Measurement,* by A. van der Ziel (Prentice-Hall, 1970). *Semiconductors,* by H. F. Wolf (Wiley, 1971), covers both the theory and practice of semiconductors and microelectronics from device manufacturers' viewpoint. *Field Effect Electronics,* by W. Gosling, W. G. Townsend and J. Watson (Butterworths, 1971), is one of the few books which covers basic theory and applications of discrete field effect transistors in detail. *Linear Microelectronic Systems,* by A. G. Martin and F. W. Stephenson (Macmillan, 1973), is a very good undergraduate text which deals well with basic principles using commercial integrated circuits by way of illustration. *Electronic Circuits: Devices, Models, Functions, Analysis and Design,* by M. S. Ghausi (Van Nostrand, 1971), is a beautifully produced book covering a wide range of topics at honours degree and post-graduate level. Very good accounts of the electronics of digital circuits are given in *Digital Circuits and Devices,* by T. Kohonen (Prentice-Hall, 1972), and *Digital Electronics,* by F. Dokter and J. Steinhauer (Macmillan, 1973). *Manual for IC Users,* by J. D. Lenk (Reston Publishing Company, 1973), also gives a good treatment of logic electronic gates.

A very good guide to practical electronic design using an academic

approach is *Amplifying Devices and Low-Pass Amplifier Design,* by
E. M. Cherry and D. E. Hooker (Wiley, 1968). *Low Noise Electronic
Design,* by C. D. Motchenbacher and F. C. Fitchen (Wiley, 1973),
gives a clear account, suitable for those who already have the basic
background, of the theory and practice of low-noise amplifier design.
Transistor Circuit Design, by L. G. Cowles (Prentice-Hall, 1972),
gives an excellent approach to design methods in small-scale circuits,
while *MOS Integrated Circuit Design,* edited by E. Wolfendale
(Butterworths, 1973), is an excellent little book on MOS/LSI principles with CAD examples. *MOS Integrated Circuits and their
Applications,* edited by M. J. Rose (Mullard, 1973), is another small
book giving a concise account of the uses of ICs, while *A User's
Handbook of Integrated Circuits,* by E. R. Hnatek (Wiley, 1973),
gives a review of fabrication methods and a guide to the selection of
the various technologies for a particular application. *MOS Integrated
Circuit Engineering,* by J. Mavor (Peter Peregrinus, 1973), gives the
theoretical background to this subject. *Transistor Circuits in Electronics,* by S. S. Haykin and R. Barrett (Iliffe, 1971), and *Applied
Electronics,* by J. F. Pierce and T. J. Paulus (Merrill, 1972), are very
well written undergraduate texts in general electronics. An unusual
and very generalised approach to electronics theory is given in
Linear Active Networks, by R. Spence (Wiley, 1970). More specialised books in active network theory are *Active Filters,* edited by L.
P. Huelsman (McGraw-Hill, 1970); *Active Integrated Circuit
Synthesis,* by R. W. Newcomb (Prentice-Hall, 1968); and *Synthesis
of RC Active Filter Networks,* by S. S. Haykin (McGraw-Hill, 1969);
and mention must be made of the classical work *Network Analysis
and Feedback Amplifier Design,* by H. W. Bode (Van Nostrand,
1945).

A good account of the application of Laplace transforms to
electronics is given in *Laplace Transforms for Electronic Engineers,*
by J. G. Holbrook (2nd edn, Pergamon, 1966). *Transients in Electronic Engineering,* by E. E. Zepler and K. G. Nichols (Chapman
and Hall, 1971), gives a good mathematical treatment of transient
phenomena. A very good post-graduate text, *Principles and Design
of Linear Active Circuits,* by M. S. Ghausi (McGraw-Hill, 1965),
gives predominantly mathematical coverage of linear circuit theory,
and *Introduction to Non-linear Network Theory,* by L. O. Chua
(McGraw-Hill, 1969), is a large book covering many aspects of non-
linear network theory with examples on practical curve tracing.

The operational amplifier has become the basic circuit in many
electronic applications and a number of small books have been
published recently in this single topic. *Elements of Linear Microcircuits,* by T. D. Towers (Butterworths, 1973), gives a good practical

account, as does *Operational Amplifiers,* by G. B. Clayton (Butterworths, 1971). *Operational Amplifiers,* by A. Barna (Wiley, 1971), gives a simple account of operational amplifier properties, and *Modern Operational Circuit Design,* by I. J. Smith (Wiley, 1971), reviews circuit design techniques based on the operational amplifier as a basic gain block. *Analogue Integrated Circuits,* edited by J. A. Connelly (Wiley, 1975), covers similar work and also can be recommended. *DC Amplifiers,* by B. Mirtes (Iliffe, 1969), gives a good theoretical account of amplifiers leading to modern commercial circuits used in analogue computers. *Operational Amplifiers: Design and Applications,* edited by J. G. Graeme, G. E. Tobey and L. P. Huelsman (McGraw-Hill and Burr Brown Research Corporation, 1971), gives a full treatment of the theory and a wide range of circuit applications.

Electronic Sensing Devices, by A. F. Giles (Newnes, 1966), reviews many physical-electronic-chemical effects and shows how they are used to solve industrial sensing problems. *Cold Cathode Tubes,* by J. B. Dance (Iliffe, 1967), *Cold Cathode Glow Discharge Tubes,* by G. F. Weston (Iliffe, 1968), and *Glow Discharge Display,* by G. F. Weston (Mills and Boon, 1972), are three excellent accounts giving the underlying principles, design, construction and performance and a wide range of applications of cold-cathode tubes.

Basic Electronic Instruments Handbook, edited by C. F. Coombs (McGraw-Hill, 1972), gives a very good survey of the subject with useful references to original papers.

Measuring Oscilloscopes, edited by J. F. Golding (Iliffe, 1971), emphasises methods and principles and is a good background introduction to the equipment handbooks, with each chapter written by staff members of Marconi Instruments. *Measurement of Transistor Parameters,* by R. Paul (Iliffe, 1969), gives comprehensive treatment of this subject. *Thick Film Microelectronics,* by M. L. Topfer (Van Nostrand, 1971), *Thick Film Circuits,* by G. V. Planer and L. S. Phillips (Butterworths, 1972), and *Thick Film Hybrid Microcircuit Technology,* by D. W. Hamer and J. V. Biggers (Wiley, 1972), give good accounts of fabrication, design and applications, with very good references to original work. *Modern Electronic Materials,* by J. Watkins (Butterworths, 1971), is a short treatment linking the theory and practical application of materials in electronics, and *Electronics Design Materials,* edited by W. F. Waller (Macmillan, 1971), gives a good sectionalised view of conductors, dielectrics, semiconductors, magnetic materials and materials techniques such as deposition, bonding and encapsulation. A very authoritative and detailed study of macro- and microelectronics practice is given in the *Handbook of Wiring, Cabling and Interconnecting for Electronics,* edited by C. A. Harper (McGraw-Hill, 1972); *Handbook of Electronic Packaging,* edited by C. A. Harper (McGraw-Hill, 1969); and *Handbook of Materials and Processes for Electronics,* edited by C. A. Harper (McGraw-Hill, 1971).

Compatibility and Testing of Electronic Components, by C. E. Jowett (Butterworths, 1972), gives a comprehensive and qualitative account of manufacturing methods and testing in microelectronics production which will be of interest to reliability, quality control, production and test engineers.

High-power applications are covered in General Electric's *Silicon Controlled Rectifier Manual* (5th edn, 1972) and *Power Diode and Thyristor Circuits,* by R. M. Davis (Peter Peregrinus, 1971). The former gives some useful practical data and the latter is an introductory treatment with a good bibliography. *Power Electronics,* by R. S. Ramshaw (Chapman and Hall, 1973), covers thyristor control of electric motors.

A very useful guide to medical electronics is given in the *IEE Medical Electronics Monographs 1 to 6,* edited by B. W. Watson (Peter Peregrinus, 1971). The treatment is purely descriptive, but there are many references quoted.

Intrinsic Safety: the Use of Electronics in Hazardous Locations, by R. J. Redding (McGraw-Hill, 1971), can be recommended for users of electronics in the chemical or petroleum industries, public utilities and equipment suppliers.

The *Electronic Circuits Manual,* by J. Markus (McGraw-Hill, 1971), is a huge book giving many practical circuits and a very good list of original references.

Transferred Electron Devices, by P. J. Bulman, G. S. Hobson and B. C. Taylor (Academic Press, 1972), and *Hot Electron Microwave Generators,* by J. E. Carroll (Arnold, 1970), cover the fundamentals of such microwave devices as Gunn diodes and give a very good list of references. *Avalanche-Diode and Microwave Oscillators,* by G. Gibbons (Oxford University Press, 1973), gives a good short account of IMPATT and TRAPATT oscillators.

Gunn-Effect Logic Devices, by H. Hartnagel (Heinemann, 1973), introduces the use of transferred electron instability to ultra-high-speed digital circuits. Several books on the laser can be recommended: *Introduction to Masers and Lasers,* by T. P. Melia (Chapman and Hall, 1967); *Laser Systems and Applications,* by H. A. Elion (Pergamon, 1967); *Laser Technology and Applications,* edited by S. L. Marshall (McGraw-Hill, 1968); and *Masers and Lasers: Physics and Design,* by J. S. Thorp (Macmillan, 1967). *Low Noise Electronics,* by W. P. Jolly (English Universities Press, 1967), gives a good short introduction to the action of microwave devices such as lasers, parametric amplifiers and travelling-wave tubes, and *Parametric Amplifiers,* by D. P. Howson and R. B. Smith (McGraw-Hill, 1970), *Analysis and Synthesis of Tunnel Diode Circuits,* by J. O. Scanlan (Wiley, 1966), and *Tunnel Diodes,* by M. A. Lee, B. Easter and H. A. Bell (Chapman and Hall, 1967), give slightly more specialised accounts of these microwave elements.

Although many of the general electronic texts already mentioned contain chapters on physical principles, these are necessarily brief, and fuller accounts of device and material fundamentals, with many references, are to be found in the following physical electronics books: *An Introduction to Physical Electronics,* by A. H. W. Beck and H. Ahmed (Arnold, 1968); *Physical Electronics,* by D. K. Ferry and D. R. Fannin (Addison-Wesley, 1971); *Introduction to Solid State Physics,* by C. Kittel (4th edn, Wiley, 1971); *Physical Electronics,* by C. L. Hemenway, R. W. Henry and M. Caulton (2nd edn, Wiley, 1967); *The Physics of Solid State Devices,* by T. H. Beeforth and H. J. Goldsmid (Pion, 1970); and *The Physics and Circuit Properties of Transistors,* by J. M. Feldman (Wiley, 1972). *Topics in Solid State and Quantum Electronics,* edited by W. D. Hershberger (Wiley, 1972), gives the theory of many new devices and many good original references. *Materials Science,* by J. C. Anderson and K. D. Leaver (Nelson, 1969), and *Science of Materials,* by T. J. Lewis and P. E. Secker (Harrap, 1966), give concise clear introductions to the physics of electronic materials.

PERIODICALS

The most comprehensive sets of periodicals produced are those published by the Institute of Electrical and Electronics Engineers (IEEE) and the Institution of Electrical Engineers (IEE). Both the IEEE and the IEE are learned societies whose duties include the dissemination of knowledge in their field. Each publishes a catalogue annually which updates their publication lists.

The primary publications of the IEEE are its *Transactions,* and these are divided up into subject groups of which the following are of interest to electronic engineers: *Aerospace and Electronic Systems*; *Circuits and Systems*; *Electron Devices*; *Consumer Electronics*; *Geoscience Electronics*; *Industrial Electronics and Control Instrumentation*; *Manufacturing Technology*; *Parts, Hybrids and Packaging*; *Reliability.*

The IEEE also publishes journals which cover a wider range of topics than the *Transactions,* i.e. *IEEE Journal of Solid-State Circuits, IEEE Journal of Quantrum Electronics.* The *Proceedings*

of the IEEE contains academic papers of long-range interest describing original research and in-depth reviews of topics of current
interest. It also contains an extensive 'Letters' section which gives
quick publication of brief reports on new research and development
topics. Several times each year the *Proceedings* and *Transactions*
are devoted entirely to an in-depth treatment of a single topic.

The periodicals of the IEE follow a similar pattern to those of the
IEEE. The *Proceedings of the IEE* appears monthly and usually
contains very academic papers on electronics, power, control and
automation and electrical science, education and management in
separate sections. These sections are also separately bound and
published individually every 3 months, and an index to the volumes
over the years 1962–1971 was published in 1972. *Electronics Letters*
is a fortnightly international periodical developed to provide speedy
dissemination of topical work in research and development. *Physics
Letters* and *Applied Physics Letters* fulfil a similar function.
Electronics and Power records and reports in a readable way on
current topics and contains much advertising copy and a reader
service for obtaining further details of products. The IEE produces
an analogous publication called *Spectrum.*

The Institution of Electronic and Radio Engineers (IERE)
publishes monthly *The Radio and Electronic Engineer,* which contains high-standard current or review papers which are much more
readable than the *Proceedings of the IEE.*

Other learned society journals which contain original and review
papers on electronics topics are *Measurement and Control*; *Journal
of Physics(C)* (*Solid State Physics*); *Onde Electrique*; *Japanese
Journal of Applied Physics*; *Royal Television Society Journal*;
Journal of the SMPTE; and *Proceedings of the Cambridge Philosophical Society.*

In addition to the publications of learned societies, there is a large
number of periodicals published by commercial publishing houses.
These contain articles of current practical interest and are written
in a very readable way. The *International Journal of Electronics*
(Taylor and Francis) contains current papers on devices, circuit
theory and applications. *Electronic Engineering* (Morgan Grampian),
Electronic Components (Bannock Press), *New Electronics* (Northwood Publications), *Electronic Equipment News* (Mercury House),
and *Electronics* (McGraw-Hill) contain many short articles of current
practical interest and reviews. *Electronics and Communications in
Japan* (Scripta Publishing Company) and *Instruments and Experimental Techniques* (Plenum), translated from Russian, give articles
and references not covered by the literature already mentioned.
Other useful commercial periodicals are: *Solid State Electronics*

(Pergamon), *Direct Current and Power Electronics* (Wynn Williams), *International Journal of Control* (Taylor and Francis), *Opto-Electronics* (Chapman and Hall), *Progress in Quantum Electronics* (Pergamon), *Optics and Laser Technology* (IPC), *Microelectronics* (Mercury House) and the *Microwave Journal* (Horizon House). Useful review articles aimed at the layman appear in the *New Scientist* (New Science Publications) and practical reviews and applications are covered in *Wireless World* (IPC Business Press).

COMMERCIAL PUBLICATIONS

Commercial companies producing electronic equipment have published much information on the background and applications of their products, as well as company journals which report on their current research work. The major publications in this category are: *RCA Review*; *Bell Laboratories Record*; *Bell System Technical Journal*; *Philips Technical Review*; *Mullard Technical Communications*; *Marconi Review*; *Marconi Instrumentation*; *Siemens Review*; *Hewlett Packard Journal*; *Acta Electronica*; *Post Office Electrical Engineers' Journal*; *IBM Technical Disclosure Bulletin*; *IBM Journal of Research and Development*; and *Commutation et Electronique*.

There are also a number of handbooks published by the following commercial firms which contain a wealth of practical data and numerous references: RCA Texas Instruments, General Electric, Motorola, Mullard, Fairchild and Signetics. Useful sources on data are *Electronics Data Services* (Pergamon), *Anglo-American Microelectronics Data* (Pergamon), *Reference Data for Radio Engineers* (Howard Sons and Company), *Microelectronics Directory* (Shaw Publishing Company) and *Microelectronics Year Book* (Shaw Publishing Company). The annual editions of *Semiconductors*, compiled by F. C. J. Finck (Kluwer, Antwerp), give a clear guide to type-designation, symbols and technical data for discrete devices.

CONFERENCE PUBLICATIONS

The IEEE, IEE and IERE sponsor many international conferences and colloquia and in most cases they publish a conference record, proceedings and digest. These give the very latest state-of-the-art in some detail and are a good source of reference material. Details of these publications are given in the annual catalogues of the Institution.

ABSTRACTS AND CURRENT PAPERS

The IEE has provided information services for the past 75 years. In 1967 these services were converted to a computer-based system called *INSPEC* (Information Services for the Physics and Engineering Community). *Electrical and Electronics Abstracts* (*EEA*) is published monthly, each containing up to six indexes providing quick access to authors, bibliographies, books, conferences, patents and reports. Twice-yearly author and subject indexes and cumulative indexes covering periods of 4–5 years are also published.

Current Papers in Electrical and Electronics Engineering is published monthly by INSPEC and provides a current-awareness publication with the title, authors, and bibliographical reference for every entry published in *EEA*, using the same classification. *Electronics and Communications Abstracts* is published monthly by Multi-Science Publishing Company. *Referativnyi Zhurnal,* the main Russian journal of abstracts, sponsored by the USSR Academy of Sciences, publishes electronic engineering abstracts under the title *Elektronika I EE Primenenie.* Other useful abstracting journals are *USSR and East European Scientific Abstracts*; *Electronics Express*; and *Bulletin Signalétique: Part 145 — Electronique.*

16

Communications

D. J. Harris

The art of communication at a distance has advanced dramatically during this century. A technique so basic to human development clearly has its roots in antiquity, with initial crude coding methods using aural and optical methods, such as the talking drum and smoke signals, but the era of modern communication commenced with the utilisation of electric current for communications in the nineteenth century. The first systems were introduced for railway signalling purposes, and by 1900 virtually world-wide telegraphic communications had been established using land-line and submarine cable, although the systems were slow and capable of transmitting only a very limited information content. The transmission of information by radio waves across the Atlantic by Marconi was a very significant step, culminating in the communication of complex data between the earth and a space probe in the vicinity of Mars in 1962, a distance of about 86 million km.

The literature available on communication reflects the general interest in the subject, the broad spectrum of specialisations that now lies within the general heading of communications and the high level of commercial activity. Vast sums are spent on communication and data transmission equipment each year, and the communications industry is one of the largest. Literature available, therefore, covers popular accounts for general reading, textbooks for professional and degree courses, specialised books giving a detailed exposition on some narrow aspect of the subject and research papers which may be of a highly abstract and mathematical nature. A distinction needs also to be made between the intensely 'practical' nature of some literature, concerned with the constructional aspects of equipment, and highly theoretical literature dealing with the basic concepts or details of the subject.

The large research and development effort, following from the commercial and military importance, results in continual innovation. The literature is thus soon outdated. It is essential, therefore, to make use of the latest publication where possible and a book which is 20 years old is invariably out of date. On the other hand, while the technology of the subject is advancing continually, basic concepts such as fields, waves and transmission line theory are as valid now as in the past and some of the older texts are still among the best.

The wide range of contemporary texts available enables background and specialised knowledge to be acquired in most areas of the subject. There are, however, a number of areas of work under development which are potentially important and which are expected to lead to major changes in the technology of communications. In some cases experience from fields not normally associated with communications has been used to innovate new techniques. Typical of these areas of development are microwave trunk waveguide systems using H_{01} circular guide; optical waveguides and optical communication systems; tropospheric scatter systems; electronic exchanges and the application of on-line computers to communication traffic control; digital systems and digital processing of communication signals; surface acoustic wave devices for signal processing; widespread use of communication satellites for direct links with groups of subscribers; and development of video-phones and conference visual displays using the telephone network. Information on such topics must be sought in professional journals and conference proceedings.

INFORMATION TO BE COMMUNICATED

The information conveyed in a communication system is usually in one of three forms. It may be aural information (speech and music); optical information defining a visual image which may be still, moving or even three-dimensional; or numeric data giving, for example, the position, speed and direction of an aircraft. The first is effectively limited in frequency by the response of the ear to less than 20 kHz, while adequate two-dimensional pictures can be obtained with a bandwidth of a few MHz. Thus, for single channels a modest frequency range only is required. There are, however, 10 million telephone exchange connections in the UK at present, and the number is increasing rapidly. Large numbers of telephone calls can be accommodated simultaneously on a single channel only by increasing the over-all frequency range of operation. For data transmission the rate of information transfer is proportional approximately to the available bandwidth, and this leads to a high operating frequency requirement.

Since different techniques are involved for different frequency ranges, the literature tends to divide itself to a large extent on a frequency basis, e.g. telephones and telephone exchanges (lower frequency ranges), radio communications (medium and high frequencies), microwave communications (wavelengths of the order of a centimetre) and optical communications (wavelength about 1μm). Space—or satellite—communications also has a technology of its own, drawing experience partly from the other fields.

TYPES OF LITERATURE AVAILABLE

Comprehensive and review-type books

This type of book used to be quite common and often designed to match the requirements of undergraduate courses and professional examinations. They can still be valuable in giving an over-all introduction to the subject, but by their nature are apt to be superficial. Books that come under this category include:

Communication, by P. Davidovits (Holt, Rinehart and Winston, 1972)

Future Development in Telecommunications, by J. Martin (Prentice-Hall, 1971)

Telecommunication System Design, by M. T. Hill and B. G. Evans (Allen and Unwin, 1973)

Telecommunications, by W. Fraser (2nd edn, Macdonald, 1967)

Telecommunications, by A. T. Starr (2nd edn, Pitman, 1958)

Telecommunications: the Booming Industry, by R. Brown (Doubleday, 1970)

The book by Hill and Evans is particularly useful in giving an up-to-date over-all view of communications techniques.

Texts dealing with specialised areas of the subject

Communication theory

The basic concepts of transmission of information have been explored extensively and come under the general heading of communication theory, or information theory.

The fundamental ideas and mathematical formalisation of the over-all processes of modulation, transmission and demodulation have been codified and analysed in detail during the past 30 years. The emphasis is on the communication channel and ways of using it, and the following texts deal with the channel, its characteristics and its use in conveying information:

Communication System Analysis, by P. B. Johns and T. Rowbotham (Butterworths, 1972)

Communication Systems: an Introduction to Signals and Noise in Electrical Communication, by A. B. Carlson (McGraw-Hill, 1968)

Communication Theory Principles, by C. W. McMullen (Macmillan, 1969)

Information Transmission, Modulation and Noise, by M. Schwarz (McGraw-Hill, 1970)

Introduction to Random Signals and Communications Theory, by B. P. Lathi (Intertext, 1968)

Introduction to the Principle of Communication Theory, by J. C. Hancock (McGraw-Hill, 1961)

Mathematical Theory of Communication, by C. E. Shannon and W. Weaver (University of Illinois Press, 1949)

Modern Communication Principles with Application to Digital Signalling, by S. Stein and J. Jones (McGraw-Hill, 1967)

Principles of Communication Systems, by H. Taub and D. L. Schilling (McGraw-Hill, 1971)

Principles of Communications Engineering, by J. M. Wozencraft and I. M. Jacobs (Wiley, 1965)

Signal Processing, Modulation and Noise, by J. A. Betts (English Universities Press, 1970)

Signals and Noise in Communication Systems, by H. E. Rowe (Van Nostrand, 1966)

Techniques of Pulse Code Modulation in Communication Networks, by G. C. Hartley (Peter Peregrinus, 1972)

Telecommunications, by J. Brown (2nd edn, Chapman and Hall, 1973)

Theory of Communication, by A. E. Karbowiak (Oliver and Boyd, 1969)

With so many good texts available it would be a difficult task to attempt an order of merit. Suffice it to say that the book by Betts is an excellent introduction, that by Taub and Schilling a very good advanced treatise, and that by Lathi perhaps the 'best buy' in terms of value for money.

Noise and its limiting effect on communications

In order for a signal to be intelligible, its magnitude must be considerably greater than the background noise that always exists. The minimum signal that can be utilised in a particular situation is then limited, since amplification will magnify the noise as well as the signal. Special techniques must be used when the received signal is necessarily small — for example, in a satellite communication system. The subject of noise is dealt with alongside communication theory in many of the books listed in the previous section. Additional books specialising in this aspect are:

Detection of Signals in Noise, by A. D. Whalen (Academic Press, 1971)
Electrical Noise, by D. A. Bell (Van Nostrand, 1960)
Noise, by A van der Ziel (Prentice-Hall, 1970)

Fields, waves, transmission lines and radiators

There is a wide range of books that introduce these subjects on an undergraduate or postgraduate text basis. The following are particularly valuable in relating to aspects of communications:

Antennas, by L. V. Blake (Wiley, 1966)
Antennas, by J. D. Kraus (McGraw-Hill, 1950)
Electromagnetic Waves and Radiating Systems, by E. C. Jordan and K. Balmain (2nd edn, Prentice-Hall, 1968)
Field Theory of Guided Waves, by R. E. Collin (McGraw-Hill, 1966)
Fields and Waves in Communication Electronics, by S. Ramo, J. R. Whinney and T. Van Duzer (Wiley, 1965)

Communication systems

There is often an overlap in the literature between the theory of the channel, as listed under communication theory, and the communication system itself. The following books deal particularly with the system and its behaviour, in some cases including the engineering aspect:

Advances in Communication Systems, edited by A. V. Balakrishnan (Vols. 1–3, Academic Press, 1965–1968)
Communication Engineering, by W. L. Everitt and G. E. Anner (3rd edn, McGraw-Hill, 1956)

Communication System Design: Line of Sight and Tropospheric Scatter, by P. F. Panter (McGraw-Hill, 1972)
Communication System Engineering Handbook, edited by D. H. Hamsher (McGraw-Hill, 1967)
Communication Systems and Techniques, by M. Schwarz, W. R. Bennett and S. Stein (McGraw-Hill, 1966)
Communication Systems Engineering Theory, by E. Sunde (Wiley, 1969)
Telecommunication System Engineering, by W. Lindsey and M. Simon (Prentice-Hall, 1972)

Telephones and automatic telephone exchanges

Exchanges rely mainly upon electromechanical and reed switches, but the pressure of increasing traffic density and complexity, with the introduction of data transmission and future introduction of video-phones, will lead to the wide-scale changeover to fully electronic switching exchanges in the near future. This change is not yet reflected in the texts, however, and information on electronic exchanges is to be found mostly in individual papers and conference reports. Texts on the general topic include:

Communication Switching Systems, by M. Rubin and C. E. Haller (Reinhold, 1966)
Introduction to Teleprocessing, by J. T. Martin (Prentice-Hall, 1972)
Principles of Telephony, by N. N. Biswas (Asian, 1971)
Telephone Exchange Systems of the British Post Office (2 vols., Pitman, 1968)
Telephony and Telegraphy, by S. F. Smith (Oxford University Press, 1969)
Telex, by R. W. Barton (Pitman, 1968)

There would appear to be a need for a new text on this subject, to include information on the latest exchange technology and traffic control.

Radio communication

Frequencies for radio communication extend from a few kHz to hundreds of MHz. Applications include the transmission of information from point to point (radio link) and the general dissemination of information over a wide area (broadcasting). At low frequencies, or for ground level line-of-sight operation, the propagation path is close to the surface of the earth, although the mechanism of propagation is different for these two situations. Long-distance communication at a frequency of a few MHz is achieved by reflection of the signal from the upper conducting layers of the atmosphere (the ionosphere) and the characteristics of this layer affect reception. Relevant texts include:

Antenna Engineering Handbook, edited by H. Jasik (McGraw-Hill, 1961)

Broadcasting Technology, Past, Present and Future (IEE, 1972)

High Frequency Communication, by J. A. Betts (English Universities Press, 1967)

Propagation of Short Radio Waves, by D. E. Kerr (McGraw-Hill, 1951)

Radio Communication, by W. F. Lovering (Longmans, 1966)

Radio Communication, by J. H. and P. J. Reyner (3rd edn, Pitman, 1972)

Radio Engineering Handbook, edited by K. Henney (5th edn, McGraw-Hill, 1959)

Radio Ray Propagation in the Ionosphere, by J. M. Kelso (McGraw-Hill, 1964)

Radio Relay Systems, by H. Carl (Macdonald, 1966)

Radio Wave Propagation at Very High Frequency and Above, by P. A. Matthews (Chapman and Hall, 1965)

Transmission and Propagation, by E. V. D. Glazier and H. R. L. Lamont (HMSO Services Text Books, Vol. 5, 1958)

Microwave communication

The extension to frequencies greater than 3000 MHz (i.e. wavelengths less than 10 cm) has resulted from the need for greater bandwidth and the advantage of short wavelengths in producing narrow beamwidths from suitable aerials of modest size. Techniques at these short wavelengths differ markedly from those at lower frequencies. There is a vast amount of literature on microwaves in general, but comparatively little on the communication aspect in particular. Some good microwave books, and those dealing with microwave communications, are as follows:

Foundations for Microwave Engineering, by R. E. Collin (McGraw-Hill, 1966)

Interference Problems Associated with Microwave Communications (HMSO, 1968)

Mechanical Engineering in Radar and Communications, edited by
   C. J. Richards (Van Nostrand, 1969)
Microwave Antennas, by A. Z. Fradin (Pergamon, 1961)
Microwave Communications, by D. J. Angelakos and T. E.
   Everhart (McGraw-Hill, 1968)
Microwave Communications, by S. Yonezawa and N. Tanaka
   (Maruzen, 1963)
Microwave Engineering, by A. F. Harvey (Academic Press, 1963)
Microwave Systems Planning, by K. L. Dumas and L. G. Sands
   (Hayden, 1967)
Radar Handbook, edited by M. Skolnik (McGraw-Hill, 1970)
Trunk Waveguide Communication, by A. E. Karbowiak (Chapman
   and Hall, 1965)

Space — or satellite — communications

The satellite communication system is a recent alternative to systems
using a submarine cable for intercontinental communications. It has
the advantage that virtually any point in sight of the satellite can be
linked to it via an earth terminal, including mobile receiving stations.
It can thus be used as a relay station to link ships and perhaps even
aircraft and other space craft or satellites. It is a very expensive
technology made necessary by the rapid build-up in international
communication requirements, including international television
coverage. The technology has developed largely over the past decade,
and while there are valuable texts, up-to-date information is to be
obtained by consulting recent articles and papers. Texts dealing with
this aspect include:

Communication Satellite Systems Technology, by R. B. Marsten
   (Academic Press, 1966)
Communication Satellites, by G. E. Mueller and E. R. Spangler
   (Wiley, 1965)
Space Communication Systems, by R. F. Filipowsky and E. I.
   Muehldorf (Prentice-Hall, 1965)
Space Communication Techniques, by R. F. Filipowsky and E. I.
   Muehldorf (Prentice-Hall, 1965)
Telecommunication Satellites, edited by K. W. Gatland (Iliffe, 1964)

Additional texts

There are texts that are relevant to the subject of communications
but do not come under the above headings. Two such books are:

Laser Communication Systems, by W. K. Pratt (Wiley, 1969)
Reference Data for Radio Engineers, edited by H. P. Westman (5th edn, Sams, 1968)

The latter is a compendium of useful information and charts across the spectrum used in communications.

Review papers giving the state-of-the-art

There are a number of journals where such papers may be found, but perhaps the most fruitful and authoritative sources are the journals of the professional institutions, such as:

Electronics and Power, published by the Institution of Electrical Engineers (IEE)
The Radio and Electronic Engineer, published by the Institution of Electronic and Radio Engineers (IERE)
Proceedings of the Institute of Electrical and Electronics Engineers (IEEE)

The following articles have appeared in *Electronics and Power* during the past 3 years and are typical of the range and type to be found:

'Technological Progress in Telecommunication Switching'	August 1970
'Intelsat — the International Telecommunication Satellite Consortium'	January 1971
'British Broadcasting and the IEE'	April/May 1971
'Telecommunications—the Grass Roots of the IEE'	April/May 1971
'Telecommunications — New Practices, Old Concepts'	September 1971
'Teleprocessing and Data Communication of the Future'	December 1971
'International Telecommunication Today'	April 1972
'What Future for the Electronic Exchange?'	August/September 1972
'The Future of Broadcasting — an Engineer's View'	October 1972
'Glass Fibres for Communication Show Promise'	December 1972
'International Telecommunication Development'	March 1973

Research and development papers and journals

The rapid development of communications equipment and systems, and all the research associated with it, has led to a continuous and almost overwhelming stream of publications. These may be research papers, papers outlining the results of development work, or details and characteristics of new equipment.

The professional institutions publish survey articles as outlined above, but the main body of their publications consists of specialised papers. Thus, the Institution of Electrical Engineers publishes the following:

Electronics and Power—papers of more general interest, including communications.

Proceedings—papers of a specialised nature, including communications.

Electronics Letters—short contributions for quick publication, which also include communication information.

The Institute of Electronic and Radio Engineers publishes:

The Radio and Electronic Engineer—papers of general interest in radio and electronic engineering, including communications.

The Institute of Electrical and Electronics Engineers publishes:

Spectrum—papers of general interest, including communications, together with a wide range of *Transactions* on various specialisations. *Transactions* of particular importance to this subject are: *IEEE Transactions on Communications, IEEE Transactions on Broadcasting* and *IEEE Transactions on Antennas and Propagation.*

There are few technical publications dealing solely with communications except those associated with particular industrial groups. The following are worthy of note:

Telecommunications International (Horizon House). This journal is available free from the publishers for qualified engineers engaged in the subject. It contains general and technical articles on telecommunications, together with a wealth of information on equipment available and technical advertisements in the field. Enquiries should be directed to Telecommunications, Computer Centre, 610 Washington Street, Dedham, Massachusetts 02026, USA.

Microwave Journal International. This is similar to *Telecommunications International,* but devoted to technical and commercial aspects of microwaves. It is also published by Horizon House.

Two journals of a somewhat practical and popular approach that carry relevant articles are *Wireless World* and *Electronic Engineering.*
Several journals published overseas are concerned with communications; for example, among those with the text in English are:

Institution of Radio and Electronic Engineers Proceedings (Australia)
Institution of Telecommunication Engineers Journal (India)
International Scientific Radio Union (URSI) Information Bulletin
Japan Telecommunication Review
Radio Science (URSI) (USA)
Telecommunication Journal (UIT Geneva)
Telecommunication Journal of Australia
Telephony (USA)

Some overseas journals are obtained in translation into English; for example, Scripta Publishing Company provides:

Electronics and Communications in Japan (from the Japanese)
Telecommunication and Radio Engineering (from the Russian)

Operating condition recommendations, regulations and reports on assemblies for co-ordinating and relating communication systems are published through URSI and CCITT.
It is common for workers in a particular field to meet at intervals to present and discuss the results of their work, often under the auspices of a professional institute. The proceedings of these conferences form a valuable collection of papers giving the state-of-the-art in the subject of the conference. Examples of such conferences are given below:

M.F., L.F. and V.L.F. Radio Propagation	IEE Conference No. 36 (1967)
International Broadcasting Convention	IEE Conference No. 69 (1970)
Trunk Communication by Guided Waves	IEE Conference No. 71 (1970)
Earth Station Technology	IEE Conference No. 72 (1970)
Digital Processing of Signals in Communications	IERE Conference No. 23 (1972)

Much valuable information is to be found in journals originating in industrial and other telecommunication groups. Enquiries for these journals should be directed to the group concerned. Typical of these journals, relevant to communications, are:

Aerial (GEC-Marconi)
Bell Laboratories Record
Bell System Technical Journal
Electrical Communication (ITT)
GEC Telecommunications
Philips Telecommunications Review
Point-to-Point Communications (Marconi)
Post-Office Telecommunication Journal
Post-Office Electrical Engineers Journal
RCA Review
Siemens Review
Systems Technology

Abstracting systems

In order to cover most of the literature in a particular subject area, it would be necessary to scan the contents list of all the available journals, a formidable task indeed. Use can be made of one of the abstracting series available to reduce the problem to one of manageable dimensions. The Institution of Electrical Engineers, for example, publishes monthly abstracts from journals under three main headings: electrical and electronics engineering, physics, and computers and control (see p. 157).

The complete range of subject topics is divided and subdivided to enable particular specialisations to be pin-pointed quickly, and the headings under which literature is classified are revised periodically to take account of changes in technology. In *Electrical and Electronics Abstracts* there is a main classification heading entitled 'Electromagnetics and Communication', under which most of the material of direct interest will be classified, but some aspects may be found under other headings if the main emphasis is, for example, on an electronic device or a type of circuit.

It is instructive and valuable to review the headings and subheadings into which 'Electromagnetics and Communication' is divided, so as to see the over-all scope of the subject and the way in which it may be classified. The headings of this major section as specified January 1975 are:

Main classification heading
3.000 Electromagnetics and Communication
Sub-classification headings
3.100 Electromagnetism
3.200 Antennas, transmission line and propagation
3.400 Information and communication theory
3.500 Telecommunication
3.600 Radar and radio-navigation
3.700 Radio, television and audio

Each of the above sub-classifications has associated a number of specialisations as follows:

3.100 Electric and magnetic fields, electrostatics; electromagnetic waves and oscillations; propagation (in unspecified medium); diffraction and scattering; interference; electromagnetic waves in plasma.
3.200 Radiowave propagation effects; optical propagation effects; antenna theory; antennas; single antennas; arrays; antenna accessories; transmission line accessories; transmission line components; waveguide theory; waveguides; optical waveguides; waveguide components.
3.400 Information theory; modulation methods; codes, speech intelligibility; signal processing and detection; communication switching theory.
3.500 Telecommunication systems; telephone systems; telegraph systems; facsimile; telemetry; other data transmission; stations and equipment; telephone stations; other stations; switching centres and equipment; electronic telephone exchanges; other telephone exchanges; other switching centres; transmission line and radio links and equipment; waveguide and coaxial cable systems; submarine cable systems; power line systems; other transmission line systems; point-to-point radio systems; mobile radio systems; satellite relay systems; space communication systems; other radio link systems; optical line systems and equipment; optical transmission lines; optical link equipment; other link systems and equipment.
3.600 Radar theory; radar systems and equipment; optical radar; radio navigation and direction finding; radio astronomical techniques and equipment.
3.700 Legislation; frequency allocation; spectrum pollution; radio and television broadcasting; radio and television transmitters; radio and television receivers; television signals, equipment and systems; audio and video recording; audio signals, equipment and systems.

One of the most valuable uses of the abstracts is to enable information on past literature in a particular field to be collated. The abstracts for previous years can be scanned and an up-to-date assessment of most of the work published in the field can be made.

Another valuable service for effecting a continuous monitor of literature in a specialist area is that offered by the Institution of Electrical Engineers under the INSPEC SDI scheme. For the payment of a fee, all literature that matches in with a particular information profile is notified to the subscriber on a weekly basis. While not inexpensive, this service can be invaluable in saving time on literature surveys and ensures that articles of interest are not overlooked.

17

Digital, analogue and hybrid computation

D. W. H. Hampshire

INTRODUCTION

Of all the topics in electrical engineering, none are advancing more rapidly than those associated with computers and computing. Although the fundamental work on digital computing goes back to Babbage, who designed a difference engine in 1835, and much of our understanding of switching systems is based upon the classical work of Boole, who formalised the treatment of logical propositions at about the same time, the explosive development of the subject has taken place since the mid-1950s.

The original theoretical ideas lay dormant for many years awaiting the development of the semiconducting diode and the transistor. These, being cheap to produce in large quantities, small in size and dissipating little power, provide the ideal means of implementing the ideas in a practical and economic manner.

The very early machines such as ENIAC and EDSAC were made using vacuum diodes and valves, but these were really prototype research tools rather than machines for commercial exploitation. Subsequently a few valve machines were produced for the market, but no large-scale production was carried out until the semiconductor machines were designed in the late 1950s. Development of computing systems then went on apace owing to the availability of convenient switching elements on the one hand, and the support of the various industrial, defence and space organisations on the other. This pattern in research and development was reflected in the form and in the quantity of the literature.

Of more recent date, the early 1970s have seen the arrival of the second revolution, that of both medium- and large-scale integration. This technique provides means for producing very large arrays of interconnected diodes and transistors on small areas of semiconductor. It has led, for example, to the production of complete arithmetic processing units in a single element—whose size of about $1\frac{1}{2}$ $\times \frac{1}{2} \times \frac{1}{16}$ in is only prevented from disappearing altogether by the need to have sufficient room for input and output pins and sufficient surface area for heat removal. This manufacturing technique has, of course, had its effects upon the literature, which now has extensive sections on large logic arrays and the interconnection, not only of switching elements, but also of whole computing units which may be either general or special purpose in nature.

This in turn has led to the digital implementation of functions which in the past have always been carried out by analogue means. A good example of this is the filter. Whereas filters have traditionally been considered in terms of linear elements, much attention is being shown at present to digital filters, which perform a similar function to the analogue but do it by means of switching elements.

There is a continuing swing towards digital systems and it is becoming increasingly difficult to say where computing ends and the processing of signals by digital means begins.

Even so, analogue computing is another branch of computing which has also played a large part in the post-war evolution of electronic and control engineering. Whereas digital computers have primarily been developed for use by non-engineers, the analogue machines were produced predominantly by engineers for engineers.

Immediately after World War II, aeronautical engineers required methods of studying the complex vibrations set up in, and by, aircraft control surfaces. The mathematical models associated with this problem consist of a set of non-linear differential equations, and so these engineers extended the ideas put forward by Lord Kelvin in the 1880s and later developed by Hartree. The aeronautical engineers turned to the electronic engineers, who during the war had developed methods for summation and integration which were used in air target predictors. The result of this marriage of engineering expertise was the general purpose analogue computer, produced, not by the electronic industry but, oddly enough, by the aircraft industry. Subsequently these machines have found increasing applications in all parts of engineering and very large machines have been built for simulation of the dynamic properties of such complex things as nuclear reactors, guided weapon and other control systems. Because they contain a large number of arithmetic units operating in parallel, they provide facilities for the fast real time simulation of complex systems. More recently the use of these analogue devices has been extended by marrying them with digital switching and computation elements to produce hybrid computing machines. This has been made practicable by the development of fast solid state signal switches.

Such machines can be used in an iterative mode in which the analogue components operate under the control of a digital processor. This is known as a parallel hybrid system. Alternatively, computers have been produced in which some processes are performed with analogue elements and others with digital elements. These are known as full hybrid systems.

Much of the literature on analogue and hybrid computing is on their application to various types of problem rather than the design of the machines themselves, and the reader needs to be wary in any literature search to separate the papers on applications from those on the systems themselves.

No computer should be viewed in isolation: it is always part of a system. It may be a special purpose machine or a dedicated general purpose machine operating as part of a control or communications system. Alternatively, it may be acting as a data processer in a system which includes the users. We then have a three-component system consisting of hardware, i.e. the machine; software, i.e. the program; and liveware, i.e. the user. No survey of the literature on computers would be complete without taking this into account. Thus a section is devoted to the programming of machines, although this is restricted to a few applications in electrical engineering.

LITERATURE SOURCES

Computer literature, already extensive, is growing rapidly. In *A Guide to Computer Literature*, by A. Pritchard (2nd edn, Bingley, 1972), the author gives a comprehensive survey of sources. He estimates that there are 40 000–50 000 items on computing published either in book form or in scientific and technical journals, conference proceedings, trade literature or research reports, and that the publishing rate is increasing by 10% per annum.

Fortunately for those in the English-speaking world, he estimates that about 66% of the items are published in English, with the next most popular language being Russian with 14%, followed by German with 7% and Japanese with 3%. It is also worthy of note that most (50%) of the authors are with industrial organisations, followed by the universities with 27% and the research organisations with 14%.

ABSTRACTING SERVICES

The would-be reader is helped by having a large number of abstracting services open to him or her. *Computer Abstracts* (Technical Information Company) is devoted principally to scientific applications and programming, and gives extensive coverage of numerical methods and algorithms. *Computer and Control Abstracts* (IEE) covers a similar field, but also takes in computer equipment and control. *Computer and Information Systems* (Cambridge Scientific Abstracts) has its main emphasis on systems and equipment, but includes only English-language sources. *Computing Reviews* (Association for Computing Machinery) aims to furnish computer-oriented persons with critical information about publications in all areas of the computing sciences. It also produces special bibliographies from time to time. Abstracts of current computer literature are also included in the *IEEE Transactions on Computers*, the coverage given being similar to that in *Computer and Information Systems*.

For further information on these and abstracting services, reference should be made to *A Guide to Abstracting Journals for Computers and Computing*, by A. R. Dorling (National Reference Library of Science and Invention, now the British Library, Science Reference Library, 1972).

PERIODICALS AND JOURNALS

Because computing systems are made mainly from electronic components, a number of papers are to be found in the standard electronics journals. A study of the relevant abstracts shows that there are contributions on computer topics in journals published by the various institutions. For example, the *Proceedings of the IEE, IEEE Spectrum* and *The Radio and Electronic Engineer.*

Specialised journals and periodicals on computer topics

A number of journals and periodicals are devoted to computer topics of various types. The following is a list of the major ones with a few brief comments.

Annales de l'Association Internationale pour le Calcul Analogique (AICA). Published quarterly by the International Association in Belgium, it contains items on analogue and hybrid computing and on simulation. Articles are written in English, French and German.

Artificial Intelligence

Computer. A news type of journal published by the IEEE Computer Group.

Computer Aided Design

Computer Bulletin. News magazine of the British Computer Society.

Computer Design. Commercial journal covering computer hardware.

Computer Journal. Research journal of the British Computer Society.

Cybernetics (US translation of the USSR research journal *Kibernetika*)

Engineering Cybernetics (English translation of the USSR *Tekhnicheskaya*)

IEEE Transactions on Computers. Contains research papers and short communications. Wide range but excludes software and numerical analysis.

IEEE Transactions on Systems, Man and Cybernetics. Research papers, mainly on systems in the wider sense.

Information and Control. Covers the theoretical aspects of logic, communication theory and automata.

Information Sciences. Research papers on a wide range of subjects including artificial intelligence, logic and formal languages.

International Journal of Man–Machine Studies. Covers pattern recognition, learning and cybernetics.

International Journal of Systems Science. Research papers on mathematical models, simulation and optimisation and control.

Journal of Cybernetics. Official journal of the American Society of Cybernetics. Research papers on automata, information theory and artificial intelligence.

Journal of Systems Engineering. Research papers on systems. Useful for computer applications.

Kibernetika (an important Russian journal on system design)

Pattern Recognition. Journal of the Pattern Recognition Society. Research papers on pattern recognition and learning processes.

SIAM Journal on Control

SIAM Journal on Numerical Analysis

Simulation. The technical journal of the Society for Computer Simulation. Research papers on digital, analogue and hybrid computer simulation.

In addition, the proceedings of two annual conferences are most important: The Fall Joint Computer Conference and the Spring Joint Computer Conference. Both are organised by the American Federation of Information Processing Societies and the proceedings are published by AFIPS Press.

House journals

Most of the major manufacturers publish a house journal which is devoted to publishing technical details of their products, or of theoretical topics related to their business. As with other journals that have already been mentioned, some of these are entirely devoted to computers when this is the main activity of the company concerned. Notable among these are:

IBM Journal of Research and Development
IBM Systems Journal
Honeywell Computer Journal

Other house journals which occasionally publish on this subject are:

Bell System Technical Journal
RCA Review
Marconi Review

The above tend to be on the electronic circuit aspects of computer design or upon the role of computers in communications.

GOVERNMENT REPORTS

A large number of reports are published each year by various government research organisations. To become informed upon those published by the US Government, NATO, RAE (Farnborough), etc., reference should be made to *Government Reports Announcements,* which is published bi-monthly by the US Government.

Of special note are the two reports which have been published in the UK for the Electrical Research Association by Ovum Ltd.:

Mini Computers (1972)
Micro-Processors: an ERA Assessment (1973)

NEWS PERIODICALS

A few weekly publications exist which provide a means of obtaining up-to-date news and comments on both activities and personalities in the computer business. A regular scan through one of these can keep one abreast of industrial and commercial development in an easily assimilated manner. *Computer Weekly* is the main UK computer newspaper, but another source of this type of information is *Electronics Weekly,* which, as its title implies, covers the whole of the electronics industry. In passing it should be noted that the *Financial Times* science and industry columns often contain items of interest on computer subjects, although these are often written more from an investor's or user's viewpoint than a technical one.

BOOKS ON COMPUTER SUBJECTS

A large number of books have been written on all aspects of computers and computing over the years, and in order to present these in some easily digested form they will be introduced in sections. This of course leads to certain difficulties, because however well the sections are chosen, there will always be a number of books which either fall equally well into a number of sections or do not fit well under any of the section headings.

The three basic divisions are digital computing, applications to particular topics, and analogue and hybrid computation. These in turn are split into a number of sub-sections, particularly in the case of digital computing, as the books on this topic far outnumber the books on the other two sections.

Digital computing

General survey texts

One of the first general surveys which gives a scholarly account of the early development and applications of both analogue and digital computers is *Calculating Instruments and Machines, by* D. R. Hartree (Cambridge University Press, 1950). There is no finer text for setting the scene against which the more modern developments have taken place.

With a first edition published in 1953 and a third edition in 1965, the partnership of Andrew and Kathleen Booth have produced *Automatic Digital Calculators* (Butterworths), which opens with a concise account of the development of digital computing 'engines' from the earliest machines up to those of the date of publication. As an early example they quote Pascal, who in 1642 designed a machine which was able to add and subtract and was originally applied to tax collection in France. This historical survey concludes with IBM's STRETCH 2 system and the ICT ATLAS—but one wonders if the applications have changed as much as the machines. As with most general texts, the authors go on to describe order codes, number representation, arithmetic units, sub-routines and compilers, and have a final chapter on applications, which include such diverse topics as X-ray crystallography, language processing, games and artificial intelligence. This book is particularly noteworthy for its bibliography, the first two editions containing an almost complete list of books and papers on digital computing machines up to 1956 occupying 22 pages. The third edition contains a list of the more significant books and papers written up to 1965 which, although selective, extends to some 15 pages.

Another text which has interests from the historical point of view is *Faster than Thought*, edited by B. V. Bowden (Pitman, 1953). This contains accounts written by the designer of every machine being built in England in 1951.

More up-to-date surveys have been written by a number of experienced authors. These general books usually start with introductory treatments on Boolean algebra and number representation in digital systems, and then go on to describe how logical elements can be put together to achieve the necessary arithmetic manipulations. This is then followed by sections which describe the operation of the general purpose machine as a system under the control of the program, which, like the data, is fed into the machine and put into the computers memory until required.

The following are representative of books coming under this heading:

Digital Computer Design, by F. G. Heath (Oliver and Boyd, 1969)
Digital Networks and Computer Systems, by T. L. Booth (Wiley, 1971)
Digital Computer Fundamentals, by T. C. Bartee (3rd edn, McGraw-Hill, 1972)
Digital Computer Principles and Applications, by A. G. Favret (Van Nostrand, 1972)

For those requiring only a simple treatment of the operation of computer systems there is: *Understanding Digital Computers,* by R. M. Benrey (Iliffe, 1965).

System description and design

Many of the texts on both logic and computer systems tend to be descriptive in nature, describing how systems work rather than how they are designed to carry out the specific tasks which they are required to perform. However, a number of authors have described the design steps either for the necessary logic or for the over-all system. A practical approach to the design of computer systems is given in *Introduction to Digital Computer Design*, by H. S. Sobel (Addison-Wesley, 1970). This also discusses the design steps in the logic necessary to implement the system design. While thinking of the practical steps in design, *Planning a Computer System*, edited by W. Buchholz (McGraw-Hill, 1962), describes the experiences of the team of designers of the IBM 7030 system. It discusses aspects of computer design from the functional point of view rather than the hardware involved.

Digital Systems: Hardware Organization and Design, by F. T. Hill and G. R. Peterson (Wiley, 1973), is a book written as much for the computer scientist and systems programmer as the electrical engineer. It is noteworthy for the introduction of A Programming Language (APL), which is a useful tool for the description of register transfers and control sequence design. The authors discuss systems from the viewpoint of information flow rather than the hardware configuration. This more theoretical type of approach is also taken in *Digital Systems Fundamentals*, by J. Motil (McGraw-Hill, 1972), who considers systems which may be deterministic, probabilistic, sequential or stochastic in their mode of operation. This text may suit those who view computing systems from a more mathematical standpoint, as it starts from the premises of set theory.

General treatment of logic systems

Although digital computers are constructed from logic elements, these elements can be used for other switching functions not necessarily concerned with computation. There is a good selection of texts which deal mainly with logic and the digital techniques which may be used in any digital system. Digital signal manipulation is becoming increasingly common, with digital methods being used in equipment in which computation as such plays no part at all.

One of the earlier texts on logic design was *Logical Design of Digital Computers*, by M. Phister (Wiley, 1959). This is an often-quoted text which has become recognised as a classic on logic design. After a treatment of Boolean algebra and its application to logic networks, the author deals with the simplification of Boolean functions. A description of the various bistable types leads to a treatment of sequential systems introducing the use of state tables and diagrams. There is a general survey of number representation and arithmetic operations, and brief treatments of memory systems and input and output equipments.

The subject cover given in Phister's book is common to a large number, although not all include the sections on storage, arithmetic and input and output equipment. Some noted texts in the field are:

Introduction to the Theory of Switching Circuits, by E. J. McCluskey (McGraw-Hill, 1965)

Introduction to Switching Theory and Logical Design, by F. J. Hill and G. R. Peterson (Wiley, 1968)

Switching and Finite Automata Theory, by Z. Kohavi (McGraw-Hill, 1970). This latter book contains sections on the design of switch contact networks and reliable systems, and also includes a discussion on suitable tests to apply for fault detection.

Logical Design of Switching Circuits, by D. Lewin (2nd edn, Nelson, 1973), provides a comprehensive treatment of both combinational and sequential logic systems, the latter having particularly good coverage; and for a very extensive comprehensive work on these topics there is *Disciplines in Combinational and Sequential Circuit Design,* by R. M. M. Oberman (McGraw-Hill, 1970). For the reader seeking a text which concentrates entirely on the design of logic circuits rather than a general discussion, *Logic Design Algorithms,* by D. Zissos (Oxford University Press, 1972) gives step-by-step description of design procedures for reliable and hazard-free combinational and sequential logic systems. It further goes on to include procedures for minimal NOR and NAND only implementations.

Further texts are:

A Survey of Switching Circuit Theory, edited by E. J. McCluskey and T. C. Bartee (McGraw-Hill, 1962)

Linear Sequential Switching Circuits, by W. H. Kautz (Holden Day, 1965)

Finite State Models for Logical Machines, by F. C. Hennie (Wiley, 1968)

Iterative Arrays of Logical Circuits, by F. C. Hennie (Wiley, 1961)

Digital Design, by R. K. Richards (Wiley, 1971)

Introduction to Digital Computer Technology, by L. Nashelsky (Wiley, 1972)

Computer Logic, by I. Flores (Prentice-Hall, 1960)

Logic of Computer Arithmetic, by I. Flores (Prentice-Hall, 1963)

Digital Computer Design, by E. L. Braun (Academic Press, 1963)

For those requiring an introduction to the subject there are:

An Introduction to Counting Techniques, by K. J. Dean (Chapman
and Hall, 1964)
Digital Techniques, by R. W. Sudweeks (Pitman, 1968)

As well as texts which may be included in the general sections, there
are a number of special interests which are worthy of consideration
in their own right. In some cases they are specialisms of a hardware
nature, and in others, of system philosophy. In each case the section
heading is sufficiently explicit to indicate the nature.

Circuit design of logic elements

Much of the computer literature deals with the interconnection of
logic elements, providing only a cursory study of the particular
circuits which are used to implement these functions. This is a gap
which is covered by *Circuit Design of Digital Computers,* by J. K.
Hawkins (Wiley, 1968), which considers both the design and the
respective performances of the various ways of implementing logic
functions and the trade-offs which are possible.

Shift register sequences

The theory of shift register sequences has found major applications
in a wide range of technological situations, such as error protection
for digital information processing, and system identification with
automatic control systems. In *Shift Register Sequences,* by S. W.
Golomb (Holden Day, 1967), there is in one volume a thorough
treatment of the theory of shift register sequences and their application together with a full bibliography. The more general case is
covered in *Linear Sequential Circuits,* by A. Gill (McGraw-Hill,
1966). Of more recent date there is on this same topic *Binary
Sequences,* by Hoffman de Visme (English Universities Press, 1971),
which contains a mathematical treatment of the properties as well
as the generation of shift register sequences.
    The structure of the logic circuits used to generate binary
sequences is extensively covered in *Error-Correcting Codes,* by W.
W. Peterson (2nd edn, MIT Press, 1972). This is a classic work of
reference for any who wish to know about the generation of
sequences and codes, whether it be for error detection, error correc-
tion or any other purpose.

Threshold logic

To date, the vast majority of digital computers use binary decision elements which perform the simple Boolean operations such as OR and AND. However, a related class of elements operate on the principle of majority decision (or, in common parlance, ballot box logic). Such elements possess greater logical power and may play an important role in the development of non-numerical processors in such areas as pattern recognition and decision-making. A volume devoted to this topic is *Threshold Logic*, by P. M. Lewis and C. L. Coates (Wiley, 1967), which develops the general theory of threshold gates and discusses synthesis procedure. There are also descriptions of realisation through such devices as the Perceptron and the Adaline. A mathematical treatment of this class of logic devices is given in *Threshold Logic,* by S. T. Hu (University of California Press, 1966).

Digital differential analysers

Braun, in his book *Digital Computer Design,* which has already been mentioned, devotes a whole chapter to this subject, but for those requiring a more detailed and comprehensive treatment of this type of machine there is *The Digital Differential Analyser,* edited by T. R. H. Sizer (Chapman and Hall, 1968). This reviews the principle of operation, the programming and a discussion of the errors in the DDA.

An alternative source of information is *Electronic Digital Integrating Computers—Digital Differential Analysers,* by F. V. Mayorov (Iliffe, 1964), which is of added interest because it gives detailed description of a Russian machine and of Russian thinking on this subject.

Analogue/digital conversion

Whenever digital computers are used in conjunction with some other hardware system, it is necessary to convert the signals from analogue to digital form at the input interface and from digital to analogue at the output. Two specialised texts on this subject which are encyclopaedic in their coverage are *Notes on Analog to Digital Conversion*

Techniques, edited by A. K. Susskind (MIT Press, 1957), and *Electronic Analog to Digital Conversions,* by H. Schmid (Van Nostrand Reinhold, 1970). These volumes discuss the theory and practical implementation in a number of different engineering situations.

Storage systems

Although most survey texts include a chapter or two on storage systems, those requiring guidance on the choice of store for various applications will find a whole volume devoted to the subject in *Digital Storage Systems,* by W. Renwick and A. J. Cole (2nd edn, Chapman and Hall, 1971). It is not a design textbook but rather a broad introductory text, but does contain an extensive bibliography for those requiring more detailed information. In a rather different vein is *Electronic Computer Memory Technology,* edited by W. B. Riley (McGraw-Hill, 1971). This is a collection of reprints from the periodical *Electronics* of articles specifically commissioned by the publishers on the technical details of representative computer memory systems.

Magnetic logic systems

In spite of the widespread use of semiconductor logic elements, there are a number of applications where for special reasons it is necessary to have alternative methods of implementation—for example, where the radiation level is high. For such situations computing elements using all-magnetic circuits have been developed. A comprehensive survey of these, together with the physical principles governing their action and an analysis of the circuits employed, is given in *Digital Magnetic Logic,* by D. R. Bennion, H. D. Crane and D. Nitzan (McGraw-Hill, 1969).

An extensive treatise on the physics of magnetic core switching is contained in *Square Loop Ferrite Core Switching,* by P. A. Neeteson (Cleaver Hume, 1964); and a collection of articles by various authors on the use of magnetic cores for switching and storage purposes appears as *Digital Applications of Magnetic Devices,* edited by A. J. Meyerhoff (Wiley, 1960). For a simple introduction to the subject there is *Square Loop Ferrite Circuitry,* by C. J. Quartly (Iliffe, 1962).

Applications to particular topics

Artificial intelligence

One of the increasing number of non-numerical applications of digital computers is that of artificial intelligence. Development and research work is at present being devoted to this subject in terms both of programming general purpose machines and of designing special purpose computers. The subject receives continuing treatment in the series *Machine Intelligence,* edited by B. Meltzer and D. Michie (Edinburgh University Press). Two texts which review the Russian position are *Pattern Recognition,* by M. Bongard (Macmillan, 1970), and *Teaching Computers to Recognise Patterns,* edited by A. G. Arkadev and E. M. Braverman (Academic Press, 1967).

Computer-aided design

Computers are being employed increasingly in the design of all types of components, devices and systems. The type of computer involvement varies from that which may be described as the automated slide rule—open loop involvement—to that in which interaction between designer and computer is very intense — closed loop involvement. With the former the designer needs to know how to program his engineering problems for the general purpose machine. With the latter the computer must be designed bearing in mind the type of problem with which the designer needs help and also the most suitable input/output facilities for the optimum flow of information and commands, whether in numerical, geometric or graphical form.

A number of texts have been written to help engineers (a) to specify their problems in a manner suitable for computer solution, (b) to choose convenient numerical methods where required and (c) to perform the programming in some suitable high-level language. The following texts are representative of those available on this aspect of the subject:

An Introduction to Digital Computing, by B. W. Arden (Addison-Wesley, 1963)

Engineering Applications of Digital Computers, edited by T. R. Bashkow (Academic Press, 1968)

Use of Digital Computers for Engineering Applications, by C. M. Haberman (Merrill, 1966)

Mathematical Methods for Digital Computers, edited by A. Ralston and H. S. Wilf (2 vols., Wiley, 1967)

Digital Computers in Engineering, by S. Seely, N. H. Tarnoff and D. Holstein (Holt, Rinehart and Winston, 1970)

Application of Computers to Engineering Analysis, by J. R. Wolberg (McGraw-Hill, 1971)

Computer Aided Decision Making, by G. K. Chacko (Elsevier, 1972)

Computer Aided Analysis and Design for Electrical Engineers, by B. J. Ley (Holt, Rinehart and Winston, 1970)

System Analysis by Digital Computer, edited by F. F. Kuo and J. F. Kaiser (Wiley, 1966)

Computer-Aided Network Design, by D. A. Calahan (McGraw-Hill, 1972)

Simulation techniques

Analogue, digital and hybrid machines are all commonly used for simulating systems of many types.

Analogue machines are normally used for simulating the dynamic behaviour of such things as control systems, chemical processes and others, which, though they may be linear or non-linear, do have continuous time responses. Digital machines, however, are used not only for this purpose, but also for systems which exhibit discontinuities such as logic, or systems in which the variable may only be known statistically. The subject is one with many devotees, as indicated by the existence of the Society for Computer Simulation in the USA, which publishes its own technical journal, *Simulation,* which provides a forum for discussing the techniques involved.

An over-all review of the subject is given in *Computer Simulation for Engineers,* by R. E. Stephenson (Harcourt Brace Jovanovich, 1971). This book discusses analogue, hybrid and digital computer simulation methods in general and also includes chapters on some of the special simulation languages which have been written to deal with electrical circuit analysis and synthesis, such as ECAP, MIMIC and COGO.

Other books which give good coverage of the subject are *Computer Simulation Techniques,* by T. H. Naylor, J. L. Balintfy, D. S. Burdick and K. Chu (Wiley, 1966), and *System Simulation,* by G. Gordon (Prentice-Hall, 1969).

More specialised texts whose titles accurately indicate their scope are *SCEPTRE: a Computer Program for Circuit and Systems Analysis,* by J. C. Bowers and S. R. Sedore (Prentice-Hall, 1971), and *LYaP AS: a Programming Language for Logic and Coding*

Algorithms, edited by M. A. Gavrilov and A. D. Zakrevskii (Academic Press, 1969).

Analogue and hybrid computation

Digital computers have long been applied to a wide range of problems from, say, that of data processing and sorting (where the volume of data is large and the scale of the arithmetic processing small) to the opposite extreme of complex mathematical manipulation involving only a small amount of data. On the other hand, analogue and hybrid machines have normally been used for a number of closely related types of problem. Because they are constructed from a large number of arithmetic units operating in parallel and also because of the ease with which integration can be performed, analogue machines are used predominantly for the simulation of dynamic systems, often of great complexity.

This difference in use is reflected in the literature, as is also the fact that analogue machines have been developed by engineers for engineers. They are operated in a 'hands on' mode with close interaction between user and the machine. The books on analogue computing concentrate in the main on two topics: firstly, on the make-up of the machines, their method of operation, sources of error and control arrangements; and secondly, on setting them up for the solution of various types of problems. They are written mainly for the user either as descriptions of the hardware and how to operate it, or on preparing problems for machine solution.

For the reader wishing to find out how to put problems on to a machine the following will give a good guide: *Systematic Analogue Computer Programming,* by A. S. Charlesworth and J. R. Fletcher (Pitman, 1967); *Analog Computer Techniques,* by C. L. Johnson (2nd edn, McGraw-Hill, 1963), and *Analog Computation in Engineering Design,* by A. E. Rogers and T. W. Connolly (McGraw-Hill, 1960).

These texts deal with the formulation of flow diagrams and the problems of amplitude and time scaling, with some discussion of solution errors. They are essentially user-orientated and are general purpose as far as the types of problem considered.

In addition, in *Principles of Analog Computation,* by G. W. Smith and R. C. Wood (McGraw-Hill, 1959), there is a useful section on the selection of rational functions for approximating to transport delay, and there is coverage of the problem of simulating given transfer functions in *Analog Computer Programming,* by M. G. Rekoff (Merrill, 1967).

On the other hand, for readers requiring information on the machines themselves two works stand out: *Analogue Computation,* by S. Fifer (4 vols., McGraw-Hill, 1961), and *Electronic Analog Computers,* by G. A. Korn and T. M. Korn (McGraw-Hill, 1956). Both are invaluable source books. The former is a mammoth treatise on all aspects of the various methods that are and have been used for integration, summation, multiplication, non-linear elements, resolving, and so on. Very extensive coverage is given. The latter text is far more concentrated but is no less encyclopaedic in its range of subject matter. It reappeared later as *Electronic Analog and Hybrid Computers,* by G. A. Korn and T. M. Korn (McGraw-Hill, 1963), a rewrite of the material which, as the title suggests, now contains a section on hybrid computing techniques. This book also contains an extremely comprehensive bibliography on the subject.

A number of texts deal with the description of various machine types and their functioning, together with discussions of the methods for preparing problems for machine solution. In general, they are not as detailed as the above but are adequate for most readers.

Introduction to Analog Computation, by J. R. Ashley (Wiley, 1963)
Analogue Computers, by M. Brand and T. E. Brand (Arnold, 1970)
The Design and Use of Electronic Analogue Computers, by C. P. Gilbert (Chapman and Hall, 1964)
An Introduction to Electronic Analogue Computers, by M. G. Hartley (Methuen, 1965)
Fundamental Analogue Techniques, by R. J. A. Paul (Blackie, 1965)
Computation by Electronic Analogue Computers, by V. Borsky and J. Matyas (Iliffe, 1968)
Design Fundamentals of Analogue Computer Components, by R. M. Howe (Van Nostrand, 1961)
Introduction to Electronic Analog Computers, by J. N. Warfield (Prentice-Hall, 1959)

Interesting in that it contains a description of a number of early machines is *Analogue Computing Methods,* by D. Welbourne (Pergamon, 1965). A corresponding book which describes some early Russian machines and the Russian outlook on applications is *Analogue Computers,* by I. I. Eterman (Pergamon, 1960).

Hybrid computation

Many of the texts already mentioned include sections on hybrid computing, but for a reader requiring a book entirely devoted to this subject there is *Hybrid Computation,* by G. A. Bekey and W. J. Karplus (Wiley, 1968). This is one of the few books entirely devoted to this subject. There are further treatments of the subject in *Analogue and Iterative Methods in Computation, Simulation and Control,* by B. R. Wilkins (Chapman and Hall, 1970), and *Analogue and Hybrid Computers,* by Z. Nenadal and B. Mirtes (Iliffe, 1968).

While considering hybrid techniques, mention should be made of a book which has a number of excellent chapters on a whole range of topics—*System Analysis by Digital Computer,* edited by F. F. Kuo and J. F. Kaiser (Wiley, 1966), which in spite of what is suggested by the title contains, in addition to material on simulation, computer-aided design and digital filters, a good chapter on full hybrid operation.

High-speed analogue computing

It is usual for analogue machines to operate at fairly low frequencies because permanent records of the output are required and these are conveniently obtained on pen recorders or X–Y plotters of limited bandwidth. Alternatively, the machine may be operating in real time in conjunction with other equipment with limited speed of response. However, this may not always be the case, and, particularly with repetitive or iterative operation, high-speed working may be desirable. There are two texts which deal with this specialised topic: *Analogue Computing at Ultra High Speed,* by D. M. Mackay and M. E. Fisher (Chapman and Hall, 1962), and *High Speed Analog Computers,* by R. Tomovic and W. J. Karplus (Wiley, 1962).

Combined texts

One of the few books which deals with both analogue and digital computers in one volume is *Electronic Computer Technology,* by N. R. Scott (McGraw-Hill, 1970), which replaces *Analogue and Digital Computer Technology* (McGraw-Hill, 1960), by the same author, which is now available in paperback form.

CONCLUSION

It is inevitable that in a survey of this kind one has had to be selective. However, if you are an author who has been overlooked, then this author offers his apologies—if your favourite book has been left out, then perhaps you would be so kind as to pass on the benefit of your experience.

18

Control engineering

N. G. Meadows

Control engineering is a multidisciplinary subject resting on a foundation of mathematical and physical principles, brought to a stage of practical implementation by engineering applications. These implementations have one important feature in common, namely the use of feedback to form a closed loop system. There are many control systems which do not involve feedback but these are of passing interest only to the control engineer. Fundamentally, those which are of interest involve the generation of an error signal related to the difference between an input reference signal and an output controlled signal. A system is then actuated by the error so that the system output is brought into correspondence with the input. Amplification of some kind is involved and this may be of current, voltage, power, torque or other physical quantity. An important feature is that the reference signal and controlled output need not be physically similar; for example, a reference voltage may be used to specify a desired pressure, flow rate, position or speed, temperature, humidity or other variable. However, for error formation the output will usually need to be transduced to a physical signal similar in kind to the input reference. From these processes we can identify the need to study the characteristics and structure of power actuators of various kinds, instruments and transducers, error-forming devices and associated hardware.

Systems are classified as open and closed loop, the latter involving error actuation. The earliest applications of control systems were usually those to regulate speed, as in a windmill or steam engine where the flyweight governor was widely used, and it is still in common use. The development of steam, gas and petrol engines was accompanied by the need to control speeds and displacements of shafts, valves and other devices. With the advent of electronics a new era of control was introduced, as evidenced by automatic gain and frequency control, voltage and current regulators and position and speed control systems.

During these developments it soon became apparent that closed loop systems could become unstable. This led engineers and physicists to the study of mathematical techniques evolved by eighteenth and nineteenth century mathematicians and their use to predict whether or not a closed loop system would be stable. In particular, Maxwell in 1868 produced his classical paper on 'Governors for mechanical systems' and Nyquist in 1932, working on electrical ones, produced a further classical paper on 'Regeneration theory' investigating stability and instability, and gradually a theoretical framework began to evolve. However, applications proceeded rapidly, particularly during World War II, and practice went ahead without theory, particularly for non-linear systems. In many cases designs were implemented based on incorrect applications of theory, and they still are, with stability being achieved by *ad hoc* methods of compensation or by running systems at a suboptimal level.

Stability can be determined from frequency domain data, as developed by Nyquist, Bode, Mikhailov, Nichols and others, provided the system is substantially linear. Extensions to non-linear systems were made via the describing function and phase-plane methods and by numerical or other solutions of non-linear differential equations. Two aspects began to dominate control development from 1945 to 1965 (approximately); these were random or stochastic signal applications and the extended theory and application of non-linear systems, although additions were still being made to linear theory. Optimal control theory was developed involving cost function algorithms or hill-climbing and other techniques for optimising outputs where interacting variables were concerned. This involved complex multivariable linear and non-linear control systems. Analogue techniques dominated early development and analogue computer simulation was extensively used to assess stability for complex systems. The advent of digital computers and general digital methods led to sampled data systems and also to computer control, either centrally or via local loops. Three-term proportional, derivative and integral controllers tended to dominate earlier process control systems, while phase lag and lead compensators were widely used in electrical and mechanical systems. These are all still extensively used but their limitations are too great for them to contribute to control other than for local loops. Process control plants, ships and aircraft require increasingly complex control schemes and the application of more sophisticated mathematical methods of analysis and synthesis.

Mathematical formulations began with linear differential equations, then to Laplace transform methods with the transfer function concept, pole and zero formulations giving links with complex variable theory and Nyquist criterion, and leading to Evans root locus method of particular use in synthesis. State-space and state-variable methods involving linear algebra with matrix formulations followed. Analytical and test methods involved frequency and time domain responses and their inter-relationships, random signal testing and the later development of pseudo-random sequences. Heavy concentration of effort was expended on the problem of identifying the dynamics of linear and non-linear systems, for mathematical or other modelling. Adaptive control systems were developed and criteria developed for non-linear system stability. Control theory and practice began to be applied to wider areas such as biomedical, economic, biological and information systems. Information and control theory were invoked to explain, and often explain away, phenomena in extensive areas, and this process is developing. The aerospace industry produced enormous advances in control instrumentation, telecommunications, information theory, stochastic processes, and system identification and adaptation. Automation was extended in industry, particularly low-cost automation, which often does not involve feedback. The development of fluidic and pneumatic devices accelerated this process, allied to hydraulic control.

The study of control systems is as prone as any other to give rise to analytic and synthetic procedures which are quickly acclaimed as the answer to engineers' unresolved problems. Norbert Wiener made some extremely wide claims for the efficacy of his work on filtering and the identification of linear and non-linear systems, failing to recognise some fundamental limitations which rendered many of his techniques valueless in real life engineering applications. Such was the reverence accorded to his work, allied to the positiveness of its expression, that engineers were arguably misled and sidetracked for several years. Of particular import has been the neglect of the effects of small non-linearities on applications linear theory, and here one may instance pseudo-random binary sequence (PRBS) testing, the use of correlation methods and other statistical identification or filter design methods which so often implicitly assume linearity but which are applied to non-linear systems. Liapounov's methods were, for a few years, accepted as another panacea; their limitations soon became apparent. And latterly there have been state–space methods; vaunted again as the answer to widely ranging problems. The author advises a degree of healthy scepticism when applying linear theory—anything resting on the principle of superposition—to non-linear systems. There are also weaknesses in the presentation of linear system theory, particularly in relation to frequency response or Nyquist loci. However, since non-linear system analysis and synthesis is difficult, identification and modelling a complex problem and optimisation in a full sense rarely achievable for other than very low-order systems, a start must be made somewhere and that start must rest in an intelligent application of linear theory, backed by an awareness of non-linear methods available and the limitations of linear and non-linear theory and practice. An enormous amount of effective design can be achieved using linear methods or successive linear design about a range of set points for non-linear systems.

However, one important fact emerges with many modern systems. It is probable that the relative success of linear theory resides in the fact that for many years control systems have often been inherently well behaved as open loop systems and unlikely to produce catastrophic disaster conditions if uncontrolled or imperfectly so. Often, too, as, for example, in electrical machines, amplifiers and oscillators, non-linearities exercise an inherent self-regulating action. However, many modern systems just will not work without automatic control. Typical examples are high-speed aircraft such as Concorde, nuclear reactors and fast reaction chemical process plants. If systems are inherently well behaved, they are unlikely to be very lively in response: the implications for aircraft are obvious. Also, if fast throughput for a process is required, quick response to changes is desirable, so that control is maintained within prescribed limits about the desired operating point. Economic viability may rest heavily on this aspect. For nuclear reactors special hazards dictate control at all times, but the economic necessity to achieve as near an optimal output as possible necessitates movement away from very safe conditions which could imply suboptimal operation. Modern large generating plant in the 500 MVA range and above are not inherently stable machines, and need to be operated as part of a complex system, but they must be pushed towards optimal conditions to justify capital and other costs.

This aspect of the control of inherently ill-behaved systems adds a new dimension to control problems. The present inability of control engineers fully to optimise multivariable systems with interacting parameters frequently results in suboptimal performance or in the installation of plant or other capacity which is larger than theoretically necessary.

The present and future clearly offer interesting and exciting challenges and many major problems are unsolved. Of new areas evolving, that of pattern recognition is becoming important and here, as in many other areas, the increased availability of digital computers can aid advances. Pattern recognition may range over such items as chromosome defect recognition, cartographic surveying and interpretation, to system identification. Other developing areas recognise not only the non-deterministic nature of many systems and of the inputs to and outputs from these, but also the imperfections of formal logic to cope with analysis and synthesis. Many control systems involve decision-making aspects, particularly but not exclusively where human operators are concerned. Modelling and control of economics systems is one area; management is another. Frequently, parameters are ill-defined and conventional logic or set theory is either inadequate or even misleading. Recently work has been directed to 'fuzzy set' theory in an attempt to encompass wider aspects of controllable systems or to attempt to bring more order to largely uncontrollable ones. Biomedical aspects are now receiving well-merited attention, and this is a wide-open field for applications and research.

Confronted with the breadth of control engineering a definitive survey of the literature would be an enormous undertaking and beyond the purpose of this reference work. As a compromise the author has chosen to present a brief survey of a number of books which can lead the student or practitioner towards an awareness of basic and well-tried techniques and extend him towards developing areas where there is still uncertainty. Reference is also made to a range of journals which can be of help in updating knowledge. The exclusion of a number of very worthy books and other sources does not imply any relative judgement on the part of the author.

BOOKS

General texts

One of the earlier books, based on differential equations and algebraic methods is *Servomechanism Fundamentals,* by H. Lauer, R. Lesnick and L. E. Matson (McGraw-Hill, 1947). This has a very good treatment of a.c. compensating networks and a useful presentation of linear analysis for low-order systems, with some non-linear aspects. A book preceding this, but of greater depth, is *Network Analysis and Feedback Amplifier Design,* by H. W. Bode (Van Nostrand, 1945). This contains fundamental electrical network theory with a specific orientation towards control theory and excellent material on feedback amplifier design and synthesis methods. Although developed in the era of thermionic devices, the techniques developed have general validity. A useful work bringing circuit theory and feedback amplifiers and simple systems more up-to-date is *Techniques of Circuit Analysis,* by G. W. Carter and A. Richardson (Cambridge University Press, 1972). This includes applications of matrices, Fourier series and transformations, complex plane analysis and synthesis, and distributed parameter systems. Extending this material is *Mathematics of Automatic Control,* by T. Takahaski (Holt, Rinehart and Winston, 1966), which links the mathematical theory of complex functions to the dynamical theory of linear control systems, including frequency and transient response, transfer function operations and stability, with some ideal filter theory and physical realisability, culminating in a section on noise suppression. A good introduction to hydraulic, pneumatic and to a lesser extent electrical systems is afforded by *Automatic Control Engineering,* by F. H. Raven (2nd edn, McGraw-Hill, 1968), which also has some useful sections on linear and non-linear theory. An interesting book which gives a very useful background to mechanical systems is *Mechanics,* by J. C. Slater and N. H. Frank (McGraw-Hill, 1947). Although not a control book, it lays useful foundations for many mechanical and theoretical aspects of control. Another valuable general book is *Principles of Control Systems Engineering,* by V. del Toro and S. R. Parker (McGraw-Hill, 1960), which included an exposition of the root locus method and self-adaptive control systems, allied to classical theory, Laplace methods, and time domain and frequency response. Other books in a McGraw-Hill series, of which this one forms a part, are as follows:

Basic Feedback Control System Design, by C. J. Savant (1958)
Control Engineers Handbook, edited by J. G. Truxal (1958)
Control System Components, by J. E. Gibson and F. B. Tuteur (1958)
Control Systems Engineering, edited by W. W. Seifert and C. W. Steeg (1960)
Feedback Control System: Analysis, Synthesis and Design, by J. C. Gille *et al.* (1959)
Feedback Control Systems, by O. J. M. Smith (1958)
Frequency Response for Process Control, by W. I. Caldwell *et al.* (1959)
Non-Linear Control Systems, by R. L. Cosgriff (1958)
Random Processes in Automatic Control, by J. H. Laning and R. H. Battin (1956)

Sampled Data Control Systems, by J. R. Ragazzini and G. F. Franklin (1958)
System Engineering: an Introduction to the Design of Large Scale Systems, by H. H. Goode and R. E. Machol (1957)

Truxal's book is very comprehensive and detailed and is strongly recommended as a standard text.

Linear and non-linear systems: theory and practice

Another early book is *Automatic Feedback Control,* by W. R. Ahrendt and J. F. Taplin (McGraw-Hill, 1951). This is a very readable introductory textbook, with a strong practical orientation. In this category comes *Textbook of Servomechanisms,* by J. C. West (English Universities Press, 1953), which is an undergraduate text written around practical systems and extended in West's second book, *Analytical Techniques for Non-Linear Control Systems* (English Universities Press, 1960). A book well worth reading is *Dynamics of Automatic Control Systems,* by E. P. Popov (Pergamon, 1962). This, among other items, has a comprehensive account of Mikhailov's methods, which are usually neglected in other works. Another very readable undergraduate text is *An Introduction to the Mathematics of Servomechanisms,* by J. L. Douce (English Universities Press, 1963), which has a large number of references to classical and other papers. A further useful textbook, particularly for electrical power systems engineers, is *Dynamic Circuit Theory,* by H. K. Messerle (Pergamon, 1965), which includes dynamically based discussions of solenoids, diaphragm microphones, transformers, separately excited shunt, series and cross-field machines, two- and three-phase machines and multiphase systems. Applicable control theory is introduced and exemplified on a wide range of physical systems.

Direct Current Machines for Control Systems, by A. Tustin (Spon, 1952), is another valuable text. *An Introduction to Process Control System Design,* by A. J. Young (Longman, 1955), introduces the special terminology and orientation of the process control engineer in a practical readable manner, covering elementary aspects of the subject at undergraduate level.

A work which covers a wider range of mainly theoretical aspects of control is *The Dynamics of Linear and Non-Linear Systems,* by P. Naslin (Blackie and Son, 1965). This incorporates linear theory with non-linear aspects, including filtered non-linear systems, sampled data systems, and topological methods giving an extensive and informative coverage of phase-plane methods and an extension to phase space. A useful section is included on the numerical and graphical computation of transients.

Another standard work is *Sampled Data Control Systems,* by E. I. Jury (Wiley, 1958). The treatment is thorough and comprehensive with useful examples. The theory developed is applicable also to sampled or discrete operation of networks, circuits and computers, and the z and modified z transforms are introduced. Design procedures for discrete compensators are developed with the analysis and study of the stability of sampled data systems. An extension of theory is given in *Linear Automatic Control Systems with Varying Parameters,* by A. V. Solodov (Blackie and Son, 1966). This contains an extended discussion of the delta function and impulse response of a linear system, block diagrams for systems with variable parameters and approximate methods for impulse response determination of systems with slowly varying parameters, and the effects of stationary and non-stationary random processes on systems. There is also a useful bibliography with this book. Of particular interest are *Introduction to the Statistical Dynamics of Automatic Control Systems,* by V. V. Solodovnikov (Constable, 1960); *Engineering Cybernetics,* by H. S. Tsien (McGraw-Hill, 1954); and *Random Processes in Automatic Control,* by J. H. Laning and R. H. Battin (McGraw-Hill, 1956), which is a comprehensive and detailed account of random signal theory and practice. Also of value is *An Introduction to the Theory of Random Signals and Noise,* by W. B. Davenport and W. L. Root (McGraw-Hill, 1958), and *Random Data: Analysis and Measurement Procedures,* by J. S. Bendat and A. G. Piersol (Wiley/Interscience, 1971), which updates work in the stochastic area. A more extended treatment is given in *An Introduction to Information Theory,* by F. M. Reza (McGraw-Hill, 1961), which also includes a very useful bibliography.

Other general undergraduate texts worthy of mention are *Feedback Theory and its Applications,* by P. H. Hammond (English Universities Press, 1958), and *Theory of Servomechanisms,* edited by H. M. James *et al.* (McGraw-Hill, 1947). This latter contains original design aspects using the Nichols chart and other detailed considerations of a fundamental nature. *Electrical Noise,* by D. A. Bell (Van Nostrand, 1960), is a useful elementary introduction to the subject, as is his earlier book, *Information Theory and its Engineering Applications* (Pitman, 1952). A useful source book, although now somewhat dated, is *Industrial Electronics Reference Book,* by the Electronics Engineers of the Westinghouse Electric Corporation (Wiley, 1948). A good introduction to state-space is given in *Linear System Theory the State Space Approach,* by L. A. Zadeh and C. A. Desoer (McGraw-Hill, 1963).

Instrumentation

Books covering practical aspects of instrumentation include *Basic Instrumentation: Industrial Measurement,* by P. J. O'Higgins (McGraw-Hill, 1966), and *Practical Instrumentation Transducers,* by F. J. Oliver (Pitman, 1972).

Modern control principles

An interesting book foreshadowing major developments in later work is *Engineering Systems Analysis,* by A. G. J. MacFarlane (Harrap, 1964). This is a precursor to some of his later work, principally published in the *Proceedings of the Institution of Electrical Engineers,* and is consistent with the author's philosophical approach, linked to generalised and specific solutions of problems in an original way. A number of interesting references are given.

A survey of more modern techniques is given in *Modern Control Theory and Computing,* edited by D. Bell and A. W. J. Griffin (McGraw-Hill, 1969). Topics include classical methods, Liapounov's methods for non-linear systems, digital and hybrid computers, random processes, identification, adaptive and multivariable systems, and optimal control. Several valuable bibliographical references are included.

Another work of modern content is *Foundations of Optimal Control Theory,* by E. B. Lee and L. Markus (Wiley, 1967). Extending this treatment is *Modal Control: Theory and Applications,* by B. Porter and R. Crossley (Taylor and Francis, 1972). Topics dealt with are the analysis and synthesis of deterministic linear time-invariant, lumped-parameter continuous time and discrete systems. There is a comprehensive matrix analysis treatment and sensitivity, as well as multivariable systems, including applications and synthesis, are covered. A very useful reference section is included. Additional material on optimal control is to be found in *Process Optimization with Applications in Metallurgy and Chemical Engineering,* by W. H. Ray and J. Szekely (Wiley/Interscience, 1973). A useful work exploring sensitivity aspects in depth is *Sensitivity Methods in Control Theory,* edited by L. Radanovic (Pergamon, 1966), which is the proceedings of an International Symposium on Sensitivity Analysis held in Dubrovnik under the auspices of the International Federation of Automatic Control (IFAC). This was a 'theme' conference; in general, all IFAC proceedings contain important source material, mainly of research or postgraduate interest.

Conference proceedings

One early conference worthy of mention led to the book *Automatic and Manual Control,* edited by A. Tustin (Butterworths, 1952). The conference was organised by the Department for Scientific and Industrial Research (DSIR), and the proceedings contain useful practical and theoretical material. Other relevant conference proceedings are *Advances in Computer Control* (IEE, 1967), which is the proceedings of the Second Control Convention of the United Kingdom Automation Council; *Stability of Systems* (*Proceedings of the Institution of Mechanical Engineers,* **178**, Part 3M, 1963–64) (and accounts of Liapounov's methods with case studies will be found here); *Proceedings of the Symposium on Computers and Automata,* edited by J. Fox (Brooklyn, Polytechnic Press, 1971), in which detailed examples of games theory, probability models, time sharing, and a variety of papers on mathematical aspects are included; *Proceedings of the Symposium on System Theory* (Brooklyn, Polytechnic Press, 1965); *Conference on Industrial Applications of Dynamic Modelling* (IEE Conference Publication No. 57, 1969); *International Conference on Centralised Control Systems* (IEE Conference Publication No. 81, 1971); and *Minicomputers in Instrumentation and Control,* edited by Y. Paker *et al.* (Polytechnic of Central London, 1972), which is the proceedings of a 3 day short course at the Polytechnic of Central London in August 1972.

Further modern developments

Remaining with modern developments, there is *State Space and Multivariable Theory,* by H. H. Rosenbrock (Nelson, 1970), which treats multivariable system design, and embraces transfer function and state-space methods as a coherent whole. This is a readable and comprehensive account which should appeal to students and engineers. As Rosenbrock points out: 'The newer state-space methods at first promised success but came in the end to seem disappointing.' This can be taken as a realistic appraisal from an author who really knows the field. As an introduction to computer control *Design of On-line Computer Systems,* by E. Yourdon (Prentice-Hall, 1972), is interesting, dealing with process control, design considerations, time sharing, applications, programs, hardware and software. The treatment is mainly descriptive and is useful for applications. *Digital*

Computer Process Control, by C. L. Smith (International Textbook, 1972), provides supplementary material, and *The Human Operator in Process Control,* by E. Edwards and F. P. Lees (Taylor and Francis, 1974), introduces the important aspect of human intervention in a control system. A comprehensive account of non-linear aspects is given in *Nonlinear Control Engineering,* by D. P. Atherton (Van Nostrand Reinhold, 1975).

Process control and instrumentation

Further industrial and measurement systems are dealt with in *Chemical Process Control: Theory and Applications,* by L. A. Gould (Addison-Wesley, 1969), and in *Principles of Industrial Process Control,* by D. P. Eckman (Wiley, 1946), which is an early work but still useful as an introductory text. The *Handbook of Applied Instrumentation,* edited by D. M. Considine and S. D. Ross (McGraw-Hill, 1964), is an extremely extensive reference book containing many applications, especially on process control. Also of value is *Handbook of Instrumentation and Controls,* by H. P. Kallen (McGraw-Hill, 1961), which gives a well-illustrated and useful treatment of practical systems such as process controllers, boilers, turbines, diesel engines, air conditioning plant, valves, actuators and flow meters. In *Instruments for Measurement and Control,* by W. G. Holzbock (2nd edn, Chapman and Hall, 1962), there is a descriptive survey of valves and controllers (pneumatic, hydraulic and electric), and a discussion of the measurement of various physical properties such as level, flow, pressure, temperature, humidity and moisture control. *Ultrasonic Transducers,* edited by Y. Kikuchi (Tokyo, Corona Publishing Company, 1969), deals with the fundamental and detailed analysis and design of piezoelectric, magnetostrictive and other electro-transducers, with a good background of theoretical ultrasonics. This is a comprehensive specialised work containing many useful references. Another relatively specialised book is *Handbook of Automatic Control Instrumentation with Isotope Sources,* by L. I. Korothkov, edited by S. Alexander (Israel Program for Scientific Translations, 1966).

Fluidics

Mainly mechanical topics are covered in the *Proceedings of the 1st International Conference on Fluid Logic and Amplification* (British Hydromechanics Research Association, 1965). *Fluid Amplifiers,* by J. M. Kirshner (McGraw-Hill, 1966), provides a mathematical treatment and contains other relevant references, while *Fluidics*, edited by E. F. Humphrey and D. H. Tarumoto (Fluid Amplifier Associates, 1965), gives a practical treatment of the subject. *Fluid Power Control*, edited by J. F. Blackburn *et al.* (Chapman and Hall, 1960), has contributions from several authors on theory and practice in control applications, with some useful analysis and a wide coverage of components and systems. *The Control of Fluid Power*, by D. McCloy and H. R. Martin (Longman, 1973), has a good bibliography and an interesting section on digital servos. It also covers sequence circuit design, testing and design aspects, logic, and fluidic, pneumatic and electrohydraulic servo-systems, and gives an up-to-date treatment of the various topics. *Fluidics: Components and Circuits*, by K. Foster and G. A. Parker (Wiley/Interscience, 1970), and *A Guide to Fluidics*, by A. Conway (Macdonald, 1971), are useful texts. The reader is warned that many of the earlier fluidic devices are not considered commercially viable. Although existing textbooks are valuable for theory, principles and applications, manufacturers' handbooks and data should definitely be sought to ensure correspondence with modern trends, e.g. in relation to wall attachment devices. This could also be important to teachers, since in the fluidics area many textbook references to particular devices are now outmoded. However, it is not easy to anticipate future trends, and these are well covered in journals and by advertisements in periodicals and other publications.

Hydraulic systems and machine tools

Hydraulic and related control systems are dealt with in *Hydraulic Control Systems*, by H. E. Merritt (Wiley, 1967), which covers fundamentals, components, linear and non-linear aspects, hydraulic, and electrohydraulic servo-systems, with a good mix of theory and practice. *Hydraulic Power and Equipment*, by W. G. Holzbock (Industrial Press, 1968), includes practical and physical aspects of valves, circuits, drives, and other devices and systems, with a good bibliography. *Hydrostatic Transmission Systems*, edited by J. Korn (Intertext, 1969), deals with practical systems and components, and is based on a series of lectures given at Enfield College of Technology in 1967. Other references will be found in books already cited in this chapter. In *Automatic Machine Tools*, by H. C. Town (Iliffe, 1968), there is a good and very well illustrated practical non-mathematical review of machine tool design and operation of automatic machine tools, including transfer machines, automatic work loading and numerically controlled machine tools. Of general interest is *Experimental Methods for Engineers*, by J. P. Holman (2nd edn,

McGraw-Hill, 1971). This book has sections on experimentation and testing methods, with measurements of flow, pressure, temperature, displacement, area, force, torque, strain and vibration, together with data acquisition and processing, which are treated briefly.

JOURNALS

Since control is such a far-ranging topic there are several hundred journals in which reference of some substance may be met. These range from journals of pure and applied mathematics where techniques relevant to control problems may be found, to specific trade journals whose main value may be in the advertisements, which clearly show trends in instrumentation, digital, analogue and other control systems, and components of a very wide variety, usually with lavish offers from manufacturers to provide details of data or handbooks. These can be of great value in a rapidly developing technology in which devices can often be as important in microcosm to the systems in which they are embodied. A representative selection is given here, covering a spectrum of international sources ranging in level from publications of principal interest to research workers to those of relevance to the technician practitioner.

a.c.m. Computing Surveys (Association for Computing Machinery)
The reader will find papers on computer algorithms, the methodology of computer selection, the development of information processing as a discipline and the responsible use of computers in a diversity of applications.
Annales de L'Institut Fourier (Université de Grenoble)
Special applications of mathematics are presented, some of which can provide a theoretical basis for control design and synthesis.
Annales de L'Institut Henri Poincaré. Section A, Physique Theorique, Section B, Calcul des Probabilités et Statistiques (Gauthier-Villars, Paris)
In a similar category to the above journal.
Annales Polonici Mathematici (Ars Polona-Research Przedmieście, Warsaw)
Research areas in the mathematics of control are featured.
Automatica (Pergamon)
This is the *Journal* of the International Federation of Automatic Control. Reference should also be made to the IFAC conference proceedings.
Automation and Remote Control (Consultants Bureau, under the editorial direction of the Instrument Society of America)
Appearing approximately 6 months after the corresponding Russian

issue, this is a translation of the journal *Avtomatika i Telemekhanika* and deals with mathematics and applications.

Bell System Technical Journal (American Telephone and Telegraph Company)
This is the General and Bell Laboratories record and contains material on fundamental research and applications.

Biomedical Engineering (United Trade Press, with the co-operation of the Biological Engineering Society)
A monthly journal which promotes communication between research workers in engineering and clinical medicine.

Bulletin: Institute of Mathematical Statistics (University of California)
A bi-monthly journal reporting research results.

Computer Survey (United Trade Press)
A statistical and analytical journal covering the electronic digital computer industry in Britain, with applications and software.

Computers and Structures (Pergamon)
Papers on special techniques in computer aided design and other areas are presented.

Control Engineering (Dun-Donnelley Publishing Corporation)
This journal contains a wealth of practical design detail and was one of the first of its kind in the field. It publishes extensive advertisements for new devices and systems, although in earlier years each issue tended to include more articles rather than advertising material and to provide valuable source material.

Cybernetica: La Revue de l'Association Internationale de Cybernétique (International Association for Cybernetics)
The aims of the Association are to ensure a permanent and organised liaison between researchers whose work in various countries is related to different sectors of cybernetics, and also to promote the development of this science and of its technical applications, as well as the propagation of results obtained in this field.

IEEE Transactions (Institute of Electrical and Electronics Engineers)
Those parts of interest to control engineers are listed below:
Aerospace and Electronic Systems
Antennas and Propagation
Audio and Electroacoustics (continued as *Acoustics, Speech, and Signal Processing*)
Automatic Control
Biomedical Engineering
Broadcast and T.V. Receivers (continued as *Consumer Electronics*)
Broadcasting
Circuits and Systems

Communications
Computers
Electron Devices
Industrial Electronics and Control Instrumentation
Industry Applications
Information Theory
Instrumentation and Measurement
Power Apparatus and Systems
Reliability
Systems, Man and Cybernetics
Although not of equal importance, control aspects feature in all of these *Transactions,* some obviously in a specialised manner, others in association with other topics, e.g. broadcasting and communications systems.

Information and Control (Academic Press)
A monthly publication mainly concerned with applicable theory, oriented towards research contributions.

International Journal of Computer Mathematics: Section A Programming Languages, Theory and Methods. Section B Computational Methods (Gordon and Breach)
Research papers form the basis of the journal.

International Journal of Control (Taylor and Francis)
New theoretical work and practical techniques are presented, mostly from research workers in universities.

International Journal of Electrical Engineering Education (Manchester University Press)
Publishes widely ranging articles on theoretical and practical topics of interest to lecturers in education and to the engineering community in general.

International Journal of Non-Linear Mechanics (Pergamon)
An interesting journal for items which may provide solutions to control engineering problems.

Proceedings of the Institution of Electrical Engineers, Electronics and Power (formerly the *Journal of the IEE*), and *Electronics Letters* (IEE)
New techniques are presented, many at a relatively high mathematical level, together with a few practical and design papers. *Electronics Letters* contains numerous short articles in each fortnightly issue.

Nuclear Engineering International (IPC Electrical-Electronic Press)
The April 1974 issue (**19**, 218, 285–299) has an interesting survey of existing world nuclear power station sites with detailed information on their progress. A wide coverage of nuclear engineering topics is provided by this journal.

Various journals and proceedings of the Institution of Mechanical Engineers, the Institution of Chemical Engineers and other institutions carry items of interest to control engineers. Professional journals from overseas afford additional material. Specialised topics will be found in *Proceedings of the Royal Society of London. Series A, Mathematical and Physical Sciences.*

Journals and review series with a special orientation towards biomedical aspects are:

Advances in Biological and Medical Physics
Advances in Biomedical Engineering
Biomedical Engineering
British Journal of Radiology
British Medical Bulletin
Computers in Biology and Medicine
Engineering in Medicine
Health Physics
International Journal of Applied Radiation and Isotopes
Journal of Biomechanics
Medical and Biological Engineering
Physics in Medicine and Biology

Journals of a mainly applied and theoretical nature, ranging from general to specific interest, are also of value to the control engineer. A selection is given below.

Computers and Electrical Engineering
Control and Instrumentation
Electronic Engineering
Electronics
IEEE Spectrum
International Journal of Electronics
Journal of Physics. E. Scientific Instruments
New Electronics
New Scientist
Review of Scientific Instruments
Scientific American

ABSTRACTING AND INDEXING SERVICES

The most commonly used English-language abstracting journal for control engineering is *Computer and Control Abstracts,* published monthly by The Institution of Electrical Engineers. The entries are arranged within a subject classification scheme under main headings which are subdivided into more specific headings, e.g. control theory is subdivided into analysis and synthesis methods, stability, optimal control, discrete systems, etc. Each issue contains author, subject, bibliography, book, conference, patent and report indexes. Cumulative indexes are available for authors and subject, and these appear at 6 monthly and 4 yearly intervals. Each abstract is numbered sequentially to facilitate easier location.

Metron is edited and published monthly by INSPEC, The Institution of Electrical Engineers, in association with Sira Institute. Each issue contains information reviews, a current titles section (over 2000 journals, conference proceedings and books are scanned) arranged under 17 headings, and subject and author indexes to current titles. Topics covered are measurement, control, automation, sensors, instruments, systems, design, manufacture and applications. Specially commissioned articles on aspects of measurement and instrument application appear regularly.

Fluidics Feedback is the only abstracts journal dealing solely with this subject, and is published every 2 months by BHRA (British Hydromechanics Research Association). The abstracts of journal articles, reports, conference proceedings, standards and patents are arranged under the following headings, which are then subdivided: applications; circuits and circuit components; general topics; moving part devices; non-moving part devices; and new products.

There are many abstracting and indexing services satisfying the needs of control engineering and related disciplines, and a selection of appropriate services is given below.

Automation Express
BECAN (Biomedical Engineering Current Awareness Notification)
Current Contents: Engineering, Technology, and Applied Sciences
Current Papers on Computers and Control
Current Papers in Electrical and Electronics Engineering
Electrical and Electronics Abstracts
Engineering Index
Fluid Flow Measurements Abstracts
International Aerospace Abstracts
International Bibliography: Automatic Control
IRBEL (Indexed References to Biomedical Engineering Literature)
ISMEC Bulletin
New Literature on Automation
Physics Abstracts
Referativnyi Zhurnal-Avtomatika, Telemekhanika i Vychislitel'naya
   Tekhnika (Automation, telemechanics and computer technology)
Referativnyi Zhurnal-Kibernetika (Cybernetics)

Referativnyi Zhurnal-Matematika (Mathematics)
Referativnyi Zhurnal-Mekhanika (Mechanics)
USSR and East Europe Scientific Abstracts-Cybernetics, Computers and Automation Technology

CONCLUSION

Certain areas omitted in the coverage of this chapter can be investigated through the bibliographies and other references in several of the books quoted. Substantial elements of other chapters in this book also complement this work, particularly Chapter 17 on Digital, analogue and hybrid computation, Chapter 14 on Electrical power systems and machines, Chapter 15 on Electronics and Chapter 16 on Communications. Further aspects occur in various mechanical engineering and other sections of the book. One important source of information which should not be overlooked resides in patent specifications; see Chapter 6. The publications of government-sponsored organisations should also be consulted, as they provide a wealth of information of interest to the control engineer.

19

Aeronautics and astronautics

G.M. Lilley, and staff of the Department of Aeronautics and Astronautics, University of Southampton

Although man has striven to fly from the time of the Greek legend of Daedalus and Icarus, his conquest of the air dates only from 1783. In that year man first travelled in balloons, but from then on steady development led to the first powered flight by the Wright brothers in 1903 and in 1961 Gagarin became the first man to orbit the earth in a satellite.

The long path to success in the conquest of flight dates from the dawn of history in the development of the arrow, the boomerang and the kite. Later came the land windmill rotor and the model helicopter. Leonardo da Vinci designed ornithopters and the parachute, and in 1783 Montgolfier designed the hot air balloon. The concept of the fixed wing came from Cayley between 1799 and 1809, and it was developed further by Henson, Stringfellow, Walker, Wenham and Langley. 1863 saw the foundation of the Societe d'Aviation, followed in 1866 by the Aeronautical Society of Great Britain. The power plant for the aeroplane had its birth in 1860 with the invention of the gas engine by Lenoir, followed by the work of Otto on a four-stroke petrol engine in 1876. The first aerodynamic experiments in wind tunnels were performed by Wenham and Brown in 1871. Lilienthal, from 1891 to 1896, was the first man to perfect controlled flight in gliders. The culmination of all these efforts came in the work of Wilbur and Orville Wright, who in 1903, with their aeroplane 'Flyer' I, powered with a 12 h.p. petrol engine, first achieved powered flight from level ground.

The science of aeronautics has its foundations in the work of Newton, Bernoulli, d'Alembert, Euler and Reynolds in their fundamental work on fluid mechanics and air resistance. Further progress came from the experimental work of Zahm, Stanton, Prandtl,

Lanchester and Rayleigh. Coordinated aeronautical research came from the establishment of the National Physical Laboratory in England in 1909 and the later foundation of the Royal Aircraft Establishment and similar establishments in Russia, Germany (Göttingen) and the USA (NACA) (National Advisory Committee for Aeronautics). The rapid advancement of knowledge in aeronautics in the last century has stemmed from the development of the wind tunnel for experimental aerodynamic research. However, following the epoch-making flights of the Wright brothers, the greatest single factor in the rapid development of aeronautics was the realisation of the aeroplane as a military weapon, and the consequent development of it during two World Wars.

The early materials of wood, steel and fabric were gradually replaced by composite structures of wood, steel and aluminium alloys, and finally in the form of metal stressed skin construction. The development of smooth surfaces and greatly improved power plants led to a rapid increase in aeroplane speed, and in 1931 Britain won the Schneider Trophy with Mitchell's Supermarine S6B at a speed of 341 mile/h and later set a world record speed of 407 mile/h. This aircraft was the ancestor of the 'Spitfire' of World War II fame. A greater boost in aircraft speed came some 10 years later from the development of the turbo-jet engine in this country, through the work of Whittle, and in Germany. Immediately after World War II, supersonic flight speeds were achieved, but it was only in 1970 that the first supersonic civil transport, Concorde, achieved its goal of $M = 2$ at an altitude of 55 000 ft.

Air transport on a world-wide basis developed rapidly following the introduction of the Douglas DC3 in 1936. Other more recent highlights in civil aviation are the Vickers Viscount, the first turbo-prop transport, the De Havilland Comet, the first turbo-jet transport, and the wide bodied transports such as the Boeing 747, the Lockheed Tristar and the Douglas DC 10, all powered with turbo-fan engines. By the early 1970s most major cities throughout the world had airports and by then air passenger transport largely replaced sea passenger transport. Community pressures had built up to reduce noise and other forms of environmental pollution. From 1970 onwards, civil transport aircraft had to be designed to be compatible with the environment in which they flew.

The development of rotary wing aircraft, the helicopter and autogyro, arose from the pioneering work of Cierva and Sikorsky. Vertical take-off of a fixed wing aeroplane was first achieved in 1953 and was fully realised in the military aeroplane, the Harrier. The flying boat, which had a long history in both civil and military aviation, eventually was discarded except for a limited number of specialised uses in favour of aircraft operating from fixed prepared runways. The airship, although full of promise for civil transportation, found the hazards of the atmosphere too great and suffered a number of tragic accidents, and little further development proceeded from 1941 onwards.

The basis of space flight was laid down in 1896 by Tsiolkovsky but the first practical experiments with liquid-fuelled rockets were in the USA by Goddard in 1926. Development of rocket-powered flight, mainly in Germany in World War II, led to the launching by Russia of 'Sputnik' I, the first artificial satellite, in 1957 and to the first manned space-flight by Gagarin in 1961. In 1969 the USA launched 'Apollo' to the moon and Armstrong became the first man to step onto its surface. Man has always striven, at the other end of the spectrum, for man-powered flight. This was first achieved in 1961 by Piggott flying 'SUMPAC', and is still a field of current world-wide interest and activity.

Thus, in 70 years the science of aeronautics and astronautics and its related technologies advanced from the achievements of the first flight of Wright at a speed of 30 mile/h to that of the orbiting satellite at a speed of 18 000 mile/h, and man-powered flight also at 30 mile/h. This science also led to the perfection of controlled flight to make all-weather operation of aircraft possible, and to the achievement of mastering the control of space rockets in flight to the moon and beyond and return to earth.

Aeronautics and astronautics has developed into one of the major branches of technology and applied science. It has exposed problems in most fields of science and has in its short life been one of the more important pacemakers for scientific achievement in the modern world. Its literature is extensive and only a brief introduction to it in the various fields of interest will be attempted below.

GENERAL BIBLIOGRAPHY

A Brief History of Flying from Myth to Space Travel, by C. H. Gibbs-Smith (HMSO, 1967)

From Low Speed Aerodynamics to Astronautics, by T. Von Karman (Pergamon, 1963)

Man-Powered Flight, by K. Sherwin (Model and Allied Publications, 1971)

Mastery of the Air, by O. G. Sutton (Hodder and Stoughton, 1965)

AERODYNAMICS

The science of aerodynamics is applied to the study of disturbances generated by the motion of a body through air. The motion of a body resulting from air-induced loads is discussed in the section on mechanics of flight. Aerodynamics as a science was formulated only a few years before the first flight of the Wright brothers and developed by a combination of theory and experiment with frequent appeal to the theorems of Bernoulli, Helmholtz, Kelvin, Rankine and the works of Reynolds, Froude, Rayleigh and Prandtl. The kinetic theory of gases (see *An Introduction to the Kinetic Theory of Gases,* by J. H. Jeans (Cambridge University Press, 1940)) and experiment (see *Modern Developments in Fluid Mechanics,* by S. Goldstein (2 vols., Oxford University Press, 1938)) confirm that air at normal pressure and temperature is a 'Newtonian' fluid and therefore its disturbed motion is governed by the Navier–Stokes equations (see *Modern Developments in Fluid Mechanics: High Speed Flow,* by L. Howarth (2 vols., Oxford University Press, 1953)). Even with the aid of modern electronic computers these equations remain unsolved, and recourse is made to a number of theoretical models involving approximations whose validity is dependent on the flow regime under investigation. Ultimately, the results from any aerodynamic theory have to be checked against experiment. The fundamentals of the subject, including details of these theoretical models, are given in a number of textbooks, including:

Aerodynamic Theory, by W. F. Durand (6 vols., Springer-Verlag, 1934–1936)
Fluid Mechanics, by L. D. Landau and E. M. Lifshitz (Pergamon, 1959)
Hydrodynamics, by H. Lamb (6th edn, Cambridge University Press, 1932)
Incompressible Aerodynamics, by B. Thwaites (Oxford University Press, 1960)
The Elements of Aerofoil and Airscrew Theory, by H. Glauert (2nd edn, Cambridge University Press, 1947)

An interesting insight into the development of the subject is given in *From Low Speed Aerodynamics to Astronautics,* by T. Von Karman (Pergamon, 1963). Typical standard textbooks for undergraduates include *Mechanics of Fluids,* by W. J. Duncan, A. S. Thom and A. D. Young (2nd edn, Arnold, 1970), and *Modern Fluid Dynamics. Vol. 1 Incompressible Flow,* by N. Curle and H. J. Davies (Van Nostrand, 1968).

The evaluation of the flow over an aeroplane from experiments made in wind tunnels is based on the principle of dynamic flow similarity (see *Physical Similarity and Dimensional Analysis*, by W. J. Duncan (Arnold, 1953)). Strict flow similarity demands that the Reynolds and Mach numbers are the same in comparisons between experimental results at model and full scale. The methods by which wind tunnels are used in aerodynamic experiments for the determination of equivalent full-scale data are given in *Wind Tunnel Technique*, by R. C. Pankhurst and D. W. Holder (Pitman, 1952), *Low-Speed Wind-Tunnel Testing*, by A. Pope and J. J. Harper (Wiley, 1966), and *Experimental Fluid Mechanics*, by P. Bradshaw (Pergamon, 1964).

The bridge between classical hydrodynamics of an inviscid fluid, in which, Lighthill has stated, phenomena can be proved but not observed, and hydraulics, in which phenomena can be observed but not proved, was spanned by Prandtl in his theory of the boundary layer (see *The Essentials of Fluid Dynamics*, by L. Prandtl (Blackie, 1952); *Boundary Layer Theory*, by H. Schlichting (6th edn, McGraw-Hill, 1968); and *Laminar Boundary Layers*, by L. Rosenhead (Oxford University Press, 1963). The study of boundary layers over wings and bodies is paramount in establishing the flow over an aeroplane and in defining its performance and stall (see *Shape and Flow: the Fluid Dynamics of Drag*, by A. H. Shapiro (Heinemann, 1964). The essential elements in boundary layer characteristics involve transition from laminar to turbulent flow and the development of turbulent flow (see *Turbulence*, by J. O. Hinze (McGraw-Hill, 1959); *The Structure of Turbulent Shear Flow*, by A. A. Townsend (Cambridge University Press, 1956); *An Introduction to Turbulence and its Measurement*, by P. Bradshaw (Pergamon, 1971); and *Turbulent Flows and Heat Transfer*, by C. C. Lin (Oxford University Press, 1959).

The study of boundary layer separation remains essentially an experimental subject, but for recent work see *Separation of Flow*, by P. K. Chang (Pergamon, 1970). The control of the boundary layer for low drag and high lift is important for good flight economics and flight safety. An account of this subject is given in *Boundary Layer and Flow Control*, edited by G. V. Lachmann (2 vols., Pergamon, 1961). A detailed summary of results of drag measurements and theory is given in *Fluid-Dynamic Drag*, by S. F. Hoerner (The Author, 1965). The study of wakes and jets is related to that of the boundary layer (see *The Theory of Turbulent Jets*, by G. N. Abramovich (MIT Press, 1963), and *Boundary Layer Theory*, by H. Schlichting (6th edn, McGraw-Hill, 1968).

The characteristics of aerofoil sections are given in *Theory of*

Wing Sections, by I. H. Abbott and A. E. Von Doenhoff (Dover, 1959), and *Handbook of Aerofoil Sections for Light Aircraft,* by M. S. Rice (Aviation Publications, 1971). The characteristics of wings and wing-body combinations are given in *Incompressible Aerodynamics,* by B. Thwaites (Oxford University Press, 1960), and *Wing Theory,* by A. Robinson and J. A. Laurmann (Cambridge University Press, 1956).

The design of propellers involves extensions to the theory of wings to allow for the rotating blades, the interference between blades and slipstream distortion. The basic theories are described in *The Elements of Aerofoil and Airscrew Theory,* by H. Glauert (2nd edn, Cambridge University Press, 1947), and *Theory of Propellers,* by T. Theodorsen (McGraw-Hill, 1948). The more general aspects of propulsion systems are dealt with in *Aerodynamics of Propulsion,* by D. Küchemann and J. Weber (McGraw-Hill, 1953).

Aerodynamic principles are also applied in the determination of the flow over land transportation vehicles, where the main distinguishing feature from aeronautical problems is the presence of the ground. This field, together with that concerning the flow over and the aerodynamic loads on tall buildings, bridges, forests, etc., is usually referred to as industrial aerodynamics. Introductory textbooks dealing with these subjects include: *Natural Aerodynamics,* by R. S. Scorer (Pergamon, 1958); *Wind Forces in Engineering,* by P. Sachs (Pergamon, 1972); and *Wind-Excited Oscillations of Structures,* by D. E. J. Walsche (National Physical Laboratory, 1972).

For recent developments in aviation, reference should be made to *Jane's All the World's Aircraft 1972–73* (S. Low, 1972). Up-to-date aerodynamic design information can be obtained from the Engineering Sciences Data Unit (ESDU).

GAS DYNAMICS

Gas dynamics is the study of compressible fluid flows. Thus subsonic, transonic, supersonic and hypersonic aerodynamics (q.v.) all essentially involve gas dynamics, while other important aeronautical applications occur in the internal flows taking place in all types of propulsion systems. Gas dynamic phenomena arise in most major components of propulsion devices, including combustion chambers, but literature specifically concerned with combustion is not covered in this section. Low-density non-continuum phenomena are important when typical flow dimensions are of similar order to the mean free path of the flow.

The fundamental effects of compressibility on fluid flows have been well understood for many decades, but the demands of high-speed flight and the related need for more powerful propulsion systems led to much more detailed study of the subject during and after World War II. An introductory text describing this new knowledge, published soon after the war, is *Introduction to Aerodynamics of a Compressible Fluid*, by H. W. Liepmann and A. E. Puckett, (Wiley, 1947). More mathematical treatments, emphasising the importance of wave processes in compressible flow, are *Supersonic Flow and Shock Waves*, by R. Courant and K. O. Friedrichs (Interscience, 1948), and *Elements of Aerodynamics of Supersonic Flows*, by A. Ferri (Macmillan, 1949), which were published at about the same time. Several other excellent general texts followed including:

Elements of Gasdynamics, by H. W. Liepmann and A. Roshko (Wiley, 1957)
The Dynamics and Thermodynamics of Compressible Fluid Flow, by A. H. Shapiro (2 vols., Ronald Press, 1954)
Gas Dynamics, by K. Oswatitsch (Academic Press, 1956)
Fundamentals of Gas Dynamics, by J. A. Owczarek (International Textbook, 1964)
Mathematical Theory of Compressible Flow, by R. Von Mises (Academic Press, 1958)

A slightly more elementary treatment emphasising internal flow applications is *Elements of Compressible Flow*, by F. Cheers (Wiley, 1963).

All the works cited above contain reviews of the ideal gas thermodynamic relationships which are employed in most compressible flow studies. A much more detailed exposition of the thermodynamic foundations of compressible flow theory may be found in *Modern Developments in Fluid Dynamics*, edited by L. Howarth (Oxford University Press, 1953), and in the first volume of the *High Speed Aerodynamics and Jet Propulsion Series* (Oxford University Press, 1955), entitled *Thermodynamics and Physics of Matter*, edited by F. D. Rossini. Other volumes in this series which treat aspects of gas dynamics in considerable depth include: *General Theory of High Speed Aerodynamics. Vol. VI*, edited by W. R. Sears (1955), and *Physical Measurements in Gas Dynamics and Combustion. Vol. IX*, edited by R. W. Ladenburg *et al.* (1955).

High-speed aerodynamics at subsonic and transonic speeds involves the development of special techniques for both theoretical and experimental studies. Accounts of these are given in *Aerodynamics of Wings and Bodies*, by H. Ashley and M. Landahl (Addison-Wesley, 1965); *A Theory of Supercritical Wing Sections*, by F. Bauer *et al.* (Springer-Verlag, 1972); and *Transonic Aerodynamics,* by C. Ferrari and F. G. Tricomi (Academic Press, 1968).

During the two decades commencing in 1950, the study of hypersonic aerodynamics developed and *Hypersonic Flow Theory,* by W. D. Hayes and R. F. Probstein (2nd edn, Academic Press, 1966), became the standard reference work. Parallel texts are *Introduction to Hypersonic Flow,* by G. G. Chernyi (Academic Press, 1961); *Elements of Hypersonic Aerodynamics,* by R. N. Cox and L. F. Crabtree (English Universities Press, 1965); and *Hypersonic Aerodynamics,* by R. W. Truitt (Ronald Press, 1959). The last author has also considered the aerodynamic heating problems arising from surface friction in *Fundamentals of Aerodynamic Heating* (Ronald Press, 1960).

Relatively few textbooks exist dealing with low-density aerodynamics, but important references are *Molecular Flow of Gases,* by G. N. Patterson (Wiley, 1956), and *Introduction to the Dynamics of Rarefied Gases,* by V. P. Shidlovsky (American Elsevier, 1967). Recent advances in this subject are best found in the various *Proceedings of the International Symposium on Rarefied Gas Dynamics* (Pergamon, 1960–).

Standard texts describing experimental facilities and techniques in supersonic and hypersonic flows are *High Speed Wind Tunnel Testing,* by A. Pope and K. L. Goin (Wiley, 1965), and, more recently, *Experimental Methods of Hypersonics,* by J. Lukasiewicz (Dekker, 1973). Much of the experimental work in hypersonic aerodynamics is carried out using shock tubes and their derivatives. A useful working handbook is the NAVORD (Naval Ordnance Publications) Report 1488 *Handbook of Supersonic Aerodynamics* (1959), and a commendably complete reference work including an extensive bibliography is *Stossrohre,* by H. Oertel (Springer-Verlag, 1966).

Gas flows involving very high temperatures are of interest in connection with entry into the earth's atmosphere, and also in hypersonic wind tunnel testing and in some propulsion systems. In these circumstances the simple ideal gas laws are no longer applicable, and must be replaced by a more complex 'real gas' thermodynamic description. An exceptionally clear introductory text describing high-temperature phenomena in gas dynamics is *Introduction to Physical Gas Dynamics,* by W. G. Vincenti and C. H. Kruger (Wiley, 1965). More detailed treatment of aspects of this topic may be found in *The Dynamics of Real Gases,* by J. F. Clarke and M. McChesney (Butterworths, 1964); *Physics of Shock Waves and High Temperature Hydrodynamic Phenomena,* by Ya. B. Zel'dovich and Yu. P.

Raizer (2 vols., Academic Press, 1966); and *Nonequilibrium Flows,* edited by P. P. Wegener (2 vols., Dekker, 1969).

PROPULSION—AIR BREATHING ENGINES

This section covers propulsion by the various types of air breathing engine, including piston engines, gas turbine engines and ram-jets. Since current development effort is centred on gas turbines, the more recent textbooks and reports tend to concentrate on this class of engine. In view of the rapid pace of progress on aero-engine developments textbooks in this field quickly date and, therefore, recent developments in aircraft engine technology can normally be found only in scientific reports and conference proceedings.

There is an excellent historical review in *Development of Aircraft Engines and Fuels,* by R. O. Schlaifer and S. D. Heron (Harvard Graduate School of Business Administration, 1950), of the organisation of the development of aero-engines by private enterprise and government organisations in the USA and the UK during the period from around 1917 to 1949. The review spans virtually the whole of the period during which both large piston engines and the first gas turbine engines were developed, and provides good material for familiarisation in this field.

Introductory technical material on piston aero-engines is available in *Aircraft Engines,* by A. W. Judge (2 vols., Chapman and Hall, 1941); *Power Plants for Aircraft,* by J. Liston (McGraw-Hill, 1953); and *Aircraft Propulsion Theory and Performance,* by A. W. Morley (Longmans, Green, 1953). Detailed information on current aero-engines can usually be obtained by application to the manufacturers. Rotary piston engines are not currently used in aeronautics but an introduction to this type of engine with references is *The Wankel Engine,* by J. P. Norbye (Bailey and Swinfen, 1971).

The ram-jet and aircraft gas turbine are well covered in the following textbooks:

Aircraft and Missile Propulsion, by M. J. Zucrow (2 vols., Wiley, 1964)
Jet Propulsion for Aerospace Applications, by W. J. Hesse and N. V. S. Mumford (2nd edn, Pitman, 1964)
Mechanics and Thermodynamics of Propulsion, by P. G. Hill and C. R. Peterson (Addison-Wesley, 1965)

These books are suitable for undergraduates and advanced course students and also form a good introduction for those entering research or industry.

Several important textbooks which concentrate on the aircraft gas turbine engine have been written in the UK and USA. Listed below is a selection of those of interest to the research worker and practising engineer.

Gas Turbine Principles and Practice, by H. Roxbee-Cox (Newnes, 1955)

High Speed Aerodynamics and Jet Propulsion Series (Oxford University Press)

Vol. II Combustion Processes, by B. Lewis *et al.* (1956)

Vol. IX Physical Measurements in Gas Dynamics and Propulsion, by R. W. Ladenburg *et al.* (1955)

Vol. X Aerodynamics of Compressors and Turbines, edited by W. R. Hawthorne (1964)

Vol. XI Design and Performance of Gas Turbine Power Plants, edited by W. R. Hawthorne and W. T. Olson (1960)

Vol. XII Jet Propulsion Engines, edited by O. E. Lancaster (1959)

Axial Flow Compressors, by J. H. Horlock (Butterworths, 1958)
Axial Flow Turbines, by J. H. Horlock (Butterworths, 1966)
Gas Turbine Theory, by H. Cohen, G. F. C. Rogers and H. I. H. Saravanamuttoo (2nd edn, Longmans, 1972)

The complex and important subject of engine control is sparsely treated in these texts. This is partly because some aspects of the control of an engine are special to that engine and not general to all engines, and partly because the art of engine control has been slow in its development. A very basic introduction to the subject is available in *Aircraft Propulsion,* by P. J. McMahon (Pitman, 1971) and a more thorough treatment is given in *Control of Aircraft and Missile Powerplants,* by A. J. Sobey and A. M. Suggs (Wiley, 1963).

PROPULSION — ROCKETS

This section is concerned exclusively with chemical rockets and primarily with their combustion characteristics (excluding pyrotechnics).

The literature on rocket technology is vast. Consequently this brief survey of the chemical rocket sphere is extremely selective and must be treated as illustrative rather than comprehensive. The history of the subject in the pre-satellite era has been briefly summarised by H. S. Seifert, for example, in *Jet Propulsion* (**25**, 594–603,

1955) under the title 'Twenty-five years of rocket development'. The subsequent history is occasionally covered in semi-popular publications, but is not conveniently reviewed at the technical level in any single source. Recent developments must be elucidated by judicious reading of the various review series mentioned later.

A good introduction to the elements of rocket theory and practice, at the undergraduate level, is given in *Rocket Propulsion Elements*, by G. P. Sutton (3rd edn, Wiley, 1963). There are also significant sections in *Mechanics and Thermodynamics of Propulsion*, by P. G. Hill and C. R. Peterson (Addison-Wesley, 1965), and in *Liquid-Propellant Rockets*, by D. Altman *et al.* (Princeton University Press, 1960), and *Solid-Propellant Rockets*, by C. Huggett *et al.* (Princeton University Press, 1960). Also note *Rocket Propulsion*, by M. Barrere (Van Nostrand, 1961).

In parallel with astronomy and a number of other scientific disciplines, rocketry has for long attracted the attention of enthusiastic amateurs, particularly in the connotation of space travel. There is an enormous wealth of published material in this vogue, and although a number of the more technical contributions are worthy of mention, one only must suffice to typify this aspect: *Rocket and Space Science Series. Vol. I Propulsion, Vol. II Propellants*, edited by B. Ketcham *et al.* (Foulsham, 1967).

Many advanced and specialised texts are devoted to propellant systems and their chemistry. Again selection permits mention of only two: *Energetics of Propellant Chemistry*, by B. Siegel and L. Schieler (Wiley, 1964), and *Propellant Chemistry*, by S. R. Sarner (Reinhold, 1966). The basic thermochemical data for performance calculations are best obtained from the *JANAF Thermochemical Tables* (2nd edn, US Department of Commerce, 1971). A range of conference proceedings also provides useful periodic collections of papers in this and other specialised areas. For example, in the AGARD Conference Series may be mentioned *Advanced Propulsion Techniques*, edited by S. S. Penner (Pergamon, 1961); *5th Combustion and Propulsion Colloquium*, edited by R. P. Hagerty *et al.* (Pergamon, 1963); and *Performance of Chemical Propellants*, by I. Glassman and R. F. Sawyer (Agardograph, 1970).

The basic theory of liquid spray combustion, both monopropellant and bipropellant, and of solid propellant deflagration is well introduced at the postgraduate level in *Combustion Theory*, by F. A. Williams (Addison-Wesley, 1965) and in *Chemistry Problems in Jet Propulsion*, by S. S. Penner (Pergamon, 1957). At a more advanced level, the *Proceedings of the 1st Rocket Propulsion Symposium* (Cranfield, 1961) and *Heterogeneous Combustion*, edited by H. G. Wolfhard, I. Glassman and L. Green (Academic Press, 1964), should be consulted. Because rockets are used to a large extent for missile propulsion, many of the details of the more sophisticated and chemically interesting propellant systems and their combustion are cloaked in the shrouds of security, but, despite this, little of basic importance is absent from the open literature.

AIRCRAFT STRUCTURES

The subject of aerospace structures covers an enormously wide field. It is concerned with the design cases, loads and environment which the structure has to withstand; with the strength, stiffness, stability and fatigue life of the structure; with the detail structural design and the materials to be used; and with the testing methods required to establish the structural integrity, life and reliability.

The design cases to be considered are laid down (for British designers) in the *British Civil Airworthiness Requirements*. Design cases for military aircraft are listed in *AVP 970*, a Government publication. Both of these are constantly being updated. Background information to some of these loading cases, especially to the atmospheric gust conditions, is contained in the AGARD monograph *Manual on Aircraft Loads*, by J. Taylor (Pergamon, 1965).

Information on the details of aircraft structures is presented in a very elementary form in *The Aeroplane Structure*, by A. C. Kermode (2nd edn, Pitman, 1964), and, with considerably more technical content, in *Analysis and Design of Flight Vehicle Structures, by* E. F. Bruhn (Tri-State Offset Company, 1965). The latter contains some aspects of detail structural design, but this is a subject on which very little is actually written. The popular journals, *Aircraft Engineering, Interavia* and *Flight International,* frequently give detailed descriptions of new aeroplanes, and new detail structural design features are sometimes discussed. For the greater part, information of this sort is transmitted by word of mouth and by practical experience in aircraft design offices. The Society of British Aircraft Constructors have produced a *Detail Design Information Handbook* (n.d.), which consists of loose-leaf instructions to draughtsmen on design features to be avoided. ESDU produces an excellent series of *ESDU Data Items* for aircraft structural design purposes, enabling strength and stiffness calculations to be rapidly undertaken. Structural design principles are not included.

The calculation of stresses within aircraft structures is a specialised branch of general stress analysis with a strong emphasis on thin shell theory. Three textbooks of a basic nature have been published or reprinted over the past 10 years, but the general science of aircraft stress analysis has advanced at such a rate that most would-be authors give up the chase. Two of the textbooks, *Aircraft Structures for Engineering Students,* by T. H. G. Megson (Arnold, 1972), and *Theory and Analysis of Flight Structures,* by R. M. Rivello (McGraw-Hill, 1969), are for undergraduates, while Bruhn's (1965) (see above) is for students and practising engineers. The advances into modern matrix methods and finite element methods are covered in such publications as *Matrix Methods of Structural Analysis,* edited by B. F. de Veubeke (Pergamon, 1964); *Recent Advances in Stress Analysis; New Concepts and Techniques and their Practical Application,* published by The Royal Aeronautical Society for the Joint British Committee for Stress Analysis (1968); and *The Finite Element Method in Structural and Continuum Mechanics,* by O. C. Zienkiewicz (McGraw-Hill, 1967). Every 3 years a conference is organised by USAF (United States Air Force), and a substantial volume of the proceedings is published (e.g. USAF Report No. AFFDL-TR-66-80 for the 1965 conference, and AFFDL-TR-68-150 for the 1968 conference, etc.). These proceedings, naturally, contain extensive lists of references to other works in this area.

Thermal stress analysis (with reference to aeronautical problems) was dealt with in *Thermal Stresses,* by B. E. Gatewood (McGraw-Hill, 1957), and in *Thermal Stress Analysis,* by D. J. Johns (Pergamon, 1965). Information on how the principles thus enunciated are developed for use in flight vehicles is largely confined to internal reports in the firms. The estimation of an aircraft's thermal environment at high speed is still carried out by the method Davies and Monahan (1952) published in the ARC (Aeronautical Research Council) *Current Paper* series. The proceedings of the 1968 symposium at Loughborough University on *High Temperature Effects on Aircraft Structures* is also a useful source of information. Volume 10 in the Pergamon series *Progress in Aeronautical Sciences,* edited by D. Küchemann *et al.* (1970), brings the theoretical position of aero-thermo-elasticity further up to date.

The subject of aero-elasticity was covered by four textbooks in the period 1951–1962, i.e. *Introduction to Airplane Vibration and Flutter,* by R. H. Scanlan and R. Rosenbaum (Macmillan, 1951); *Introduction to the Theory of Aeroelasticity,* by Y. C. Fung (Wiley, 1955); *Aeroelasticity,* by R. L. Bisplinghoff, H. Ashley and R. L. Halfmann (Addison-Wesley, 1955); and *Principles of Aeroelasticity,* by R. L. Bisplinghoff and H. Ashley (Wiley, 1962). Of these, *Aeroelasticity,* by Bisplinghoff, Ashley and Halfmann is now regarded as a standard work, and *Principles of Aeroelasticity,* by Bisplinghoff and Ashley, as a more advanced treatise. The other two are good introductions for students. The AGARD *Manual of Aeroelasticity* constitutes a handbook for workers in industry and contains useful information on all aspects of the subject. It is continually being revised and updated. On the other hand, the textbooks mentioned are tending to become somewhat dated. The aerodynamic aspects of aero-elasticity require extensive knowledge of unsteady flow. This is covered in several advanced texts, e.g. *The Potential Theory of Unsteady Supersonic Flow*, by J. W. Miles (Cambridge University Press, 1957), and 'Unsteady airfoil theory', by A. I. van de Vooren, in *Advances in Applied Mechanics. Vol. V*, edited by H. L. Dryden *et al.* (Academic Press, 1958). The AGARD *Conference Proceedings 80* (1970) provides a good cross-section of recent developments, both theoretical and experimental, in this aerodynamic field.

Fatigue of aeroplanes and their components has caused a proliferation of literature in the last 20 years. The principal information sources are conference proceedings, AGARD publications and reports from the national aeronautical research institutes. A regular series of bi-annual conferences is organised by the International Committee on Aeronautical Fatigue, and the proceedings contain both reviews of recent work and new papers on particular topics. The American Society for Testing and Materials (ASTM) produces annual bibliographies of fatigue publications generally, and periodic special publications on aircraft fatigue and crack propagation. The AGARD proceedings of the *Symposium on Random Load Fatigue* (1972) is one of the most recent sources. Section 4 of *International Conference on Structural Safety and Reliability,* edited by A. M. Freudenthal (Pergamon, 1972), is quite comprehensive on aircraft fatigue. The *ESDU Data Items* present information on fatigue for practical design purposes.

AEROSPACE MATERIALS

A broad view of this subject may be gained from books such as *Space Materials Handbook,* edited by C. G. Goetzel *et al.* (Addison-Wesley, 1965), or *Materials in Space Technology,* edited by G. V. E. Thompson and K. W. Gatland (Iliffe, 1963). The four volumes of *Aerospace Structural Metals Handbook,* from the Mechanical Properties Data Center, Belfour Stulem Inc., Michigan, USA, give much information and are being continuously updated. Much space in these volumes is devoted to the aluminium alloys and steels which still form the bulk of aerospace structural materials. Information on titanium is contained in *Aircraft Designers' Handbook for Ti and Ti Alloys* (USAF Report AFML-TR-67-142).

Up-to-date lists of material properties for designers are found in the AGARD *Materials Properties Handbook* and in the ASTM *Data Series*. Most aircraft manufacturers have their own information handbooks of material properties. Composite materials (i.e. fibre-reinforced plastics) are well introduced in L. N. Phillips' chapter on 'Composites' in *Polymer Science*, edited by A. D. Jenkins (North-Holland, 1972). Information on carbon fibres is contained in the *Proceedings of the First Carbon Fibre Conference* (Plastics Institute, 1971). *Design in Composite Materials* (IMechE, 1973) covers the design philosophy. Plastics of many different types are used in aerospace engineering. *Polymers in Space Research*, edited by C. L. Segal *et al.* (Dekker, 1970), covers thermally stable polymers, polymers at low temperature and solid propellants. Likewise, *Chemistry in Space Research*, by R. F. Landel and A. Rembaum (Elsevier, 1970), covers much the same ground. Many monographs of the Plastics Institute, the *Encyclopedia of Polymer Science and Technology*, edited by H. F. Mark *et al.* (16 vols., Wiley, 1964–1973), and the *Polymer Science and Technology Series* (Interscience) have sections relevant to aerospace technology. Annual symposia are held by the Society of the Plastics Industry, Society of Plastics Engineers, Society of Aerospace Material and Process Engineers, and normally contain papers of interest to aerospace engineers. In particular, see the *Proceedings of the Symposium on Reinforced Plastics for Aerospace Applications* (Plastics Institute, 1973).

V/STOL AIRCRAFT

V/STOL aircraft is a sub-system within aeronautical engineering. The topic naturally breaks into two sections, namely STOL (Short Take-Off and Landing) aircraft, which operate with a take-off run which is shorter than for conventional aircraft, and VTOL (Vertical Take-Off and Landing) machines, which have the capability of rising vertically from the ground. In the latter category, the helicopter is the only widespread example, but other classes of machines such as the direct jet lift aircraft of the Harrier class are now gaining increasing importance. Obviously these two classes of aircraft are not mutually exclusive in that almost all VTOL aircraft are being designed and built to perform a short take-off and landing at overload weight.

Design requirements for both of these classes of aircraft are to be found in the Civil Aviation Authority (CAA) publication entitled *British Civil Airworthiness Requirements*, or, for military aircraft, in *AVP 970*. In many cases, particularly appropriate to VTOL machines, the regulations have not been fully framed and in these cases the designer has to agree the limiting cases with CAA in the case of British machines.

STOL aircraft differ from conventional aircraft mainly in the lift systems. Development of more sophisticated mechanical flaps has enabled the designer to reduce take-off length to the order of 2500 ft. Performance of these flap systems is not published in any one reference, but an indication of their performance may be gained by extrapolation of the *ESDU Data Items* on high lift devices and by consulting scientific reports.

A great deal of work in recent years has been concentrated on integrating propulsion with lift systems, to produce higher lift coefficient. Such work has resulted, for example, in the development of the jet flap and the augmentor wing. A useful introduction to such topics is given in *Boundary Layer and Flow Control,* edited by G. V. Lachmann (2 vols., Pergamon, 1961). A further discussion of some of these topics is contained in *Aerodynamics of V/STOL Flight,* by B. W. McCormick (Academic Press, 1967). The importance of the propulsion system to the efficiency and integrity of these integrated systems cannot be over-stressed. No single publication deals comprehensively with the problems of engine lift system integration.

In the VTOL case, most literature is naturally associated with the rotary winged aircraft, of which the helicopter is the principal example. An excellent introduction to the helicopter is given in *Aerodynamics of the Helicopter,* by A. Gessow and G. C. Myers (Macmillan, 1952). Unfortunately, this book has not been revised recently, but it is extremely accurate and very readable. A later publication, *Helicopter Dynamics and Aerodynamics,* by P. R. Payne (Pitman, 1959), has a very extensive review of the literature, but unfortunately is marred by a large number of errors. The translation of the Russian book *Helicopters,* edited by M. L. Mil (1957), has been undertaken by NASA (National Aeronautics and Space Administration) and two volumes are available (NASA TTF-494 and TTF 519). These latter publications are extremely comprehensive and well worth study by the serious student.

The engineering of the helicopter has been dealt with in two publications, *Helicopter Engineering,* by R. A. Young (Ronald Press, 1955), and *Principles of Helicopter Engineering,* by J. S. Shapiro (Temple Press, 1955).

Conferences are organised regularly on V/STOL subjects. The main organising bodies are The Royal Aeronautical Society, AIAA (American Institute of Aeronautics and Astronautics), American Helicopter Society, ICAS (International Council of Aeronautical

Sciences, NASA, AGARD and NATO (North Atlantic Treaty Organisation). Many of the papers given at these conferences may be obtained from the sponsoring bodies in published form subsequent to the conference.

MECHANICS OF FLIGHT

Flight mechanics may be defined as the dynamics and statics of an aeroplane as a flexible body having finite strength being propelled and manoeuvred in the atmospheric environment. It ranges from the selection and implementation of flight profile specifications and requirements, through vehicle layout and design to the identification of critical manoeuvre and other loading cases used in structural design.

It is possible to distinguish between the following areas of flight mechanics: performance; stability and control. The literature survey is included under these headings. Although they overlap with flight mechanics, the following subjects have been specifically excluded since they are considered in the other sections: celestial and orbital mechanics; aeroelasticity; component loading or loading actions.

A large body of information on flight mechanics is available in the form of textbooks, manuals, requirements and specifications, journals, conference proceedings, lecture series, reports, current papers, memoranda, etc. However, it should be appreciated that many industrial and government organisations will have developed in-house methods for solving problems of flight mechanics, which may or may not have general availability in the form of reports. This process has been intensified by the use of large-scale digital computer programs in such areas as layout, design and performance optimisation and sensitivity studies. The literature surveyed here should therefore be regarded as introductory, but also as useful in establishing, understanding and developing the principles of flight mechanics.

General literature

The following bibliography contains a selection of books which describe aircraft, their design philosophies and their behaviour with, on the whole, a minimum resort to mathematical formulations:

An Introduction to Aeronautical Engineering, by A. C. Kermode (7th edn, Pitman, 1962)

An Introduction to the Dynamics of Airplanes, by H. N. Abramson
(Ronald Press, 1958)
Flight Handbook: Theory and Practice of Powered Flight, edited by
W. T. Gunston (Aero., 1972)
Flight Without Formulae, by A. C. Kermode (3rd edn, Pitman,
1960)
Mastery of the Air: an Account of the Science of Mechanical Flight,
by O. G. Sutton (Basic Books, 1966)
The Anatomy of the Aeroplane, by D. Stinton (Foulis, 1966)
The Science of Flight, by O. G. Sutton (2nd edn, Penguin, 1955)

Performance literature

The basic principles of performance estimation are contained in *The
Performance of Civil Aircraft,* by F. B. Baker (Pitman, 1950), and
Airplane Performance, Stability and Control, by C. D. Perkins and
R. E. Hage (Wiley, 1949). Some aspects have been updated and
included in a broad coverage of aeronautical topics in *Aerodynamics
for Engineering Students,* by E. L. Houghton and A. E. Brock
(Arnold, 1960). At this time, the most modern and authoritative work
is *Aircraft Performance: Prediction Methods and Optimization*
(NATO, 1973).
  Detailed information and data required in performance estimation
and reduction are contained in *Airplane Design,* by G. Corning
(Bailey and Swinfen, 1953), and in the following *ESDU Data Items
Sub-Series:*

Aerodynamics	(11 vols.)
Transonics Aerodynamics	(2 vols.)
Performance	(3 vols.)

  Information about propulsive units can be obtained from the
engine manufacturers in the form of brochures, specifications and
performance characteristics. It may be specific to an individual
power plant or more general in nature.
  Performance bibliography also includes *Airplane Aerodynamics,*
by D. O. Dommasch, S. S. Sherby and T. F. Connolly (4th edn,
Pitman, 1967); *Aerodynamic Theory Vols. V and VI,* edited by W.
F. Durand (Springer-Verlag, 1936); *Theory of Flight,* by R. Von
Mises (Dover, 1959); and *Aerodynamics of V/STOL Flight,* by B.
W. McCormick (Academic Press, 1967). Flight testing aspects of
performance are discussed in volumes I, III and IV of the *AGARD
Flight Test Manual* (4 vols.) (NATO, various datings). Performance

requirements are presented in the *British Civil Airworthiness Requirements* (Air Registration Board, now the Civil Aviation Authority (CAA)), and their military equivalent.

Stability and control literature

Also included under this heading are subjects such as gust and turbulence response, handling qualities and stability augmentation. The latter topic has an interface with automatic control and the inherent flight vehicle dynamic behaviour.

All the textbooks in this field are excellent and some are outstanding. They range in approach from the applied *Aircraft Stability and Control for Pilots and Engineers,* by B. Dickinson (Pitman, 1968), to the mathematical *Flight Stability and Control,* by T. Hacker (Elsevier, 1970). There are differences between American and British notations for quantities such as stability derivatives. Books using the US notation are:

Airplane Performance, Stability and Control, by C. D. Perkins and
　R. E. Hage (Wiley, 1949)
Automatic Control of Aircraft and Missiles, by J. H. Blakelock
　(Wiley, 1965)
Dynamics of Atmospheric Flight, by B. Etkin (Wiley, 1972)
Modern Flight Dynamics, by W. R. Kolk (Prentice-Hall, 1961)
Stability and Control of Airplanes and Helicopters, by E. Seckel
　(Academic Press, 1964)

The UK notation introduced by Gates is used in the following:

Aircraft Stability and Control, by A. W. Babister (Pergamon,
　1961)
*An Introduction to the Longtitudinal Static Stability of Low-Speed
　Aircraft,* by F. G. Irving (Pergamon, 1966)
ESDU Data Items. Aerodynamics Sub-Series. Vol. 3
The Principles of the Control and Stability of Aircraft, by W. J.
　Duncan (Cambridge University Press, 1952)

The rationalised systems of notation of Hopkin are described in the *ESDU Data Items. Dynamics Sub-Series.*

Other sources of information on stability and control are *High Speed Aerodynamics and Jet Propulsion. Vols. VII and VIII,* edited by A. F. Donovan and H. R. Lawrence (Princeton University Press, 1957), and the AGARD *Flight Test Manual* (1959). Specialist material is contained in various AGARD series — for example, AGARD *CP 17 Stability and Control* (2 vols., NATO, 1967).

AVIONICS

In modern aircraft there is a marriage between electronics and aviation which has led to the coining of the word 'avionics'. This has become an important part of total aircraft technology and is a field which is developing as rapidly as the reliability of the hardware and the understanding of the man–machine interface allow. The trend towards automatic flight relies upon the development of airborne equipment and ground equipment. Avionics equipment will appear in the form of navigation, air traffic control and landing aids on the ground, while in the air there are, in addition to the associated airborne computers, inertial navigators, data displays etc.

With this very wide coverage, the literature on avionics is dispersed through the media dealing with the constituent parts, viz. aeronautics, electronics and automatic control. Up-to-date literature is best obtained by consulting the relevant abstracting and indexing journals mentioned in the final section of this chapter and in Chapter 11.

The learned societies publishing papers on avionics are those most concerned with the separate disciplines, i.e. AIAA, Royal Aeronautical Society, IEEE (Institute of Electrical and Electronics Engineers), IEE (Institution of Electrical Engineers), IERE (Institution of Electronic and Radio Engineers).

Reference should also be made to conference proceedings published by the above societies and by AGARD and IFAC (International Federation of Automatic Control).

Textbooks relevant to avionics are largely dealt with in the specialist subjects, and, in these, applications to aircraft are sometimes given. A relevant text on automatic control is *Automatic Control of Aircraft and Missiles*, by J. H. Blakelock (Wiley, 1965).

ASTRONAUTICS

The scientific and technical communications explosion of the 1960s is particularly apparent in the field of astronautics, which benefited from the stimulus of massive US Government funding and imaginative and challenging technical goals. More recently, however, the changing political and economic climates have led to a shift in emphasis from planetary exploration to near-earth operations and a change in launch strategy to one of reusability. The commercial and scientific exploitation of satellite technology for communications and surveillance, both scientific (earth resources, astronomical) and military, and a consolidation of the techniques acquired during the Apollo programme, notably in space medicine, are anticipated. Textbooks as a primary literature source, have not, in general, adjusted to this change in environment.

The field perhaps least influenced by the changes suggested is that shared with astronomy, notably celestial mechanics. The traditional mathematical treatment is illustrated by *An Introduction to Celestial Mechanics,* by F. R. Moulton (2nd edn, Macmillan, 1958). Some of the newer mathematical techniques and an awareness of the artificial satellite role, with corresponding emphasis on mission and navigational applications orbit prediction and determination from observation, can be found in *Astrodynamics,* by S. Herrick (2 vols., Van Nostrand, 1971–1972).

Of the very many texts which have attempted to cover a broad spectrum of techniques having space applications, from, for example, space medicine to propulsion and guidance albeit superficially, useful introductory examples are provided by *Space Technology,* edited by H. S. Seifert (Wiley, 1959), and *Sourcebook on the Space Sciences, by S. Glasstone* (Van Nostrand, 1965). Illustrative of the diversity of topic and volume of literature published in the last decade are the 34 volumes in the *Space Technology Series* published by Prentice-Hall. Each is devoted to a comparatively narrow specialist area—for example, *Space Communications Systems,* by R. F. Filipowsky and E. I. Muehldorf (1965). In view of the pace of technical development, as well as changing aspirations, textbooks cannot provide information on the current status of technologies and should be viewed essentially as background material sources. Only the published proceedings of specialist conferences, research agency reports and journals will provide current awareness.

Fundamental astronomical data for observational and navigational purposes is available in the annual publications:

American Ephemeris and Nautical Almanac (USGPO)
Astronomical Ephemeris (HMSO)

Specialist conference proceedings appear regularly in the form:

Advances in Astronautical Sciences
Progress in Aeronautics and Astronautics
AGARDograph series
COSPAR Space Research
International Astronautical Congress

REPORTS AND JOURNALS

The lists of textbooks in aeronautics and astronautics is very extensive and only a few of the more important ones are listed and referred to above. However, many recent developments are not found in textbooks, and reference must be made to scientific and technical report series and journals.

In searching for reports and papers on current work, reference should be made to the following bibliographical sources:

Scientific and Technical Aerospace Reports (STAR)
International Aerospace Abstracts (IAA)
R & D Abstracts

Advance information on *STAR* and *IAA* is provided by the *NASA/SCAN Notification* under numbered categories. A computer search of the literature is available from the ESRIN Space Documentation Service, Via Galileo Galilei, 000444 Frascati, Italy. In the UK, the terminal link at the Technology Reports Centre may be used for search requests.

Scientific report series dealing with aeronautics and astronautics include:

Aeronautical Research Council (ARC) London. Reports and Memoranda (R & M) and Current Papers (CP)
National Aeronautics and Space Administration, Washington, *NASA (NACA up to 1958)*. Technical reports, notes and memoranda. Translations, Contractors Reports and Special Publications including bibliographies, data compilation and conference proceedings.
Royal Aircraft Establishment, National Gas Turbine Establishment and Rocket Propulsion Establishment. Technical reports, notes and memoranda. Distributed by the Technical Reports Centre at St Mary Cray, Orpington, Kent, BR5 3RF.
The Advisory Group for Aeronautical Research and Development (AGARD) issue state-of-the-art reviews (*AGARDographs*) and *Conference Proceedings (CP)* AGARD publications in the UK are distributed by the Defence Research Information Centre at St. Mary Cray.
The International Civil Aviation Organisation (ICAO) issues reports on its conventions dealing with the regulation of international air transport and navigation. It also issues digest and statistical information on air traffic etc.
The Institut du Transport Aerien (ITA) publishes a regular *Bulletin* and issues reports in the field of socio-economics of air transport.

Aeronautical report series are also available from the countries listed below.

Australia	Aeronautical Research Laboratories (ARL)
	Weapons Research Establishment (WRE)
	Commonwealth Scientific and Industrial Research Organisation (CSIRO)
Canada	National Research Council (NRC)
	National Aeronautical Establishment (NAE)
France	Office National d'Etudes et de Recherches Aerospatiales (ONERA)
	Centre National de la Recherche Scientifique (CNRS)
Germany	Deutsche Forschungs—Und Versuchsanstalt für Luft und Raumfahrt (DLR)
Holland	Nationaal Lucht—en Ruimtevaartlaboratorium (NLR)
Sweden	Flygtekniska Forsoksanstalten (FFA)
India	National Aeronautical Laboratory (NAL)
	Council for Scientific and Industrial Research (CSIR)
	Aeronautical Research Committee (ARC)
Japan	National Aerospace Laboratory (NAL)
South Africa	Council for Scientific and Industrial Research (CSIR)
Belgium	Von Karman Institute for Fluid Dynamics (VKIFD)
Russia	Academy of Science. Union of Soviet Socialist Republics (ASUSSR)

Reports are also issued by most aeronautical departments in universities throughout the world.

Basic research in aeronautics and astronautics is kept under review in the following serial publications:

Advances in Aeronautical Sciences (American Institute of Aeronautics and Astronautics, 1958–   )

Advances in Space Science and Technology (Academic Press, 1959–   )

Advances in the Astronautical Sciences (American Astronautical Society, 1957–   )

Annual Review of Fluid Mechanics (Annual Reviews Inc., 1969–   )

International Symposium on Combustion (Combustion Institute, 1954–   )

Progress in Aerospace Sciences (Pergamon, 1961–), formerly *Progress in Aeronautical Sciences*

Progress in Astronautical and Aeronautics (Academic Press, 1960–). Some issues published under the title *Progress in Astronautics and Rocketry.*

Journals dealing with the topics considered in the previous sections include:

Aeronautical Society of India. Journal
AIAA Journal
Aircraft Engineering
American Helicopter Society. Journal
Astronautica Acta
Aviation Week and Space Technology
BIIL/Biuletyn Informacyjny Instytutu Lotnictwa
British Interplanetary Society. Journal
Canadian Aeronautics and Space Journal
CASI Transactions
Combustion and Flame
Combustion Science and Technology
Composites
Computers and Structures
DFVLR-Nachrichten
ESRO/ELDRO-CERS/CECLES Bulletin
Flight International
Gas and Oil Power
Helicopter and VTOL World
ICAO Bulletin
Interavia—World Review of Aviation
International Journal of Solids and Structures
Japan Society for Aeronautical and Space Sciences. Journal
Journal of Aircraft
Journal of Fluid Mechanics
Journal of Mechanical Engineering Science
Journal of Spacecraft and Rockets
Materials Engineering
Physics of Fluids
Plastics and Polymers
Recherche Aerospatiale
SAMPE Journal
Soviet Aeronautics
The Aeronautical Journal
The Aeronautical Quarterly

Transactions of the ASME. Series A. Journal of Engineering for Power

Transactions of the ASME. Series E. Journal of Applied Mechanics

Transactions of the ASME. Series F. Journal of Lubrication Technology

20

Chemical engineering

L. B. Cousins

During this century there has been a tremendous increase in a fourth primary technology, namely chemical engineering, based on chemistry and classical physics, which has led to the scientific design of process engineering equipment. The other primary engineering technologies—civil, mechanical and electrical engineering—were already established before the end of the nineteenth century. Two papers by Gregory (1972) and Peck (1973) trace the history of the chemical industry and provide further reading.

Danckwerts (1972) defines chemical engineering as 'that branch of engineering concerned with processes which change the composition or properties of matter in bulk'. Chemical engineers are involved in the control of chemical and physical changes that occur during manufacture, and these factors determine the type of engineering process and equipment to be installed. They are concerned with design procedures, so that full-scale plant or equipment can be designed by calculation, and with improvements in the efficiency of processes or process plant. The more precisely the design procedures can be developed, the more precisely it is possible to optimise the design of plant.

Hougen (1965) shows that the functional activities of the chemical engineer are very wide. Chemical engineers are employed in research and development, process design and process engineering, computer programming, pilot plant experiments, selection and design of equipment, process control, plant operation, sales, marketing, management and administration. Much of chemical engineering science is concerned with scaling-up, automation and the production of a theoretical model of a process or flow in a plant, thus eliminating expensive pilot plant experiments.

The interests of chemical engineers are very varied. Some of their major topics of interest are:

methods for the estimating thermodynamic quantities in systems;

interfacial phenomena, particularly with relation to the formation and behaviour of foams, drops and bubbles, etc., in mass transfer equipment;

catalysis—particular with relation to the design and operation of a chemical reactor;

fluid mechanics—the study and the flow of single-phase, two-phase and multiphase gas/liquid/solid systems in various shapes and sizes of plant;

fluidisation;

non-Newtonian fluids;

mixing of fluids;

in heat and mass transfer processes, e.g. distillation adsorption of gases, desalination;

solids handling—including crushing, grinding, and mechanics of flow of fine powders;

application of computers in design and control theory;

development of procedures for optimising the design of single items of plant and of complete process plant;

design of chemical plant and machinery, e.g. process vessels and pumps.

In the past, most chemical engineers have been employed in the chemical and petroleum industry. They are now making inroads into other specialised branches of science and technology. According to Danckwerts (1972), the proportion of chemical engineers employed in the non-chemical industries in Britain has risen to over 25%. Chemical engineers are required for work on combustion engineering, gas engineering, food processing and extraction metallurgy, in electrical power generation and civil engineering and on environmental problems, such as pollution, the treatment of sewage and sludge and the reduction of noise.

A paper by Constan (1973) previews the 75th national AIChE meeting held in June 1973. The three main fields of interest that dominated this meeting were the environment, food and bioengineering. On the environment, the complete spectrum of water, air and solids was well covered.

Marshall (1973) in a plenary address to First Pacific Chemical Engineering Congress, October 1972, Kyoto, Japan, has predicted problems of society which challenge chemical engineering in the last three decades of the twentieth century. He sees the main problems confronting engineers as: environmental pollution control; the energy crisis; urban transport and emission control; and food production, health care, crime control and urban renewal. He considers that chemical engineers are well equipped with their knowledge of separation processes, reaction kinetics, analysis and synthesis of processes to overcome some of these problems.

Thus, the chemical engineering profession has a wide range of disciplines and a chemical engineer can be employed in a variety of posts from a research scientist to marketing, sales and management. Similarly, the information in which the chemical engineer is interested is dispersed very widely throughout the scientific and technological literature. A good proportion of the published literature is very readily recognised as being on chemical engineering topics, but there is also a large proportion of other literature which contains some articles which are of interest to chemical engineers. For instance, many articles of interest are to be found in the mechanical engineering literature, particularly that presented by the various national mechanical engineering societies.

The report literature is often a valuable source of data for the practising engineer and scientist, because many company and research reports contain all of the reliable experimental data obtained in tabulated form or in good graphical form. Often, when a report is published in a journal, the experimental and numerical data are condensed to meet the editor's requirement. However, many company reports are written confidentially and are not therefore available for general use.

PRIMARY SOURCES OF BRITISH LITERATURE

The bulk of the published British literature is produced by individual organisations, i.e. by publishing houses, by the research organisations, by various institutes and by the universities.

Two journals which cover marketing and company matters are *European Chemical News* and *Chemical Age International*. *European Chemical News* is published weekly by the International Publishing Corporation (IPC) and provides details of marketing information on chemicals. It summarises new chemical plant projects, product distribution and future contracts, and presents topical news items regarding new technical products and company finance. In it are listed European chemical prices, new plant and equipment and some sources of supply of chemical equipment. *European Chemical News* often includes surveys of chemical plant manufacture, chemical plant contractors and principal chemical companies in a particular country. Also with this journal is published an occasional supplement in its *Chemscope* series. For instance, on 15th June 1973 a survey was published on the Japanese chemical industry and the manner in which it is adapting to a changing world. This survey included the role of the big trading companies in Japan, petrochemicals, finance and the recovery of profits, export of engineering services and plants, industrial relations, research and development, energy resources and the outlook for chemical engineering. *Chemical Age International*, which is published weekly by Benn, includes news items on the chemical and processing industry, information on products, technology, company markets, company finance, plant under construction and new patents on chemical processes, and short reviews on progress in a particular industry. It regularly includes surveys on the projects under construction in the petroleum refinery, the chemical and the allied industries in specific countries. See, for example, the survey on Italian Projects in the 27 April 1973 issue. This survey includes the total value of new plant, the expansion and investment. Annually the journal surveys the top 200 companies world-wide.

The technical page of the *Financial Times* also produces news items about new products and new ideas which are of interest to chemical engineers.

Another journal which includes company and industrial news is *Chemistry and Industry,* which is published twice monthly by the Society of Chemistry and Industry, London. This journal publishes technical articles which are primarily on chemical subjects, but many of the articles, news items, equipment notes and technical sales literature are of interest to chemical engineers. It also lists recent publications and sales literature.

Process Engineering, which recently incorporated *Chemical and Process Engineering,* is published monthly by Morgan & Grampian. It is produced primarily for technical personnel responsible for the construction, operation and maintenance of process plant, control equipment and systems. It includes news items, information on processes, projects, equipment control services, materials, new technical literature and longer articles. An annual review on heat transfer is published in August each year under the title of *CPE Heat Transfer Survey.*

A similar journal to *Process Engineering* is *Processing,* which is published monthly by IPC. This publication has recently incorporated the journal *Process Technology International* (formerly *British Chemical Engineering*) and publishes monthly scientific and technical papers in chemical engineering and process engineering. It includes special reports and papers on processes, plant, management and technical data. Many issues contain a nomogram of use to chemical engineers. New patents and some abstracts from international journals covering many chemical engineering topics are listed. The journal also contains a diary of world events in the field, news items and comment.

The other journals reviewed in this section are primarily more scientific in nature than the previous ones mentioned.

Two very good journals published monthly by Pergamon are *Chemical Engineering Science* and the *International Journal of Heat and Mass Transfer*. *Chemical Engineering Science* primarily publishes papers on research, dealing with the application of chemistry, physics and mathematics to chemical engineering. However, the journal is not necessarily concerned with pure science. Topics dealt with range from general principles to particular industrial processes, and include the development of new chemical processes and plant design. Some of the areas covered include: chemical engineering aspects of fluid mechanics, applied reaction kinetics, process control mechanics and statistics of mixing processes, mechanics of process equipment and machinery, optimisation and computer-aided design. The journal includes shorter communications, book reviews and, occasionally, current Soviet papers of interest. Papers are published in English, French or German. Descriptive articles concerning existing industrial processes and plant are not considered to be appropriate to the journal. The *International Journal of Heat and Mass Transfer* only covers part of the chemical engineering field. The aims and scope of the journal are to provide a means for exchange of basic ideas in heat and mass transfer between research workers and engineers located throughout the world. Emphasis is placed on original research, both analytical and experimental, to increase the basic understanding of transfer processes and their application to engineering problems. The journal regularly contains reviews and bibliographies on recent papers in heat and mass transfer, and reviews new books. Papers are usually in English, but the journal also includes articles in French or German or Russian, with abstracts in each of these four languages.

Pergamon also produce an occasional publication, *Progress in Heat and Mass Transfer,* which is primarily devoted to selected papers from important conferences. It is a major reference source of heat and mass transfer and closely related fields.

Another particularly relevant journal which is published every other month by Elsevier is *The Chemical Engineering Journal*. This periodical prints papers on original research and development work, interpretative reviews and discussions on the latest developments in chemical engineering. It includes short communications and book reviews. Elsevier also publish the international quarterly

Environmental Pollution. It contains mainly research papers containing original results, surveys of major pollution topics and reviews of new books in the field.

The monthly international journal *Powder Technology* publishes articles on all aspects of the formation of particles and their characterisation and on the study of systems containing particulate solids. Articles include research papers, reviews, short communications, letters and book reviews.

The Institution of Chemical Engineers produce three separate periodicals: *The Chemical Engineer,* which is published monthly; *The Transactions,* which are published quarterly; and *The Symposium Series,* which is published as required. *The Chemical Engineer* includes topical news and notes about the Institution, about industry and about general activities. It deals with the practical application of chemical engineering science and informs the engineer of changes and progress in his technology and profession. Papers published in the *Transactions* are original contributions to chemical engineering knowledge. They include results of research or experimental work and new developments of plant or processes. The *Symposium Series* records the papers and discussion at meetings organised by the Institution.

The mechanical engineering literature also provides a useful source of information for chemical engineers. The Institution of Mechanical Engineers produce annually a number of volumes entitled the *Proceedings of the Institution of Mechanical Engineers;* the biannual journal *Heat and Fluid Flow,* consisting of selected papers sponsored by the Thermodynamics and Fluid Mechanics Groups of the Institution; and the bi-monthly *Journal of Mechanical Engineering Science,* which contains some theoretical papers of interest to many chemical engineers.

Some British universities produce occasional publications. One which often provides useful information is the quarterly *Birmingham University Chemical Engineer.*

Other British journals which are of fringe interest are:

Process Biochemistry, which incorporates *Biochemical Engineering* and is useful therefore to biochemical engineers.

Journal of the Institute of Fuel, Combustion and Flame, Combustion Science and Technology and *Energy Digest* (incorporating the *Journal of Fuel and Heat Technology*), which are all very relevant for chemical engineers who have an interest in combustion.

Cryogenics, for those with interests in refrigeration and low temperatures.

Corrosion Science and *The British Corrosion Journal*

Desalination, which includes useful contributions on heat and mass transfer topics, evaporation, reverse osmosis and other techniques. The various nuclear engineering journals.

In addition to the published literature, the research organisations in the United Kingdom make many of their research reports available for general use. These research organisations include the various UKAEA research centres at Harwell, Winfrith, Culham, Windscale, Risley and Dounreay. Harwell produce reports on the various aspects of chemical engineering, process technology and pollution. Winfrith, Windscale and Dounreay are more inclined to heat and mass transfer in nuclear reactors, and Culham are interested in fusion techniques.

The laboratories of the Departments of Trade and Industry, i.e. The National Engineering Laboratory, the Warren Springs Laboratory and the National Physical Laboratory, all produce reports in this field. The work of interest to chemical engineers, carried out by the National Engineering Laboratory, is primarily in heat transfer and fluid flow and in flow measurements and rheology. That of the Warren Spring Laboratory of interest is in fluid–solids handling, in rheology and environmental pollution, while the National Physical Laboratory are providing physical property data on many pure inorganic and organic compounds used by the chemical and metallurgical industries.

The various research associations also produce and abstract a large amount of appropriate literature. Prominent among these are the British Hydromechanics Research Association, which produces a monthly abstract journal, *Civil Engineering Hydraulics Abstracts* (formerly *Channel*), and various journals on fluidics, hydraulics and fluid engineering, as well as books and bibliographies; the Heating and Ventilation Research Association, which produces the periodical *Heating and Ventilation Engineer* and the abstract journal *Thermal Abstracts;* and the Water Research Centre.

Private industry and government-owned industries, such as the Gas Council, CEGB, the National Coal Board and British Steel, produce many interesting reports, but only a few of them are made available for general use. However, a great deal of very applicable information can be found in the 'trade' literature.

A compilation of chemical research in progress in the UK has been written by Poll (1968). This review indexes the work undertaken by the universities, the research associations, government laboratories and other organisations. Poll has recently completed a new compilation of this research and this is being considered for publication by the Institution of Chemical Engineers, London.

NORTH AMERICAN LITERATURE

The chemical engineering literature published in the USA is dominated by the major societies with interests in this field, namely: the American Institute of Chemical Engineers, the American Society of Mechanical Engineers and the American Chemical Society.

The American Institute of Chemical Engineers (AIChE) is probably the most prolific source of literature on the subject, publishing information which covers the whole spectrum of chemical engineering science. *Chemical Engineering Progress* is published monthly with the intention of recording the latest developments in the field with articles written by practising engineers. These articles, together with a news section, are designed to keep working chemical engineers abreast of new technology and practices. It includes articles on chemical marketing and lists future meetings and courses. This Institute also produces every other month the *AIChE Journal*, which is a fundamental and research-orientated publication devoted to chemical engineering. In addition to these periodicals, the AIChE publish a series of soft-cover books in their symposium and monograph series. Some recent publications in these series are:

Air Pollution and its Control (1972)
Food and Bio-Engineering—Fundamental and Industrial Aspects (1971)
Crystallisation from Solution (1972)
Recent Advances in Separation Techniques (1972)
Vacuum Technology at Low Temperatures (1972)
Fluidized Bed Fundamentals and Applications (1973)
CEP Technical Manual—Sulphur and SO_2 Developments (1971)
Drag Reduction in Polymer Solution (1973)
CEP Technical Manual—Cooling Towers (1971)
Chemical Engineering Computing (1971)
Water (1972)
Industrial Process Design for Pollution Control (1971)

Also, the AIChE have produced a series of volumes in a series entitled 'Loss Prevention' which are a series of articles on cause of failure in process plant, prepared by the editors of *Chemical Engineering Progress*.

The publications of the American Society of Mechanical Engineers (ASME) provide a wealth of knowledge for chemical engineers who are primarily interested in heat transfer, fluid dynamics and power generation. They publish a monthly journal *Mechanical Engineering*, which contains one or two technical review-type papers, usually on power generation or mechanical engineering. This journal gives news items of the Society and includes abstracts of papers to be given at forthcoming symposia. ASME also produce journals for basic research and fundamental studies in their *Transactions of the ASME* series. Many of the papers reproduced in these journals have previously been presented at ASME Symposia. These journals include: *Journal of Heat Transfer, Journal of Applied Mechanics, Journal of Engineering for Power, Journal of Engineering for Industry and Journal of Fluids Engineering*, which replaced *Journal of Basic Engineering* in March 1973. All of these are published quarterly. In addition to these journals, the ASME produce a monthly journal, *Applied Mechanics Reviews*. This periodical often contains a review article, but foremost it is an abstracting journal. The abstracts are extremely well presented, but the literature covered is by no means comprehensive.

The AIChE and the ASME organise each year, jointly and independently, a large number of seminars and symposia. Many preprints of papers given at these symposia are available on sale from these two organisations.

The American Chemical Society (ACS) publish many journals of interest to the chemical engineering profession. *Chemical and Engineering News* is the official publication of this Society. It is designed to keep members informed of policies and activities of the Society and of events and trends in the chemical world. Weekly issues contain short news stories, longer trend stories and features on topics of broad interest related to the chemical world. A bound annual index is available.

The ACS monthly *Chemtech* is orientated towards the chemist and the chemical engineer. It not only covers preparation, characterisation and use of materials, but also engineering design operation and distribution, and it is concerned with the problems of the technologist.

Scientists and engineers engaged in the study and maintenance of the natural environment through the use of chemical principles would be interested in the monthly journal *Environmental Science and Technology*. In this journal papers on fundamental research and technology in water, air and waste chemistry, news reports and political industrial aspects of environmental management are printed.

Of interest to chemical engineers in the food industry is the *Journal of Agricultural and Food Chemistry*. The subjects covered in this bi-monthly journal include pesticides, chemistry of food processing, biochemistry of nutrition, chemistry of flavours and compounds isolated from food products.

The three most significant journals of interest to chemical engineers which are published by the ACS are the quarterly publications in the Industrial Engineering Chemistry (I and EC) series, namely: *I and EC—Process Design and Development, I and EC—Product R and D* and *I and EC—Fundamentals*. The *I and EC Process Design and Development* reports orginal work on design methods, concepts and applications to the development of process and process equipment. It includes empirical or semi-theoretical correlations of data, determination of design parameters, methods of integrating systems analysis and process control into process design and development, scale-up procedures and other experimental process development techniques *I and EC—Product R and D* publishes papers applicable to the preparation and properties of chemicals with emphasis on reactions and mechanism studies relevant to the preparation of chemical products. The *I and EC—Fundamentals* is devoted to original scientific papers dealing with the frontiers of chemical engineering science. Those aspects of physical and chemical phenomena are covered which can lead to improved design, more precise mathematical description of future events and more profitable engineering. Papers are selected on the basis that they contain significant conclusions and not just recorded data.

Another publication in the Industrial and Engineering Chemistry series of interest is the *Annual Reviews of Industrial and Engineering Chemistry*, which is published in book form by the American Chemical Society. The first such review, dated 1970, was published in 1972. Chapters in this review covered a broad range of topics in industrial chemistry and chemical engineering from heat and mass transfer, distillation, applied mathematics and process control to plastics and crystallisation.

Two ACS quarterly journals are concerned with chemical and physical data. The *Journal of Chemical and Engineering Data* includes experimental data of interest in the fields of chemistry and chemical engineering, while the *Journal of Physical and Chemical Reference Data* presents critically evaluated compilations on physical and chemical properties.

McGraw-Hill produce fortnightly the very practical journal *Chemical Engineering*. This journal is subtitled 'Chemical Technology for Profit-Minded Engineers', which fully describes its outlook. As well as providing a forum for technical papers on the whole aspect of chemical engineering and articles of careers in chemical engineering, it contains a commentary on events and news in chemical engineering. It includes international news, news features, new products and services, engineering features, manufacturers' literature, operation and maintenance of plant and many useful advertisements. An occasional supplement is published with this journal. It is entitled

Chemical Engineering DESK BOOK and is devoted to one aspect of chemical engineering technology. For instance, the issue dated 18 June 1973 was devoted to a survey of Environmental Engineering and included sections on the Laws and Regulations, Pollution Control Technology, Inplant Pollution Control and a Buyers Guide. The section on Pollution Control Technology contained an article on information sources on environmental pollution by Bennett (1973).

A journal of considerable value to the practising process and design engineer is *Hydrocarbon Processing,* which is published monthly by the Gulf Publishing Company. This journal contains many practical articles of interest to chemical engineers working in or for the hydrocarbon processing industry. Special reports, other technical features, management guidelines and safety guidelines are included. It also contains many helpful advertisements and useful physical property data.

The Canadian Journal of Chemical Engineering is published bimonthly by the Canadian Society for Chemical Engineering. It publishes original research, new theoretical interpretation and critical reviews in the science or industrial practice of chemical engineering or applied chemistry. Full-length articles, communications, notes or letters are published in English or French.

Other North American journals of specific interest to some chemical engineers are:

ASHRAE Journal
Advances in Cryogenic Engineering
Cryogenics and Industrial Gases
*Journal of Research of the National
    Bureau of Standards, Part B,
    Chemistry and Physics*
} all of use to engineers working in the low-temperature or refrigeration field

Anti-Corrosion
Corrosion
Materials Protection and Performance
Protection of Metals
Wear
} for those engineers with corrosion problems

Combustion
} for those with interests in power systems

Pipeline and Gas Journal
Oil and Gas Journal
Oil, Gas and Petrochem Equipment
Journal of Petroleum Technology
} for those working in the oil, gas and petroleum industry

American Nuclear Society Transactions
Advances in Heat Transfer
Annual Review of Fluid Mechanics
Proceedings of the Heat Transfer and
   Fluid Mechanics Institute
} for those with interests in heat transfer and fluid mechanics

Food Manufacture
Journal of Applied Polymer Science
Journal of Acoustical Society of
   America
Optics and Laser Technology
} for those with more specialised interests

Research establishments, universities and some industrial companies in the USA and Canada produce many reports in specific areas of chemical engineering science which are generally available. Some of the major sources of information include: Atomic Energy of Canada Ltd, Atomics International, Dow Chemicals, General Electric, Goodyear Atomic Corporation, Gulf Atomics, Massachusetts Institute of Technology, NASA, National Bureau of Standards, Noyes Data Corporation, Oak Ridge National Laboratory, Union Carbide, University of California, University of Cincinnati, University of Michigan, University of Houston, etc., and the United States Atomic Energy Commission.

OTHER PRIMARY SOURCES OF PUBLISHED LITERATURE

There are a vast number of publications which could be included in this section. However, only those which the author has personally found most relevant to the subject have been included. For those who wish to delve further, details of many of the other publications can be found in some of the abstracting literature and in extensive lists of published information.

The main German chemical engineering publication is *Chemie Ingenieur Technik*, which is published monthly. Articles are in German with abstracts in German and English. This journal covers the whole range of industrial chemistry and chemical engineering science.

The leading journal of the German chemical trade and industry is *Chemische Industrie*. It is directed to the needs of the executive as well as the chemical engineer. It provides information on home and international market trends, production procedures and processes. The same company publishes the quarterly English-edition journal *Chemische Industrie International*, which is also devoted to the market of chemical goods, auxiliaries and production aids. It includes regular world market feature articles on the principal chemical and plastics products, advanced processes and manufacturing techniques.

For those chemical engineers who are working on energy power generation and heat transfer, a very useful journal is *Brennstoff – Wärme–Kraft*, which publishes, monthly, the articles in German and each abstract in English and German. Another journal of interest to the heat transfer and fluid dynamics expert is *Wärme–und Stoffübertragung*, which is published monthly. This journal covers recent knowledge on scientific principles of the transport processes of heat and mass as well as allied material properties. It often includes articles on condensation, heat exchanger design, boiling, drying and natural convection. Articles are in English or German.

The monthly *Chemische Technik* publishes papers of original scientific work, training, selection of equipment, costs, automation and reliability of chemical plants. Those with interests in cryogenics and refrigeration should note the journals *Luft–und Kaltetechnik* and *Kaltetech Klimatisierung*.

The major French journal in this field is *Chimie et Industrie— Genie Chimique,* which is published bi-monthly. Most articles are in French. This journal covers the whole field of chemical engineering and chemical industry. A French journal of interest to the combustion energy and thermal industry is *Revue Generale de Thermique*. This journal is published monthly; articles are in French with abstracts in six languages. Chemical engineers who are working in refrigeration and food technology would be interested in the publications of the Institut International du Froid, who publish a regular *Bulletin* and an *Annexe au Bulletin,* which includes proceedings of meetings and organised by this Institute. For those engineers with interests in petroleum and natural gas, the journal *Revue Institut Francais du Petrole* is very useful.

From Poland the journals *Pregemysl Chemiczmy, Bulletin Acad. Polon. Science* and *Inzynieria Chemiezna* all contain the odd article of interest to chemical engineers in each issue. The Czechoslovakian journal *Collection of Czechoslovak Chemical Communications,* written in English, includes a chemical engineering section in most issues.

The Japanese journals that include a large number of scientific and technical papers are *Journal of Chemical Engineering of Japan* which covers all aspects of chemical engineering science, and *Bulletin of the JSME,* which includes articles on thermal and fluid transport. From India the publications of interest are *Indian Chemical Engineer, Indian Journal of Technology* and *Chemical*

Engineering World. The publication from Australia worthy of inclusion is *Mechanical and Chemical Engineering Transactions,* which is published by the Australian Institution of Engineers.

There are very many Russian, East European and Japanese journals published in the field of chemical engineering. Many of the articles published in these countries are translated into English. Some are cover-to-cover translations, while others contain articles selected from various sources. The major publications in this translation category are:

International Chemical Engineering, which is published quarterly by the American Institute of Chemical Engineers. It is devoted in the main to selective translations of current engineering literature from the Soviet Union, the countries of Eastern Europe and Asia.

Thermal Engineering, the cover-to-cover translation of *Teploenergetika,* which is devoted to the comprehensive coverage of research and practice in the power industry, including district heating.

Chemical and Petroleum Engineering, which is the cover-to-cover translation of *Khimicheskoe i Neftyanoe Mashinostroenie.* This is a practical journal of value to the process engineer.

Heat Transfer—Soviet Research, Heat Transfer—Japanese Research, Fluid Mechanics—Soviet Research and *Fluid Mechanics —Japanese Research.* All of these consist of translations of papers selected from the Russian or Japanese literature, some of which may have been translated by another source.

The cover-to-cover translation *Soviet Physics — Doklady.*

Research establishments, universities and some industrial companies outside North America and Britain produce many reports in particular areas of chemical engineering. Some of the main sources are: Australian Atomic Energy Commission; Aktiebolaget Atomenergi in Sweden; AB Atomenergi Studsvik Sweden; AEK Denmark; CISE Milan; CEN Grenobel; CNEN Italy; CEA France; Euratom; Israel Atomic Energy; Kernforschungs, Karlsruhe and Julich; Royal Institute of Technology, Sweden; and Skoda Works, Czechoslovakia.

SECONDARY SOURCES OF CHEMICAL ENGINEERING LITERATURE

Current research activities are reported in journals and technical reports long before they appear in books. This research material may be located through a variety of indexing services. Some of these are quite comprehensive, some collective, while others give a comprehensive coverage to a group of journals. The published literature is reasonably well indexed by the available indexing and abstracting services. It is possible to locate general or specific information on most aspects of chemical engineering in which research has been performed.

The Institute for Scientific Information reproduce the tables of contents from more than 700 journals in *Current Contents Engineering Technology & Applied Sciences.* Many of the chemical engineering journals are listed in this weekly publication.

One abstract source for the chemical engineering literature is *Engineering Index,* which produces annotations of selected articles from journals, engineering societies, associations, some university and industrial research reports and technical papers, some conferences and some books. Patents are not included.

Probably the widest secondary source of the published chemical engineering literature is provided by the *Chemical Abstracts* service. A list of the subject areas of chemical engineering covered by this service was given in a paper by Dickman and O'Dette (1971). Patents are indexed as scientific and technical information documents. *Chemical Abstracts* service covers the science of a particular discipline, e.g. chemical engineering, but not all the technology of all related subjects. Much of the chemical engineering information in *Chemical Abstracts* is concentrated in sections 47 to 64, the sections on 'Applied Chemistry and Chemical Engineering', although useful information can be found in some of the other sections. An analysis by Baker (1971) of the national origin of papers abstracted showed that *Chemical Abstracts* has a wide coverage of literature from journals produced in the USSR, Japan, Eastern Europe, and SE Asia. *Chemical Abstracts* service provides several current awareness services by computer processing techniques. The relevant computer services to chemical engineers are *Chemical Titles, CA Condensates, Polymer Science and Technology—Journals* and *Polymer Science and Technology—Patents.*

An abstract journal devoted solely to chemical engineering is *Theoretical Chemical Engineering Abstracts,* which is issued bimonthly. Very good summaries of journal articles, some British reports and US Government research reports are reproduced. Contents include: fluid dynamics, heat transfer, mass transfer, heat/mass transfer, kinetics and thermodynamics, chemical reactor engineering, design control, mixing, physical separation, grinding, crushing, pumps, compressors, economics, optimisation, nomograms, general and books.

Some other secondary sources of information which contain sections of interest to chemical engineers are: *Government Reports Announcements, International Aerospace Abstracts, STAR, Nuclear Science Abstracts, Science Abstracts, Dissertation Abstracts* (which abstracts recent North American theses), *Pollution Abstracts, R & D Abstracts, Food Science and Technology Abstracts, BLLD Announcement Bulletin* (which lists British report literature, translations produced by British organisations and some university theses) and *Heat Bibliography* (produced annually by the National Engineering Laboratory at East Kilbride and listing references on all aspects of technology associated with heat and heat transfer).

An important source of technical and company information of value to chemical engineers can be found in the patent literature. Many of the patent abstracting journals contain sections on the subject.

Some recent recommended books

During the past 30 years many very good books have been written on all aspects of chemical engineering science and technology and some of them have become reference books. In the section below are recorded some recommended books published more recently which supplement or even supersede the recognised classical texts.

In the field of chemical reaction engineering three recommended books are *Chemical Reaction Engineering,* by O. Levenspiel (2nd edn, Wiley, 1972); *Chemical Reaction Engineering,* a supplement of Chemical Engineering Science (Pergamon, 1971); and *Chemical Reactor Theory—an Introduction,* by K. G. Denbigh and J. C. R. Turner (2nd edn, Cambridge University Press, 1971). The last book covers the main types of idealised reactor, a review of chemical kinetics, behaviour and design of tubular reactors and continuous stirred reactors, residence time distribution, mass transfer factors, chemical factors, thermal characteristics of reactors and their stability and operation.

For those in the separation field there are four books worthy of note: *Industrial Processing with Membranes,* edited by R. E. Lacey and S. Loeb (Wiley, 1972), which discusses electrodialysis, reverse osmosis and ultrafiltration; *Industrial Filtration of Liquids,* by D. B. Purchas (2nd edn, L. Hill, 1971), a practical book which provides a systematic survey with a treatment of the basic problems to allow selection of the most suitable equipment; and two books on crystallisation—*Crystallisation,* by J. W. Mullin (2nd edn, Butterworths,

1972), which deals with plant and processes of bulk crystallisation and the necessary theoretical background, and *Industrial Crystallisation* (Institution of Chemical Engineers, 1970), which records the proceedings of the symposium organised by this Institution in April 1969.

For engineers with interests in drying, there have been four noteworthy practical books on this subject. A book designed to provide the process engineer with a reasoned technical assessment of the type of dryer most likely to meet his needs is *Industrial Drying,* by A. Williams-Gardner (L. Hill, 1971), *Drying—Principles and Practice,* by R. B. Keey (Pergamon, 1972), considers the principles of drying of solids and gives an engineering description of 13 classes of drying. The drying of ceramics, wool, foodstuffs and textiles is also discussed. Another book on this subject is *Drying of Solids in the Chemical Industry,* by G. Nonhebel and A. A. H. Moss (Butterworths, 1971). This book includes a large amount of operating data relating to full-scale performance of various dryers. In addition, theoretical aspects of drying and plant selection are discussed. A more specialised book in this field is *Spray Drying,* by K. Masters (L. Hill, 1972), which is a practical approach to principles, operation and application of the subject.

A text which covers the whole subject field and collates results of recent research work is *Fluidisation,* edited by J. F. Davidson and D. Harrison (Academic Press, 1971). The proceedings of a seminar held in India in January 1971 are recorded in *Particle Technology,* edited by D. Venkateswarke and A. Prabhakara Rao (Indian Institute of Technology, Madras, 1972).

In the field of heat transfer and fluid dynamics, there have been several very good books. A text which covers many aspects of heat transfer and has an emphasis on practical applications is entitled *Heat Transfer,* by F. A. Holland, R. M. Moores, F. A. Watson and J. K. Wilkinson (Heinemann, 1970). On two-phase flow there have been several authoritative texts by well-known research workers. These include: *Annular Two-Phase Flow,* by G. F. Hewitt and N. S. Hall-Taylor (Pergamon, 1970), which covers a wide range of practical flow and boiling situations, including heat transfer and burnout; *One Dimensional Two-Phase Flow,* by G. B. Wallis (McGraw-Hill, 1969), on the basic techniques for analysing this type of flow and their application to a wide variety of problems; *Convective Boiling and Condensation,* by J. G. Collier (McGraw-Hill, 1972), a practical approach to the existing state-of-the-art in the area of boiling and condensation; and *The Flow of Complex Mixtures in Pipes,* by G. W. Govier and K. Aziz (Van Nostrand Reinhold, 1972), which considers the flow of two-phase mixtures and non-Newtonian fluids.

Another practical book on non-Newtonian technology is *Flow Properties of Polymer Melts*, by J. A. Bryson (Iliffe, 1970). A book written by a number of contributors which deals with many aspects of low-temperature technology is *Cryogenic Fundamentals*, edited by G. G. Haselden (Academic Press, 1971). On the high-temperature side, a book which discusses heat, mass and momentum transfer in flames is *Combustion Aerodynamics*, by J. M. Beer and N. A. Chigier (Applied Science, 1972).

Chemical engineers working in extraction processes should note the books *Recent Advances in Liquid–Liquid Extraction*, edited by C. Hanson (Pergamon, 1971), and *Counter-Current Extraction*, by S. Hartland (Pergamon, 1970). The first discusses the theory and application of liquid/liquid extraction and concentrates on advances made in the last decade, while the second is concerned with rationalising the many equations and design procedures in the chemical engineering literature that are available for determining the number of stages or transfer units required to carry out certain mass transfer operations.

An up-to-date review of the existing knowledge of catalysis is given in *Mass Transfer in Heterogeneous Catalysis*, by C. N. Satterfield (MIT Press, 1970).

On process plant design, development and operation three books have been noted. These are: *Optimisation: Theory and Practice*, by G. S. G. Beveridge and R. S. Schechter (McGraw-Hill, 1970); *Phase Equilibrium in Process Design*, by H. R. Null (Wiley, 1971); and *Process Plant Design*, by J. R. Backhurst and J. H. Harper (Heinemann, 1973). A major part of the last book is devoted to the design and optimisation of major plant items, including heat exchangers and both tray and packed towers. A section deals with costs and the presentation of costing data.

Modelling techniques are discussed in *Process Analysis by Statistical Methods*, by D. M. Himmelblau (Wiley, 1970), a book on development and analysis of empirical mathematical models of process and modelling of specific processes; and *Modelling and Simulation in Chemical Engineering*, by R. G. E. Franks (Wiley, 1972), which has an emphasis on digital simulation.

A standard reference work on industrial solvents with particular emphasis on those used in cellulose laquers in *Solvents*, by T. H. Durrans (8th edn, Chapman and Hall, 1971).

A practical approach to the design of high-pressure equipment at pressures up to 45 000 lbf/in^2 and more is given in the book by W. R. D. Manning and S. Labrow entitled *High Pressure Engineering* (L. Hill, 1971).

Many aspects of water pollution are covered in *Physicochemical*

Processes for Water Quality Control, by W. J. Weber (Wiley, 1972). There have been several books on chemical engineering practice. These books primarily intended for the student are: *An Introduction to Thermodynamics,* by R. S. Silver (Cambridge University Press, 1971); the translation *Physical Principles of Chemical Engineering,* by P. Grassman (Pergamon, 1971); and *A Handbook of Unit Operations,* by D. A. Blackadder and R. M. Nedderman (Academic Press, 1971), for aiding the student with graphical and analytical calculation methods which form the groundwork of chemical engineering. The proceedings of the international conference held in Melbourne and Sydney in August 1970 cover most aspects of chemical engineering. They appear in *Chemeca '70* (Butterworths and the Institution of Chemical Engineers, 1971). The final book noted is *The Soviet Chemical Industry,* by G. Hemy (L. Hill, 1971). It includes chapters on the chemical industry, chemical engineering, planning, capital investment, manpower, operating costs and materials, services, profits, economics and foreign trade.

Handbooks and encyclopaedias

Handbooks provide a ready reference source of data and design information. The most useful texts of this type are concise and they critically evaluate many aspects of a particular subject. So that they can be really valuable sources of information, it is important that only the most reliable and most accurate material is included in them.

Probably the most useful book which covers many aspects of chemical engineering in some detail is *Chemical Engineers' Handbook,* edited by R. H. Perry and C. H. Chilton (5th edn, McGraw-Hill, 1973). Many specialists have contributed to this edition of a classic work which relates chemical principles with operation practices. This book provides the facts, figures, methods and data needed for solving a wide variety of problems which the chemical engineer encounters in his work.

A comprehensive reference book of the current knowledge of heat technology is the *Handbook of Heat Transfer,* edited by W. M. Rohsenow and J. P. Hartnett (McGraw-Hill, 1973). It includes a large collection of thermophysical property data as well as the various types of heat transmission, conduction, convection and radiation. The effects of chemical reaction, electrical fields and magnetic fields on heat transfer are discussed. Boiling, condensation, mass transfer and heat exchanger design are all well covered.

Information on pollution is given in *Industrial Pollution Control Handbook,* edited by Herbert Lund (McGraw-Hill, 1971). This gives guidance on water quality, odours, reducing waste load, heat recovery from solid waste incinerators and gaseous pollutants. It provides details on plant layout, equipment and materials of construction and a training manual and gives reasons for inaccurate unreliable data of emissions.

The Materials Handbook, edited by George S. Brady, (10th edn, McGraw-Hill, 1971), gives the engineer an understanding of a wide variety of materials, the applications of each, problems that may occur in procurement and suitable substitutes. Coverage includes: metals and alloys, abrasives, plastics, synthetic resins, industrial chemicals, petroleum products, fuels, refractories, minerals, high-heat plastics, high-energy fuels, aerospace materials, nuclear shielding materials and pollution control materials.

A valuable source of the present state of research and technology in the field of chemical plant and equipment is the *ACHEMA Jahrbuch,* which is published by DECHEMA, the German Chemical Engineering Society. Every three years DECHEMA organises the ACHEMA exhibition and congress, which is located in Frankfurt. In 1973 more than 2200 firms exhibited their products of chemical process equipment, processes and process plant. The 1971/73 edition of the *Jahrbuch* was published in three volumes. Volume I, entitled *Chemical Engineering Research and Education in Europe,* contains 360 reports from university departments and research establishments on their activity in the field of chemical engineering and its fundamentals; in particular, industrial chemistry, process engineering, physical chemistry, materials science and instrumentation. Volume II, entitled *Technical Developments in Chemical Plant and Equipment,* contains reports from manufacturers of chemical equipment, machinery and plant for laboratory and full-scale operation. Volume III is the *Guide to Chemical Plant and Equipment in Europe.* It contains not only an index of the firms, institutions and organisations appearing in the yearbook and participating in ACHEMA, but also gives details of which firms produce a particular product and, hence, where further knowledge regarding the function and possible application of the product in question is likely to be available. An index of the trade names is included.

Details of the UK chemical manufacturers, contractors, plant manufacturers, industrial research organisations, professional and trade organisations and consultants can be found in the *Chemical Industry Directory,* produced by *Chemical Age* and published by Benn.

Two encyclopaedias which often provide a good starting point on a particular aspect of chemical engineering are *Chemical Engineering Practice*, by H. W. Cremer and T. Davies (Butterworths, 1956–1960), in 12 volumes, and the multi-volume *Encyclopaedia of Chemical Technology* of Kirk-Othmer (2nd edn, Wiley, 1963–).

Physical property information sources and data sheets

Chemical engineers require transport and thermophysical property data for calculation of process plant design and efficiency. There is a vast amount of data, some good and some bad, on very many substances scattered throughout the literature, but there are very many gaps in the knowledge. Many industrial organisations have built up their own data banks or have their own calculation methods for the compounds and mixtures with which they usually work. Unfortunately, these data sources are not often generally available to others.

There have been many attempts by individuals and organisations to produce compilations of physical properties and now there are several organisations offering information from data banks or from calculation methods using the computer. Some organisations are offering for sale complete data banks and computer programs.

To name all the available sources is inefficient in terms of space and guiding the reader, so only a brief review of the considered major sources is given.

The Properties of Gases and Liquids, by R. C. Reid and T. K. Sherwood (2nd edn, McGraw-Hill, 1966), presents a critical review of the various estimation procedures for a limited number of properties and fluids. Properties include: critical properties, *P–V–T* and thermodynamic properties, vapour pressures, latent heats, heat capacities, surface tensions, viscosities, thermal conductivities, diffusion coefficients, heats of formation and free energies of formation. Recommendations are made regarding the best methods of estimating each property and of extrapolating available data.

A ready reference book of some physical and chemical data is given in *Handbook of Chemistry and Physics*, edited by R. C. Weast (56th edn, CRC Press, 1975). This book is revised every year to incorporate new data. Two other valuable sources of data are Landolt-Bornstein's *Tables* and *Gmelins Handbuch der Anorganischen Chemie*, both multi-volume German compendia. (See *The Use of Chemical Literature* for details of these and other compilations.)

Jean Timmermans has compiled two sets of volumes on much of the early published data in tabulated form. These are entitled *The Physico-Chemical Constants of Binary Systems in Concentrated*

Solution (Interscience, 1959), in four volumes, and *The Physico-Chemical Constants of Pure Organic Compounds* (1965), in two volumes.

Tables of numerical data on physical properties of numerous substances, liquids, solids, gases and metals were compiled in *International Critical Tables* (McGraw-Hill, 1926–1933), in seven volumes plus an index.

In past years, R. W. Gallant has produced a most useful series of papers on the physical and transport properties of many hydrocarbons for the journal *Hydrocarbon Processing*. The entire series, entitled *Physical Properties of Hydrocarbons*, has since been published in two volumes by the Gulf Publishing Company, New York. Volume 1 covers paraffinic hydrocarbons, alcohols, oxides and glycols. Volume 2 covers other oxygenated hydrocarbons, nitrogen-containing compounds, sulphur-containing compounds and aromatics. The properties of about 170 compounds are included.

CODATA, the Committee on Data for Science and Technology, promote and encourage, on a world-wide basis, the production and distribution of compendia and other forms of critically selected numerical data substances. The National Bureau of Standards (NBS) has carried out work on all the elements, inorganic compounds, C1 and C2 organics and aqueous solutions of important acids. Selected data have appeared in NBS circulars and publications. They have also produced selected values of properties of hydrocarbons and related compounds. This later programme is closely associated with the Thermodynamic Research Centre (TRC) at Texas A and M University. TRC covers many inorganic and organic compounds and substances of importance to the chemical industry. Their publications have included *Thermophysical Properties of High Temperature Solid Materials,* edited by Y. S. Touloukian (Macmillan, 1967), in eight volumes, *Selected Values of Properties of Chemical Compounds* (TRC, 1968), *Thermophysical Properties Research Literature Retrieval Guide,* in three volumes, edited by Y. S. Touloukian (2nd edn, Macdonald, 1967). Properties included in the last-mentioned are thermal conductivity, specific heat, viscosity, thermal radiative properties, diffusion coefficient, thermal diffusivity and Prandtl number. Supplement 1, 1964–1970, to this retrieval guide has appeared in six volumes (IFI/Plenum, 1973).

Between 1970 and 1972 TRC has produced in 13 volumes a massive work entitled *Thermophysical Properties of Matter,* edited by Y. S. Touloukian and C. Y. Ho (Plenum). These volumes reproduce data for the thermal conductivity, specific heat, thermal expansion and thermal radiative properties of metallic elements, alloys and non-metallic solids; the thermal conductivity of liquids and gases; thermal radiative properties of coatings; thermal diffusivity and viscosity. These volumes are available as a complete set, as a sub-set or individually.

TRC are about to produce new tables of the thermodynamics and other properties of mixtures of non-electrolytes. This series is entitled *International Data Series Selected Data on Mixtures,* in which experimental data is presented in tabular and graphical form on a single sheet.

Some of the work carried out by TRC is closely associated with the American Petroleum Institute Research Project. This Institute has compiled the physical and thermodynamic data and correlations of substances of most interest to petroleum refiners, for process evaluation and equipment design, in two volumes entitled *Technical Data Book—Petroleum Refining* (American Petroleum Institute, 1970).

The Engineering Sciences Data Unit (ESDU), London, produce a wide range of selected data for application in engineering in the form of data sheets and memoranda. The institutions currently supporting this work are the Royal Aeronautical Society, the Institution of Mechanical Engineers, the Institution of Chemical Engineers and the Institution of Structural Engineers. The data sheets of most value to chemical engineers have been produced on physical data and reaction kinetics, heat transfer and fluid mechanics. ESDU has produced a very comprehensive index to all its publications.

Somewhat similar data sheets to those of the ESDU, on heat transfer, covering a wide range of problems, are published in the *VDI Wärmeatlas* (Verein Deutscher Ingenieure, Verlag GMBH, Düsseldorf). Another source of data of a somewhat similar type are the *Heat Transfer and Fluid Flow Data Books* from General Electric Company, Schenectady, New York.

During the past few years several physical property data services, which use computer methods, have become established. The American Institute of Chemical Engineers wrote a program called 'APPES' (*A*merican *I*nstitute of *C*hemical *E*ngineers *P*hysical *P*roperty *E*stimation *S*ystem). This program was bought by the National Engineering Laboratory near Glasgow and they are able to provide rapid estimates of the physical properties of gases and liquids (Martin, 1970). The Institution of Chemical Engineers, London, in collaboration with BP Chemicals International Ltd and the Computer Aided Design Centre, have a program PPDS, *P*hysical *P*roperty *D*ata *S*ervice, for the physical properties of pure components which calculates the properties of mixtures using recommended mixing rules. In June 1973 the data bank contained 350 compounds.

At the 1973 ACHEMA exhibition the German company Friedrich

Uhde demonstrated their program package for calculating the thermophysical properties of pure chemical substances and mixtures (liquid or gaseous). This program package is the basis of the DECHEMA thermophysical properties service. Computer programs for the thermodynamic and transport properties of cryogenic fluids have been written at the NASA-Lewis Research Centre in a joint venture with the National Bureau of Standards.

Sponsored research organisations

To aid government and private industry, several sponsored research organisations have been established. The advantages of these organisations are that they are able to provide industry with expertise in a particular field and they have suitable research facilities available. They are, therefore, able to carry out industrial research for one company or for a group of companies at a fraction of the cost that the company would pay if it did its own research. These sponsored research organisations are a source of very sophisticated information in that they provide their subscribers with very advanced reports on topics of interest. Some of them provide consultancy and advanced retrospective information facilities for their subscribers. There are several well-known research centres which give this type of service to the chemical, process and petroleum industries.

One of the first organisations of this type to become established was the Battelle Memorial Institute. Its major research centres are at Columbus, Pacific Northwest Laboratories, Frankfurt and Geneva. Battelle is a multi-disciplinary research, development and educational organisation which carries out contract research, laboratory work, pilot plant work, economic development and planning studies. It uses income from its own endowment to support a substantial research and education programme. Battelle has carried out quite a lot of work in the field of chemical engineering on, for example, polymers, mechanical and thermal process engineering, refrigeration and food technology, internal combustion engineering and environmental research. Recently the Geneva Research Centre in collaboration with the Frankfurt centre produced a forecast of plant and equipment demand in the chemical and oil refinery industry. This *Process Plant* study gives forecasts of the demand, up to 1974 and 1980, for five major items of equipment. During the study a computerised storage and retrieval system was developed for information on individual chemical plants throughout the world. This data bank, known as *Chemplant Data Bank*, is now available as a service to industry.

Two organisations which provide a service for designers, users and manufacturers of heat transfer equipment are Heat Transfer Research Inc. (HTRI), in the USA, and the Heat Transfer and Fluid Flow Service (HTFS), at the Atomic Energy Research Establishment, Harwell, UK. Both of these organisations produce reports, design methods and computer programs which are proprietary to the member subscribing companies. The cost of the individual services is distributed among the sponsoring companies. Although these two organisations have similar basic objectives, their work has not overlapped to any great extent and the two services often complement one another. As part of their service, HTFS give a limited amount of consultancy and an extensive information retrieval service. In 1967 HTFS set up its own special library for the world-wide heat transfer and fluid flow literature. To date 21 000 documents are held in this library. New additions are listed in the current awareness bulletin *HTFS Digest* which is distributed monthly to both the HTFS sponsoring companies and non-sponsors. All the literature held in this library has been indexed using keywords formulated by HTFS, and a computerised storage and retrieval system has been developed. The library also stores a vast quantity of physical property information. Literature searches on all this information are carried out on request. In addition to the co-operative subscription service, HTFS provides a rechargeable confidential consultancy and testing contract service for any individual company or group. The HTFS and HTRI services are backed up by extensive underlying and application-orientated research programmes by each organisation on many aspects of heat transfer and fluid flow.

In the separation field, Fractionation Research Inc. (FRI) was founded in 1952 by 30 companies to test the performances of fractionation devices on a commercial scale. Now 80 companies subscribe to the work. FRI carries out co-operative contract research for a group of sponsoring companies. They have made a systematic study of tray designs and packing materials. Monthly progress reports, topical reports, plant test data and a tray design handbook have been presented. Recently they have completed a computer program package for rating sieve trays.

Another organisation in this field, which is in the embryo stage, is The Separation Processes Service at AERE Harwell. This project is organised in a similar manner to HTFS. Reports, consultancy and information will be provided to a group of subscribers and rechargeable contract research will be carried out. This work will be backed up by an extensive research programme.

AERE at Harwell has set up many multi-disciplinary technical services for clients in industry and other outside organisations. In the field of chemical engineering and process technology, the Heat Transfer and Fluid Flow and Separation Processes Services have already been mentioned. In addition, there is a large amount of information on biochemical and microbiological techniques, corrosion, desalination and water treatment, disposal of industrial waste, electrohydraulic crushing, high-temperature processes, mathematical modelling and optimisation of chemical plant and processes, reverse osmosis, solvent extraction, tracer techniques and treatment of sludges. The Hazardous Materials Service provides a consultancy service on the treatment and disposal of toxic and non-toxic wastes. They have a storage and retrieval of information system based upon the methods used by HTFS and produce monthly the current awareness communication *Industrial Wastes Information Bulletin.*

Two newly proposed organisations are the Design Institute for Multiphase Processing, which is sponsored by the American Institute of Chemical Engineers, and Fluid Property Research Incorporated, which will be supported in the same manner as HTRI.

Further details of some of these and other contract research organisations were given in a paper by Woodward (1972).

Other sources of information

For chemical engineers requiring details of where and how to obtain information from US Government sources, a comprehensive review was given in a paper by Elizabeth Biggert (1972). The areas covered included technical and scientific, economic, legislative and geographical sources.

The Cryogenic Data Centre was established in 1958 at the National Bureau of Standards, Boulder, Colorado, to organise the low-temperature literature from all over the world (Johnson, 1968). This centre has a very large collection of literature on the properties of materials at low temperatures.

An information retrieval service called CPI DATA is now available from *Chemical Week* (McGraw-Hill). Data collected from 1000 US and foreign publications include production and consumption figures, construction activity, new processes, etc. An SDI service and retrospective retrieval of abstracts on the subject during the previous 12 months are available.

In Britain, the Institution of Chemical Engineers Library and Information Service has compiled a thesaurus of about 650 keywords, for which abstracts of published papers on chemical engineering have been indexed and stored using feature cards. Over 12 000 papers have been keyworded; the annual intake to the system is

2000. Retrieval of this information is available to clients on a subscription basis. The Institution is also involved with the Physical Property Data Service which was described earlier. In addition, they are carrying out an analysis of COMPENDEX, which is a version of *Engineering Index* on magnetic tape, for Selective Dissemination of Information.

In the field of particle science and technology, an information service has been organised at the University of Technology, at Loughborough. This service produces the monthly current awareness bulletin *Particle Technology* and carries out literature searches.

Another university organised service is the Information Centre on High Temperature Processes (Brain, 1973), in the Department of Fuel Science at the University of Leeds. This department has built up an index of over 30 000 items. They will make a search of the index on request.

The International Food Information Service at Reading carries out literature surveys based primarily on its own extensive *Food Science and Technology Abstracts*.

The Chemical Engineering Societies of Canada, Germany, France, Japan, UK, USA, USSR and Yugoslavia have founded the International Centre for Heat and Mass Transfer. Its aim is to promote and foster international co-operation in research and application in the field of heat and mass transfer. The Centre acts as a focal point for topical international information on the subject and produces a regular newsletter. It also organises an annual seminar on specialised topics from the field.

REFERENCES

Baker, D. B. (1971). 'World's Chemical Literature Continues to Expand', *Chemical and Engineering News*, July 12, 37–40
Bennett, G. F. (1973). 'Information Sources', *Chemical Engineering Deskbook Issue*, **80**, 39–47
Biggert, E. C. (1972). 'Federal Government Information Sources', *Chemical Engineering*, **79**, (11), 103–114
Brain, M. E., Livesey, J. B. and Williams, A. (1973). 'A Specialised Information Centre for High Temperature Processes', *Aslib Proceedings*, **25** (5), 186–190
Constan, G. L. (1973). 'Detroit to Host 75th National AIChE Meeting', *Chemical Engineering Progress*, **69** (3), 73–76, 85
Danckwerts, P. V. (1972). 'Chemical Engineering Science', *Chemical Engineer*, (262), 222–225
Dickman, J. T. and O'Dette, R. E. (1971). 'Chemical Engineering Content of *Chemical Abstracts*', *Chemical Engineering Progress*, **67** (11), 79–81
Gregory, S. A. (1972). 'Chemical Engineering as a Profession in the UK', *Chemical Engineer*, (260), 130–134
Hougen, O. A. (1965). 'Chemical Education in the USA', *Chemical Engineer*, (191), CE 222–231
Johnson, V. J. (1968). 'Development and Operation of a Specialised Technical

Information and Data Centre', *Journal of Chemical Documentation*, **8** (4), 219–224

Marshall, W. R. (1973). 'Chemical Engineering in the Last Three Decades of the 20th Century', *Chemical Engineering Progress*, **69** (2), 25–32

Martin, C. N. B. (1970). 'Using a Comprehensive Physical Property Estimation System', *Chemical Engineer*, (241), CE 285–288

Peck, W. C. (1973). 'Early Chemical Engineering', *Chemistry and Industry*, (11), 511–517

Poll, A. (1968). 'Chemical Engineering Research in Progress in the UK—No. 4', *Chemical Engineer*, (221), CE 279–326

Woodward, F. N. (1972). 'The Role of Consultants', *Chemistry in Britain*, **8** (4), 154–157

21

Fuel technology

L. S. Brown and M. E. Horsley

INTRODUCTION

Fuel technology is the study of the derivation, processing and utilisation of combustible materials, and includes the design and manufacture of burners and the equipment or apparatus in which the fuels are used. This is clearly a very wide field and in the present context, therefore, it is confined to the study of fossil fuels—coal, oil, gas—and their use in the industrial furnace and boiler field.

This wide definition shows the topic to be truly interdisciplinary and having connections with, for example, chemistry, chemical and process engineering and mechanical engineering. For this reason, an information search must be spread over topic titles other than that of fuel technology.

GENERAL OR BACKGROUND INFORMATION

For the non-specialist pursuing a general query, there are several potential sources of information on fuel technology matters. A good encyclopaedia or technical dictionary is a sensible starting point and the majority of company libraries, however small, will contain yearbooks, handbooks or general textbooks which, owing to the interdisciplinary nature of the subject, often contain relevant information. A survey of some such books has shown that, for example, 18% of the text of thermodynamics books is devoted to fuel technology, while the percentage for engineering handbooks/yearbooks/reference manuals is about $4\frac{1}{2}\%$. Similarly, books related to mining, minerals extraction, heating and ventilating, chemical engineering practice, etc., may prove fruitful.

BASIC TEXTBOOKS

Many of the textbooks in the subject originated several years ago and, while revisions have been incorporated in the recent editions, some books show an historical bias towards solid fuels and do not necessarily cover the important recent advances. Amongst the useful library texts are the following:

Fuel: Solid, Liquid and Gaseous, by J. S. Brame and J. G. King (6th edn, Arnold, 1967)

An Introduction to the Study of Fuel, by J. C. MacRae (Elsevier, 1966)

Elements of Fuel Technology, by G. W. Himus (2nd edn, L. Hill, 1958)

Fuel Science, by J. H. Harker and D. A. Allen (Oliver and Boyd, 1972)

These are general 'student style' textbooks containing a wide range of fundamental information.

Fuels and Fuel Technology, by W. Francis (2 vols., Pergamon, 1965). A wide range of information is presented in data sheet style, with many references for further reading.

Efficient Use of Fuel (2nd edn, HMSO, 1958). Originally produced in the war years as an aid to fuel efficiency, this very practical book nevertheless contains much fundamental information.

Technical Data on Fuel, edited by H. M. Spiers (British National Committee, World Power Conference, 1961). Essential data for industrial and research purposes are tabulated in this book.

Dictionary of Fuel Technology, by A. Gilpin (Newnes–Butterworths, 1969). Essential data on fuels are presented, in a layout which the title suggests.

Efficient Use of Energy (IPC, 1975)

SPECIALISED TOPICS

The foregoing paragraphs have indicated the likely sources of general information which may be required by the undergraduate or the plant engineer. When specific studies or investigations are to be pursued, the more specialised literature has to be consulted, and the standard starting points are bibliographies, abstracts, indexes and journals.

It will be necessary to consult a bibliography of bibliographies (for example, *Index Bibliographicus, Vol. 1. Science*) to determine the appropriate bibliographies, but it must be borne in mind that these range from (a) relatively brief ones carried in data books, through (b) clearly relevant and listed ones, to (c) fringe topic bibliographies, because of the broad nature of the subject. Examples of these three are (a) the list of about 250 items presented in *Technical Data on Fuel* (see above); (b) *Efficient Use of Fuels: Bibliographic Index* (United Nations Economic Commission for Europe, 1955); and (c) the scattered references in *Chemical and Process Engineering—Unit Operations, a Bibliographical Guide,* by K. Bourton (Macdonald, 1967).

Indexes, usually cumulated annually, list references over a broad spectrum and are not peculiar to the fuels field. Typical indexes include *Applied Science and Technology Index* (The H. W. Wilson Company), *British Technology Index* (Library Association) and *Engineering Index* (Engineering Index Inc.), and it is stressed again that the literature search must range over several subject headings. For example, in *Applied Science and Technology Index,* a heading 'Fuel' exists but relevant information lies also under other headings. Thus, in one particular volume the 'Fuel' heading carried 9 sub-headings such as 'Fuel Economy' and 'Fuel Research'; and referred the reader to 18 other headings such as 'Coal' (with 16 sub-headings), 'Propellants' (5 sub-headings), 'Blast Furnaces—Fuel' and 'Gas Turbines, Aircraft—Fuel'. Additionally, if the topic investigated may lie under a separate heading, e.g. 'Heat Transfer' would be a reasonable heading to survey for furnace design, then this also must be pursued. A further value of an index survey is that the indexes list the journals surveyed, which in turn are sources of current information, as outlined below.

Abstracts give both a bibliographical reference to a publication and a synopsis of varying length, often saving much literature search time by clarifying ambiguous titles. The abstracts, published at regular intervals, represent the most useful source of current information, and the premier publication in the fuel technology context is *Fuel Abstracts and Current Titles* (generally called *FACTS*), published monthly by the Institute of Fuel, London. This covers the full range of fuel technology by sub-headings and cross-references, drawing for its information on original sources and other abstracting services of national and international repute. The subject and author index of each issue is presented also as an annual index, thus providing a useful guide to past references.

There are many other abstracting services of use in this context, some quite specialised and some with a wide range of topics, produced commonly by learned bodies, government agencies and industrial organisations on a national and international basis. *FACTS* itself carries a list of organisations which produce appropriate abstracts, and there are standard lists, such as that of the British Library, Lending Division, which must be widely consulted, as previously indicated, because of the nature of the topic. Confining this summary, for brevity, to the ones of immediate interest, examples include *Coke Review* and *NCB Abstracts A*, which are clearly of a specialised nature, and *Air Pollution Abstracts* and *CEGB Digest*, which will contain relevant but diluted information.

While the abstracting and indexing services are very valuable, they inevitably present their information several months after the publication of the paper or text, simply because the service has to obtain the prime source and then reprocess it, a situation which may be aggravated where secondary services are involved. This in no way detracts from the value of abstracting or indexing services, for they cover much more ground than the individual researcher can, but it does emphasise the need for the individual to peruse, on publication, a representative range of relevant journals. Incidental benefits of journal reading include the presence of trade information and advertisements, which may help with the updating of practical application and knowledge, plus advance notice and reports of lecture meetings on relevant topics.

Journals can be split broadly into four groups:

Learned journals containing mainly research-style papers and of regular wide interest to the fuel technologist—for example, the *Journal of the Institute of Fuel*; *Fuel*; *Energy Digest* (incorporating the *Journal of Fuel and Heat Technology*); *Journal of Engineering for Power. ASME Transactions Series A*; *Revue Générale de Thermique* (France); *Thermal Engineering (Teploenergetika,* Russia); *Brennstoff–Wärme–Kraft* (Germany). Some of the overseas magazines are available in translated form, but others require access to a translation service.

Publications of a regular but specific rather than wide interest. These are considered below under separate topic headings.

Industrial journals and house magazines, containing review articles, progress reports and product appraisals. Examples include *Energy World*; *Oil and Gas Firing*; *Power* (USA); *Combustion* (USA); *Heat Engineering* (Foster Wheeler); *Brown Boveri Review* (Switzerland).

Fringe interest publications, which yet again emphasise the broad nature of the subject. The researcher will not confine himself to publication titles of immediately apparent interest, but will consider other headings such as chemistry, mechanical engineering and chemical engineering and the sub-divisions thereof. Examples in this group include the *Transactions of the Institution of Chemical Engineers*; the *American Institute of Chemical Engineers Journal*; *International Journal of Heat and Mass Transfer*; *Journal of Heat Transfer*. *ASME Transactions Series C*; the *Industrial and Engineering Chemistry* group of journals (USA); *Chemistry and Industry*; *Chemical Process Engineering*.

There is a very large number of scientific and technical journals available (for example, Butterworths' *World List of Scientific Periodicals* contains about 50 000 entries), but continued pursuance of references will soon highlight those of major importance to the individual. Researchers seeking shorter lists of periodicals should consult their company or institution librarian and the indexes noted previously.

For further consideration of specialised literature, the subject may conveniently be broken down according to the aspects of the work to be studied and the headings used here are: solid fuels and by-products; liquid fuels; gaseous fuels; steam raising and usage; furnaces and refractories; space heating; combustion and flames; air pollution; and analysis, testing and measurement. These topics are not rigidly compartmented; hence, the references and conference proceedings mentioned under one heading are sure to contain material of value to the other headings. The lists can be by no means exhaustive in the space available and, in particular, no individual papers have been mentioned. The major journal is the *Journal of the Institute of Fuel,* which must be consulted whatever the topic, and is not therefore included under the topic headings. It is implicitly understood that, whenever possible, the overseas equivalent journals will be consulted.

Much of the relevant investigational work is pursued in research establishments which may or may not issue external reports. It is suggested that the information officer or librarian of such establishments be contacted to indicate the available range of published material.

Solid fuels and by-products

Authoritative works on the coal substance include *Coal: Its Formation and Composition,* by W. Francis (Arnold, 1961), and *Coal; Typology, Chemistry, Physics and Constitution,* by D. W. van Krevelen (Elsevier, 1961). *Chemistry of Coal Utilisation,* edited by H. H. Lowry (Wiley, 1963), with contributions by a number of international experts, covers the wide fields implied by its title. *Coal, Coke and Coal Chemicals,* by P. J. Wilson and J. H. Wells (McGraw-Hill, 1950), is a useful work with ample descriptive content. Conference reports published by the Institute of Fuel include *Pulverised Fuel* (3 vols., 1947); *Pulverised Fuel* (2nd Conference, 1958); *Use of Small Coal—Today and Tomorrow* (1958); and *Science in the Use of Coal* (1958).

The British Carbonisation Research Association publishes reviews, reports and abstracts, including foreign translations and, in similar vein, the National Coal Board (Technical Intelligence Branch) and the US Bureau of Mines are well-recognised.

Liquid fuels

The Petroleum Handbook (5th edn, Royal Dutch/Shell Group, 1966) gives a broad picture of all aspects of the oil industry, and *Petroleum Refinery Engineering,* by W. L. Nelson (4th edn, McGraw-Hill, 1958), is a standard handbook in its field. A concise account of liquid fuels production and utilisation is provided by *Liquid Fuels,* by D. A. Williams and G. Jones (Pergamon, 1963), while *The Atomisation of Liquid Fuels,* by E. Giffen and A. Muraszew (Wiley, 1953), is recommended.

The *Institute of Petroleum, (Report) IP* (formerly the *Journal of the Institute of Petroleum*) and the *Proceedings of the American Petroleum Institute* contain highly relevant material, while commercial periodicals such as *Hydrocarbon Processing and Petroleum Refiner* or *Oil and Gas Firing* could also be consulted.

Within the range of specialist conferences, those of the Institute of Fuel are notable. For example, the researcher may well be expected to consult the *Proceedings of the 3rd Conference on Applications of Liquid Fuels* (Torquay, 1966).

Gaseous fuels

Gas Making and Natural Gas (British Petroleum Company Ltd, 1972) contains, inter alia, useful descriptions of gasification processes from both solid and liquid bases. *Natural Gas: a Study,* by E. N. Tiratsoo (2nd edn, Scientific Press, 1973), is a comprehensive treatment of its subject. *Gas Engineers Handbook,* edited by C. G. Segeler (Industrial Press, 1965), and *Fuel Gasification (Advances in Chemistry, 69),* by F. C. Shora (American Chemical Society, 1967), may provide helpful references.

Institute of Fuel conference reports include *Gasification Processes* (with the Institution of Gas Engineers, 1962); *Industrial Gas Symposium* (1967); and *Utilisation of Natural Gas* (Eastbourne, 1968).

Clearly it is implicit that the British Gas Corporation will be contacted for specialist information, while among the serial publications which should be consulted are the *Journal of the Institution of Gas Engineers*; *American Gas Association Monthly*; and *Oil and Gas Firing*. Textbooks, periodicals, etc., on solid and liquid fuels usually have substantial sections or references to gaseous fuels derived from non-gaseous sources.

Steam raising and usage

Works which provide a thorough descriptive treatment of boiler plant include *Steam Generation,* by J. N. Williams (4th edn, Allen and Unwin, 1969), and *Steam: its Generation and Use* (38th edn, Babcock and Wilcox Company, 1972). *Large Boiler Furnaces,* by R. Dolezal (Elsevier, 1967), deals with theory, construction and control. *Boiler House and Power Station Chemistry,* by W. Francis (4th edn, Arnold, 1962), has descriptive and analytical parts and its coverage reflects the preponderance of solid fuel usage in power stations. In view of the element of empiricism in boiler design, *Pilot Plants, Models and Scale-up Models in Chemical Engineering,* by R. E. Johnstone and M. W. Thring (McGraw-Hill, 1967), could be of considerable use. In the field of steam utilisation, Oliver Lyle's *Efficient Use of Steam* (HMSO, 1963) remains a classic work.

The Institute of Fuel's reports on the *Symposium on Chimney Design* (1966) and the *Symposium on Combustion in Marine Boilers* (1968) cover specialised topics, as do the *Proceedings of the Institution of Mechanical Engineers,* the *Transactions of the Institute of Marine Engineers* and the *Journal of Engineering for Power. ASME Transactions Series A,* from time to time. In the range of commercial periodicals, the *Heating and Air Conditioning Journal* (formerly the *Steam and Heating Engineer*) is useful for background information.

Furnaces and refractories

Among those books which provide a satisfactory blend of theory and practice are *Modern Furnace Technology,* by H. Etherington and G. Etherington (3rd edn, Griffin, 1961), and *Science of Flames and Furnaces,* by M. W. Thring (2nd edn, Chapman and Hall, 1962).

Concise studies by J. D. Gilchrist include *Furnaces* and *Fuels and Refractories* (Pergamon, 1963). *Industrial Furnaces,* by W. Trinks and M. H. Mawhinney (Vol. 1, 5th edn, Wiley, 1961, and Vol. 2, 4th edn, Wiley, 1967), is a standard work, particularly strong on detailed description. *Refractories and their Uses,* by K. Shaw (Applied Science, 1972), deals with selection, types and uses, while *Steel Plant Refractories—Testing, Research and Development,* by J. H. Chesters (2nd edn, United Steel, 1957), is a more specialised work.

Institute of Fuel conference reports include *Waste Heat Recovery from Industrial Furnaces* (1961); *Fuel and Energy for Various Steelmaking and Ironmaking Processes* (1962); and *Incineration of Municipal and Industrial Waste* (1969).

Relevant serial and other publications include the *Journal of the Iron and Steel Institute;* the *Journal* and *Reports* of the British Cast Iron Research Association; *Reports* of the British Iron and Steel Research Association; the *Refractories Journal* of the Refractories Association of Great Britain; and research papers and monographs of the British Ceramic Research Association.

Space heating

A considerable amount of information is contained in the *IHVE Guide,* published by the Institution of Heating and Ventilating Engineers (4th edn, 1970), and the *Handbook of Fundamentals,* published by the American Society of Heating, Refrigerating and Air Conditioning Engineers (1968). These works have a preponderance of tabulated data with a minimum of discussion. Other sources are the *Heating Handbook,* by R. H. Emerick (McGraw-Hill, 1964), and *Heating and Hot Water Services in Buildings,* by D. Kut (Pergamon, 1968).

Periodicals with relevant content include the *Building Services Engineer;* the *Journal* and *Transactions of the American Society of Heating, Refrigerating and Air Conditioning Engineers; Heating and Ventilating Engineer;* and *Heating and Air Conditioning Journal.*

Thermal Abstracts, published by the Heating and Ventilating Research Association, is a valuable source of information.

Combustion and flames

This is probably the most widely and deeply researched area of fuel technology and the investigator here is faced with many interdisciplinary references. No attempt can be made in this summary to be exhaustive.

Combustion, Flames and Explosions of Gases, by B. Lewis and G. von Elbe (2nd edn, Academic Press, 1961), is a very thorough study of its subject. *Combustion Aerodynamics*, by J. M. Beer and N. Chigier (Applied Science, 1972), stresses the value of theory to the practical designer, and *Fuels and Combustion*, by M. L. Smith and K. W. Stinson (McGraw-Hill, 1952), maintains a satisfactory balance between theory and practice. *Flames: their Structure, Radiation and Temperature*, by A. G. Gaydon and H. G. Wolfhard (3rd edn, Chapman and Hall, 1970), is an authoritative study with ample coverage of experimental method, while F. J. Weinberg's *Optics of Flames* (Butterworths, 1963) is a more specialised work. *Heat Transmission*, edited by W. H. McAdams (3rd edn, McGraw-Hill, 1954), is deservedly well known, as is *Heat Transfer*, which is volume 7 of *Chemical Engineering Practice*, edited by H. W. Cremer (Butterworths, 1963) and which covers solids drying and space heating aspects.

Publications of the Institute of Fuel under this topic heading include reports of the conference *Pulverised Fuel Flames* (1960) and their four symposia *Flames and Industry*. The work of the International Flame Research Foundation, based at IJmuiden, deserves particular mention. Its reports and published work (such as is available from the Institute of Fuel) are most significant.

The various *Symposia on Combustion*, published by the Combustion Institute, are essential sources of reference, being the proceedings of regular international conferences. Other sources are *Combustion and Flame*; the *Transactions of the Institution of Chemical Engineers*; the *Journal of the American Chemical Society*; and the *International Journal of Heat and Mass Transfer*. Useful publications of the National Engineering Laboratory include an annual *Heat Bibliography* and various reports and research summaries.

Air pollution

Comprehensive coverage of this topic is provided by *Air Pollution: a Comprehensive Treatise*, edited by A. C. Stern (3 vols., 2nd edn, Academic Press, 1968); *Atmospheric Pollution: Its Origins and Prevention*, by A. R. Meetham (Pergamon, 1964); *Combustion-Generated Air Pollution*, edited by E. S. Starkman (Plenum, 1971); and *Industrial Pollution Control Handbook*, edited by H. F. Lund (McGraw-Hill, 1971).

The annual conference proceedings of the National Society for Clean Air contain highly relevant material, while periodical sources include the *Journal of the Air Pollution Control Association*; *Environmental Science and Technology*, published by the American Chemical Society; *Atmospheric Environment*; and the *Air Pollution Abstracts*, published by Warren Spring Laboratory.

Analysis, testing and measurement

The necessity for standard methods for testing fuels, etc., is obvious and various organisations promulgate appropriate specifications. Reference to the *BSI Yearbook* of the British Standards Institution under its subject index produces the appropriate specification, e.g. *BS 845: 1961 Code for Acceptance Tests for Industrial Type Boilers.*

Standards of the American Society for Testing and Materials are contained in 47 parts (i.e. volumes) published annually. Each part deals with a specific group, e.g. *Part 26: Gaseous Fuels, Coal and Coke.*

The current Institute of Petroleum's *Standards for Petroleum and its Products Part 1*, deals with methods for analysis and testing. An example is No. 219/67, 'Standard Method of Test for Cloud Point of Petroleum Oils'.

Information on established methods of instrumentation can be obtained from publications such as *Instrument Engineers Handbook*, edited by G. G. Liptak (2 vols., Chilton, 1970). In the rapidly advancing field of instrumentation, however, reference to periodical literature is advisable, examples being the *Journal of Physics E. Scientific Instruments*, published by the Institute of Physics and Physical Society; the *Transactions of the Instrument Society of America*; the *IEEE Transactions on Instrumentation and Measurement*; and the *Transactions of the Society of Instrument Technology.* Company publications such as *Marconi Instrumentation, Hewlett Packard Journal,* and *Kent Technical Review* contain much useful and up-to-date information, while background information is obtainable from journals such as *Control and Instrumentation.*

22

Production engineering

J. O. Cookson

INTRODUCTION

The scope of production engineering can be elucidated by considering the IProdE (Institution of Production Engineers) definition of a production engineer as 'one who is competent . . . to determine the factors involved in the manufacture of commodities, and to direct the production processes to achieve the most efficient co-ordination of effort, with due consideration to quantity, quality and cost'. The current IProdE *Guide to the Grade of Member* informs us that professional production engineers work in the manufacturing industries, which range from electronics to pharmaceuticals and from plastics to food processing, and that they are not limited as in earlier years to metal cutting only. However, while the techniques applied to the organisational, control and management functions of the production engineer may be common throughout this wide field, the specific technologies will differ considerably, and for the present purpose we will confine attention to the manufacturing technologies used in the traditional engineering industries manufacturing machinery, motor cars, aeroplanes, domestic appliances and all the other capital equipment and consumer goods. A common feature of these manufacturing industries is their use of machine tools, defined in the broadest sense as 'a power-driven machine, not portable by hand whilst in operation, which works by cutting, forming, physico-chemical processing, or a combination of these techniques'. There will also be a fairly widespread use of methods of joining and fabricating, such as welding or riveting, and of various techniques for fitting parts together and for marshalling, handling, and assembling of materials and parts during the manufacturing processes.

In addition to the actual manufacturing processes, the associated measurement, inspection and quality control techniques must be included, together with a range of production control and planning functions.

INDEXING AND ABSTRACTING PUBLICATIONS

To begin to find one's way about in the literature of production engineering one can consult the indexing and abstracting publications, dealing in the main with journals and conference proceedings. Some of the major ones covering wide areas of engineering are useful. An essential source is *Engineering Index* for abstracts. *British Technology Index* and *Applied Science and Technology Index* give titles only, but include many smaller useful items not included in *Engineering Index.* Some sections of *Computer and Control Abstracts* cover the more sophisticated machine tool applications of numerical control and its developments. The much more recently developed INSPEC publication, *ISMEC,* has yet to prove its usefulness as a current awareness device, in relation to its price. Its cover, in terms of number of journals, is not as complete as might be desired. Other abstracting publications with narrower, specialised interests, either in particular technical subjects or in cover of geographical areas, include *Industrial Diamond Review—Abstracts* and *Japan Science Review—Mechanical and Electrical Engineering. Production Technology Abstracts and Reports from Eastern Europe* ceased publication at the end of 1973, but much of the material included in it is available from the Russian language *Referativnyi Zhurnal.*

Such abstracting and indexing publications are, because of their cost, likely to be available only in larger libraries and thus not accessible to practising production engineers in many—indeed, most—companies. Indeed, practising production engineers are not likely to have the time or inclination to use them and their main use will be to research workers or by information and library personnel on behalf of engineers.

There are, however, a number of smaller, more specialised publications available more likely to be used by engineers. One, the *PERA Bulletin,* is primarily available to members of the Production Engineering Research Association (PERA) free of charge six times a year, but is republished with a time lag in *The Production Engineer,* covering the whole range of production engineering topics from a variety of English-language journals. Another, the *Library Bulletin* of The Machine Tool Industry Research Association (MTIRA) is, again, available to members free of charge, but is also generally available on an annual subscription basis. In this case the cover is more restricted to the manufacturing techniques of production engineering and to machine tools, but also covers new books, standards and catalogues. A major advantage is fast publication, a few days after the end of each month for which material is included.

STANDARDS

The *British Standards* which will be used from time to time by production engineers include a wide range of those concerned with ferrous and non-ferrous materials and engineering products common to many engineering subjects, but some of special interest include:

BS308: Parts 1, 2 and 3: 1972 Engineering Drawing Practice. To define the basic methods of presenting engineering information in the form of drawings.
BS3800: 1964 Methods for Testing the Accuracy of Machine Tools
BS4640: 1967 Classification of Metal Working Machine Tools by Types
BS4656: 1971– The Accuracy of Machine Tools and Methods of Test. In various parts for specific types of machine.
BS4813: 1972 Method of Measuring Noise from Machine Tools
BS Handbook No. 18: 1972 Metric Standards for Engineering. Includes a selection covering common items used in the workshop, including cutting tools.

The *BSI Yearbook* should be consulted for the range of relevant standards, including cutting tools and their fixing points on machine tools, many of which are now in agreement with the world-wide *ISO Standards.*
A wide-ranging study which includes references to many standards and literature items is *Specifications and Tests of Metal Cutting Machine Tools,* published by UMIST (University of Manchester Institute of Science and Technology). Schlesinger limits for acceptance tests are still in use, after 40 years, although never official standards in their originally published form, but they are gradually being superseded as *British Standards,* based on *ISO Standards,* are written. They can be found in *Testing Machine Tools,* by G. Schlesinger (7th edn, Machinery Publishing Company, 1966).
A full discussion of the relevant standards likely to be encountered in the workshop as metric units and workpieces and machine tools

to metric standards are introduced is continued in a manual, *Metrication in the Machine Shop* (HMSO, 1970).

Much work on standards directly concerned with machine tools is coordinated by the Machine Tool Trades Association and some of their draft standards are generally available.

PATENTS*

The sections *Abridgments of Patent Specifications* of most interest are likely to be B3 Metal working, B4 Cutting; Hand tools, and B5 Working non-metals; Presses.

JOURNALS

Two journals likely to be read by most production engineers, to keep abreast of current news and announcements of new machines and the various tools of their trade, are *Machinery and Production Engineering* and *Metalworking Production*. Of these, the latter is aimed towards supervisory and managerial levels; the other includes a more practical, technical bias. Other journals having a similar, non-academic appeal include *Iron Age International, Tooling* and, with wider engineering interests, *Engineer, Engineering* and *Engineers' Digest*.

The professional journal is *The Production Engineer,* the journal of the Institution of Production Engineers. The Institution of Mechanical Engineers, in *The Chartered Mechanical Engineer,* also has some interest in production engineering. *BNCS News,* the journal of the British Numerical Control Society, has a specialised interest for users and designers of numerically controlled machines.

For various metal forming interests in particular, the relevant specialised journals are *Metallurgia and Metal Forming* and *Sheet Metal Industries*. The specialised interests of welding are catered for in the journal *Metal Construction and British Welding Journal.*

Metrology and Inspection covers interests in measurement and quality control.

Journals likely to be of prime importance to research and development engineers in production and in machine tool design include *International Journal of Machine Tool Design and Research, International Journal of Production Research* and, especially, *Annals of CIRP,* for very valuable contributions from a wide range of overseas research centres.

* See Chapter Six.

Useful journals from American publishers include *American Machinist*; *Iron Age*; *Manufacturing Engineering and Management*; *Modern Machine Shop*, for general cover; *Abrasive Engineering* and *Automation*, for specialist interests; and the *Transactions of the ASME*. *Journal of Engineering for Industry*, for research papers. The reading of foreign-language journals is likely to be restricted among production engineers, but research and development engineers would do well to scan German journals such as *Industrie-Anzeiger*, which includes regularly articles reporting research carried out in the machine tool laboratory of the Technical High School, Aachen, and *Werkzeug Maschine International*, which includes contributions from a number of different countries. *Werkstattstechnik*, the professional journal of the production engineering group of the VDI (Verein Deutscher Ingenieure), carries much material on the development of production methods and organisation. *Werkstatt und Betrieb* is another German-language journal including a wide range of technical articles on metal cutting and forming machines and processes. Collectively, these German journals carry much more detailed technical data and engineering analysis than is customary in British and American journals.

Two relevant Russian journals available in English, as cover-to-cover translations, are *Machines and Tooling (Stanki i Instrument)*, which is entirely concerned with machine tools, cutting tools and related processes, and *Russian Engineering Journal (Vestnik Mashinostroeniya)*, which includes some production engineering among general mechanical engineering.

The Dutch journal *Metaalbewerking* is also of quite high standard from the point of view of original material, particularly on presses and sheet-metal working, and includes some contributions originating from research carried out by the TNO (Organisatie V. Toegepast Natuurwetenschappelijk Onderzoek) research organisation.

Some machine tool manufacturers issue journals—for example, Alfred Herbert Ltd with *Machine Tool Review, Wickman News* and the German *VDF Information,* although publication of some can be irregular. *Machine Tool Engineer* is published in India by Hindustan Machine Tools Ltd.

CONFERENCES

The annual International Machine Tool Design and Research Conference, organised in alternate years by UMIST and Birmingham University, proceedings published for 1960–1970 under the title

Advances in Machine Tool Design and Research (Pergamon) and from 1971 as *Proceedings* . . . (Macmillan) results in a considerable number of papers on machine tool design, metal cutting and metal forming appearing in a series of bound volumes. Other relevant published conference proceedings emanate, from time to time, from the Institution of Production Engineers and the Institution of Mechanical Engineers. A very prolific American source of conferences is the Society of Manufacturing Engineers, but the papers may not always be published subsequently in collected form. Retrospective searches will be confusing unless the previous titles of this organisation are remembered—ASTE, ASTME. The *Proceedings* of the Annual Meeting and Technical Conference of the Numerical Control Society, in the USA, provide forward-looking papers on the development and application of control techniques.

Notable conference publications in recent years include *Machinability* and *Materials for Metal Cutting,* published by the Iron and Steel Institute as *ISI Reports,* and the proceedings of the International Grinding Conference, published as *New Developments in Grinding* (The Grinding Institute, 1972). Occasional MTIRA conferences provide information on aspects of machine design, noise and trends in production processes, and although the conferences may be restricted to members, the proceedings are usually for sale.

Other British sources of conference proceedings are the Birniehill Institute and the Welding Institute.

DIRECTORIES

The directory most commonly useful for production engineers seeking suppliers of the machines, tools, equipment and services used in their factories is *Machinery's Annual Buyers' Guide.* In its 44th year after seeing others attempt more detailed, expensive, machine specification manuals or microfilm catalogue collection systems, it is an invaluable general-purpose aid to the production engineer. An additional feature which is especially useful is a guide to trade names and agents for foreign machines. While this guide does include presses and sheet-metal working, a useful additional source is the *Sheet Metal Industries Year Book* (Industrial Newspapers), which provides a good buyers' guide covering materials, plant and equipment for all types of press work, not solely for sheet-metal working, as well as much technical information on materials and processes. *The Engineer Buyers Guide* (Morgan Grampian, annual) contains some similar material.

The production engineer may need to organise the supply of

castings and forgings, and for this the *Foundry Directory and Register of Forges* (Standard Catalogue Company, annual) will be of great value. The *Engineering Components, Materials Index* (Technical Indexes Ltd. includes suppliers of materials and semi-finished products which will be of interest to production engineers, as well as mechanical and electrical components and equipment required by the machine tool manufacturer. This product data book is used in conjunction with a microfilm cassette system to provide access to catalogue material.

Apart from the *Telephone Directory-Yellow Pages* there is scarcely another directory of any great relevance, although this is not to say that there will not be difficulties sometimes in locating machinery or services. Then, reference can be made to the enquiry services provided by technical journals—in particular, *Production Engineering* and *Machinery* — and by the research associations PERA and MTIRA.

DICTIONARIES

Practitioners in particular technical fields do not often feel the need for definitions of the terms they use, except perhaps in rapidly developing technologies where glossaries may appear as appendices in books and in technical journals. Mechanical engineering diction-aries for technical terms are relatively scarce; those for production engineering, almost non-existent. *A Dictionary of Machining,* by E. N. Simons (Muller, 1972), covers the process of metal-cutting. Some of the multilingual dictionaries useful for assistance in translation also provide useful definitions and explanations of terms. The most outstanding of these *Spanende Werkzeugmaschinen,* by H. G. Freeman (W. Giradet, 1963), gives German-to-English translation and is truly encyclopaedic, although the text is in German. Others include *The Machine Tool. An Interlingual Dictionary of Basic Concepts,* by E. Wüster (Technical Press, 1968), with an English–French master volume giving definitions of terms, including parts of machines; *Technical Dictionary of Production Engineering,* edited by R. Walther (Pergamon, 1972), in two volumes, English/German and German/English; and *Toolroom Machinery in Four Languages,* by H. E. Horten (CR Books, 1966), in English, German, French and Spanish. A more specialised language dictionary is *English–Russian Dictionary of Metal-Working and General Engineering Shop Terms,* edited by A. L. Zaigevsky (Soviet Encyclopaedia Publishing House, 1969).

Various processes are dealt with separately in *Dictionary of*

Production Engineering (W. Giradet, 1960–1969), German–English–French in four slim volumes dealing with *1. Forging and Drop Forging* (1960), *2. Grinding and Surface Roughness* (1963), *3. Sheet Metal Forming* (1965) and *4. Fundamental Terms of Cutting* (1969), all edited by expert committees from CIRP (College Internationale Recherche Production).

Cutting tools and cutting processes are also the subject of *Wörterbuch Werkzeuge*, by H. G. Freeman (W. Giradet, 1960), a German/English dictionary; and *Illustrated Technical Dictionary of Metal Cutting Tools*, by T. Heiler (Blackie, 1964), in five languages — English, German, French, Italian and Spanish.

More specialised dictionaries not likely to be completely up to date in a rapidly developing subject are *Numerical Control of Machine Tools. Dictionary in Four Languages*, by H. E. Horten (Hinkel, 1969), and *NC Lexicon*, by Y. H. Attiyate (IAMI NC Lexicon, 1971).

BOOKS

Production engineering subjects are dealt with at a variety of levels, varying from the academic or theoretical to the practical—the latter sometimes largely unrecorded by the written word, being trade or craft practice. It is proposed to mention here a selection of those books which might be expected to be useful to production engineers already educated in their subject and those likely to be in supervisory, managerial positions, thus eliminating craft training texts.

Apart from general introductory material, it is convenient to divide the subject so as to follow the specialist interests of various groups of people as follows: management, production control, quality control, inspection, production planning, method study, ergonomics, handling and assembly, metal cutting and forming, welding; design of machine tools.

General, introductory and descriptive books

For those, possibly librarians and information scientists, who may have to find their way to data or reading material on a subject encompassed within production engineering, or for those production engineers venturing into a new part of their field, some explanatory, descriptive and introductory sources may be required.

Modern Machine Tools, by F. H. Habicht (Van Nostrand Reinhold, 1963), *Machine Tools*, by H. D. Hall and H. E. Linsley

(Industrial Press, 1957), and *Fundamentals of Manufacturing Processes and Materials,* by C. Edgar (Addison-Wesley, 1966), are readable, descriptive texts. A recently published clear, concise, less detailed explanation of basic machine tools and metal cutting processes is given in the well presented and illustrated *Machine Tools, Metals and Cutting Fluids* (1972), only slightly biased towards the consideration of cutting oils because of its source, the BP oil company. A general text aimed at undergraduate level, covering processes and machines, is *Modern Workshop Technology,* by H. Wright Baker (3rd edn, Macmillan, Part 1, 1966; Part 2, 1969), and another, *Manufacturing Technology,* by M. Haslehurst (English Universities Press, 1970), includes some consideration of the planning and organisational activities.

For more technical detail, it is difficult to match, in a single volume, the subject cover and usefulness of the *Tool Engineers' Handbook,* edited by F. W. Wilson (2nd edn, McGraw-Hill, 1959), despite the fact that it has not been revised since the second edition of 1959 and is exclusively American in its references to standards and materials. For the more sophisticated, and newer, numerically controlled machines and their programming, the *Numerical Control Users Handbook,* edited by W. H. P. Leslie (McGraw-Hill, 1970), can be consulted. A wide range of metal forming processes is covered in *Chipless Machining,* by C. H. Wick (Industrial Press, 1960).

Management, production control

Management in its widest sense includes the organisational control of many activities within the total company structure—policy, product design, marketing, etc.—which are outside the scope of production engineering. However, management activities are carried on within the production engineering function and some textbooks are written from this standpoint, with a technical bias. Managers need to be familiar with the methods available to them and to be aware of the implications of 'modern' terms such as MBO (management by objectives), management by exception, etc. These are most likely to be included in recent general books such as *Engineer's Handbook of Management Techniques,* edited by D. Lock (Gower Press, 1973), or more specific books such as *Management by Objectives in Action,* by J. W. Humble (McGraw-Hill, 1970). Standard texts such as *Modern Production Management,* by E. S. Buffa (4 th edn, Wiley, 1973), and *The Management of Production,* by J. D. Radford and D. P. Richardson (3rd edn, Macmillan, 1972), are likely to concentrate on information on the systems for organising and controlling production, also dealt with in handbooks on the organisation of production such as *Production Handbook*, edited by G. B. Carson (3rd edn, Ronald Press, 1972), and *Tool Engineers' Handbook*, edited by F. W. Wilson (2nd edn, McGraw-Hill, 1959), which cover all aspects. *Elements of Production Planning and Control*, by S. Eilon (Collier/Macmillan, 1962), has a more British base, and is particularly valuable for its many references. Other standard British textbooks include *The Principles of Production Control*, by J. L. Burbidge (3rd edn, Macdonald, 1971). At a more basic level, *Modern Production Control*, by A. W. Willsmore (3rd edn, Pitman, 1963), and *Production Control in Practice*, by K. G. Lockyer (Pitman, 1966), provide guidance on the simpler, practical systems and the associated paper work. The particular application of critical path analysis to production control and planning is dealt with in *An Introduction to Critical Path Analysis, by* K. G. Lockyer (3rd edn, Pitman, 1970), and *Critical Path Analysis by Bar Chart*, by C. W. Lowe (2nd edn, Business Books, 1969).

As computers come increasingly into use for handling management information, including production control, source books such as *Factfinder 3: Production Control Packages. A Survey of Computer Applications for Production Control in the UK*, by G. K. Holden (National Computing Centre, 1973), will be useful.

Financial control is a function often separate from production engineering with the company, but since some knowledge of it is essential, some books specifically written for the production engineer exist; examples are *An Engineer's Guide to Costing* (IProdE, 1969) and *Cost Control for Production Management*, by W. A. Reynolds and J. B. Coates (Machinery Publishing Company, 1970).

Quality control, inspection

Of the textbooks on metrology, a selection such as *Handbook of Industrial Metrology*, by the American Society of Tool Manufacturing Engineers (ASTME) (Prentice-Hall, 1967), for a general introduction, and *Engineering Metrology*, by K. J. Hume (3rd edn, Macdonald, 1970). *Metrology for Engineers*, by J. F. W. Galyer and C. R. Shotbolt (Cassell, 1969), and *Dimensional Metrology for Engineers*, by G. G. Thomas (Butterworths, 1974), for more practical detail, should answer most queries. A less commonly available book, *Foundations of Mechanical Accuracy*, by W. R. Moore (Moore Special Tool Company, 1970), provides a unique detailed exposition of the measuring and manufacturing techniques associated with accuracies of the order of millionths of an inch.

The general principles of quality control and the associated management systems can be found in, for example, *Elements of Production Planning and Control,* by S. Eilon (Collier Macmillan, 1962), and *Production Handbook,* edited by G. B. Carson (3rd edn, Ronald Press, 1972), while general management inspiration is to be found in *Contribution of Inspection to Quality Assurance,* by S. Weinberg (Industrial and Commercial Techniques, 1970), and *Managing to Achieve Quality and Reliability,* by F. Nixon (McGraw-Hill, 1971).

Production planning and methods

The general information on standard manufacturing methods and machines is contained in the handbooks referred to earlier. *Manufacturing, Planning and Estimating Handbook* (ASTME, 1963) deals comprehensively with its subject and is overlapped to some extent by *Tool Engineers' Handbook.* Books on individual processes and machines include *Broaching,* by C. Monday (Machinery Publishing Company, 1960); *Modern Honing Practice,* by A. J. Cox (Machinery Publishing Company, 1969); *Copy Turning,* by K. J. Downes (Machinery Publishing Company, 1961); *Jig Boring,* by R. S. Connell (Machinery Publishing Company, 1963); and many more books from the same publisher, including *Machinery's* Yellow-Back Series of smaller books. Other useful books include *A Treatise on Milling and Milling Machines* (Cincinnati Milling Machine Company, 1959); *Mechanical Presses,* by H. Makelt (Arnold, 1968); and *Hydraulic Presses,* by G. Oehler (Arnold, 1968).

The detailed techniques for the design and use of equipment and tools for the manufacture of individual components is dealt with in *Production Engineering: Jig and Tool Design,* by E. J. H. Jones and H. C. Town (8th edn, Butterworths, 1972), and *Tool Design,* by C. Donaldson, G. H. Le Cain and V. C. Gould (3rd edn, McGraw-Hill, 1973), and, for press tools, dies and moulds, *Die Design Handbook* (2nd edn, McGraw-Hill, 1965).

For a concept attracting attention in the literature in more recent years, involving classification of workpieces and their manufacture in similar groups, a useful introduction is *Readings in Group Technology,* by G. A. B. Edwards (Machinery Publishing Company, 1971). This is a method for batch-production; mass-production is dealt with in *Mass-production Management. The Design and Operation of Production Flow-Line Systems,* by R. Wild (Wiley, 1972).

The concepts, applications and implementation of work study, activity sampling, work measurement, time study, time and motion study and such like, as aids to the development of productivity, are well covered in a sensible, practical text, *Work Study and Related Management Services,* by D. A. Whitmore (Heinemann, 1968). Other standard texts include *Work Study,* by R. M. Currie (3rd edn, Pitman, 1972), and *Introduction to Work Study* (International Labour Office, 1966).

Ergonomics

Ergonomics can be regarded as the study of factors influencing the efficiency of human work, or the relationship between man and his environment, and is thus of importance to production engineers and has obvious links with specific subjects such as work study, design of machine controls, etc. A general text, providing data on the capabilities of operatives, design of equipment, the working environment, measurement of work and work organisation, is *Ergonomics,* by K. F. H. Murrell (Chapman and Hall, 1965). By the same author, *Ergonomics, Fitting the Job to the Worker* (British Productivity Council, 1960) is a shorter text for the non-specialist.

An allied field is the acquisition of skills, and *Industrial Skills,* by W. D. Seymour (Pitman, 1966), provides both theoretical and practical information. For training schedules related to specific engineering skills, the many publications of the Engineering Industries Training Board should be consulted.

Handling and assembly

Much of the literature on materials handling and conveying is concerned with warehouses, moving of bulk materials or other situations not directly applicable to production engineering, but *Materials Handling in the Machine Shop,* by C. Hardie (Machinery Publishing Company, 1970), is one of the very few books to deal with the production engineer's situation directly. A particular problem in handling is that of dealing with the surplus material removed from the components when they are machined. The only book treating this as a subject in its own right is *Swarf and Machine Tools,* edited by P. J. C. Gough (Hutchinson, 1970).

Handling of component parts is one step on the way to automated assembly, and two authoritative texts on this subject are *An Introduction to Mechanical Assembly,* by W. V. Tipping (Business Books, 1969), and *Mechanised Assembly,* by G. Boothroyd and A. H.

Redford (McGraw-Hill, 1968). The former book is particularly strong on practical applications, while the latter is rather more scholarly and includes examination of the scientific basis for the performance and design of machines. A multi-part publication, *Production Engineering Data Memoranda* (IProdE), provides 12 parts on automated assembling, which illustrate just about all the available mechanisms and devices employed for the purpose.

Books on the currently developing subject of so-called 'robot' devices, which can be used for loading and unloading machines, are mainly academic and theoretical, with the exception of *Industrial Robots—A Survey,* edited by G. Lundstrom, L. Lundstrom, A. Arnstrom and B. W. Rooks (International Fluidics Services, 1972).

Metal cutting and forming

The basic principles are discussed for both cutting and forming in *Manufacturing Properties of Materials,* by J. M. Alexander and R. C. Brewer (Van Nostrand Reinhold, 1967), while *Introduction to the Principles of Metalworking,* by G. W. Rowe (Arnold, 1965), concentrates on forming and theories of metal deformation and *Fundamentals of Metal Machining,* by G. Boothroyd (Arnold, 1965), on cutting. All give a good guide to the literature at the date of publication. *The Performance of Metal Cutting Tools,* by R. Tourret (Butterworths, 1958), is a compilation from hundreds of references providing a summary of research and testing on cutting processes, and provides a basis for predicting the effects of changes in work-piece material, tool material, cutting conditions, etc., on the performance.

Two books from the ASTME manufacturing data series provide a thorough understanding and sound recommendations on the subjects in question. These are *Cutting Tool Material Selection,* edited by H. J. Swinehart (ASTME, 1968), and *Cutting and Grinding Fluids: Selection and Application,* edited by R. K. Springborn (ASTME, 1967).

The actual data on how to cut metal can be difficult to pin down in the published literature for any but the most common operations on widely used materials, e.g. turning of mild steel, drilling En8, and even then the recommendations are apt to be imprecise and general, to allow for the unpredictable variations in workpiece material and the unspecified details of tool angles or cutting fluid. Once again American publications, such as the previously mentioned *Tool*

Engineers' Handbook, provide the most comprehensive single-volume collections of data, and the effort must be made to interpret them and translate the material specifications into British equivalents. *Machining Data Handbook* (2nd edn, Metcut Research Associates, 1972) is perhaps not widely known, but includes neatly tabulated up-to-date data for a wide range of machining operations and workpiece materials. *Metals Handbook. Vol. 3 Machining* (8th edn, American Society for Metals, 1967) can be used to supplement this and is a mix of tabulated data (some of it from *Machining Data Handbook*) and case studies. *Machining Science and Applications: Theory and Practice for Operation and Development of Machining Processes,* by M. Kronenberg (Pergamon, 1966), is a veritable mine of information, but most of it from German sources, with attendant difficulties in relating references to material specifications.

The equivalent British information, so far as it exists in detailed factual form, is scattered and not well presented in book form. The books mentioned at the start of this section can be supplemented by exploration for data amongst those included in the 'mainly descriptive' category in the earlier general section.

For metal forming, *Chipless Machining,* by C. H. Wick (Industrial Press, 1960), gives information on a range of metal forming processes, including some of the less commonly known, and updating can be provided in some fields from *Developments in High-Speed Metal Forming,* by R. Davies and E. R. Austin (Machinery Publishing Company, 1970). More tabulated data and practical working formulae are given in *Sheet-Metal Industries Yearbook* (Fuel and Metallurgical Journals Ltd., 1974) in relation to the cutting, bending and drawing of sheet and plate materials, while *Metals Handbook. Vol. 4, Forming* (8th edn, American Society for Metals, 1969) provides both case studies and data for a wide range of processes. Information on the carrying out of operations on various types of press is given in *Mechanical Presses,* by H. Makelt (Arnold, 1968); *Hydraulic Presses,* by G. Oehler (Arnold, 1968); *Metal Forming Handbook,* by L. Schuler (4th edn, L. Schuler, 1966); and *Bliss Power Press Handbook* (E. W. Bliss Company, 1950); the last two books being sponsored by press manufacturers. Data on a specific press technique, which has been increasingly developed and applied in recent years, is excellently provided in another manufacturer's book, *Fine Blanking Practical Handbook* (Feintool A. G., 1972).

There is a range of 'newer' machining processes, i.e. processes developed the past 20 years or so, including spark, electrochemical, ultrasonic machining methods and others, which are described in, for example, *Non-Traditional Machining Processes,* edited by R. K. Springborn (ASTME). One of the more important of these processes is described in more detail in *Electrochemical Machining,* edited by A. E. De Barr and D. A. Oliver (Macdonald, 1968), and in *Fundamentals of Electrochemical Machining,* edited by C. L. Faust (Electrochemical Society, 1971).

Welding

A comprehensive treatment of welding processes is given in *Welding Handbook,* by the American Welding Society (Macmillan) in its six volumes, updated at intervals: *Part 1: Fundamentals of Welding* (6th edn, 1968); *Part 2: Gas, Arc and Resistance Processes* (6th edn, 1969); *Parts 3A and 3B: Welding, Cutting and Related Processes* (6th edn, 1971); *Part 4: Metals and their Weldability* (6th edn, 1972); and *Part 5: Applications of Welding* (6th edn, 1973). A more specifically British review, which includes some practical detail, is provided by *The Solid Phase Welding of Metals,* by R. F. Tylecote (Arnold, 1968), which is less academic in style than its title might indicate. It also includes more recent developments such as friction welding, explosive welding and ultrasonic welding. Another modern technique is dealt with in *Electron Beam Welding,* by A. H. Meleka (McGraw-Hill, 1971).

A practical book describing methods and devices for improving production in welding shops is *Production Welding,* by P. F. Woods (McGraw-Hill, 1971). An important consideration in welding practice is dealt with in *Control of Distortion in Welded Fabrications* (2nd edn, Welding Institute, 1968) and, in fact, many other authoritative texts on specific welding processes are available from The Welding Institute.

The techniques for some related joining processes are described in *Soldering and Brazing Technology,* by S. C. Churchill (Machinery Publishing Company, 1963).

Design of machine tools

Good, authoritative textbooks on machine tool design are not common. Some, such as *The Theory of Machine Tools,* by J. W. Browne (Cassell, 1968), *Design and Construction of Machine Tools,* by H. C. Town (Iliffe, 1971), and *Design of Machine Tools,* by O. A. Johnson (Chilton, 1971), are intended for students up to about HNC level.

A standard text on the general subject of the design of machine tools is *Design of Metalcutting Machine Tools,* by F. Koenigsberger

(Pergamon, 1964). The English translations of the Russian books *Machine Tool Design,* translated by N. Weinstein (4 vols., Central Books, 1970), are very complete and unique since they include much descriptive material on the mechanisms of individual types of machine. For particular aspects of design there are *Machine Tool Structures,* by J. Tlusty and F. Koenigsberger (Pergamon, 1970); *Vibration of Machine Tools,* by G. Sweeney (Machinery Publishing Company, 1971); and *Machine Tool Dynamics,* by D. B. Welbourn and J. D. Smith (Cambridge University Press, 1970).

There are some descriptive texts which could be of assistance in building machines for specific purposes, where the idea for the mechanism is all-important—e.g. *Transfer and Unit Machines,* by E. D. Lloyd (Machinery Publishing Company, 1969); *Automatic Machine Tools,* by H. C. Town (Iliffe, 1968); and *Special-Purpose Production Machines,* by W. Charmon (Crosby Lockwood, 1968)—but further detailed assistance in design is limited to the journal and conference literature, which may be of some sporadic help to the designer unless the publications of the Machine Tool Industry Research Association and Production Engineering Research Association are available. These are initially restricted to members of the respective associations, but some become generally available after a period of time. In particular, items from the series of design procedures *Notes for Designers,* published by MTIRA and generally available, include some on the design of spindles. Research report subjects include the use of cast-iron in machine tools, variable-speed drives, adaptive control, hydrostatic bearings and ergonomics.

For those capable of reading in German, two books giving an insight into German research applied to machine tool design and production engineering are *Moderner Werkzeugmaschinenbau,* by H. Opitz (W. Giradet, 1971), and *Moderne Produktionstechnik,* by H. Opitz (W. Giradet, 1970).

Of course, the information required for the design of many machine tool details is similar to that used by other designers throughout mechanical engineering, since bearings, gears, belt drives, etc., are all commonly used.

HISTORICAL ASPECTS

The history of production engineering, at least so far as the published literature is concerned, revolves mainly around the development of the machines and the biography of the men who designed and made them. No attempt will be made to provide a guide to the original source material here, but a selection of the more modern books on the subject will provide reference to the source material. A readable general account of the history of machine tool development combined with biographical treatment is *Tools for the Job,* by L. T. C. Rolt (Batsford, 1965). *A History of Machine Tools 1700–1910,* by W. Steeds (Oxford University Press, 1969), is detailed and factual and contains many illustrations. More detailed treatment of individual machine types is available from a series of monographs, *History of the Lathe,* by R. S. Woodbury (MIT Press, 1961), *History of the Milling Machine* (MIT Press, 1960), *History of the Grinding Machine* (MIT Press, 1959) and *History of the Gear Cutting Machine* (MIT Press, 1958), by the same author.

Various publications from the Science Museum, including the catalogue to the machine tool collection, are of interest.

Issues of *Machinery* (New York), *Transactions of the ASME,* for the USA, and the *Proceedings of the Institution of Mechanical Engineers,* for UK, from the latter half of the nineteenth century onwards, include much fascinating material, including lively contributions to discussions and letters to the editor.

23

Design

C. A. Murfin

INTRODUCTION

This chapter is limited to consideration of the literature which deals with the activity of design. The publications referred to are those which may be found to be of use in obtaining an understanding of the discipline of design and the influence exerted by factors not directly in the engineering field. Those publications concerned with conventional engineering analysis, the design of a specific product or design for a particular function are not included. Exclusion of these is based on the premise that an appreciation of the discipline of design and the many design methods and techniques available forms the base on to which specialised detailed knowledge can be added for design in a particular field.

The chapter starts with the literature which relates to systematic design, design methodology and design methods or techniques, and then proceeds to give an introduction to the literature dealing with some of the influences of major importance in design. Where reference to a publication is made through a single topic, it is not implied that the book is limited to consideration of that topic only.

THE ACTIVITY OF DESIGN

Design. A word which is interpreted and employed by each user in his own special way. Definitions and descriptions abound in the literature, where there are as many as there are authors. Here engineering design is the subject matter, but even this qualified term does not have a short and exact definition. The reason for this is the scope and complexity of the designing activity which incorporates a range of technical and non-technical factors in a combination of the sciences, mathematics and arts carried out by man acting individually or in groups. Designing in the sense given in *Sciences of the Artificial*, by H. A. Simon (MIT Press, 1969), is to a limited degree an artificial science, but the activity of design is a discipline in its own right. An understanding of the discipline of design combined

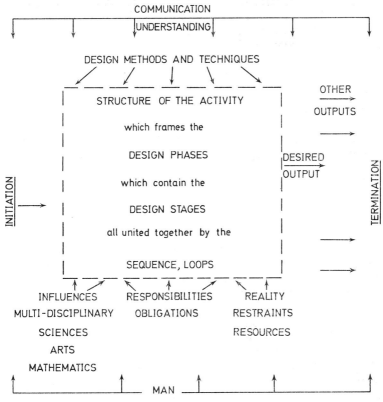

COMMUNICATION

UNDERSTANDING

DESIGN METHODS AND TECHNIQUES

STRUCTURE OF THE ACTIVITY

which frames the

DESIGN PHASES

which contain the

DESIGN STAGES

all united together by the

SEQUENCE, LOOPS

OTHER OUTPUTS

DESIRED OUTPUT

INITIATION

TERMINATION

INFLUENCES   RESPONSIBILITIES   REALITY
MULTI-DISCIPLINARY   OBLIGATIONS   RESTRAINTS
SCIENCES   RESOURCES
ARTS
MATHEMATICS

MAN

Figure 23.1   A model of the activity of design

with an appreciation of the many design methods which can be employed within the discipline provides the foundation from which all designers work regardless of their particular fields. The discipline, together with the applied methods, forms the common ground for design in all branches of architecture, civil engineering, electrical engineering and mechanical engineering.

In the last few decades considerable effort has been directed towards the study of design and designing. The impetus to these studies has been provided by two sources. One of these is the expressed dissatisfaction with the standard or performance of contemporary design. The other is the growth in magnitude and complexity of engineering design with the need to integrate the efforts of many individuals. Publication of the Feilden Report *Engineering Design* (Council for Scientific and Industrial Research) (HMSO, 1963) focused attention upon the problem. It is through an understanding of the over-all activity of design in terms of its structure, contents, interactions and sequences, combined with an appreciation of man's ability, that the opportunity to improve design will be made available.

In the studies of the design activity it is possible to classify the main areas of interest in the following manner:

The framework in which the nature of the design activity and the structure of the problem may be examined.

The procedural aspects of the activity to define the sequences and the main phases through which the design passes from initiation to termination.

The design process by which work is progressed through each phase in steps and stages.

Design methods and techniques which can be of use in tackling particular difficulties at any stage.

The most excellent book *Design Methods. Seeds of Human Futures,* by J. C. Jones (Wiley/Interscience, 1970), provides a very readable introduction to all the interest areas given above. Part I reviews the design activity from craft evolution to the systematic procedures of the present and part II is a directory of some 35 design methods. A classical example of early work by a design theorist can be found in an impressive book by the erstwhile champion of a systematic approach to design, *Notes on the Synthesis of Form,* by C. W. Alexander (Harvard University Press, 1964). The approach adopted in this text was to consider the nature of design as a science and then to reduce the problem to small parts which could be solved and then synthesised to a whole. A similar procedure can be found in *Introduction to Design,* by M. Asimow (Prentice-Hall, 1962), where design is broken down into major phases each containing a number of steps which have to be followed through in sequence. A practical example of this type of approach, but incorporating parts of Matchett's Fundamental Design Method with a model programme and check lists, can be found in *Problem Analysis by Logical Approach,* by R. W. Latham (Atomic Weapons Research Establishment, UKAEA, 1965). The systems approach to design for problems in situations where inputs and outputs of the system may be clearly identified is presented in *A Methodology for Systems Engineering,* by A. D. Hall (Van Nostrand, 1962). These procedures may be compared with that proposed in *Technological Innovation—A Methodology,* by L. B. Archer (Inforlink, 1971), for continuous development and modification of models and analogies of the problem with successive iterations converging on a final solution. *The Discipline of Design,* by V. K. Handa, P. H. Roe and G. N. Soulis (Allyn and Bacon, 1967), together with *Science of Engineering Design,* by P. H. Hill (Holt Rinehart and Winston, 1970), are both useful in part for their individual interpretation of the activity. A varient on the work of Asimow is presented in the form of a textbook for the engineering student in *Introduction to Engineering Design,* by T. T. Woodson (McGraw-Hill, 1966). *Engineering Design,* by J. H. Faupel (Wiley, 1965), and *Engineering Design,* by R. Matousek (Blackie, 1964), have a tendency to concentrate attention upon the detail phase of the activity. For the interested observer *The Design of Design,* by G. L. Glegg (Cambridge University Press, 1969), is a well prepared and very readable little book. The student about to start design will find that *Design Engineering,* by J. R. Dixon (McGraw-Hill, 1966), with its sections on inventiveness, analysis and decision making, and *An Introduction to Engineering and Engineering Design,* by E. M. Krick (2nd edn, Wiley, 1969), are very effective texts. An example of systematic development, as opposed to modification of existing design, from inception to commencement of detail design, can be found in *Design Synthesis,* by T. H. Ellinger (2 vols., Wiley, 1968). The engineer will find rewarding some aspects, particularly the description of design techniques and procedures, of *Design in Architecture,* by G. Broadbent (Wiley, 1973). An interesting example of an interrogation method, similar to the standard work study critical examination approach, can be found in *The Rational Manager,* by C. H. Kepner and B. B. Tregoe (McGraw-Hill, 1965).

The most important journal in the design field is the *Engineering Designer* (Institution of Engineering Designers), which includes articles of current interest and design abstracts. The *'DMG: DRS' Journal—Design Research and Methods,* of the Design Research Society (UK) and the Design Methods Group (USA), is the source for viewing current thinking in the theoretical field. *Mechanical Design—Reference Sources,* by J. N. Siddall (University of Toronto Press, 1967), is a now somewhat dated reference source to the literature in mechanical engineering design.

Many conferences and meetings concerned with engineering design have been held in recent years. Contributions to these have been fairly large in quantity, but very variable in quality. Contained within them are a significant number of papers which are of value and when interpreted by the individual designer for his particular situation can be of direct use. There are, however, too many papers which are not easy to read or understand and for which practical implementation of the concepts is even more difficult. It is unfortunate that many papers are prepared with clarity and intelligibility given a very low merit rating by the authors. In design, where clear unambiguous communication is essential, and is achieved with drawing at the detail design phase, it is somewhat contradictory to find that some authors do not prepare their papers with ease of understanding a primary requirement. The following publications contain the contributions to the more important conferences which have covered the whole spectrum of the activity of engineering design:

Conference on Design Methods, edited by J. C. Jones and D. G. Thornley (Pergamon, 1963)

The Design Method, edited by S. A. Gregory (Butterworths, 1966)

Design Methods in Architecture, edited by G. H. Broadbent and A. Ward (Lund Humphries, 1969)

Emerging Methods in Environmental Design and Planning, edited by G. T. Moore (MIT Press, 1970)

Design Participation, edited by N. Cross (St. Martins, 1973)

Papers and articles on engineering design may be found in a number of journals and periodicals. The following list provides the titles of those which are the more prominent:

Proceedings of the Institute of Electrical and Electronics Engineers
Proceedings of the Institution of Civil Engineers
Proceedings of the Institution of Mechanical Engineers
ASME Transactions. Journal of Engineering for Industry
Architects Journal
Mechanical Engineering
Machine Design
Engineering

To these may be added the periodicals which contain information which may be of use to the designer. These include:

Design Engineering
Design News
Engineers Digest
Engineering Materials and Design
Product Engineering

DECISION THEORY

Design has been defined as an iterative decision-making process in *Introduction to Design,* by M. Asimow (Prentice-Hall, 1962). As decisions have to be made at all levels and stages within the over-all activity, a formalised approach as provided by decision theory, from the science of operational research, may be of considerable use. A sensible introduction to the concept may be found in *Elementary Decision Theory,* by H. Chernoff and L. E. Moses (Wiley, 1959),

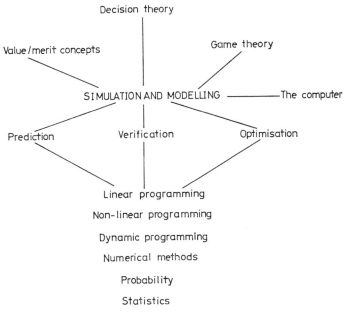

Figure 23.2   Decision theory

or in the more recent *Elements of Decision Theory,* by B. W. Lindgren (Collier Macmillan, 1971). The key to the application of decision theory is the allocation of value or merit to the design variables. A general approach to this can be found in *The Study of Values,* by G. W. Allport, P. E. Vernon and G. Lindzey (Houghton Mifflin, 1960), and more specific studies relating to the decision process in *Design for Decision,* by I. J. Bross (Macmillan, 1953), and *Decision and Value Theory,* by P. C. Fishburn (Wiley, 1965). The over-all strategy to be adopted in the decision-making situation of our competitive world falls within the theory of games. Readable explanations of this are given in *Game Theory. A Non-Technical Introduction,* by M. D. Davis (Basic Books, 1970), and *The Compleat Strategyst,* by J. D. Williams (McGraw-Hill, 1966), with the more technical applications in *Games and Decisions,* by R. D. Luce and H. Raiffa (Wiley, 1957). Establishing a realistic model in a large design exercise can be difficult, and some guidance may be found in *The Art of Simulation,* by K. D. Tocher (English Universities Press, 1963). Simulation, a technique from operational research, combined with the speed and capacity of the computer, is well described in *The Design and Use of Computer Simulation Models,* by J. R. Emshoff and R. L. Sisson (Macmillan, 1970), with *Computer Modelling and Simulation,* by F. F. Martin (Wiley, 1968) typical of the whole range of books available on the subject. Analogue and digital methods for system design and performance optimisation is competently presented in *Computer Simulation of Dynamic Systems,* edited by R. J. Kochenburger (Prentice-Hall, 1972), and *System Simulation,* by G. Gordon (Prentice-Hall, 1968). A good example in the use of decision theory can be found with *Prediction and Optimal Decision,* by C. W. Churchman (Prentice-Hall, 1961), and for the design situation with *Product Design and Decision Theory,* by M. K. Starr (Prentice-Hall, 1963), and *Analytical Decision Making in Engineering Design,* by J. N. Siddall (Prentice-Hall, 1972). This last book contains a good coverage of optimisation analysis with a fair range of references dealing with the mathematical techniques employed. Further consideration of optimisation may be found in *Introduction to Optimisation Techniques,* by M. Aoki (Collier Macmillan, 1971), and *Optimisation: Theory and Practice,* by G. S. G. Beveridge and R. S. Schechter (McGraw-Hill, 1970). A typical example on the application of optimisation methods in design using the computer can be found with *Optimisation Methods for Engineering Design,* by R. L. Fox (Addison-Wesley, 1971), which can be compared with an optimisation technique applied to simple mechanical engineering components provided in *Optimum Design of Mechanical Elements,* by R. C. Johnson (Wiley, 1961), and *Mechanical Design Synthesis,* by R. C. Johnson (Van Nostrand Reinhold, 1971).

ELEMENTS OF SYSTEM DESIGN

The process of systems design is one of the design procedures which can be usefully adopted in particular situations. The primary intent of the process is to use a logical and systematic approach to decision-making which will encourage compatibility between the hardware, human and environment elements of the total system. An adequate introduction to the subject matter can be found in *The Design of Engineering Systems,* by W. Gosling (Heywood, 1962), with further expansion of the methods in *Systems Engineering Methods,* by H. Chestnutt (Wiley, 1967), and *A Methodology for Systems Engineering,* by A. D. Hall (Van Nostrand, 1962). Consideration of design from the systems viewpoint reveals the cybernetic problem of information flow, communication and control. With the human element an

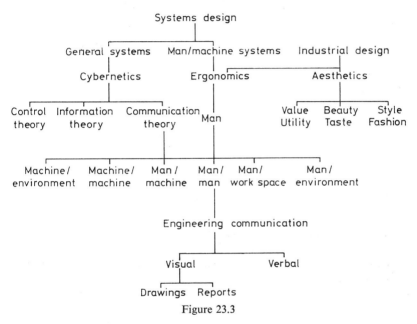

Figure 23.3

integral part of the system, as stressed by the contributions in *The Human Operator in Complex Systems,* edited by W. T. Singleton, R. S. Easterby and D. Whilfield (Taylor and Francis, 1967), then ergonomics and the behavioural sciences, sociology, psychology and anthropology are of importance. Including man the user as an essential part of the system introduces industrial design, which covers aesthetics as well as other social and human factors.

CYBERNETICS

Cybernetics is the science concerned with communication, information manipulation and control in any system which responds to its environment. It has been defined as 'Control and communication in the animal and the machine' in the first of the two classic texts *Cybernetics*, by N. Wiener (2nd edn, Wiley, 1961), and *The Human Use of Human Beings*, by N. Wiener (Sphere, 1969). For the beginner, *An Introduction to Cybernetics*, by W. R. Ashby (Chapman and Hall, 1961), is a rewarding text and *What is Cybernetics?*, by G. T. Guilbaud (Heinemann, 1959), provides a readable answer to the question posed. A review of the subject is contained in *Survey of Cybernetics*, by J. Rose (Iliffe, 1969), and a somewhat dated bibliography in *Cybernetics: A Subject Guide*, by F. and H. Alam (New Science Publications, 1968). A thorough grasp of the basic concepts in communication theory and information theory is necessary before control theory can be applied to a cybernetic system in which there may be man to man, man to machine, machine to machine and man/machine to environment contact. A reasonable survey of modern communication theories is contained in *Human Communication*, by J. L. Aranguren (Weidenfeld and Nicolson, 1967), while *Symbols, Signals and Noise*, by J. R. Pierce (Hutchinson, 1962), provides a good exposition of the essentials of both information and communication theory. Sound introductions to information theory can be found in *The Psychology of Human Communication*, by J. Parry (University of London Press, 1967), and *On Human Communication*, by C. Cherry (2nd edn, MIT Press, 1966).

COMMUNICATION MAN/MAN

A simple discourse on communication as a personal activity within the design process is succinctly presented in *Engineering Communications*, by A. B. Rosenstein, R. R. Rathbone and W. F. Schneerer (Prentice-Hall, 1964). The designer has always to communicate the results of his endeavours to others through drawings and design documents. A large number of books are readily available which deal adequately with graphics, drawing and draughting at all levels. Typical of the American publications which tend to cover the ground from how to hold a pencil, through the principles of draughting to technical illustrating are *Design Drafting*, by J. H. Earle (Addison-Wesley, 1972), and *Basic Graphics*, by P. E. Luzadder (2nd edn, Prentice-Hall, 1968). The most important document for the British mechanical engineer is *BS 308: 1972 Engineering Drawing Practice*, which contains all the information necessary for good draughting practice. For the presentation of the written report, *Communicating Technical Information*, by R. R. Rathbone (Addison-Wesley, 1966), provides a modest little text which is easy to read and includes a section dealing with the more common errors. However, an author with perceptive experience as a report recipient, a willingness to view his own text from the reader's position and an awareness of individual style has a realistic attitude for the production of a readable, informative and persuasive document. The primary aids to the author can then be limited to a good dictionary such as the *Shorter Oxford English Dictionary* (3rd edn, Oxford University Press, 1973); a means to an enhanced vocabulary with *Roget's Theaurus of English Words and Phrases,* edited by R. A. Dutch (Penguin, 1970); advice on how plain words should be used with *Complete Plain Words,* by Sir Ernest Gowers (Penguin, 1970); and the conventions of grammar and vocabulary provided by *Fowler's Dictionary of Modern English Usage,* revised by Sir Ernest Gowers (2nd edn, Oxford University Press, 1965). The most important group of formal documents in the over-all design activity is the specifications. The intent of a specification is to inform the addressee of what has to be done and the limits within which it must be achieved. It has, therefore, to be prepared with considerable care to ensure that all that is essential is included and any unnecessary restraint or ambiguity is excluded. A useful guide which is very general and needs critical interpretation for a specific case can be found in *PD 6112: 1967 Guide to the Preparation of Specifications* (British Standards Institution).

ERGONOMICS

All systems require the attention of man during the periods of manufacture, use and maintenance. The more effectively man can carry out these activities, the more efficient and productive will be the system. The primary aim in a man/machine system design approach is to include the human operator as an integral part of the system throughout the design activity. To do this, the engineering designer needs to have an understanding of human characteristics and performance. It is within ergonomics that the designer will find anatomical, physiological and psychological data relating man to his working activity, equipment and environment. A sound introduction to the subject matter can be found in *Ergonomics—Man in his Working Environment,* by K. F. H. Murrell (Chapman and Hall, 1965), and *The Biology of Work,* by O. C. Edholm (Weidenfeld and Nicolson, 1967). The psychological aspects are thoroughly expanded in *Systems Psychology,* by K. B. de Greene (McGraw-Hill, 1970), and *Psychological Principles in System Development,* edited by R. M. Gagne (Holt, Rinehart and Winston, 1962). Equipment designers will find that the following books provide extensive data: *Human Engineering Guide to Equipment Design,* by C. T. Morgan, J. S. Cook, A.

Chapanis and M. W. Lund (McGraw-Hill, 1963); *Human Engineering Guide for Equipment Designers,* by W. E. Woodson and D. W. Conover (2nd edn, University of California Press, 1964); *The Human Body in Equipment Design,* by A. Damon, H. W. Stoudt, and R. A. McFarland (Harvard University Press, 1966); and two sections of *Anthropometrics for Designers,* by J. Croney (Batsford, 1971). Guidance and recommendations on a number of detailed topics are provided in some *British Standards,* reference to which can be found in the current *BSI Year Book.* A most effective information, abstracting and photocopying service is provided by the Ergonomics Information Analysis Centre in the Department of Engineering Production of the University of Birmingham. The quarterly journal *Applied Ergonomics* (IPC Business Press) is a useful source for current information and case histories on the application of ergonomic data. An engineer looking for a single exposition will find very rewarding a series of articles published in this journal between December 1969 and September 1971 in the form of 15 chapters from an applied ergonomics handbook.

INDUSTRIAL DESIGN

The total value of a marketed design will include an element of aesthetic value attributed to it by each potential user. Judgement of the aesthetic worth to be ascribed to a product during its design cycle and how it is to be achieved requires prediction of the personal desires of people other than those concerned with the design. Each individual has a concept of beauty and taste, and exhibits a particular emotional and intellectual response which may be influenced by current fashion and style. It is these considerations which make it difficult to subject aesthetics to direct analysis and are a reason why it is one of the less well developed disciplines involved in design. In the social sciences, where people are the focal point, techniques have been developed for the collection of data, establishment of case law and prediction of trends and probabilities; see *Subject Bibliography of the Social Sciences and Humanities,* by B. M. Hale (Pergamon, 1970). It is possible that these procedures could be extended into the field of aesthetics in engineering design, but little evidence of this is available. The literature dealing with the direct application of aesthetic considerations to an engineering design is very limited in both quality and quantity. The two books *The Aesthetics of Engineering Design,* by F. C. Ashford (Business Books, 1970), and *Industrial Design for Engineers,* by W. H. Mayall (Iliffe, 1967), need to be read in conjunction with *Art and Alienation,* by

H. Read (Thames and Hudson, 1967), and *Art and Illusion*, by E. H. Gombrich (3rd edn, Phaidon, 1968), to obtain some real insight to the problem. The Council of Industrial Design provides a very useful technical and information service through its records, index and selection service of practising designers in Britain, while its exhibitions and the monthly magazine *Design* present current aspects of industrial design. A stimulating presentation on the morality of industrial design, the role of the industrial designer and the problem of relating design work to the good of mankind can be found in *Design for the Real World*, by V. Papanek (Thames and Hudson, 1972). Further consideration of social awareness and political consciousness can be found in *Design, Nature and Revolution*, by T. Maldonado (Harper and Row, 1972).

COMPUTER—AIDED DESIGN (CAD)

The computer has become a most powerful tool for use throughout the activity of engineering design wherever detailed, complex or extensive analysis has to be performed. In cases where lengthy numerical work has to be carried out for purposes such as data reduction, justification, planning and control, its use is essential and has now become commonplace. Entry to the literature for this type of application can be readily made through the specific subject matter and the appropriate mathematical techniques. Here attention is confined to a particular interpretation of the term 'computer-aided design' rather than to the general case of the use of computers in engineering. Basic appreciation of design in conjuction with the computer can be found in *Computer Aided Design*, by B. Herzog (Industrial and Commercial Techniques, 1969), and *An Introduction to Computer-Aided Design*, by C. R. Mischke (Prentice-Hall, 1968). For really effective computer-aided design there must be direct contact between the designer and the computer to permit a dialogue or interaction between them. This area of man–computer graphics is well presented in *Interactive Graphics for Computer Aided Design*, by M. D. Prince (Addison-Wesley, 1971), although the examples are drawn from the American aircraft industry. The Computer Aided Design Centre, Cambridge, which is directed by the Departments of Trade and Industry, provides computer, library and consultancy services as described in *Computer Aided Design for Industry* (HMSO, 1972). Specific application of CAD in the circuits field can be found in publications such as *Computer Aided Design of Electronic Circuits*, by E. Wolfendale (Iliffe, 1968), *Computer Aided Design of Magnetic Circuits*, by A. Kusko and T. Wroblewski (MIT

Press, 1969), and *Computer Aided Integrated Circuit Design*, by G. J. Herskowitz (McGraw-Hill, 1968).

MAN — THE DESIGNER

The activity requires the synthesising of many influences which are not considered to be directly in the field of engineering and yet are factors upon which design decisions have to be made. The designer who must play a creative role in the activity has to take into account the 'non-engineering' factors when he is attempting to generate solutions. If he is to show due cognizance of these factors, then a measure of familiarity with them is essential. Reference to some of these factors is given in this and the following sections.

Technical competence
(Numeracy, Literacy, Numerical dexterity)

Economic comprehension	Value recognition
Aesthetic perception	Creative ability

THE DESIGNER

Environment appreciation	Human understanding
Ethical standards	Legal awareness
Political consciousness	Social cognition

Publications which deal directly with the designer as a human being and a participant in the activity of design are noticeable by their paucity. An attempt has been made in *Human Problem Solving*, by A. Newell and H. A. Simon (Prentice-Hall, 1971), to model the human problem solver and formulate a framework for the designer's approach to the work. At a personal level, books such as *Personal Knowledge*, by M. Polanyi (Routledge and Kegan Paul, 1962), and *The Study of Man*, by M. Polanyi (University of Chicago Press, 1959), discuss the nature of human knowledge. These books, by a physical scientist, discuss knowledge as something essentially personal and postulate that the process of achieving knowledge is made up of the same essential elements whatever field of knowledge may be under exploration. The maintenance of individual responsibility and mutual human respect together with improving communication between people and peoples is the subject matter of *Personal Values in the Modern World*, by M. V. C. Jeffreys (Penguin, 1962). Problems in the field of moral philosophy, from the objectivity of values to the freedom of the will, are lucidly presented in *Ethics*, by P. H. Nowell-Smith (Penguin, 1954). This book contains valuable sections on words such as 'good', 'right', 'ought' and 'duty', which are suitable for understanding by the layman.

CREATIVITY

Creativity is the most important characteristic which must be exhibited by any designer for both his own and society's benefit. There is nothing in engineering which can aspire to the intense personal satisfaction which results from having had a significant creative part in a new design. The pleasure of achievement over-whelms the pain and anguish which may have been experienced in generating the ideas and pushing them through to practical realisa-tion. The creative ability of being able to see new relationships, produce unusual ideas or deviate from traditional patterns is demon-strated to different degrees by all designers. That the standard of design is sometimes low can be attributed in part to the accent placed upon routine analysis rather than encouragement of innovation in many educational and industrial establishments. It may be postulated that creativity cannot be taught or learnt, but any innate ability can certainly be matured and developed to reach a higher level. Approaches to reducing the creativity problem in engineering design rest between the two extremes of stimulating multi-directional random or inspirational thought by overcoming mental blocks/ inhibitions and encouraging thorough systematic examination of existing patterns to establish new relationships.

Insight	Preparation	Sequential thought
Lateral thought		Strategic thought
Perseverance	TO CREATE	Concentration
Withdrawal		Incubation
Perception	Motivation	Immersion

From the many publications which deal in a general manner with the psychological aspects of creativity, some of the more readable for the engineer are *Creativity. Selected Readings,* edited by P. E. Vernon (Penguin, 1970); *Contemporary Approaches to Creative Thinking,* edited by H. E. Gruber (Atherton Press, 1967); *Imagination and Thinking,* by P. McKellar (Basic Books, 1957); and *Anatomy of Judgement,* by M. L. J. Abercrombie (Hutchinson, 1960), which stresses the importance of perception and insight. The provocative use of information and challenge to accepted concepts to bring about insight and creativity is termed 'lateral thinking', as opposed to 'vertical thinking', in *Lateral Thinking. A Textbook of Creativity,* by E. de Bono (Ward Lock Educations, 1970), and the associated *Use of Lateral Thinking,* by E. de Bono (Cape, 1967). The proposals outlined in the former constitute a way of thinking to avoid the problem of mental blocks which can occur with logical thought applied to particular creative situations. The approach may be found by the individual reader to be useful for re-structuring his thinking to gain flexibility of mind. The book is well presented and the very useful first portion includes techniques of alternatives search, assumptions challenge and suspended judgement leading to fractionation, reversal, brainstorming, analogies, random stimuli and polarisation. Further useful reading can be found with *Source Book for Creative Thinking*. by S. J. Parnes and H. P. Harding (Scribner, 1962).

Identification of the psychological processes involved in creativity and explanation of the mechanisms by which creative potential can be released and directed to the solution of specific problems is contained in *Synectics*, by W. J. J. Gordon (Harper and Row, 1961). The examination of the operational mechanisms with the synectic process of making the strange familiar and the familiar strange, together with application of the personal, direct, symbolic and fantasy analogies, is thorough. The bibliography is extensive and pertinent. The importance of imagination and use of creative ability from problem orientation to evaluation is well brought out in *Applied Imagination*, by A. F. Osborn (3rd edn, Scribner, 1963). This book, by the originator of brainstorming, contains an extensive discussion on the principles of creativity, then leads on to ways of developing creative abilities and includes sections which deal with perceptive, emotional and cultural blocks, together with deferred judgement. In *Professional Creativity*, by E. K. Von Fange (Prentice-Hall, 1959), an attempt is made to explain creative thinking and the processes behind it, based upon the author's considerable experience, and a systematic plan is proposed for creativity. A book which is very easy to read and understand is *Creative Engineering Design*, by H. R. Buhl (Iowa State University Press, 1960). It is set out in the conventional design process pattern, with chapters on need, recognition, definition, preparation, analysis, synthesis, evaluation and presentation, and contains some references and a short bibliography. The problem of 'set' or mental block as an inhibiting factor in the generation of new ideas is well illustrated. Some readers may find useful parts of *Introduction to Creative Design*, edited by D. H. Edel (Prentice-Hall, 1967).

The presentation and style of *Morphological Creativity*, by M. S. Allen (Prentice-Hall, 1962), are somewhat unusual, but careful personal interpretation can make the contents useful for a designer. The technique presented of creating new ideas by analysis of the form and structure of existing ones and changing the relationship between them owes much to the work of F. Zwicky. A number of methods which may be of use in developing design concepts, including morphological analysis, with an expression of the need for adequate planning and scheduling, are presented in *Creative Synthesis in Design,* by J. R. M. Alger and C. V. Hays (Prentice-Hall, 1964). The forced relationship technique of combining elements of a design in different ways in the search for a better concept is well described in *Creative Thinking,* by C. S. Whiting (Reinhold, 1958). The somewhat similar technique of attribute testing as a base for the systematic search for alternatives or variations to an existing arrangement is included in *The Techniques of Creative Thinking,* by R. P. Crawford (Hawthorn Books, 1964). Another systematic technique, in this instance for aiding synthesis by examination of the interdependencies between elements of the design problem, is given in *AIDA: Technique for the Management of Design,* by J. R. Morgan (Institute for Operational Research, 1972).

Many aspects of the very broad subject of creativity are discussed in the contributions to the following three publications: *Creativity and its Cultivation,* edited by H. H. Anderson (Harper and Row, 1959); *Scientific Creativity. Its Recognition and Development,* edited by C. W. Taylor and F. Barron (Wiley, 1963); and *Creativity: Progress and Potential,* edited by C. W. Taylor (McGraw-Hill, 1964). Creativity considered as an expression of innovation to provide that which is of social worth is presented in *Creativity and Innovation,* by J. W. Haefele (Van Nostrand Reinhold, 1962). A very good reference source for creativity in the engineering field can be found in *Creativity and Innovation in Engineering,* edited by S. A. Gregory (Butterworths, 1972). This book contains sections which deal with the definition and measurement of creativity, education, problem-solving and management.

ECONOMICS AND COST

An appreciation of the over-all economics of technology and an understanding of the methods for allocating and controlling costs is necessary for many engineers. Unfortunately, the range of books available on these subjects contains little that has been prepared to meet the needs of the engineer rather than the accountant. At first sight, *Engineering Economy,* by E. P. DeGarmo and J. R. Canada (5th edn, Collier Macmillan, 1973), appears most useful, as it is intended for engineers with little prior knowledge of economics or accounting. Although it covers economic principles and accounting concepts, the very scale of the work, with the presentation of many techniques and methods for particular situations, tends to obscure the fundamentals. A much more limited approach has been used for the *Introduction to Engineering Economics* (The Institution of Civil

Engineers, 1969). A number of techniques have been proposed for examining the relationship between the cost of a project and the benefits which accrue or the effectiveness of the end product. These may be found in publications such as *Cost Effectiveness,* by J. M. English (Wiley, 1968); *Cost Benefit Analysis,* edited by M. G. Kendall (English Universities Press, 1971); and the early chapters of *Cost Benefit Analysis,* by E. J. Mishan (Allen and Unwin, 1972), in which it is argued that society as a whole should benefit from a particular project, not merely those directly engaged upon it. Perhaps the easiest to read from an engineer's viewpoint is the short but explicit *Cost Benefit Analysis and Public Expenditure,* by G. H. Peters (2nd edn, The Institute of Economic Affairs, 1968), which presents the concepts together with some applications. The economics of engineering are discussed in *Cost Engineering Analysis,* by W. R. Park (Wiley/Interscience, 1973), and *Modern Cost Engineering Techniques,* by H. Popper (McGraw-Hill, 1970). The reduction of costs is a continuous exercise throughout a product life cycle, and some of the procedures which can be adopted are described in the *Manual of Cost Reduction Techniques,* by M. Radke (McGraw-Hill, 1972), and in *Value Management: Value Engineering and Cost Reduction,* by E. D. Heller (Addison-Wesley, 1971). The well-known organised approach for finding the lowest cost of meeting the required functions is described in *Techniques of Value Analysis in Engineering,* by L. D. Miles (2nd edn, McGraw-Hall, 1972), and *Value Analysis,* by W. L. Gage (McGraw-Hill, 1967).

A very brief but precise explanation of cost, cost control and the influence of the various elements of expense upon them can be found in *An Engineer's Guide to Costing* (The Institution of Production Engineers and The Institute of Cost and Works Accountants, 1969). Cost data for direct application to design is never readily available and data sheets are not published in the same way as material or performance figures are quoted. Estimating at the design stage is difficult, although assistance can be found with publications such as *Mechanical Estimating Guidebook,* by J. Gladstone (McGraw-Hill, 1971), and *Guide to Estimating for Heating and Ventilating Contractors* (Heating and Ventilating Contractors' Association, 1971).

DESIGN MANAGEMENT

Hidden in the midst of the very extensive literature dealing with 'management' are a few books which attempt to direct attention towards the special case of creative design. Management, which here includes organisation and planning, of design work on both the individual and interactive group level relies heavily upon conventional techniques tempered by experience. One of the earlier books, *The Management of Innovation*, by T. Burns and G. Stalker (Tavistock Publications, 1961), illustrates clearly the management problem of communication in their example taken from the electronics industry, while *Managing Creative Scientists and Engineers*, by E. Raudsepp (Collier Macmillan, 1963), attempts a wider viewpoint. A short but explicit book which discusses design and the business in terms of the design task, decision planning and organisation is *Managing Design*, by B. T. Turner (Mantec Publications, 1973), which forms a companion to the earlier work *Management of Design*, by B. T. Turner (Industrial and Commercial Techniques, 1968). Two other modern books, which may be found to be useful upon individual interpretation, are *Management, Innovation and System Design*, by I. G. and M. E. Wilson (Auerbach Publications, 1971), and *Management of Engineering Design*, by D. J. Leech (Wiley, 1972).

An important element in managing a design task is planning the over-all activity. It is necessary to plan for the effective allocation of time, resources, facilities and manpower to specified stages in the activity. As design in the industrial environment is always a group exercise, an appreciation of the interaction and dependencies between the individual parts which serve to make up the whole is desirable. For simple tasks the conventional techniques of milestones and bar charts, which may incorporate check points and related events, are effective tools. With more extensive and complex projects, techniques such as monitor and network analysis become necessary. Network analysis may be used to advantage at the individual level, but is more usually associated with large projects and the use of a computer. Care is always necessary with the computerised systems to ensure that the results justify the effort and cost. A good general introduction to network analysis can be found in *Network Analysis for Planning and Scheduling*, by A. Battersby (3rd edn, Macmillan, 1970), and in *Planning by Network*, by H. S. Woodgate (2nd edn, Business Books, 1967). Typical examples of the many publications which deal with critical paths and program evaluation and review are *Project Management with CPM and PERT*, by J. J. Moder and C. R. Philip (2nd edn, Van Nostrand Reinhold, 1971), and *Planning and Control with PERT/CPM*, by R. I. Levin and C. A. Kirkpatrick (McGraw-Hill, 1966).

Current safety legislation is an important consideration for design management. The Health and Safety at Work . . . Act (HMSO, 1975) is the fundamental text, with Section 6 essential reading for all designers.

24

Automotive engineering

W. G. Stevenson

The vast size of the world's automotive industry, the variety of its products and the social transformation brought about in many countries by their use (involving large road construction and town planning projects) have resulted in a copious literature of interest to the automotive engineer, the oil technologist, the economist, the legislator and several other kinds of specialist. We are concerned here with the literature requirements of the automotive engineer only. He may be an inventor, a research worker, a development engineer, a designer, a production engineer, a fleet operator, and so on. It is therefore necessary to include a fairly broad range of subject matter, out of the total available, in order to be sure of meeting the needs of any kind of specialist within the field of automotive engineering.

The automotive engineer may be concerned with any matters within the fields of vehicle design, manufacture and operation. In the past, design criteria were largely customer requirements. In the case of cars, for example, consideration was given by the manufacturer to styling, comfort, convenience of control, fuel economy, reliability, durability, silence, cost, and more. While the needs of the customer are still as urgent, legal factors are now controlling vehicle design and operation to a much greater extent than hitherto, and this influence is on the increase. Environmental considerations are largely at the root of vehicle legislation in many countries, and the three most important examples of it at the present time concern vehicle requirements for safety on the road, air pollution and noise prevention. It may be necessary also for the automotive engineer to consider possible alternatives to petroleum-based fuels as a source of power, or, at least, to consider the possibility of a more economical use of the present fuels; and this introduces the question of some new type of power unit and the considerable amount of research and development required to establish its feasibility.

The foregoing remarks are given to illustrate the considerations determining the choice of literature which now follows.

SOURCES OF ORIGINAL RESEARCH REPORTS

Professional institutions

Society of Automotive Engineers

The object of the Society of Automotive Engineers, which is located in Warrendale, Pa., is 'to promote the Arts, Sciences, Standards and Engineering Practices connected with the design, construction and utilisation of self-propelled mechanisms, prime movers, components thereof, and related equipment'. It pursues this object through meetings of its members. Of the main meetings, each of which is repeated each year, six are of interest mainly to automotive engineers, as follows:

International Automotive Engineering Congress. This is held in January, and is by far the largest meeting of the six, with between 200 and 300 papers given. These are mostly divided into short symposia on topical subjects which may relate to any aspect of vehicle design and performance. The papers are based on work carried out by the manufacturer in laboratory or on proving ground, by government agency, by universities and technical colleges, by consulting engineers and others. While the bulk of the papers are of US origin, European countries and Japan have also been represented.
National Automobile Engineering Meeting. This is held in May. It is on similar lines to the foregoing meeting but the number of papers is less—about 100.
National West Coast Meeting. This is held in August. Again, it is on similar lines to the previous two meetings, with no apparent restriction of subject matter, but the number of papers is between 20 and 30.
Combined National Farm, Construction, Industrial Machinery and Powerplant Meeting. This is one of the larger meetings, with over 100 papers presented. It is held in September. While a number of papers deal with researches, the bulk of them appear to describe particular models of vehicles, such as farm tractors, and components, such as hydraulic pumps and circuits.
National Commercial Vehicle Engineering and Operations

Meeting. This is a specialised meeting held in October, and comprises between 20 and 30 papers. A limited amount of research is reported on, but the bulk of the papers are devoted to design considerations for trucks, including power requirements, descriptions of truck components, and advice to operators. The effect of legislation on design is amply treated.

National Fuels and Lubricants Meeting. This is held at the end of October and the programme lists about 20 papers. All papers appear to be based on research. They cover exhaust-emission control, whether affected by fuels or not, and all other aspects of the application of fuels and lubricants to vehicle engines.

In June 1973 a conference entitled *Fleet Week 73* was held. This combined the SAE Powerplant and Commercial Vehicle Engineering and Operations Meetings with seven other conferences; and a total of about 60 papers were presented. Whether this will be a permanent arrangement, within the series of annual SAE meetings, remains to be seen. Programmes of all the meetings, listing the complete set of papers at each, can be obtained from the Society of Automotive Engineers, 2 Pennsylvania Plaza, New York, NY 10001, and all papers are available for sale.

Stapp Car Crash Conferences. These are held annually, in November. The Society of Automotive Engineers acts as administrative sponsor and publishes the proceedings of each conference. The papers are sponsored by the Biomechanics Research Center of Wayne State University, The University of Michigan, and the University of California at San Diego, some of the papers being contributed by research staff at these three universities. All aspects of passenger safety in road accidents are considered, purely medical research being involved in some cases, and the vehicle structure, passenger restraint systems, safety glass, and additional vehicle factors being correlated with passenger injuries in others.

The Institution of Mechanical Engineers

The Institution of Mechanical Engineers, Automobile Division, organises meetings at which single papers or, at most, two or three papers are given. These papers cover a variety of researches carried out by manufacturing firms, universities, technical colleges, etc. In addition, the Institution organises conferences on broad subjects, the papers of any one conference together providing a comprehensive treatment of the subject chosen. The conference papers are also reports of researches. Neither meetings nor conferences appear to be held regularly, but the total output of papers directly on automotive subjects in a year (together with other papers which sometimes have a bearing on automotive problems) makes this a useful source of information to the specialist in this field. Information on availability of papers can be obtained from the Institution of Mechanical Engineers, Birdcage Walk, London, SW1H 9JJ.

Periodicals

Avtomobilnaya Promishlennost. This Russian periodical is published monthly, and carries reports of research from a large number of institutes in the USSR. Its range appears to be the whole of automotive technology. Of the total number of organisations contributing to the contents of this journal, the following six are most frequently represented: Scientific-Research Institute for Vehicles and Vehicle Engines; Moscow Higher Technical Institute in the name of Bauman; Moscow Vehicle/Road Institute; Moscow Institute of Vehicle Mechanics; Scientific-Research Institute for the Tyre Industry; and State Design-and-Construction Bureau for Buses.

Bulletin of JSAE and *Bulletin of the JSME.* Both of these Japanese journals are published in English, the former being issued annually by the Society of Automotive Engineers of Japan, and the latter monthly by the Japan Society of Mechanical Engineers. As its title indicates, the *Bulletin of JSAE* is devoted exclusively to automotive subjects. The authors appear to be either from motor manufacturers or universities. The topical subjects of safety, air pollution and noise predominate in the latest issue, but several papers on comfort, as related to suspension characteristics and shocks in the transmission system, and one paper on the bending fatigue strength of truck frames are also published in this issue.

The authors contributing to the *Bulletin of the JSME* are also from industry and the universities. The papers of automotive interest are few in number, and appear to be mainly concerned with engine and engine component problems. As reports of research, however, they are of equal value with the *Bulletin of JSAE*. Papers of both these Japanese journals are apparently in the Japanese language originally, and a considerable time—possible 2 years—may elapse before they are available in English.

Automotive Engineering. This journal is published monthly by the Society of Automotive Engineers, and its range of subject matter includes everything relevant to progress in automotive technology (with a limited interest in power units for applications such as in aircraft). The main articles, although authoritative, are apparently intended for automotive engineers in general and not just for specialists in the particular subjects of the articles. They are in most cases based on papers given at the SAE meetings mentioned earlier in this chapter, and their choice appears to be based on a few principles of selection, such as the probability of their appealing to a large number of readers, their connection with matters of public concern, e.g. air pollution and road safety, and their bearing on future developments.

The Journal of Automotive Engineering* (JAE). *JAE* is published monthly by the Institution of Mechanical Engineers, Automobile Division. Each issue carries a number of articles either written specially for *JAE* or taking the form of complete or abridged versions of papers delivered at meetings or conferences. The main theme of these articles appears to be all aspects of vehicle and component design, in which the results of research may be called on to aid calculations or to determine design principles. Legislation in its relation to design is also treated. When the article covers an actual research project, this is normally found to be biased towards the application of its results to the design of some component, with a computer being used to integrate the relevant parameters.

Automotive Design Engineering*. This is a monthly journal intended for 'designers and development engineers of automotive vehicles and equipment'. Most of the articles therefore deal with design principles, trends in design and the latest design developments of the vehicle as a whole, its components and sub-assemblies. In addition, there is the occasional article in the form of a report on research or development work and, more frequently, articles describing test equipment used in this field.

Specialist Journals. In addition to the periodicals which set out to describe research in potentially any part of the field of automotive technology, there are the specialist journals dealing with particular components or systems of the vehicle, or aspects of vehicle performance. A comparative newcomer to this class is *Vehicle System Dynamics,* described by the publishers as an 'international journal of vehicle mechanics and mobility'. It is published not more than four times annually. The editors aim at a rigorous treatment of problems, both theoretical and practical, in the researches described,

* Continued as *Automotive Engineer* (1975–)

and the scope is stated to include 'the kinematics and kinetics of vehicle systems and their components, and model studies of the vehicle/passenger system'. The *Journal of Sound and Vibration,* the official medium of publication for the British Acoustical Society, is issued monthly, and deals extensively with the research work of the Institute of Sound and Vibration Research at Southampton University. Although problems of vehicle noise and vibration form only a part of the subject matter, the research work related to this field is of high quality, and includes both objective and subjective aspects.

The specialist journals derive their material from industrial and academic bodies, which may act as publishers in addition to carrying out the research. Reports of such work may, however, be issued by publishing houses, professional societies and others. For example, the American Society for Testing and Materials produces a quarterly journal called *Tire Science and Technology,* and defines its objectives in doing so as the production of 'a multidisciplinary journal acting as a focal point for the significant literature of mechanics and dynamics, traction, wear, strength, and durability of tires'. Emphasis is laid on evaluating standard methods of tyre testing to be used both on the road and in the laboratory: the importance of such standards for economic and safety reasons is stressed. The criteria to be observed in judging papers offered for publication give some indication of the value of this journal. They are stated as follows: 'Contributions are judged on the basis of: (1) the paper contributes to the permanent literature; (2) the paper includes experimental data or mathematical derivations, or both, either as original work as is needed to confirm the theoretical conclusions of proposals; (3) the substance of the paper has not been published in the open literature.'

Other examples of specialist journals of interest to the automotive engineer are *Technical Aspects of Road Safety,* an international quarterly publication dealing with road safety from the points of view of both vehicle and traffic, and *Bosch Technische Berichte,* published two to four times each year by the Bosch organisation, and dealing with development and production matters in addition to fundamental research. In this latter journal, fuel injection equipment is featured prominently in the basic research, but other subjects, such as electric motors for battery-driven vehicles and the spacing of vehicles in traffic by an automatic control system, are also to be found. The Bendix Corporation in the USA publishes the *Bendix Technical Journal* three times a year. The intention in doing so is to describe important scientific and engineering investigations carried out by members of the Bendix staff; and this periodical is stated to 'complement rather than be competitive with other scientific and technical society journals'. As far as possible, each issue is limited to a single topic—for example, automotive braking, or the control of vehicles in novel public transport systems—and this policy ensures a fairly comprehensive discussion of current problems within each subject selected. As a final example of a specialist journal, the irregularly published *Lucas Engineering Review* might be cited. Its aim is stated among the editorial contents as follows: 'It records the engineering and thinking behind the products of the Lucas Organisation rather than the products themselves. In this way, it is hoped to make a contribution to scientific and engineering knowledge'. Its subject matter is mainly the vehicle electrical system.

Automobiltechnische Zeitschrift (ATZ) and *Motortechnische Zeitschrift (MTZ)*. These two German monthly periodicals are from the same publisher. As their titles indicate, *ATZ* covers all parts of the automotive vehicle, *MTZ* the engine only (this latter journal also dealing with power units for railway, marine and stationary applications). Both treat their subject matter comprehensively and include development, design, and production in addition to research. The sources of their contribution are mainly industry, universities and technical colleges, and these contributions may come additionally from Eastern European or Balkan countries and Japan; but the language of publication is invariably German. In most cases, the reports of research appear to have been specially contributed, but in others they have been first presented at conferences, and are later reprinted in *ATZ* or *MTZ*. Most of the articles in any one issue may form a symposium on a single, though possibly broad, aspect of vehicle construction or operation.

Deutsche Kraftfahrtforschung und Strassenverkehrstechnik. This is an irregular publication of the Verein Deutscher Ingenieure, and is unusual among the automotive periodicals in carrying only one lengthy article in each issue. The subject matter of these articles can be broadly described as vehicle operating characteristics, chiefly of a dynamical nature. For example, tyre friction and rolling resistance, suspension, steering and braking, the problems of vehicle and trailer combinations, including stability during braking and the strength of couplings, are dealt with theoretically and practically. Current environmental problems, such as the vehicle in relation to road safety and to its exhaust gas emissions, are given due prominence. The research reported is of a high standard, and the articles appear to be written specially for this journal.

Ingénieurs de l'Automobile. This monthly French journal is an official publication of the Société des Ingénieurs de l'Automobile (SIA). It covers all aspects of automotive technology, including production. Some of its articles are reports of researches carried out in industry and elsewhere, but these are not necessarily published originally in the SIA journal. Research carried out in France is among the most likely to be first reported in this publication.

ATA. This is the official journal and proceedings of the Italian Associazione Tecnica dell' Automobile, and is published monthly. All aspects of the automotive vehicle are potential subjects for articles, of which usually two or three report on research. These reports do not necessarily receive their first publication in *ATA*, however, even if they refer to Italian research. For example, it is possible that some such work might first be reported in English at an SAE meeting, and only several months later in *ATA* in its language of origin. The investigations described are carried out in industry, technical colleges and research establishments. English abstracts of research articles are presented on detachable cards in each issue.

Co-operative organisations

British Technical Council of the Motor and Petroleum Industries (BTCMPI)

BTCMPI is composed of automotive engineers and petroleum technologists. As Council members, they represent their respective industries in the effort to correlate the properties of petroleum products with vehicle requirements, fuels and lubricants being the main petroleum derivatives of interest. A meeting of the full Council, of the Fuels or the Lubricants Committee, or of any of the project groups, provides a forum for the discussion of subjects such as cam and tappet wear, piston ring and gudgeon pin scuffiing, exhaust emissions from petrol engines, diesel engine exhaust smoke and carburettor icing. Such discussions may lead to new research projects or may be for the purpose of assessing progress in existing researches. At the end of the research comes the report. Almost all reports from this source are available for general sale, and may be ordered from the BTCMPI Secretariat, c/o The Motor Industry Research Association, Watling Street, Nuneaton, Warwickshire, CV10 OTU.

Coordinating European Council (CEC)

The CEC derives its membership from the United Kingdom, West Germany, France, Belgium, Holland, Switzerland, Sweden, Spain and Italy. Its function is to promote the development and standardisation of performance tests of fuels and lubricants for automotive vehicles (also for marine and other kinds of engines) in the participating countries. For this purpose, research projects are planned and carried out in laboratories in these countries, and the results are correlated in the process of devising internationally acceptable test procedures. The work of the Council in the various countries is controlled by the corresponding national committees, which also supply members of the Council. For example, in the United Kingdom the appropriate committee is the British Technical Council of the Motor and Petroleum Industries. Details of reports available, and copies required, can be obtained from the Executive Secretary, Coordinating European Council, 61 New Cavendish Street, London W1M 8AR. The CEC does not formulate specifications for fuels and lubricants. Its publications comprise descriptions of test methods for assessing the properties of fuels and lubricants, accepted by Council members for application in their respective countries, and reports of investigations into such questions as the consequences of removing lead from petrol and the scientific basis of exhaust smoke evaluation.

Coordinating Research Council (CRC).

The CRC performs a similar function in the United States to the BTCMPI in the United Kingdom. It is formed by the co-operation of the American Petroleum Institute and the Society of Automotive Engineers. The annual output of reports lies between five and ten. In subject matter they are concerned chiefly with the development of methods of test for fuels and lubricants, the determination of current vehicle needs, such as for octane value of fuel, and investigations into aspects of the exhaust emission problem in petrol and diesel engines. Examples of researches which have formed the subjects of reports are as follows: a study of the characteristics of full-flow light-extinction smokemeters for diesel engine exhaust studies; the influence of leaded and unleaded fuels on octane requirement increase in 1971 cars; an assessment of recommended practices for constant-volume sampling of vehicle exhaust gases; a survey of the octane number requirements of 1971 cars; a study of the effect of humidity of the intake air on nitric oxide formation in diesel engine exhaust gases; and an evaluation of techniques for estimating hydrocarbons also in diesel engine exhaust gases. The CRC will apparently notify any bona fide enquirer whenever reports are available for purchase. To obtain this service, it is necessary to write to the following address: Coordinating Research Council, Inc., 30 Rockefeller Plaza, New York, NY 10020. The actual reports are obtainable from the Society of Automotive Engineers, the address of which is given above.

The Motor Industry Research Association (MIRA)

MIRA is financed partly by its member companies of the British and European Motor Industries and partly by income from sponsored research. Investigations carried out at MIRA are generally aimed at the improvement of automotive vehicles and their components. Much research has been done and is still being done in the fields of vehicle durability and operating economy. Within recent years, the problems of road safety, as affected by vehicle design and noise and exhaust gas pollutants of the atmosphere, have become prominent subjects of investigation. In the past, practically all researches were financed out of general funds. Nowadays there is an increasing amount that is privately sponsored, with no restriction as to the type or location of the sponsor. Reports arising out of individually financed work are the property of the sponsors and can only be made generally available with their permission. Reports accruing from general-fund work are normally put on sale 1 year after publication. All MIRA reports of research which are available for sale are listed in a special section of *Automobile Abstracts* (see below).

Government agencies

National Highway Traffic Safety Administration (NHTSA)
The NHTSA is a section of the US Department of Transportation, and sponsors research projects on various aspects of road safety. They are carried out in universities and elsewhere, and the results are published in reports which are available for purchase. These are listed in a semi-monthly abstracting journal called *Highway Safety Literature,* which may be obtained, free of charge, on application to the National Highway Traffic Safety Administration, Office of Administration, Washington DC 20590. Many of these reports deal with road safety from the point of view of the vehicle or the vehicle in relation to driver and passenger.

Transport and Road Research Laboratory (TRRL)

The work of TRRL, which comes under the Department of the Environment, is divided between vehicles and roads, each of the two sections including a very wide range of subjects. On the vehicle side are the main environmental problems of noise, air pollution, and safety. Work on the first two of these appears to be mainly limited to the effect of the vehicle population and the topography of its immediate environment on noise and air pollution. The safety research is split into physical and psychological factors such as tyre/road interaction or the effectiveness of protective helmets on the one hand, and driver viewing time for instruments or the value of extra brake lights on the other. Reports of this work are published from time to time, and copies can be obtained by arrangement with the Transport and Road Research Laboratory, Crowthorne, Berkshire.

International conferences

The results of research are sometimes presented initially at international conferences which arise out of the co-operation of national bodies in the countries represented. It should be pointed out, however, that conferences of this kind do not always attract first reports of original work, and it will occasionally be found that an author has already contributed essentially the same paper to another conference or to a periodical.

The Fédération Internationals des Sociétés d'Ingenieurs des Techniques de l'Automobile (FISITA), which was founded in 1947, has constituent bodies in 14 countries — for example, the Institution of Mechanical Engineers in the UK and the SAE in the USA. It organises a conference every second year in one of the countries, the official languages of these conferences being English, French and German. Proceedings of each conference are published, and the representative body in the country where a particular one is held is responsible for issuing its papers.

The NATO Committee on Challenges to Modern Society is the sponsor of a joint effort by the UK, France, Holland, Sweden, Germany, Italy, Japan and the USA on an experimental safety vehicle. Each country carries out a separate piece of research, and all report at an annual conference. Details of papers presented can be obtained from TRRL.

SECONDARY SOURCES OF INFORMATION ON AUTOMOTIVE ENGINEERING

Automobile Abstracts (formerly *MIRA Abstracts*). The Motor Industry Research Association issues a monthly journal known as *Automobile Abstracts*. Each number carries one or more reviews of recent literature on some aspect of automotive research or development, or some piece of experimental equipment. The word 'development' may relate to the work carried out in producing a new prototype vehicle, component or sub-assembly, or a discussion of design principles underlying the finally manufactured unit. Materials used in vehicle construction, or in vehicle operation, such as fuels and lubricants, also form part of the subject matter of *Automobile Abstracts*. Economic and organisational aspects of manufacture are excluded, but a production process will be of interest if correlated with the performance of the finished product.

The above definitions also apply to the abstracts proper, which form the bulk of this publication, and vary from 50 to 300 words in length. A list of references to literature of marginal interest concludes each issue, together with an author index. There is also an annual index. Every effort is made to include only recently published material. Some of the abstracts refer to complete translations into English made by MIRA from the original articles, and, every quarter, a special section lists MIRA reports, etc., available for sale. Offprints of this section are available free of charge, but *Automobile Abstracts* itself is available on subscription (currently £40 p.a.) from MIRA, Watling St., Nuneaton, Warwicks., CV10 0TV.

Bulletin Mensuel de Documentation. This French monthly abstract journal is published by La Société Auxiliaire Technique de l'Automobile, du Motocycle et du Cycle, and may be had in single parts or by annual subscription. It is comprehensive in its coverage of vehicle construction and operation, including, for example, legislation applicable to this field. The material is up to date, and reasonably well classified, but the abstracts are too short to be of much value, or even nonexistent in some cases. The absence of an annual index greatly impairs its usefulness as a reference tool.

Intereurope Service of Automotive Regulations. This service covers regulations affecting the construction and operation of automotive vehicles from a number of countries, including the USA. These are translated into English when originally in another language. The service is available on subscription from Intereurope, 54a/56a Peach Street, Wokingham, Berks., RG11 1XZ. It is in the form of loose-leaf folders, and is regularly updated by additions to the sheets. Filing instructions and a subject index are also provided.

BIBLIOGRAPHY

Pigott, W. D. (comp.). *A Checklist for Automobile Engineering Libraries.* Motor Industry Information Group (1973). Available from Mrs J. Pearson, Commercial Information Officer, GKN Group Technological Centre, Birmingham New Road, Wolverhampton, WV4 6BW.

This bibliography provides a fairly comprehensive, roughly classified, unannotated list of individual works on automotive engineering and cognate subjects, and forms a useful supplement to the foregoing more generalised treatment of the literature. Specific headings are followed by lists of textbooks and symposia, and reference works include polyglot dictionaries, a professional directory, and data books on a variety of subjects. Abstracting and indexing services are included, as are sources of standards. Short lists are given of periodicals and of bodies organising conferences.

25

Thermodynamics and thermal systems

J. B. Arnold

Thermodynamics is one of the basic sciences which may be applied to a wide range of scientific and technological problems. It is concerned with the states and energies of substances and their interactions with external systems as heat and work processes. Thermodynamics as a science was developed in the nineteenth century as a result of the work of Carnot, Clausius, Kelvin, Joule, Maxwell, Gibbs and Planck. Today two distinct approaches to thermodynamics exist. The classical approach deduces the general laws of thermodynamics from experimental observation of complete systems which consist of a very large number of molecules, i.e. a macroscopic approach. The other approach, statistical thermodynamics, argues the same thermodynamic laws by considering the statistical behaviour of the individual molecules, i.e. a microscopic approach. Classical thermodynamics leads to relatively easy solutions to many mechanical engineering problems, particularly in the fields of power generation and energy transfer. Statistical thermodynamics has unique application to chemical reactions, to electrical/electronic devices and to systems operating in extreme conditions of low or high pressures and temperatures. In recent years attempts have been made to unify classical thermodynamics, statistical thermodynamics and quantum mechanics. A major work on the unified approach is *Principles of General Thermodynamics,* by G. N. Hatsopoulos and J. H. Keenan (Wiley, 1965). In the foreword of this book a short authoritative historical review is given, with references to the works of early thermodynamicists.

CLASSICAL THERMODYNAMICS

Two famous books on classical thermodynamics, *Thermodynamics,* by J. H. Keenan (Wiley, 1941), and *Heat and Thermodynamics,* by M. W. Zemansky (4th edn, McGraw-Hill, 1957), expressed the thermodynamics theories in such a rigorous and progressive manner that they strongly influenced subsequent writers. Textbooks written on thermodynamics express the individual viewpoints of their authors and the variety of needs of the readers. Some aim to present the academic theory only, while others act as a course text for students studying for first or higher degrees and may contain numerous worked examples.

Popular undergraduate course texts are: *Engineering Thermodynamics, Work and Heat Transfer (SI Units),* by G. F. Rogers and Y. R. Mayhew (2nd edn, Longmans, 1967), which has a particularly good section on gas and vapour power cycles and includes worked examples to illustrate the text. *Applied Thermodynamics for Engineering Technologists (SI Units),* by T. D. Eastop and A. McConkey (Longmans, 1970), covers the same ground as the previously mentioned text, but aims at a lower level of student by providing many worked examples. A rigorous first-year undergraduate text is *Engineering Thermodynamics (SI Units),* by D. B. Spalding and E. H. Cole (3rd edn, Arnold, 1974).

Advanced Engineering Thermodynamics, by R. S. Benson (Pergamon, 1967), covers certain thermodynamic topics not fully treated in normal texts. Of the American textbooks, which are usually written in British units, *Fundamentals of Classical Thermodynamics,* by G. J. Van Wylen and R. E. Sonntag (2nd edn, Wiley, 1965), and *Engineering Thermodynamics,* by J. B. Jones and G. A. Hawkins (Wiley, 1960), are well-written and well-produced course texts, concentrating mainly on thermodynamic fundamentals. *Thermodynamics,* by J. P. Holman (McGraw-Hill, 1969), presents a more generalised approach and is easier to digest than the major work *Principles of General Thermodynamics,* by G. N. Hatsopoulos and J. H. Keenan (Wiley, 1965). *A Course in Thermodynamics,* by J. Kestin (Ginn Blaisdell, Vo. 1, 1966; Vol. 2, 1968), has an individualistic approach and extends thermodynamics into the field of electrical phenomenon, mechanics and materials science.

A selection of other textbooks follows.

Available Energy and Second Law Analysis, by E. A. Bruges (Butterworths, 1959)
Basic Engineering Thermodynamics, by M. W. Zemansky and H. C. Van Ness (McGraw-Hill, 1966)

Second Law of Thermodynamics, by S. R. Montgomery (Pergamon, 1966)

Thermodynamics, by W. C. Reynolds (McGraw-Hill, 1965)

Thermodynamics, An Auto-Instructional Text, by M. Mark (Prentice-Hall, 1967)

Thermodynamics, An Introduction, by R. Battino and S. E. Wood (Academic Press, 1968)

STATISTICAL THERMODYNAMICS

An early work in this field is *Statistical Thermodynamics,* by R. H. Fowler and E. A. Guggenheim (Cambridge University Press, 1939). Statistical thermodynamics, once solely of interest to physicists and chemists, is now entering the course curricula of engineering degree courses. Quantum mechanics is treated in *Mathematical Foundations of Quantum Mechanics,* by J. Von Neumann (Princeton University Press, 1955), and its integration with thermodynamics in *Principles of General Thermodynamics,* by G. N. Hatsopoulos and J. H. Keenan (Wiley, 1965). A selection is as follows:

A Course in Statistical Thermodynamics, by J. Kestin and J. R. Dorfman (Academic Press, 1971)

Equilibrium Statistical Mechanics, by F. C. Andrews (Wiley, 1963)

Generalised Thermodynamics, by L. Tisza (MIT Press, 1966)

Statistical Mechanics, by R. H. Fowler (2nd edn, Cambridge University Press, 1936)

Statistical Thermodynamics, by J. F. Lee, F. W. Sears and P. L. Turcotte (Addison-Wesley, 1963)

Statistical Thermodynamics, by E. Schrodinger (2nd edn, Cambridge University Press, 1952)

Statistical Thermodynamics, by C. L. Tien and J. H. Lienhard (Holt, Rinehart and Winston, 1971)

Thermal Physics, by C. Kittel (Wiley, 1969)

Thermodynamics of Electrical Processes, by M. McChesney (Wiley, 1971)

THERMODYNAMIC PROPERTIES OF FLUIDS

It would be highly convenient for engineers if an international standardising committee were to produce and maintain a publication of the thermodynamic and other properties of fluids. Unfortunately, a publication of this type does not exist and engineers have to rely on data from a variety of sources, such as national standardising bodies, individual researchers and the manufacturers of the fluids, notably ICI and Du Pont.

In 1956 a series of publications from Butterworths entitled *Thermodynamic Functions of Gases,* edited by F. Din, were started. To date three volumes have been issued: Vol. 1, 1956 covers ammonia, carbon dioxide, carbon monoxide; Vol. 2, 1956 covers air, argon, acetylene, ethylene and propane; Vol. 3, 1961 covers ethane, methane and nitrogen. This series is the official publication of the Thermodynamics Committee of the Mechanical Engineering Research Board of the British Government.

In 1963 the Sixth International Conference on the Properties of Steam (*Proceedings* published by the American Society of Mechanical Engineers in October 1963) agreed to a skeleton set of steam tables. The conference also set up an International Formulation Committee, which issued in 1967 the 'IFC Formulation'. This provides the equations and constants for computing the properties of steam (IFC Secretariat, Verein Deutscher Ingenieure, 4 Dusseldorf 1, Postfach 1139, Graf-Recke-Strasse 84, Germany). The UK Committee on the Properties of Steam has issued the *UK Steam Tables* (Arnold, British Units, 1967; SI Units, 1970). These tables agree with the skeleton tables of the 6th International Conference and include the 'IFC Formulation'. Data of the normal thermodynamic properties, viscosity, thermal conductivity and surface tension are presented. For many years, Keenan and Keyes steam tables have been extensively used. The recent edition of *Steam Tables, Thermodynamic Properties,* by J. H. Keenan, F. C. Keyes, P. G. Hill, J. G. Moore, published by Wiley, is available in British (1969) or SI Units (1970). The thermodynamic properties of gases at low pressure may be readily computed from thermodynamic relationships and equations expressing the variations of specific heat in terms of temperature. The theory is outlined in many thermodynamic textbooks and fully explained in *Gas Tables,* by J. H. Keenan and J. Kaye (Wiley, 1948), which also presents the results for air, carbon dioxide and combustion products.

The Abridged Thermodynamic and Thermochemical Tables, by F. D. Hamblin (Pergamon, 1971), is available in British or SI units and presents data for steam and other common fluids.

The *Proceedings of the Joint Conference on the Thermodynamic and Transport Properties of Fluids,* published by the Institution of Mechanical Engineers (1957), covers a wide range of fluids and units.

Thermodynamic Atlas, by I. Kolin (Longmans, 1967), provides temperature–entropy and enthalpy–entropy diagrams and other data for air, water, ammonia, Freon 12, carbon monoxide and carbon

dioxide in metric units. Also presented are diagrams and tabulated data of the performance of steam power cycles, refrigerators, and air compressors.
Other references are:

Handbook of Chemistry and Physics, by R. C. Weast (56th edn, CRC Press, 1975)
Tables of Physical and Chemical Constants, by G. W. C. Kaye and T. H. Laby (14th edn, Longmans, 1973)
Thermodynamic Data for the Calculation of Gas Turbine Performance, by D. Fielding and J. E. C. Topps (HMSO, 1959)
Seventh International Conference on the Properties of Steam (ASME, 1968)
Fourth (1968) and Fifth (1970) Symposiums on Thermophysical Properties (ASME)
Tables of Thermal Properties of Gases, by J. Hilsenrath (NBS Circular 564, 1955)
Thermodynamic and Transport Properties of Gases, Liquids and Solids (ASME, 1959)

See also the section on Heat Transfer and the chapter on Fuel Technology.

TEMPERATURE

Temperature is perhaps the most difficult measurable property to define conceptually. It is defined by means of the laws of thermodynamics in order to link with other thermodynamic properties and theories. However, thermodynamic temperature is a theoretical concept and an internationally defined scale is used for calibration and practical purposes. The temperature scale is defined in *The International Practical Temperature Scale of 1968* (HMSO). A major reference work, *Temperature, Its Measurement and Control in Science and Industry,* edited by C. M. Herzfeld is issued in three parts by the Instrument Society of America (Vol. 4, 1972). The British Standards Institution and the National Bureau of Standards have numerous publications relating to temperature measurement.
Other references are:

The Calibration of Thermometers, by C. R. Barber (HMSO, 1971)
Temperature Measurement in Engineering, by H. D. Baker *et al.* (2 vols., Wiley, 1953 and 1961)
Thermocouple Temperature Measurement, by P. A. Kiuzie (Wiley, 1973)

HEAT AND MASS TRANSFER

Whereas thermodynamics is only concerned with the magnitude of heat and mass transfer and its effects on the system, the subject 'heat and mass transfer' is concerned with the mechanism and the rate at which the energy transfers. Although heat and mass transfer are distinctly different processes, the similarity between the theory and the interactive nature of the processes imply similar reference sources. Division of the topic usually occurs under the headings of conduction, convection, radiation, mass transfer and combined processes and applications. Heat and mass transfer is usually a subject in the curriculum of first degree courses in mechanical engineering, while a number of higher degree courses specialise in the topic.

General textbooks

A large number of textbooks are available in heat and mass transfer and the following books have been the authoritative texts for a number of years: *Analysis of Heat and Mass Transfer,* by E. R. G. Eckert and R. M. Drake (McGraw-Hill, 1972); *Heat, Mass and Momentum Transfer,* by W. M. Rohsenow and H. Y. Choi (Prentice-Hall, 1961); *Heat Transfer,* by M. Jakob (2 vols., Wiley, 1949 and 1957).

For an introductory undergraduate course in heat transfer, *Heat Transfer,* by J. P. Holman (3rd edn, McGraw-Hill, 1972), *An Introduction to Engineering Heat Transfer,* by J. R. Simonson (McGraw-Hill, 1967), *Heat Transfer Engineering,* by H. Schenck (Longmans, 1960), and *Principles of Heat Transfer,* by F. Kreith (2nd edn, International Textbook, 1965), are recommended. A case-study, problem-solving approach is adopted in *Heat Transfer,* by F. A. Holland (Heinemann, 1970). Of the newer general textbooks, *Momentum, Energy and Mass Transfer,* by J. C. Slattery (McGraw-Hill, 1972), and *Heat Transfer,* by F. J. Bayley, J. M. Owen, A. B. Turner (Nelson, 1972), are interesting.

Books on specific topics and applications

Conduction of Heat in Solids, by H. S. Carlsaw and J. C. Jaeger (Clarendon Press, 1959), is a definitive work in conduction, while *Analytical Methods in Conduction and Heat Transfer,* by G. E. Myers (McGraw-Hill, 1971), *Boundary Value Problems of Heat Conduction,* by M. N. Ozisik (International Textbook, 1968), and *Computer Aided Heat Transfer Analysis,* by J. A. Adams and D. F.

Rogers (McGraw-Hill, 1973), concentrate on numerical methods for solving conduction problems by digital computers.

Convective Heat and Mass Transfer, by W. M. Kays (McGraw-Hill, 1966), is a satisfactory teaching text, and *Heat and Mass Transfer in Boundary Layers,* by S. V. Patankar and D. B. Spalding (International Textbook, 1970), is a useful postgraduate text. Two informative books on radiation heat transfer are: *Radiative Transfer,* by H. C. Hottel and A. F. Sarofin (McGraw-Hill, 1967), and *Thermal Radiation Heat Transfer,* by R. Siegel and J. R. Howell (McGraw-Hill), 1972). Recommended specialist books on mass transfer are *Mass Transfer Operations,* by R. E. Treybal (2nd edn, McGraw-Hill, 1968), and *Convective Mass Transfer,* by D. B. Spalding (Arnold, 1963). A practical approach is adopted by T. Hobler in his book *Mass Transfer and Absorbers* (Pergamon, 1966). Three advanced books on specialised topics are *Unsteady State Heat Transfer,* by Y. V. Kudryavtsev (Iliffe, 1966), *Heat and Mass Transfer in Recirculating Flow,* by A. D. Gosman (Academic Press, 1969), and *Extended Surface Heat Transfer,* by D. Q. Kern and A. Kraus (McGraw-Hill, 1972). *Compact Heat Exchangers,* by W. M. Kays and A. L. London (2nd edn, McGraw-Hill, 1964) is a definitive text.

Other specialised texts are:

Fundamentals of Aerodynamic Heating, by R. W. Truitt (Ronald Press, 1960)
Heat and Mass Transfer in Capillary-Porous Bodies, by A. V. Luikov (Pergamon, 1966)
The Heat Pipe, by D. Chisholm (Mills and Boon, 1971)

Design Data

For many years *Heat Transmission,* by W. H. McAdams (3rd edn, McGraw-Hill, 1954), has been a popular source of data for the design of heat transfer equipment. However, the *Handbook of Heat Transfer,* by W. M. Rohsenow and J. P. Hartnett (McGraw-Hill, 1973), presents a complete summary of the methods and data. A major source of industrially verified design data is *Heat Transfer and Fluid Flow,* produced by The General Electric Company (in the UK from Rowse Muir Ltd., Ascot).

Articles

Heat and mass transfer, because of its complexity and wide

application, attracts a large number of researchers. Their works appear in numerous journals and at conferences held throughout the world. The *International Journal of Heat and Mass Transfer*, published monthly by Pergamon, is a major reference for research papers and annually presents a review of recent literature and progress. Articles presented at the triennial *International Heat Transfer Conference* were published by the American Institution of Chemical Engineers in 1967 and by the Elsevier Publishing Company in 1970. *Annual Proceedings of the Heat Transfer and Fluid Mechanics Institute* are published by the Stanford University Press. The American Society of Mechanical Engineers publishes quarterly the *Journal of Heat Transfer. ASME Transactions, Series C,* while the British Institution of Mechanical Engineers, besides issuing separate papers, periodically sponsors conferences or symposia, generally on a specific topic, and publish the proceedings in book form.

Advances in Heat Transfer (Academic Press, annual) aims to bridge the gap between the textbook and the research articles by providing reviews of specialised topics.

Other research publications are:

Canadian Journal of Chemical Engineering
Journal of Fluid Mechanics
Journal. Physical Society of Japan
Journal of Engineering for Power. ASME Transactions, Series A
    Thermal Engineering
Heat Transfer-Soviet Research
Warme–und Stoffubertragung
National Engineering Laboratory Reports

See also the chapters on Fuel Technology and Chemical Engineering.

THERMAL SYSTEMS

Prime sources of reference in this section are the engineering institutions, notably the Institution of Mechanical Engineers, the American Society of Mechanical Engineers and the Society of Automotive Engineers, who regularly publish research reports and reviews and periodically hold conferences on specific topics followed by publication of the proceedings. Unlike theoretical thermodynamics, which is well documented in textbooks, the applications of thermodynamics are constantly changing in design methodology and performance, with the result that textbooks dealing with applications quickly become dated. Also, few first degree courses specialise in sufficient depth in any specific topic to justify the publication of a wide range of modern textbooks. However, a professional engineer with a good knowledge of thermodynamics should be able to proceed directly to recent research publications and follow his specific interest.

A major source of information related to thermodynamic applications is the Society of Automotive Engineers (2 Pennsylvania Plaza, New York), who publish separate reports, specifications, books and index covering the topics of IC engines, gas turbines, energy conversion, aerospace, motor cars, industrial plant, fuels, lubricants, noise, pollution, safety, materials, production, etc. Their publications deal with engineering applications, design and experiences, in addition to analytical research reports. Research and technical reports are available as separate copies and are announced in the SAE's monthly magazine *Automotive Engineering.* Annual *Transactions of the SAE* are available in book or microfiche form; the library edition contains some 200 selected papers, while the complete transactions of approximately 700 papers is available only on microfiche. The *SAE Handbook,* published annually, contains specifications for materials, components, hardware, safety and performance of engineering devices. The Society also publishes the *Co-ordinated Research Councils (CRC) Reports* on fuels and the *American National Standards Institute Reports.* All the SAE publications over the past 2 years are indexed in an annual checklist. Single reports are indexed by number, report titles, author's name and organisation, while the books and booklets are also provided with a synopsis of the contents. Periodically the Society issues a library index of the technical reports published over the past 5 years. The Society also publishes the *Aerospace Standards (AS), Aerospace Recommended Practices (ARP)* and *Aerospace Information Reports (AIR) Index* and the *Aerospace Material Specifications (AMS) Index.* Motor vehicle safety standards and regulations are issued and updated by the SAE's safety standards service.

The American Society of Mechanical Engineers covers the broader aspects of mechanical engineering and as such overlaps the more specific interests of the SAE. ASME publishes a monthly general interest magazine entitled *Mechanical Engineering* and also the monthly *Applied Mechanics Reviews,* which covers books, films, journals, as well as research papers, and as such provides an excellent reviewing service for practising engineers. The quarterly *Transactions of the ASME* are divided into eight journals which include the *Journal of Engineering for Power,* the *Journal of Heat Transfer* and the *Journal of Fluids Engineering.* Three bound volumes of the transactions are available annually. Other publications are the *American National Standards* and the *ASME Standards,* but of particular interest to thermal systems are the *Performance Test Codes* (which cover all types of power plant) and the books and booklets on heat transfer and energy/power. A major service provided by ASME is the *Boiler and Pressure Vessel Code* (1974) and its addenda service, with a revision appearing every 3 years.

Publications from the British Institution of Mechanical Engineers (IMechE) (1 Birdcage Walk, London) are a monthly journal entitled *Chartered Mechanical Engineer (CME)* containing general interest articles and the monthly newspaper, *Mechanical Engineering News.* Other journals catering for specialised interests are the monthly *Journal of Automotive Engineering** and the bi-monthly *Railway Engineering Journal,* which are both application-orientated, while *Heat and Fluid Flow,* published semiannually, and the *Journal of Mechanical Engineering Science* are research-orientated. Individual research reports are available separately and when bound together constitute the *Proceedings of the Institution of Mechanical Engineers.*

Some years ago most IMechE research reports were presented at evening meetings of the Institution followed by publication of the report and critique. Today a large number of the research reports are presented at conferences held on specific topics, which are followed by the conference proceedings printed in book form.

Reviews and abstracts of research reports relating to thermodynamics, heat and mass transfer and their applications are provided by *Heat Bibliography,* published annually by HMSO; *Engineering Index*; *Applied Mechanics Reviews*; *the Motor Industry Research* Association's *Automobile Abstracts*; and *International Aerospace Abstracts.*

POWER GENERATION AND CONVERSION

Reciprocating engines

A number of textbooks aimed at the general interest, maintenance, and user market describe the mechanism and design details of engines and their auxiliaries. A. W. Judge has written a series of books of this type, all published by Chapman and Hall. Also in this category are *Diesel Engines Principles and Practice* (2nd edn, Newnes, 1962) and *Marine Diesel Engines* (5th edn, Newnes–Butterworth, 1972), by C. C. Pounder, and *The Modern Diesel,* edited by

* Continued as *Automotive Engineer* (1975–   )

D. S. D. Williams (14th edn, Butterworths, 1972). Of interest to undergraduates are *The Internal Combustion Engine*, by F. A. Z. Schmidt (Chapman and Hall, 1965), and *Combustion Processes*, by L. C. Lichty (6th edn, McGraw-Hill, 1967), which may satisfy some course requirements, while *Engines*, by D. H. Marter (Thames and Hudson, 1962), is useful elementary background reading for engineering students. Of interest to the engine specialist is *High Speed Internal Combustion Engine*, by Sir H. R. Ricardo and J. G. G. Hempson (5th edn, Blackie, 1969), which describes the experiences of the famous consulting company. The two volumes by C. F. Taylor entitled *The Internal Combustion Engine in Theory and Practice* (2nd edn, MIT Press, 1966) are probably the most quoted textbooks referred to by modern researchers. Volume 1 deals with engine performance and volume 2 with combustion and mechanical design details. Although these textbooks were written before the use of the digital computer in the synthesis of IC engines, the analytical approach is well documented and the volumes contain a considerable amount of data correlated from a wide range of experimental and commercial engines. *Rotary Piston Machines*, by F. Wankel (Iliffe, 1965), describes a wide range of rotary engine mechanisms, while the performance and kinematics of the Wankel engine itself is treated in *The Wankel RC Engine: Design and Performance*, by R. F. Ansdale (Iliffe, 1968).

The majority of British research papers concern the compression ignition engine, while American reports concentrate on spark ignition engines. A significant contribution to the use of computers in IC engine design and performance prediction has been achieved by a team of British universities headed by Professor Benson of UMIST. Their work regularly appears at conferences and in reports from the IMechE and the British Ship Research Association. Notable work on diesel engine combustion has been achieved by Professor Lyn of the University of London and by the CAV Company. This work is published in the *Lucas Engineering Review* and by the IMechE. Dr. Karim's (Imperial College of Science and Technology) work concentrates on the more fundamental aspects of engine combustion. Dr. Annand of UMIST specialises in engine heat transfer, while Professor Priede of the Southampton University Institute of Sound and Vibration works on combustion-induced noise and its radiation from engines. Recent major conferences organised by IMechE relating to diesel engines are: *Diesel Engines: Breathing and Combustion* (1966), *Thermal Loading of Diesel Engines* (1965), *Diesel Engine Combustion* (1970) and *Computers in Internal Combustion Engine Design* (1968). Other IMechE conference proceedings on engines are *Mechanical Design of Diesel*

Engines (1967), *Accuracy of Electronic Measurements in IC Engine Development* (1966), *Transport Engines of Exceptionally High Specific Outputs* (1969), *Some Unusual Engines* (1975), *Critical Factors in the Application of Diesel Engines* (1971) and *Reliability of Diesel Engines and Its Impact on Cost* (1974).

Basic research of engine performance carried out by Ricardo and Company has been recorded in the publications of IMechE and SAE; W. M. Scott's work on using high-speed photography of diesel engine combustion is published by the SAE and is entitled *Looking in on Diesel Combustion.* The 1951–1952 Part III *Proceedings of the Automobile Division of the IMechE* contained a classic report on SI engine combustion by D. Downs and R. W. Wheeler entitled 'Recent developments in knock research'.

Annually the SAE collects a number of papers dealing with the practical aspects of running and designing engines and publishes them under the title *Engineering Know-How in Engine Design.* The series, now at Part 21 (1973), deals mainly with SI engines. The SAE publishes annually approximately 40 individual reports relating to SI engines and approximately 20 reports on diesel engines. The *Proceedings* of the annual International Conference of the International Federation of Societies of Engineers in Automobile Technology are another major source of automotive engine research papers.

Magazines dealing with the applications, availability, and accessories of power generators are: *Power* (McGraw-Hill), *Power Engineering* (Technical Publishing Company), *Gas and Oil Power* (Whitehall Press Ltd), and *Diesel and Gas Turbine Progress* (Diesel Engines Inc.). All cater for the interests of engine users.

Engine emissions

Owing to recent public concern about the pollution of the environment, a vast amount of research has been performed to look into the problem of emissions from IC engines and to control the obnoxious gases. The SAE publishes approximately 60 articles a year on the subject and their major reports have been collected into three volumes entitled *Vehicle Emissions.* Part I covers the period 1955–1963, part II the period 1963–1967 and part III the period 1967–1970. The IMechE have held two conferences entitled *Motor Vehicle Air Pollution Control* (1968) and *Air Pollution Control in Transport Engines* (1972). A recent textbook on engine emissions is *Emissions from Combustion Engines and their Control,* by N. A. Henein and D. J. Patterson (Ann Arbor, 1972), and is recommended reading for engineers initially studying the problem.

GAS TURBINES AND ROCKETS

Since their inception, gas turbines and rocket propulsion have been areas of advanced technology concerned with national defence. This has resulted in a vast amount of research having been performed by government organisations such as the National Gas Turbine Establishment, the Royal Aircraft Establishment and the National Aeronautics and Space Administration. In addition, research and development contracts have also been placed with a few of the larger companies. Because of the confidentiality of the reports of this research and because engineers working in these organisations have ready access to the design data within their company, the quantity of freely available data is far less than the research performed would suggest.

The treatment of gas turbine performance and systems design is well established and is covered in a number of textbooks. *Gas Turbine Theory,* by H. Cohen, G. F. C. Rogers and H. I. H. Saravanamutto (2nd edn, Longmans, 1972), in SI units, is a revised version of an earlier text of the same title. This book concentrates on basic theory and is suitable for courses specialising in gas turbines.

Many of the textbooks in gas turbine technology were written in the late 1940s and the 1950s, and while the thermodynamic principles have not changed the methods and design data have been superseded. *Gas Turbine Principles and Practice,* edited by one of the early workers in gas turbines, Sir H. Roxbee-Cox (Newnes, 1955), was a much quoted reference. Two volumes in the series *High Speed Aerodynamics and Jet Propulsion* are basic references: volume XI, entitled *Design and Performance of Gas Turbine Power Plants,* edited by W. R. Hawthorne and W. T. Olsen (Oxford University Press, 1960) and volume XII, entitled *Jet Propulsion Engines,* edited by O. E. Lancaster (Oxford University Press, 1959). A more recent text is *Aircraft Gas Turbine Engine Technology,* by L. I. E. Treager (McGraw-Hill, 1970). *Gas Turbine Engineering Handbook,* edited by J. W. Sawyer (3 vols., 2nd edn, Gas Turbine Publications Inc, 1972, is a series of summaries on gas turbine component design, combustion, materials, and user information, and not a design data book. The long list of references following each summary is useful. An American translation of a book of Russian origin entitled *Theory of Jet Engines,* by A. L. Klyachkin (1969), is available from US Department of Commerce (reference number FTD-MT-24-123-70).

Other textbooks, some of which are out of print, are:

Cycle and Performance Estimation, by J. Hodge (Butterworths, 1955)

Gas Turbine, Analysis and Practice, by B. H. Jennings and W. L. Rogers (McGraw-Hill, 1953)

Gas Turbines, by H. A. Sorensen (Ronald Press, 1951)

Gas Turbine Power, by G. M. Dusinberre and J. C. Lester (2nd edn, International Textbook, 1958)

Principles of Jet Propulsion and Gas Turbines, by M. J. Zucrow (Wiley, 1948)

The aerodynamic design of the turbomachines in gas turbines is covered in two textbooks by J. H. Horlock entitled *Axial Flow Compressors* (Butterworths, 1958) and *Axial Flow Turbines* (Butterworth, 1966). *Centrifugal Compressors* is printed by the SAE (1961) and *Advanced Centrifugal Compressors,* by ASME (1971).

Most of the textbooks on rocket motors deal with the construction details and principle of operation rather than in-depth methodologies.

The Chemistry of Propellants, by S. S. Penner and J. Ducarne (Pergamon, 1960)

Handbook of Astronautical Engineering, edited by H. H. Koelle (McGraw-Hill, 1961)

Internal Ballistics of Solid Fuel Rockets, by R. N. Wimpress (McGraw-Hill, 1950)

Liquid Rockets and Propellants, edited by L. E. Bollinger (Academic Press, 1960)

Rocket Propulsion Elements, by G. D. Sutton (3rd edn, Wiley 1963)

Research papers in gas turbines and rocket technology are regularly published by the professional institutions: IMechE, ASME, SAE, the Royal Aeronautical Society, the American Institute of Aeronautics and Astronautics, L'Association Aeronautique et Astronautique de France and the American Rocket Society. However, probably the main sources of reference are the government research agencies NASA (formerly NACA) and the Aeronautical Research Council.

Three books from the IMechE are a selection of research papers dealing with *Axial Radial Turbomachinery* (1970), *Internal Aerodynamics* (1970) and *Technical Advances in Gas Turbine Design* (1969).

AIAA produce semi-monthly the *International Aerospace Abstracts,* and NASA issue fortnightly the *Scientific and Technical Aerospace Reports.*

STEAM PLANT

Most of the published information on steam plant is concerned with the reliability of the plant to operate efficiently and safely for long periods. Of these data, emphasis is placed on plant for electrical power generation, although a wide range of steam plant is used in process and environmental heating. The use of steam as the thermodynamic working fluid brings its own problems of corrosion, boiling and condensation, and two-phase flow and its measurement. The design of steam turbines is technically similar to the aerodynamic design of gas turbine turbomachinery except that low-pressure steam turbines frequently have to contend with water droplet impingement and subsequent blade erosion.

A major work in eight volumes prepared by the Central Electricity Generating Board (CEGB), published by Pergamon (2nd edn, 1971), entitled *Modern Power Station Practice,* deals comprehensively with all aspects of generating and controlling electrical power from steam heated by fossil and nuclear fuels. This series covers the thermodynamic and materials design of all plant components along with instrumentation and control. In a three volume series produced by the editors of *Power* and published by McGraw-Hill (1967), volume 1 entitled *Power Generation Systems* deals with the complete system, whilst volume 2 entitled *Plant Energy Systems* looks at specific components of the plant, i.e. compressors, fans, refrigeration and lubrication. *Steam, Its Generation and Use,* produced and published by the Babcock and Wilcox Company (38th edn, 1972), deals with the design details of large steam generators and superheaters. Specialised design and performance details of cooling towers are discussed in *Cooling Towers: Principles and Practice,* by W. Stanford and G. B. Hill (2nd edn, Carter Thermal Engineering, 1970), and *The Industrial Cooling Tower,* by K. K. McKelvey and M. Brooke (Elsevier, 1959). The *ASME Boiler and Pressure Vessel Code* and its addenda provides comprehensive data for the design, use, maintenance and repair of boilers and pressure vessels.

Research papers relating to steam plant are primarily published by the IMechE and the ASME, with the SAE publishing those papers relating to small steam engines for automotive applications. The ASME in its *Journal of Engineering for Power* publishes individual reports, while papers on the fluid design of turbomachinery and pumps, prior to 1973, appeared in the *Journal of Basic Engineering* and are now published in the *Journal of Fluids Engineering.*

Research papers from the IMechE are sponsored by the Steam Plant Group and the Thermodynamics and Fluid Mechanics Group. Individual papers are published, but a large number of reports appear as bound volumes of the proceedings of conferences. Notable recent titles are: *Modern Steam Plant Practice* (1971), *Steam Plant Operation* (1973), *Steam Plant for the 1970's* (1970), *Advanced Class Boiler Feed Pumps* (1971) and *Glandless Pumps for Power Plant* (1970). Papers relating to the specific problems of operating with wet steam appear in three volumes: *Wet Steam* (1966), *Wet Steam 2* (1970) and *Wet Steam 3* (1971).

ENERGY CONVERSION

Due to man's insatiable demand for energy, the conservationists' lobby and the foreseeable exhaustion of the fossil fuel supply, research is now taking place on a variety of new energy sources and new methods of producing work.
Textbooks on the subject are:

Direct Energy Conversion, by S. W. Angrist (2nd edn, Allyn and Bacon, 1971)
Direct Energy Conversion, by G. W. Sutton (McGraw-Hill, 1966)
Energy Conversion, by S. L. Chang (Prentice-Hall, 1963)

The SAE publish, annually, the *Proceedings of the Intersociety Energy Conversion Engineering Conference,* and in the 1971 edition there is a 4-year author and subject index and a 3-year *IECEC Proceedings* abstracts.

REFRIGERATION, HEATING, VENTILATING AND AIR CONDITIONING

Thermal environmental engineering is primarily concerned with designing suitable systems to provide a specific environment to suit the needs of an industrial process or to provide human comfort. Generally, the task involves maintaining the air within a closed space or building within specified levels of temperature, humidity and/or purity.
The American Society of Heating, Refrigerating and Air-Conditioning Engineers (ASHRAE) (345 East 47 Street, New York) and the Institution of Heating and Ventilating Engineers (IHVE) are the major sources of design data. The ASHRAE *Handbook and Product Directory* (formerly the ASHRAE Guide) is divided into three sections, *Systems* (1973), *Applications* (1971) and *Equipment* (1972), which along with the *Handbook of Fundamentals* (1972)

provide an excellent information source of products, methods and data. Revision of the series occurs over a 4-year cycle. The *IHVE Guide* (1973), now in SI units, concentrates solely on design data. *Specification* (Architectural Press is an annual publication giving, in addition to other data, the heat transfer coefficients of a wide range of commercially available building materials and structures. Other data sources are: *Handbook of Air Conditioning System Design,* prepared by the Carrier Air Conditioning Company (McGraw-Hill, 1966), *Handbook of Heating, Ventilating and Air Conditioning,* by F. and J. Porges (6th edn, Butterworths, 1971) and *Trane Air Conditioning Manual* (Trane Company, 1965).

Research papers and articles are published in the *ASHRAE Journal* and *Transactions*; in *Building Services Engineer,* formerly the *Journal of the IHVE*; and in the *Bulletin de l'Institut International du Froid* (papers in English and French).

Conference proceedings on *Heat and Mass Transfer in Refrigeration Systems* and *Air Conditioning* are published by the International Institute of Refrigeration (1963). *Heating and Ventilating for a Human Environment* (1967) is available from the IMechE.

Other sources of research papers are *Heating and Ventilating Engineer and Journal of Air Conditioning,* a monthly magazine from Technitrade Journals; *Refrigeration World and Air Conditioning Review,* available alternate months from the Scientific Press; and the monthly *Refrigeration and Air Conditioning,* published by Refrigeration Press. Many other articles appear in magazines and journals which specialise in other fields, but are users of the thermal systems, e.g. food technology, hospitals, scientific laboratories, motor cars, aeroplanes, spacecraft.

Most of the available textbooks adopt an applications-orientated approach aimed at practising engineers. However, the textbook *Thermal Environmental Engineering,* by J. L. Threlkeld (2nd edn, Prentice-Hall, 1970), deals with thermodynamic fundamentals and is therefore suitable as a course text for undergraduates.

A selection of textbooks covering a range of interests are:

Air Conditioning Engineering, by W. P. Jones (Arnold, 1967)
Cryogenic Engineering, by J. H. Bell (Prentice-Hall, 1963)
Cryogenic Fundamentals, by G. G. Haselden (Academic Press, 1971)
Cryogenic Technology, edited by R. W. Vance (Wiley, 1963)
Environmental Engineering, Analysis and Practice, by B. H. Jennings (International Textbook, 1970)
Modern Air Conditioning, Heating and Ventilating, by W. H. Carrier, R. E. Cherne, W. A. Grant, and W. H. Roberts (Pitman, 1959)

Principles of Refrigeration, by R. J. Dossat (Wiley, 1961)

Refrigeration and Air Conditioning, by W. F. Stoecker (McGraw-Hill, 1958)

Refrigeration at Sea, by R. Munton and J. R. Stott (Maclaren, 1967)

Trane Refrigeration Manual (Trane Company, 1956)

26

Marine technology

R. C. Kahler and M. J. Shields

Current developments in most branches of science and technology are outstripping the terminology available to describe them, and the marine field is no exception. The traditional terms 'Marine Engineering' and 'Naval Architecture' are already becoming difficult to define, some overlap with subjects of interest to mechanical engineers and 'Offshore Technology' begin to find increasing mention in the literature, it is becoming more and more difficult to decide where the overlap occurs. We have, therefore, decided to adopt the term 'Marine Technology' for our heading, since this best covers the subjects of primary interest within our field.

Most marine subjects overlap into other fields. Diesel engines, steam and gas turbines, and pumps are used as commonly on land as at sea, and although the construction of a ship may be regarded as a specialised subject, it is basically only an assembly process using the same techniques of plate handling, forming, cutting and welding as any other type of structural engineering. The design processes, too, make use of structural analysis techniques which are encountered in civil engineering and aeronautical engineering, while on the hydrodynamics side there are many affinities with aerodynamics. In offshore work, drilling rigs, for example, borrow from their land-based counterparts, and many operations such as dredging, piling and reclamation can really be regarded as part of civil engineering. The ship itself is a self-contained community, and its design and operation require the support of such disciplines as electrical and electronic engineering, acoustics, paint and corrosion technology, and marine biology, while even such subjects as interior decor and furnishing are not beyond the scope of the well-stocked marine library. Finally, of course, economics and management sciences are becoming increasingly important in all fields of engineering.

Any information worker in the marine field must, therefore, be fully aware of the sources of information in all these varied subjects. Most of them will, however, be covered elsewhere in this book, and so, having defined the extent of the problem, we propose to concentrate here on sources having a specifically marine bias.

CLASSIFICATION SOCIETIES

Classification Societies are a peculiarly marine institution, at least in their origins. Their primary function is to establish rules governing the construction and equipment of ships and their machinery, and to inspect them during the course of construction, and at intervals during their service life, in order to ensure that they are built and maintained in a safe condition, and thus form an acceptable insurance risk. Each of the Societies maintains an active research department, mainly concerned with investigations into structural strength and metallurgical matters. The publication policies of the Societies vary, some producing periodicals which include technical articles based on the Society's work, while others publish detailed reports of their investigations.

Most of the major maritime countries of the world have a national Society, but all the Societies operate at least to some extent on an international basis, and do not restrict themselves entirely to ships owned in their country of origin. The oldest-established and most widespread is Lloyds Register of Shipping (commonly confused with Lloyds Insurance, with which it has no connection other than the coffee-house of the same name in which both had their origin), and this maintains surveyors in most parts of the world, and classifies many ships of foreign construction and ownership. Other major Societies are the American Bureau of Shipping, Bureau Veritas (France), Germanischer Lloyd (again no relation), Det norske Veritas (Norway), Registro Italiano Navale, Nippon Kaiji Kyōkai (Japan) and the USSR Register of Shipping. All Societies publish their *Rules* in the form of annual updated volumes which, while being very similar in their coverage, do differ considerably in detail, so that it is not sufficient to assume that because, for example, Lloyds Register covers a constructional detail in a particular way any of the other Societies will have an identical requirement. Copies of the *Rules* are bulky and rather expensive, and some of the foreign Societies publish only in their own language, so that it is often not practicable for the smaller marine library to maintain complete sets of even the most commonly encountered Classification Society *Rules*. Complete sets of the major Societies of the world are held by such libraries as that of the Department of Trade (Marine Division), the British Ship Research Association and Lloyds Register itself. Probably they would not be willing to lend these, but in many cases they could be consulted or sections could be photocopied under the usual conditions. It is worth noting that the *Rules* of all the countries of the Soviet Bloc have now been unified, and that those of the USSR Register of Shipping are available in a Russian/English edition.

Lloyds Register does not produce any reports available to the general public, but publishes an irregular serial entitled *100A1* which includes articles that describe the work of the Society without, however, going into any great technical detail. Technical reports are issued by the research departments of Det norske Veritas and Registro Italiano Navale, while periodicals such as *The Surveyor* (ABS), *Bulletin Technique du Bureau Veritas* (BV) and *Veritas* (DnV) are also available. Ships have also to comply with rules and regulations issued by the Department of Trade, and by similar bodies overseas. These deal mainly with such matters as the welfare and safety of crew members and also the safe operation of ships.

LEARNED SOCIETIES

The remarks of the overlap of marine technology into other fields made at the beginning of this chapter cannot be overemphasised when we consider the learned societies. It is probably true to say that there can be few papers published by any of the major technological societies of the world which could not in some way be relevant to problems in the marine field. This would be particularly true, for example, of the Institution of Mechanical Engineers or the Institution of Electrical Engineers, or of such American societies as the ASME, the IEEE or the American Society of Civil Engineers, whose *Transportation Engineering Journal of ASCE* and *Journal of the Waterways, Harbors and Coastal Engineering Division* are obviously of major interest. There are nevertheless several societies directly concerned with the marine field, and any marine library will undoubtedly find it necessary to obtain their transactions or proceedings.

To take the British societies first, the oldest is the Institution of Naval Architects, established under that title in 1860, and becoming the Royal Institution of Naval Architects (RINA) in its centenary year. Early papers given to the Institution covered marine engineering as well as naval architecture, and although marine engineers formed a separate Institute in 1889, papers on this subject are still occasionally to be found in the *RINA Transactions*. These transactions in their complete form can be obtained only as an annual bound volume, but the quarterly journal *The Naval Architect* carries the complete texts (with discussions) of papers likely to be of general interest, in addition to original articles and news items. Other more specialised papers are published in *Supplementary Papers*, which appears once or twice a year.

The Institute of Marine Engineers again publishes its *Transactions* annually and, in addition, a monthly journal known as the *Marine Engineer's Review*. This contains the usual news items and, in addition, some longer technical articles which are, however, usually not based on papers read at meetings of the Institute. There are also abstracts of articles from the current technical press.

Two further British societies which reflect the strong regional bias of the shipbuilding industry are North-East Coast Institution of Engineers and Shipbuilders (NECIES), founded at Newcastle-upon-Tyne in 1884, and the Institution of Engineers and Shipbuilders in Scotland (IESS). Strictly speaking, the latter can claim with some justification to be the senior British learned society in the field, since it owes its origin to the Institution of Engineers in Scotland, which was founded in 1857. The Scottish Shipbuilders Association was founded in 1860, and the two societies amalgamated in 1865, adopting the present title 10 years later. Both the North-East Coast and the Scottish Institution publish *Transactions,* with parts available at intervals throughout the year.

The Society of Naval Architects and Marine Engineers (SNAME) is the principal society in our field in the United States, and, as its name implies, it covers the whole field of marine technology. It also fulfils the functions of the regional institutions in Britain in that it has many more or less autonomous local sections spread throughout the United States and Canada. The annual *SNAME Transactions* contain papers presented at the annual meeting of the main society, but not the papers presented at the many meetings which each of the local sections arranges throughout the year. Some of the papers read at local meetings, but by no means all, are reprinted in the quarterly journal *Marine Technology,* which is published by the main society in New York. In addition, this includes a news section which lists the papers to be read at forthcoming meetings of the local sections, and in many cases it is possible to obtain copies of these papers from the secretaries of the local sections concerned. The *Journal of Ship Research,* which is also published quarterly by SNAME, contains a high proportion of papers relating to the hydrodynamic aspects of ship design. There are also occasional publications of various types from this Society, and altogether it is without doubt the most important source in the United States, and perhaps the most useful single source in the world.

Another useful serial produced in the USA is the bi-monthly *Naval Engineers Journal,* published by the American Society of Naval Engineers. This is, in fact, the transactions of the Society under a new guise, with some additions. The subject coverage is not limited to marine engineering, but the emphasis tends to be on naval rather than merchant marine matters. Finally, so far as the USA is concerned, there is the Marine Technology Society, which holds an annual conference and publishes a bi-monthly journal, concentrating mainly on ocean engineering and allied subjects. In Europe learned societies do not contribute anything like such a large fraction of the available marine technological literature as they do in Britain or the USA. The Schiffbautechnische Gesellschaft (STG) in Germany and the Association Technique Maritime et Aéronautique (ATMA) in France hold an annual meeting each at which major technical papers are read, and these are later published in an annual volume of proceedings. The Scandinavian countries have a supranational 'society' known as the Nordic Ship-Technical Meeting (NSTM). This holds an annual meeting at a venue in one or other of the contributing countries (Norway, Sweden, Denmark, and Finland), and since it is titled in the language of the host country, the initials NSTM represent something slightly different on the various occasions; hence, the use of the rather clumsy English title above. Four papers, usually lengthy and of a very high technical standard, are read at each of these meetings, each paper by a representative of, and in the language of, one of the Scandinavian countries, although thankfully the Finnish contributors have up to now presented their papers in Swedish. Little if anything is available in the form of papers or transactions from learned societies elsewhere on the Continent (although we shall meet some again as publishers of periodicals). There is an Italian Society of Naval Architects, while in Spain naval architects belong to a branch of the National Institution of Civil Engineers. In both cases at least some of the papers read are to be found as articles in the periodical press.

So far as countries of the Soviet bloc are concerned, the learned society, at least in the technological field, seems to blend imperceptibly into the state-supported research institute, and will therefore be dealt with under that heading. The main exception appears to be the Kammer de Technik of East Berlin, which is a general purpose technical society with a shipbuilding section. It is also a major East German publisher of technical books and periodicals.

There are several learned societies in Japan publishing a great volume of very important papers on all aspects of marine technology, the difficulty being that they do so almost exclusively in Japanese. Only the Japan Society of Mechanical Engineers publish regular proceedings in English, all other societies contenting themselves with at best a title and abstract in English of various levels of comprehensibility. Moreover, both translations and transliterations of the titles of these societies and their journals vary considerably, so that it is worth taking a little time here to discuss this.

The words 'Nihon' and 'Nippon', for example, are variant readings of the same characters in Japanese (meaning 'Japan'), and the choice between them is entirely a matter of personal preference. During and before the last war Nippon was fashionable, but Nihon has tended to take over, and some societies have changed the transliteration of their titles to suit.

The national society is the Society of Naval Architects of Japan, also referred to as the Japan Society of Naval Architects. The Japanese title of this organisation was formerly Zōsen Kyōkai, but from volume 123, June 1968, of their *Journal* they have been known as Nihon Zōsen Gakkai, and the journal itself as *Nihon Zōsen Gakkai Rombun-Shu*. The journal is published twice yearly, in June and December, each issue having a separate volume number, and it contains a large number of papers on all aspects of naval architecture, including both hydrodynamics and structures. Equal to the national society, at least in output, is the unambiguously entitled Society of Naval Architects of West Japan (Seibu Zōsen Kai) whose journal *Seibu Zōsen kai Kaihō,* is again published twice a year, in March and July. Also based in West Japan (at Osaka University) is the Kansai Society of Naval Architects (Kansai Zōsen Kyōkai), which publishes a quarterly journal.

All aspects of ship operation and traffic engineering are covered by the Nautical Society of Japan, also known as the Japan Institute of Navigation, whose *Journal* is published twice yearly. This Society also has a second publication, published three times a year, known as *Navigation.* The *Journal* is known in Japanese as *Nippon Kōkai Gakkai- Tō* and the other publication as *Kōkai.* The two publications are similar in format and are easily confused. A useful point is that *Navigation* has, over recent issues, been publishing in serial form an English–Japanese dictionary of marine terms which, when complete, will be very extensive.

One final organisation worth noting is the Society for Materials Science, Japan (Nippon Zairyō Gakkai), whose monthly journal

Zairyō covers all aspects of the technology of both metals and non-metals, corrosion, biological degradation and many other subjects of interest to the marine technologist.

The above journals, together with many others in the fields of engineering and industrial chemistry, have been described in a recent Aslib publication, *A Classified Directory of Japanese Periodicals, Engineering and Industrial Chemistry,* compiled by P. C. R. Mason (1972). In view of the confusion of Japanese names and titles mentioned above, this booklet should be a valuable addition to the reference section of any marine or technical library. It is available from Aslib at £3.00 per copy.

RESEARCH INSTITUTES

Under this heading we must perforce deal with a great variety of institutions of one sort or another which are not easily classifiable under clear-cut headings. They include direct state-controlled research establishments, state-financed university research institutes and departments, industrial research associations (perhaps with some indirect government aid), national or regional research and test laboratories, and all other similar organisations which publish reports or other printed information, or from whom technical information can be obtained. A list of such bodies is given at the end of this chapter.

In general, private industry has not been included in this section. Individual firms may, of course, have research departments, but the work is usually of a confidential nature, and they are often unwilling to publish except through learned societies. In any case, most trade directories will provide all the guidance necessary to firms whose main interest is in the marine field, and direct enquiry is usually the only way of ascertaining whether or not a particular firm is willing to divulge the results of its research, and at what price. However, there are exceptions here as well. There are some American companies, such as Litton Industries, Oceanics Inc., and Hydronautics Inc., which are wholly or partly devoted to marine research, and in fact are often major sources of information, their non-confidential reports being available through the National Technical Information Service of the USA. They have therefore been included in our list, as have some of the national ship design companies of the Soviet bloc whose function is indistinguishable from that of the national research laboratory or research association in western countries.

Another noticeable omission is the standards institute. There is no single body in the world exclusively devoted to the production of marine standards or specifications. They are usually produced by the national standardisation body of the country concerned, perhaps in collaboration with some marine-oriented research institute. In the UK, for example, the British Ship Research Association is collaborating with the British Standards Institution in the production of a set of *British Marine Standards,* and is also represented on the appropriate ISO panels. The BSI is the best source of information on British and foreign standards, and in fact holds some collections of foreign standards in its library.

Addresses of foreign standards institutions are given in the *BSI Yearbook* in the section entitled 'Complete Sets of British Standards Maintained in Overseas Countries'.

The direct government research establishments in the UK that are of greatest relevance to the marine field include a number under the Ministry of Defence, Navy Department. These establishments usually retain the now-obsolete term 'Admiralty' in their title, an example being the 'Admiralty Experiment Works'. A list is given at the end of this chapter, but in fact there is no need to apply directly to any of them for information, since such of their reports as are available to the general public may be obtained through the Defence Research Information Centre (DRIC) at Orpington, Kent. These, and other reports from government research stations, are announced in *R & D Abstracts,* available free of charge from DRIC. A similar situation obtains in the USA, where all work performed for the government that is not classified is notified in the abstract journal *Government Reports Announcements,* available on subscription from the National Technical Information Service (NTIS), Springfield, Virginia. All reports cited in these two publications are available in microfiche form from the British Library, Lending Division at Boston Spa. The other main British government research establishment of interest to those engaged in the marine field is the National Physical Laboratory, which has two departments whose work is relevant, namely Ship Division and Maritime Science Division. Both issue reports, although not all are freely available, and much of the work also forms the basis for papers to the learned societies.

Just as government security classifications obstruct the availability of information from some state research establishments, so commercial confidentiality affects the industrial research associations. For obvious reasons most of the European ship research associations are prepared to disclose the results of their current work only to their own member firms. However, the British Ship Research Association acts as the European clearing house for such information as is available, and abstracts it in its *Journal of Abstracts of the*

British Ship Research Association, which will be fully described in a later section. Generally speaking, the BSRA is prepared to help and advise in all matters apertaining to information work in the marine field, and to offer its extensive information facilities to non-members at reasonable rates, and is therefore probably the best method of keeping in touch with research activities in the marine field.

Outside Britain, Europe and America, there are few countries with national ship research institutions, the chief exceptions being the USSR, with the Leningrad Shipbuilding Institute and the Central Research and Scientific Institute of the Merchant Marine, and Japan, with the Sempaku (or Senpaku) Gijutsu Kenkyushō (the Ship Research Institute of Japan), all three of which publish only in their original languages. Central research facilities do exist in other countries, such as East Germany (Rostock) and Bulgaria (Varna), but their publications are not always easy to obtain, and in many cases are of limited relevance to the western world. Countries which do not as yet have extensive research facilities, but which may well establish them in the near future, are the 'coming nations' of ship-building: South Korea, India, Brazil and Spain.

Many of the most important research institutes are attached to universities. The Norwegian equivalent of the British Ship Research Association, the Ship Research Institute of Norway (Norges Skipsforskningsinstitutt (NSFI), for example, is based at the Technical University of Norway at Trondheim, and the Danish Ship Research Institute at the National Technical University at Lyngby. At the other end of the scale, virtually any university or polytechnic has departments dealing with mechanical engineering, strength of materials or metallurgy—all subjects of interest in the marine field—and it is therefore extremely difficult to decide where to draw the line when selecting universities for inclusion in a list such as we have given. For the sake of brevity, it has been decided to eliminate all universities which do not have active interests in the subjects most directly concerned with this chapter, namely naval architecture and marine engineering. Even so, the list is extensive, and it is worth noting that the more important of these university research departments are: in the UK, those of Glasgow, Newcastle-upon-Tyne and Southampton; in the USA, the University of Michigan and the Stevens Institute of Technology; and in the Soviet Union, the Universities of Leningrad and Moscow. European universities with extensive marine interests are those of Hamburg, Hanover, Rostock, Berlin, Delft, Zagreb and Naples, while the Japanese research effort is concentrated in the Universities of Tōkyō, Tōhoku and Kyushu.

PERIODICALS

A full listing of all periodicals relevant to shipping and shipbuilding would be far beyond the scope of this chapter. For example, the library of the British Ship Research Association subscribes to some 500 periodicals, all of which have some relevance to its work, and the reader is therefore referred, not only to our own listing of the more directly relevant marine literature, but also to periodicals listed eleswhere in this volume, as well as to directories, catalogues and periodicals.

There are some periodicals, of course, which deserve special mention. No marine library should be without the weekly *Marine Week* (formerly *Shipbuilding and Shipping Record*) or the monthly *Shipping World and Shipbuilder*. Slightly more specialised are *Motor Ship, Tanker and Bulk Carrier* and *Ship and Boat*, which deal respectively with diesel-engined ships, large bulk cargo vessels and the smaller craft, while *Fairplay* is possibly the best weekly periodical for market information, together with the dailies *Lloyds List* and the *Journal of Commerce*.

All the above are published in Britain. Most of the American journals are published by learned societies, and have already been mentioned. The only two commercial periodicals of any consequence coming from the USA are *Maritime Reporter* and *Marine Engineering/Log*, both of which are primarily useful for industrial news items rather than technical articles. The latter is perhaps the better in this respect, but those familiar with it 60 years or so ago will be disappointed if they expect to find the same standards of technical writing.

There are also a number of English-language magazines published in other countries, notably *International Shipbuilding Progress*, published in Rotterdam, which attracts papers by leading naval architects from all over the world. Two important and informative English-language journals from Japan are *Zōsen* and *Japan Shipbuilding and Marine Engineering*, while from Scandinavia there is *Norwegian Maritime Research*, until recently known as *European Shipbuilding*.

Foreign-language periodicals always present problems, and usually expensive ones at that, but they cannot be ignored in the marine field. Some of the most important information is contained in these journals, perhaps more so than in foreign-language learned society transactions. Chief among them, both in informative content as well as in relative ease of translation, are the West German periodicals *Hansa* and *Schiff und Hafen*; also from West Germany are *Forschungshefte für Schiffstechnik*, and a periodical recently introduced by the Verein Deutscher Ingenieure (VDI) that deals mainly with ocean engineering and that the VDI have elected to call *Meeres-technik/Marine Technology*. The English version of the title, if it is ever used, will clearly lead to confusion with the well-established and excellent periodical of that name published by the Society of Naval Architects and Marine Engineers to which reference has been made earlier. The use of a bilingual title reflects the growing trend in the German technical press to publish some articles in English, which can only be welcomed. Eastern Germany also produces useful journals, the principal ones from our point of view being *Seewirtschaft* and *Schiffbauforschung*. France is surprisingly poor in marine technical journals, *Navires, Ports et Chantiers* and *Journal de la Marine Marchande* being the only two of note.

Italy also produces two journals, *La Marina Italiana* and *Tecnica Italiana*, while *Ingenieria Naval* is the main contribution from Spain. News from the Netherlands may be obtained from *Schip en Werf*, while Scandinavia is well represented by *Svensk Sjøfarts Tidning* (the Swedish Shipping Gazette) and by three Norwegian periodicals, *Norwegian Shipping News* (with contents largely in English), *Skip* and *Skipsteknik*, together with other specialised titles dealing with the coasting and fishing industries. Satisfactory coverage of the countries of the Soviet bloc, other than East Germany, may be obtained by taking *Brodogradnja* (Yugoslavia) and *Budownictwo Okretowe* (Poland), while the Soviet Union itself produces *Morskoi Flot* and *Sudostroenie*. Finally, the Japanese journal *Sempaku* can be very useful to those with access to Japanese translation facilities, but not otherwise, since it contains no English-language material whatever, even the titles being in Japanese.

House journals are as many and varied as the firms that publish them, and since they exist chiefly to publicise the products of the company, they are often of limited usefulness. However, the Japanese firms of Mitsubishi, Hitachi Zosen and Ishikawajima Harima publish extremely good English-language versions of their respective *Technical Reviews,* and these are well worth noting.

ABSTRACTING SERVICES

The abundance of periodicals that may contain possibly relevant matter makes the need for abstracting services especially acute in the marine field. Unless a marine information service is prepared to acquire a wide range of serials from all over the world, with the attendant costs and linguistic problems, a subscription to a comprehensive marine abstracting service becomes an absolute necessity. Unfortunately, however, while there are many well-known general-purpose technical abstracts services and reference works available, there are relatively few which specialise in marine subjects. Undoubtedly the largest marine abstracting service in the world is the *Vodnoe Transport* (Water Transport) section of *Referativnyi Zhurnal,* but this has several disadvantages apart from the obvious one of being exclusively in Russian. Although its coverage is very wide, it includes many obscure Soviet sources which are usually unavailable to western readers. Coverage of other sources is no better than that of western abstracting journals and is, in fact, relatively late. Moreover, such abstracts as are given, in addition to being in Russian, are so brief as to be of little practical value.

The oldest-established and widest-ranging publication in the west is the *Journal of Abstracts of the British Ship Research Association,* now in its 31st year. It covers, in English, all the major sources of marine information throughout the world, and reproduces about 3000 abstracts a year. Although it does include some title-only references, most are informative abstracts. The *BSRA Journal* also contains lists of available publications, such as translations, bibliographies and derestricted BSRA reports, and it is backed up by a comprehensive library and information service which is offered to non-members of the Association at reasonable rates. Despite this extensive coverage, it is obtainable at the relatively low subscription rate of £20 sterling per annual volume of 12 monthly issues and index.

Until 1973 the *BSRA Journal* was the only abstracting service covering the marine field in the west. With the advent of two new journals this is, however, no longer the case.

The first of these, to be produced by the Institute of Marine Engineers, will in fact be an extension of an existing service. For many years the *Transactions of the Institute of Marine Engineers,* and more recently the *Marine Engineer's Review,* has contained an abstracts section chiefly as a service to members of the Institute. Since 1971, when the *Transactions* were transformed into the *Review,* these abstracts have also been reprinted quarterly under their own cover. Now, in collaboration with Lloyds Register of Shipping, the Institute is to publish another quarterly, the *Journal of Technical References.* In the past the *Marine Engineer's Review* has produced about 700 quite lengthy informative abstracts per year, some reprinted from the *BSRA Journal.* There is at the time of writing no indication as to what the present or future coverage of these journals will be in terms of numbers of periodicals scanned or of the extent to which the number of abstracts published is to be increased. The quoted subscription rate is £10 for the *Journal of Technical References* and £6 for the *Abstracts,* or £15 for a combined subscription to both.

The other newcomer is again an extension of an existing service, but here it is one which has up to now been limited to Scandinavia. It is, however, somewhat less of an unknown quantity than the *Journal of Technical References,* since samples of earlier Norwegian-language issues are available, and these can be assessed and compared with the first of the English-language issues which have recently come to hand at the time of writing. This publication, formerly *Artikel Indeks før Skip,* then the *Scandinavian Ship Abstract Journal,* and now known as *Ship Abstracts* is produced by the Norwegian Centre for Informatics with the collaboration of the Norwegian and Swedish Ship Research Institutes and the Finnish Association of Metal and Engineering Industries. The Danish Ship Research Institute, which contributed to the Scandinavian edition, seems to have withdrawn from the association with the changeover to English. This publication is produced by computer, which at least ensures that each issue contains its own index, with a cumulative index being supplied at the end of the year. For some reason the publishers have decided to produce ten issues per year, which they claim will contain about 3000 abstracts. In fact, however, the POLDOC computer program used for processing the information reprints abstracts under alternative section headings, allotting them new numbers, instead of producing 'see also' references, so that this figure is an exaggeration. Independent checks on both the Scandinavian and English-language versions indicate that about 15% of abstracts printed are repeats, so that the true total of abstracts is nearer 2500 than 3000. Because of the need to minimise computer storage, the abstracts are extremely brief, usually no more than two or three lines. Although English is used for all abstracts, and for most foreign-language titles, including those from Scandinavia, the German and Dutch titles retain the original languages, which, in view of the brevity of the abstracts, does not make for easy understanding. Computer production does not appear to have had any appreciable effect in speeding up publication. The lag between original publication date and the abstract journal is about the same as for the *BSRA Journal* and is in fact slightly greater than for abstracts produced in the *Marine Engineer's Review. Ship Abstracts* is by far the most expensive of the three services.

Most other marine abstracting services are trivial in comparison with any of the above, or are merely extensions to periodicals. An example of this is the Norwegian periodical *Skip,* which since January 1973 has been printing each month Norwegian translations of a selection of the BSRA abstracts. The German periodicals *Hansa* and *Schiff und Hafen* reprint lists of titles of recent articles, which are supplied by the Documentation Service of the STG, and the Jugoslav periodical *Brodogradnja* also lists Serbo-Croat translations of titles of recent articles taken directly from a number of sources that are available locally, but relying on the *BSRA Journal* for its wider coverage.

TRADE DIRECTORIES AND YEARBOOKS

There are relatively few directories which apply solely to the marine field, but those which do are of great value. Chief among them is the so-called 'Blue Book', the *Directory of Shipowners, Shipbuilders and Marine Engine Builders,* which lists not only commercial undertakings but also consultative and research bodies on a worldwide scale, and is very comprehensively indexed. It is produced annually by the publishers of the periodical *Marine Week,* IPC Business Press. Its main rival, with very similar coverage and indexing, is the *International Shipping and Shipbuilding Directory,* published by Benn, which has the advantage of listing by country in certain of its sections. A considerable volume of useful information on docks, harbours and available facilities is contained in *Ports of the World,* another Benn publication, or the very similar *Port Dues, Charges, and Accommodation,* published by George Philip and Son. Finally, the Japanese *Zōsen Yearbook,* produced by the publishers of the periodical *Zōsen,* is also worth mentioning for the extra information on Japanese organisations which it contains.

Other yearbooks will, of course, also be necessary if the librarian wishes to cover firms and organisations whose business is not strictly marine.

One further class of yearbook which has not been mentioned is the *Register Book,* which each of the Classification Societies produces annually, and which contains details of each ship classed by that society. The Lloyds Register volumes also include vessels classed by other societies, so that they may be regarded as an up-to-date listing of the world's merchant fleet.

APPENDIX: A SELECTED LIST OF SOURCES OF INFORMATION IN MARINE TECHNOLOGY

Classification societies

American Bureau of Shipping
Bulgarian Register of Shipping
Bureau Veritas
China Corporation Register of Shipping
Czechoslovak Register of Shipping

(East) German Ship Inspection and Classification Society (DSRK)
Germanischer Lloyd
Hellenic Register of Shipping
Hungarian Ministry of Transport and Communications Shipping Administration
Jugoslavenski Registar Brodova
Korean Register of Shipping
Lloyds Register of Shipping
Nippon Kaiji Kyōkai
Norske Veritas
Polish Register of Shipping
Register of Shipping of the USSR
Registro Italiano Navale
Rumanian Register of Shipping

Learned societies

American Society of Civil Engineers (NY)
American Society of Mechanical Engineers (NY)
American Society of Naval Engineers (Washington)
American Welding Society (Miami)
A. N. Krylov Scientific and Engineering Society of the Shipbuilding Industry (Leningrad)
British Joint Corrosion Group (London)
British Society for Underwater Technology (Poole)
European Federation of Corrosion (Frankfurt/Main)
Federation of Engineers and Technicians in Yugoslavia (Belgrade)
German Society of Engineers (VDI) (Dusseldorf)
Institute of Electrical and Electronic Engineers (NY)
Institute of Marine Engineers (London)
Institution of Engineers and Shipbuilders in Scotland (Glasgow)
Institution of Mechanical Engineers (London)
International Cargo Handling Coordination Association (London)
Japan Association of Underwater Technology (Tokyo)
Kansai Society of Naval Architects (Kansai Zōsen Kyōkai) (Osaka U.)
Nautical Society of Japan (Jap. Inst. of Nav.) (Tokyo U. of Merch. Mar.)
North-East Coast Institution of Engineers and Shipbuilders (Newcastle-upon-Tyne)
Norwegian Society of Professional Engineers (Oslo)
Permanent International Association of Navigation Congresses (Brussels)
Royal Institute of Navigation (London)
Royal Institution of Naval Architects (London)
Scientific and Engineering Society of Water Transport (Moscow)
Shipbuilding Society of China (Shanghai)
Shipbuilding Technical Society (STG) (Berlin)
Society for Shipbuilding Technology (Shanghai)
Society of Automotive Engineers (NY)
Society of Consulting Marine Engineers and Ship Surveyors (London)
Society of Naval Architects and Marine Engineers (NY)
Society of Naval Architects of Japan (Tokyo)
Society of Naval Architects of West Japan (Fukuoka)
Swedish Association of Engineers and Architects (Stockholm)
Technical Maritime and Aeronautical Association (ATMA) (Paris)
Welding Institute (Abingdon, Berks.)

Research institutes and universities

Admiralty Experiment Works (Haslar)
Admiralty Materials Laboratory (Poole, Dorset)
Admiralty Research Laboratory (Teddington, Middlesex)
Alexandria Institute of Oceanography and Fisheries
All-Union Research Institute of Marine Fishing Industry and Oceanography (Moscow)
Association of Naval Engineers (Spain) (Madrid) affiliated to the Institution of Civil Engineers
Association for the Investigation of Naval Construction (Madrid)
Battelle Memorial Institute (Columbus, Ohio)
Belgian Maritime Research Centre (Ceberena) (Brussels)
Booz Allen Applied Research Inc. (Bethesda, Md)
British Ship Research Association (Wallsend)
Central Scientific Research Institute of the Merchant Marine, USSR (Leningrad)
Central Ship Design Bureau (Gdansk)
Centre for Advanced Marine Studies (Marseille)
Centre for Research and Oceanographic Studies (Paris)
Centre for Technical Naval Studies (CETENA) (Genoa)
Chalmers University of Technology (Göteborg)
Delft University of Technology
Department of Mechanics and Control Processes (ACAD SC USSR) (Moscow)
Far-Eastern V.V. Kuibyshev Polytech. Inst. (Vladivostock)
Food and Agriculture Organization (FAO) (Rome)
Foundation for Nuclear Propulsion of Merchant Ships (The Hague)
Galati Polytechnic Institute (Balati)
Gdansk Technical University
Gorky Institute of Water Transport Engineers
Gothenburg University
Hamburg Shipbuilding Test Establishment (HSVA) (Hamburg)
Higher Institute of Electrical and Mechanical Engineering (Varna)
Hydrodynamic Test Channel (Madrid)
Hydronautics Inc. (Laurel, Md.)
Indian Institute of Technology (Kharagpur)
Institute for Maritime Sciences (Split)
Institute for Shipping Research (Bergen)
Institute for the Use of Nuclear Energy in Shipbuilding and Navigation (Hamburg)
Institute of Marine Research (Helsinki)
Institute of Marine Technology (Quebec)
Institute of Naval Construction (Genoa)
Institute of Naval Construction (Naples)
Institute of Shipping Economics (Bremen)
Institute of Transport, Maritime Affairs and Communications (Zagreb)
Intergovernmental Marine Consultative Organization (IMCO) (London)
International Maritime Committee (Antwerp)
Israel Ship Research Institute (Haifa)
Japan Shipbuilding Industry Foundation (Tokyo)
Kyushu University (Fukuoka)
Leningrad Admiral S.O. Makarov Higher School of Marine Engineering
Leningrad Institute of Water Transport
Leningrad Shipbuilding Institute

Litton Ship System Inc. (Los Angeles, Calif.)
Maritime Construction Research Institute (IRCN) (Paris)
Maritime Institute (Gdansk)
Massachusetts Institute of Technology (Cambridge, Mass.)
National College of Naval Engineering (Paris)
National Gas Turbine Establishment (NGTE) (Farnborough)
National Enginering Laboratory (Glasgow)
National Institute of Naval Architecture Studies and Experiments (Rome)
National Institute of Shipping and Shipbuilding (Rotterdam)
National Naval Architecture Research and Testing Center (Rome)
National Physical Laboratory (Teddington)
National Ports Council (London)
National Research Council of Canada (NRCC)
Naval Construction Research Establishment (Dunfermline)
Naval Ship Research and Development Center (Washington)
Netherlands Ship Model Basin (NSMB) (Wageningen)
Netherlands Ship Research Centre (Delft)
Norwegian School of Economics and Business Administration (Bergen)
Novosibirsk Institute of Water Transport Engineers
Oceanics Inc. (NY)
Odessa Higher School of Marine Engineering (USSR)
Odessa Institute of Water Transport Engineers
Pusan National University (Pusan)
Research And Design Institute of Shipbuilding (Varna)
Research Institute for Shipbuilding (Berlin)
Research Institute for Shipbuilding for Inland Navigation (Duisburg)
Rhine-Westphalia Technical University (Aachen)
Royal Institute of Technology (Stockholm)
Shanghai Shipbuilding Institute
Shipbuilding and Navigation Study Unit TNO (The Hague)
Shipbuilding Research Association of Japan (Tokyo)
Shipbuilding Research Centre of Japan (Tokyo)
Shipbuilding Research Institute (Vienna)
Ship Design (Brodoproject) (Zagreb)
Shipowners Refrigerated Cargo R.A. (Cambridge)
Ship Research Institute (Tokyo)
Ship Research Institute of Norway (Trodheim)
Skibsteknisk Laboratorium (Lyngby)
Stanford Research Institute (Menlo Park, Calif.)
Stevens Institute of Technology (Hoboken, NJ)
Swedish Ship Research Foundation (SSF) (Göteborg)
Swedish State Shipbuilding Experimental Tank (SSPS) (Göteborg)
Technical University of Berlin
Technical University of Denmark (Lyngby)
Technical University of Hanover
Technical University of Helsinki
Technical University of Istanbul
Technical University of Norway (Trondheim)
Technical University of Vienna
United States Coast Guard Academy (New London, Conn)
United States Merchant Marine Academy (Kings Pt., NY)
United States Naval Academy (Annapolis, Md.)
United States Naval Applied Science Laboratory (Brooklyn, NY)
United States Naval Civil Engineering Laboratory (Pt. Hueneme, Calif.)

United States Naval Postgraduate School (Monterey, Calif.)
United States Naval Research Laboratory (Washington)
United States Naval Ship Systems Command (NAVSHIPS) (Washington)
United States Navy Bureau of Yards and Docks (NAVDOCKS) (Washington)
University of Alexandria
University of Belgrade
University of California (Berkeley)
University of Genoa
University of Glasgow
University of Hamburg
University of London
University of Madrid
University of Michigan
University of Newcastle-upon-Tyne
University of Osaka Prefecture
University of Rostock
University of Southampton
University of Strathclyde (Glasgow)
University of Trieste
University of Tokyo
University of Wales Institute of Science and Technology (Cardiff)
University of Zagreb
Webb Institute of Naval Architecture (Long Island)
White Fish Authority (Edinburgh)

Periodicals

ASCE Proceedings. Journal of the Waterways, Harbors and Coastal Engineering Division (New York)
Brodogradnja (Zagreb)
Budownictwo Okretowe (Warsaw)
Bulletin Technique du Bureau Veritas (Paris)
Canadian Shipping and Marine Engineering (Toronto)
Cargo Handling and Shipbuilding Quarterly (East Bentleigh, Victoria, Australia)
Central Scientific Research Institute of the Merchant Marine, USSR (Leningrad): *Transactions*
Chalmers University of Technology (Gothenburg, Sweden): *Transactions (Shipbuilding Section)*
Commercial Fishing (Fleetwood, Lancs.)
Dock and Harbour Authority (London)
Fairplay (London)
Fish Industry Review (WFA) (Edinburgh)
Fishing News (London)
Fishing News International (London)
Hansa (Hamburg)
Hitachi Zosen News (Osaka)
Holland Shipbuilding and Marine Engineering (Rotterdam)
Hovering Craft and Hydrofoil (London)
ICHCA, Journal (London)
IHL Engineering Review (Tokyo)
Indian Shipping (India)
Institut für Seeverkehrswirtschaft (Bremen): *Statistik der Schiffahrt*
Institute of Marine Engineers (London): *Transactions*

Institution of Engineers and Shipbuilders in Scotland (Glasgow): *Transactions*
International Shipbuilding Progress (Rotterdam)
Israel Shipping Research Institute (Haifa): *Journal*
Japan Shipbuilding and Marine Engineering (Tokyo)
Journal de la Marine Marchande (Paris)
Journal of Hydronautics (Easton, Pa.)
Journal of Maritime Law and Commerce (Silver Spring, Md.)
Journal of Ship Research (New York)
Journal of Technical References (London)
Journal of Transport Economics and Policy (London)
Leningrad Shipbuilding Institute (Leningrad): *Transactions*
Marina Italiana (Genoa)
Marine Engineering/Log (New York)
Marine Engineers' Review (London)
Marine Observer (London)
Marine Pollution Bulletin (London)
Marine Safety Council of the United States Coast Guard (Washington): *Proceedings.*
Marine Technology (New York)
Marine Technology Society (Washington): *Journal*
Marine Week (London)
Mariner's Mirror (London)
Maritime Institute (Gdansk): *Transactions*
Maritime Reporter (New York)
Meerestechnik (Dusseldorf)
Mitsubishi Heavy Industries Technical Review (Japan)
Morskoj Flot (Moscow)
Motor Ship (London)
Nautica (Amsterdam)
Nautical Society of Japan (Tokyo): *Journal*
Naval Architect (London)
Naval Engineers Journal (Washington, DC)
Navigation (Tokyo)
Navires, Ports et Chatiers (Paris)
Navy International (London)
Netherlands Ship Model Basin (Wageningen): *Reports*
Neubauten (New Ships) (Hamburg)
North East Coast Institution of Engineers and Shipbuilders (Newcastle-upon-Tyne): *Transactions*
Northern Offshore (Oslo)
Norwegian Fishing and Maritime News (Bergen)
Norwegian Shipping News (Oslo)
Norwegian University of Technology (Trondheim): *Ship Model Tank Reports*
NSFI-NYTT (Trondheim, Norway)
Ocean Industry (Houston, Texas)
Offshore (Houston, Texas)
Offshore Services (Kingston, Surrey)
Offshore Technology (London)
Ports and Dredging (The Hague)
Reed's Marine Equipment News (Sunderland)
Royal Institute of Navigation (London): *Journal*
Royal Institution of Naval Architects (London): *Transactions*
Safety at Sea International (London)
Scandia (Hamburg)

Scandinavian Shipping Gazette (Copenhagen)
Schiff und Hafen (Hamburg)
Schiffahrt International (Hamburg)
Schiffbauforschung (Rostock, East Germany)
Schiffstechnik: Forschungshefte fur Schiffbau und Schiffsmaschinenbau (Hamburg)
Schip en Werf (Rotterdam)
Seatrade (Colchester, Essex)
Seewirtschaft (E. Berlin)
Sempaku (Tokyo)
Ship and Boat International (London)
Shipbuilding and Marine Engineering International (London)
Shipbuilding and Transport Review International (The Hague)
Shipbuilding, Ship Repair, and Services (Sydney)
Shipping Digest (New York)
Shipping Statistics and Economics (London)
Shipping World and Shipbuilder (London)
Skip (Oslo)
Skipsteknikk (Oslo)
Smaaskipsfart (Oslo)
Society of Naval Architects and Marine Engineers (New York): *Transactions*
Society of Naval Architects of Japan (Tokyo): *Journal*
Society of Naval Architects of West Japan (Fukuoka): *Journal* (also known as *Journal of Seibu Zōsen Kai*)
South African Shipping News and Fishing Industry Review (Cape Town)
Sudostroenie (Leningrad)
Surveyor (New York)
Svensk Sjofarts Tidning (Gothenburg)
Swedish State Shipbuilding Experiment Tank (Gothenburg): *Reports*
Tanker and Bulk Carrier (London)
Tecnica Italiana (Trieste)
Transport (Brussels)
Transportaion Engineering Journal of ASCE (New York)
Veritas (Oslo)
World Dredging and Marine Construction (USA)
World Fishing (London)
World Ports and Marine News (Washington)
World Shipbuilding Bibliography (Gdansk)
Zōsen (Yokyo)

Abstracting journals

British Ship Research Association. *Journal of Abstracts* (Wallsend)
British Technology Index (London)*
Brodogradnja (Zagreb)†
Budownictwo Okretowe (Warsaw)†
Engineering Index (New York)*
Government Reports Announcements (Springfield, Virginia)*
Hansa (Hamburg)†
Journal of Technical References (London)
Marine Engineers Review (London)†

*General technical coverage, not restricted to marine subjects
†Contains abstracts or lists of titles in addition to normal editorial matter

Navires, Ports et Chantiers (Paris)†
Referaten – Dienst Wasserfahrzeuge, Schiffbau (Rostock)
Referativnyi Zhurnal – Vodnoe Transport (Moscow)
Revista de Informacion Marina Mercante (Madrid)†
Schiff und Hafen (Hamburg)†
Ship Abstracts (Oslo)
Skip (Oslo)†
World Shipbuilding Bibliography (Gdansk)

Recent textbooks and reference books

Basic Ship Theory, by K. J. Rawson and E. C. Tupper (Longmans, 1968)
Automation in Merchant Ships, by J. A. Hind (Fishing News, 1968)
Marine Corrosion, by T. H. Rogers (Newnes, 1968)
Technology of Ship Repairing, by D. D. Benkovsky *et al.* (MIR, 1967)
Basic Naval Architecture, by K. C. Barnaby (5th edn, Hutchinson, 1967)
Naval Architecture, Examples and Theory, by B. Baxter (Griffin, 1967)
The Economics of Sea Transport, by C. O'Loughlin (Pergamon, 1967)
Applied Naval Architecture, by R. Munro-Smith (Longmans, 1967)
Automation on Shipboard, by G. J. Bonwick (Macmillan, 1967)
Centralised and Automatic Controls in Ships, by D. Gray (Pergamon, 1966)
Principles of Naval Architecture, by J. P. Comstock (SNAME, 1967)
Shipbuilding Technology, by V. K. Dormidontov *et al.* (MIR, 1966)
Ship Construction, by D. J. Eyres (Heinemann, 1972)
Schiffbautechnisches Handbuch (5 vols., VEB Verlag, 1961)
Background to Ship Design and Shipbuilding Production, by J. A. Hind (Temple Press, 1965)
Ship Design and Construction, edited by A. M. D'Arcangelo (SNAME, 1969)
Marine Steam Boilers, by J. H. Milton (3rd edn, Newnes–Butterworths, 1970)
Practical Construction of Warships, by R. N. Newton (3rd edn, Longmans, 1970)
Nuclear Ship Propulsion, by R. F. Pocock (Allan, 1970)
Creative Naval Architecture, by G. N. Hatch (Thos. Reed, 1971)
Marine Diesel Engines, by C. C. Pounder (5th edn, Newnes–Butterworth, 1972)

†Contains abstracts or lists of titles in addition to normal editorial matter

27

Metallurgy

W. J. D. Jones

INTRODUCTION

Metallurgy is the study of all aspects of metals: the chemical treatment of metal ores to release the metal (known as extraction metallurgy); the study of metal and alloy atomic structures and microstructures to facilitate the formulation of alloys with desirable properties (physical metallurgy); the fabrication of metals into desired shapes and structures (production or fabrication metallurgy); their final preparation for service (metal finishing); their response to complex engineering service conditions (engineering metallurgy); and, finally, their deterioration in service, which, if allowed to occur, would lead to the conversion of the metal back into a chemically compounded form, the state in which it naturally exists and from which it was initially extracted (corrosion and protection). The study of this cycle of events is metallurgy.

If the reader is not familiar with the subject and requires a broad introduction to all its aspects, the best work to consult is *An Introduction to Metallurgy*, by A. H. Cottrell (Arnold, 1967). This is a very good introduction written by the doyen of the profession. It consists of short chapters on almost every aspect of metallurgy, each terminating with a list of references for further reading to take the reader on to a more advanced understanding. This book is not the most elementary available and it is biased towards the science of metals rather than their engineering aspects. A more elementary science-based approach is found in the books mentioned below on materials science. A more elementary engineering metallurgy approach is found in such books as *Metallurgy for Engineers*, by E. C. Rollason (4th edn, Arnold, 1973), or the more elementary two-

426

volume work *Engineering Metallurgy*, by R. A. Higgins (English Universities Press, Vol. 1, 4th edn, 1973, Vol. 2, 2nd edn, 1970).

The research techniques and the physical concepts and theories which have evolved, mainly in studies of metals, have been found to apply to other solid materials such as ceramics, carbon, glasses, polymers, etc., which have recently become the subject of much research and development. This has led to a broadening of the study of metallurgy into the more general materials science and materials technology. Some of the information sources for metallurgy, therefore, have titles bearing the word 'materials' and they contain information on materials other than metals and their alloys.

The present trend in education, particularly its early phase, is towards replacing metallurgy wherever possible by a unified study of all materials. This method of approach is shown in *The Elements of Materials Science*, by L. Van Vlack (3rd edn, Addison-Wesley, 1975), which is the most widely used elementary text. A slightly more advanced approach is given in the four-volume work *The Structure and Properties of Materials*, by J. Wulff *et al.* (Wiley, 1964–1966).

DICTIONARIES

The technology of metals is fraught with industrial jargon and proprietary names for patented processes and alloys. These words are explained in *A Dictionary of Metallurgy*, by A. D. Merriman (Macdonald and Evans, 1958), or the later *Concise Encyclopaedia of Metallurgy* (1965), by the same author and publishers. A briefer but similar work is *Dictionary of Metallurgy*, by D. Birchon (Butterworths, 1965).

More recently published are *Dictionary of Metallurgy*, by W. E. Clason (Elsevier, 1967), which contains metallurgical terms in English, French, Spanish, Italian, Dutch and German, and *Dictionary of Metal Finishing and Corrosion*, by H. W. Dettner (Elsevier, 1971).

English-speaking metallurgists are well served with translations of important works from other languages. Many foreign research journals are regularly published in cover-to-cover English translations (see p. 55).

The following organisations produce translations of important papers into English. They usually have an index of available translations for sale and will translate to order.

British Industrial & Scientific International Translation Service, The Metals Society, 1 Carlton House Terrace, London, SW1Y 5DB

Henry Brutcher Translations, PO Box 157, Altadena, Calif. 91001, USA

National Translations Center, John Crerar Library, 35 W 33rd St., Chicago, Illinois, 60616, USA

DATA SOURCES

A large body of data exists in this subject consisting of all the known properties of metals, alloys, some of their important compounds and sundry allied materials such as refractories, glasses, carbon and polymers, which are used either in the metallurgical industries or in conjunction with metals. Many readers consult metallurgical texts merely to learn the value of some parameter for a particular alloy. For this purpose the following books are useful. Some contain very little explanatory text and concentrate on providing tables of values or other data, while others additionally give a more detailed explanation of the underlying principles or practical details of the metallurgical processes.

The best-known works of reference in general metallurgy are the following two. *Metals Reference Book*, by C. J. Smithells (5th edn, Butterworths, 1975) consists mainly of data covering the complete range of metallurgy. In contrast, *Metals Handbook* (American Society for Metals (ASM), Ohio) is of a more encylopaedic nature, with explanations of each topic considered. At present this work consists of nine volumes written by expert committees of the ASM:

Vol. 1 *Properties and Selection of Materials* (1961)
Vol. 2 *Heat Treatment, Cleaning and Finishing* (1964)
Vol. 3 *Machining* (1968)
Vol. 4 *Forming* (1969)
Vol. 5A *Forging* (1972)
Vol. 5B *Melting and Casting* (1972)
Vol. 6 *Welding and Brazing* (1971)
Vol. 7 *Atlas of Microstructures of Industrial Alloys* (1972)
Vol. 8 *Metallography and Phase Diagrams* (1973)
Vol. 9 *Fractography and Atlas of Fractographs* (1974)

Most other reference books are more limited in their scope, with respect to either the metals or the properties considered. However, if one's interests fall within their confines, then the following reference books may be as useful as the former.

The most comprehensive collection of data on British steels is found in *The Mechanical and Physical Properties of the British*

Standard En Steels (BS970–1955), by J. Woolman and R. A. Mottram (Pergamon):

Vol. 1   *En 1 – En 20*   (1964)
Vol. 2   *En 21 – En 39*   (1966)
Vol. 3   *En 40 – En 363* (1969)

Engineering Data on Cast Iron and many other publications on various aspects of cast irons are published by the British Cast Iron Research Association, Alvechurch, Birmingham B48 7QB. Information on stainless steels (or any other type of steel) may be obtained from the User Advisory Service of the British Steel Corporation, London. Data and other information on the use of some individual non-ferrous metals may be obtained from the following bodies:

Aluminium Federation, Broadway House, Calthorpe Road, Five Ways, Birmingham, B15
Cobalt Information Centre, 7 Rolls Buildings, Fetter Lane, London, EC4A 1JA
Copper Development Association, Orchard House, Mutton Lane, Potters Bar, Herts
Gold—Johnson Matthey and Company Ltd., Hatton Garden, London, EC1
Lead Development Association, 34 Berkeley Square, London, W1X 6AJ
Magnesium Elektron Ltd., Clifton Junction, Manchester
Climax Molybdenum, Villiers House Strand, London, WC2N 5JS
International Nickel Company Ltd., Thames House, Millbank, London, SW1
Niobium—Murex Ltd., Rainham, Essex
Precious metals generally—Johnson Matthey and International Nickel Company
Rare earths—Johnson Matthey
Silver—Johnson Matthey
Tantalum—Murex Ltd
Tin Research Institute, Fraser Road, Perivale, Greenford, Middlesex
Titanium—IMI Kynoch Works, Witton, Birmingham 6
Tungsten—Murex Ltd
Zinc Development Association, 34 Berkeley Square, London
Zirconium—IMI

The following books are also useful sources of data. Note that books of American origin give alloy names and specification designations which are different from those used in Britain.

Encyclopaedia of Engineering Materials and Processes, edited by
H. R. Clauser *et al.* (Van Nostrand Reinhold, 1963)
Engineering Alloys, Names, Properties, Uses, by N. E. Woldman
and R. C. Gibbons (5th edn, Van Nostrand Reinhold, 1973)
Engineering Materials Handbook, by C. L. Mantell (McGraw-Hill,
1958)
Handbook of Lattice Spacings and Structures of Metals and Alloys,
by W. B. Pearson (2 vols., Pergamon, 1958 and 1967)
Materials Data Book for Engineers and Scientists, by E. R. Parker
(McGraw-Hill, 1967)
Metal Bulletin Handbook (6th edn, Metal Bulletin Ltd, 1973). This is
an annual compilation of data relating to the costs and production
of metals and alloys in the UK and other principal markets.

Many data books are published which do not deal principally with
metals, but which are frequently useful to the metallurgist, such as
the following:

Handbook of Chemistry and Physics, edited by R. C. Weast (56th
edn, CRC Press, 1975)
The Handbook of Thermophysical Properties of Solid Materials, by
A. Goldsmith, T. E. Waterman and H. J. Hirschhorn (Macmillan,
1961). This is a five-volume work in which values of density,
melting-point, latent heats of fusion, vaporisation and sublimation,
specific heat, thermal conductivity, thermal diffusivity, emissivity
and reflectivity, linear thermal expansion, vapour pressure and
electrical resistivity are collected. Volume 1 elements, volume 2
alloys and volume 3 (including) cermets and intermetallics, contain
metallurgical data.

STANDARD SPECIFICATIONS

The use of most metals in industry is usually voluntarily governed
by codes of practice which may lay down the (minimum) properties
or metallurgical condition or recommended procedures for use, etc.,
of metals in a particular application. This is necessary to establish
standards of safety and efficient performance and to establish an
acceptable working basis for a contractual relationship between
manufacturers and users of alloys.

The alloys specified for use in a particular application may be
different in different countries; they may even be different in
different industries within a country. Nevertheless, these specification documents contain a wealth of information on metallurgical

properties and give a strong indication of the alloys which should be used in particular circumstances.

The standard specifications for general engineering in the UK are established by the British Standards Institution (2 Park Street, London, W1), which publishes a number of documents on metal usage.

Some industries do not use the general engineering BS system, but have their own, e.g. the aircraft industry. The North Sea oil industry growing at the present time chooses to work to the standards of the American Petroleum Institute.

In America, many bodies publish standard specifications. These are found listed in the *Index of US Voluntary Engineering Standards,* edited by W. J. Slattery (NBS Special Publication No. 329, 1971) and *Supplement 1* (1972), by the same editor and publisher. The most important metal specifications are produced by the American Society for Testing and Materials (ASTM) (1916 Race St., Philadelphia, Pa.), the Society of Automotive Engineers (SAE) and American Iron and Steel Institute (AISI).

The *ASTM standards,* issued annually, constitute 47 parts in 1974 covering the use of all materials in a wide range of applications. They are very detailed regarding the test procedure for establishing the suitability of a metal for a particular purpose or determining data.

Most countries have evolved their own systems of specifications. If it is necessary to establish the identity of a metal referred to by a foreign specification designation or to establish the equivalent British specification, this may be done with the aid of *Metallic Materials Specification Handbook,* by R. B. Ross (2nd edn, Spon, 1972).

METAL MONOGRAPHS

Any reader who has a query on a particular metal would naturally look for a book bearing its name as a title. He would be likely to find one, as the majority of metals have at some time been the complete or joint subject of a book which here will be called a metal monograph. This type of book commonly attempts to give a complete account of the metal from its occurrence in nature through its extraction, alloying and fabrication and including its properties and uses (unless otherwise stated in the title), dealing mainly with the well-established facts available on that metal. After a metal has been in use for some time, its extraction procedures usually change less than its physical and engineering aspects, which continue to grow owing to the continuing development of better alloys and the increasing knowledge of the mechanical properties of the alloys in a wider range of engineering service conditions. When second or later monographs appear, they may deal mainly with these latter developments.

The scope of a metal monograph is too large to give recent research results on every aspect of the metal and its alloys, and they are published too infrequently to be consistently up-to-date. Nevertheless, this type of book may form a suitable starting point for a more detailed search.* It is, therefore, useful to have a list of recently published metal monographs such as is given here. Only some of the most recently published books are listed and they commonly contain references to previously published books and other publications on that metal.

Books on iron alloys are listed separately because the amount of information available on iron and its alloys is much greater than that on any other metal and would not be easily contained within one book. Iron is difficult and expensive to prepare in a high-purity form and there is no great demand for it in this form, because it is mechanically weak. The vast majority of iron used is in the form of its alloy with carbon known as steel. Steel, cast-iron and other iron-based alloys are collectively known as ferrous alloys and all other metals, by contrast, are non-ferrous alloys. This demarcation correctly reflects the relative magnitude of the various metals used.

Non-ferrous metals

Aluminum Alloys: Structure and Properties, by L.F. Mondolfo (Butterworths 1976)
The Technology of Aluminium and Its Alloys, by P. C. Varley (Butterworths, 1970)
Antimony, by C. Y. Wang (3rd edn, Griffin, 1952)
Beryllium, by G. E. Darwin and J. H. Buddery (Butterworths, 1960)
Boron: Preparation, Properties and Applications, edited by G. K. Gaule (Plenum, 1965)
Cadmium, by D. M. Ch'zhikov (translated from Russian by D. E. Hayler) (Pergamon, 1966)
Calcium Metallurgy and Technology, by C. L. Mantell and C. H. Hardy (Reinhold, 1945)
Caesium—See Rubidium
Chromium, by A. H. Sully and E. A. Brandes (2nd edn., Butterworths, 1967)
Columbium (US terminology)—see Niobium (European name)
Cobalt, Its Chemistry, Metallurgy and Uses, by R. S. Young (Reinhold, 1960)
Copper, edited by A. Butts (Collier Macmillan, 1970)
Germanium, by V. I. Davydov (Gordon and Breach, 1967)

* A comprehensive account and bibliography of the properties of all pure metals and some alloys, their ores and extractive metallurgy is given under each element in *Gmelins Handbuch der anorganischen Chemie* (see chapter 8 of *The Use of Chemical Literature* (2nd edn. ,1969)).

Gold; Recovery Properties and Applications, edited by E. M. Wise (Van Nostrand, 1964)

Lead and Lead Alloys, by W. Hofman (Springer-Verlag, 1970)

Principles of Magnesium Technology, by E. F. Emley (Pergamon, 1966)

Manganese, by A. H. Sully (Butterworths, 1955)

Molybdenum, by L. F. C. Northcott (Butterworths, 1956)

For Molybdenum see also Niobium

The Winning of Nickel; Geology, Mining and Extraction Metallurgy, by J. R. Boldt and P. Quenau (Longmans, 1967)

Engineering Properties of Nickel and Nickel Alloys, edited by J. L. Everhart (Plenum, 1971)

'Niobium (Colombium)' in *Columbium and Tantalum*, edited by F. T. Sisco and E. Epremian (Wiley, 1963)

Niobium, Tantalum, Molybdenum and Tungsten, edited by A. G. Quarrell (Elsevier, 1961)

Tantalum and Niobium, by G. L. Miller (Butterworths, 1969)

Palladium. Recovery, Properties and Uses, by E. M. Wise (Academic Press, 1968)

International Conference on Plutonium 1965, edited by A. E. Kay and M. B. Waldron (Chapman and Hall, 1967)

Rhenium, by B. W. Gonser (Elsevier, 1962)

Rubidium and Caesium, by F. M. Perel'man (2nd edn., Pergamon, 1965)

The Technology of Scandium, Yttrium and the Rare Earth Metals, by E. V. Kleber and B. Love (Macmillan, 1963)

Selenium and Selenides, by D. M. Chizhikov and V. P. Shchastlivyi (Collets, 1968)

Silver: Economics, Extraction and Use, edited by A. Butts and C. D. Coxe (Van Nostrand, 1967)

Tantalum—see Niobium

Tellurium and Tellurides, by D. M. Chizhikov and V. P. Shchastlivyi (Collets, 1970)

Thorium—see Uranium

Tin and Its Alloys, edited by E. S. Hedges (Arnold, 1960)

Extractive Metallurgy of Tin, by P. A. Wright (Elsevier, 1966)

Titanium, by A. D. and M. K. McQuillan (Butterworths, 1955)

Titanium Science and Technology, edited by R. I. Jaffee and H. M. Burte (4 vols., Plenum, 1973)

Tungsten, by K. C. Li and C. Yu Wang (3rd edn, Reinhold, 1955)

Tungsten: Metallurgy, Properties and Applications, by C. J. Smithells (Chemical Publishing Company, 1958)

For Tungsten see also Niobium

Uranium, by J. H. Gittus (Butterworths, 1963)

Extraction and Metallurgy of Uranium Thorium and Beryllium, by R. G. Bellamy and N. A. Hill (Pergamon, 1963)

The Metallurgy of Vanadium, by W. Rostoker (Wiley, 1958)

Yttrium—see Scandium

Zinc, edited by C. H. Mathewson (Reinhold, 1959)

Zirconium, by G. L. Miller (2nd edn, Butterworths, 1957)

Iron and its alloys

Alloying Elements in Steels, by E. C. Bain and H. W. Paxton (ASM, 1961)

Aspects of Modern Ferrous Metallurgy, edited by J. S. Kirkaldy and R. G. Ward (University of Toronto Press, 1964)

The Making, Shaping and Treating of Steels, by H. E. McGannon (8th edn, US Steel Corporation, 1964)

The Manufacture of Iron and Steel, by G. R. Bashforth (4 vols., 3rd edn, Chapman and Hall, 1964)

Martensite: Fundamentals and Technology, by E. R. Petty (Longmans, 1970)

Recent Advances with Oxygen in Iron and Steel Making, edited by W. J. B. Chater and J. L. Harrison (Butterworths, 1964)

Stainless Iron and Steel, by J. H. G. Moneypenny (2 vols., Chapman and Hall, 1951 and 1954)

The Structure of Alloys of Iron, by W. Hume-Rothery (Pergamon, 1966)

The Technology and Properties of Ferrous Alloys for High Temperature Use, by M. G. Gemmill (Butterworths, 1966)

The Technology of Ferrous Alloys for Ambient and Low Temperature Use, by T. F. Pearson (Butterworths, 1968)

The latest developments in ferrous metallurgy are well covered in many of the volumes of the continuing AIME (American Institute of Mining and Metallurgical Engineers) Metallurgical Society Conference Series and the continuing Special Reports Series of the Iron and Steel Institute.

TEXTBOOKS

Any question relating to a particular metal is not always answered by reference to a book devoted to that metal. Questions of detailed behaviour are often better answered in a text dealing with a particular property or the principles underlying the behaviour of all or many metals. In addition to having a bibliography of metal monographs it is necessary to have one on subjects.

There is no continuous series of volumes (comparable with the *Handbuch der Physik: Encyclopedia of Physics* (54 vols., Springer-Verlag, 1955–)) which covers the whole field of metallurgy topic by topic. To have a comprehensive coverage of the principles of metallurgy, it is necessary to list books from a wide range of publishers. This is done below for each part of the subject. Wherever possible a recently published book is given which reviews the field and lists other more detailed books.

Metallurgy is a practical science. Several books have been published on the techniques commonly used, such as *The Physical Examination of Metals,* edited by B. Chalmers and A. G. Quarrel (2nd edn, Arnold, 1960), and more recently *Physical Metallurgy Techniques and Applications,* by K. W. Andrews (2 vols., Allen and Unwin, 1973), and *Tools and Techniques in Physical Metallurgy,* by F. Weinberg (2 vols., Dekker, 1970). These volumes give a brief introduction to each of a number of techniques. Much more detailed information is given in a series of volumes under the general editorship of R. F. Bunshah and the general title of *Techniques of Metals Research* (Wiley). The individual books are arranged as follows (the parts referred to are separate books):

Vol. 1 *Techniques of Materials Preparation and Handling, Part 1* (1968), *Part 2* (1968), *Part 3* (1968)
Vol. 2 *Techniques for Direct Observation of Structures and Imperfections, Part 1* (1968), *Part 2* (1969)
Vol. 2A *The Stereographic Projection and Its Application* (1969)
Vol. 3 *Modern Analytical Techniques for Metals and Alloys, Part 1* (1970), *Part 2* (1970)
Vol. 4 *Physiochemical Measurements in Metals Research, Part 1* (1970), *Part 2* (1970)
Vol. 5 *Measurement of Mechanical Properties, Part 1* (1971), *Part 2* (1971)
Vol. 6 *Measurement of Physical Properties, Part 1* (1971), *Part 2* (1972)
Vol. 7 *Techniques Involving Extreme Environment, Non-Destructive Testing, Computer Methods in Metals Research and Data Analysis, Part 1* (1972), *Part 2* (1972)
These books are more than handbooks of experimental methods; they give good reviews of their subject.

Extraction metallurgy

Books on this topic are usually either descriptive of the processes of extraction of metals singly or in groups, or are concerned with the scientific understanding of the chemical principles regulating the processes. An example of the first type is *An Outline of Metallurgical Practice,* by C. R. Hayward (3rd edn, Van Nostrand, 1952). Examples of the second type are *Extraction Metallurgy,* by J. D. Gilchrist (Pergamon, 1967), and *Physical Chemistry of Iron and Steel Manufacture,* by C. Bodsworth and H. B. Bell (2nd edn, Longmans, 1972).

Many other examples of both types of books could be given, but it is more fruitful to draw attention to a bibliography of books and research papers in extraction metallurgy which has been published recently in *Principles of Extraction Metallurgy,* by F. Habashi (4 vols., Gordon and Breach, 1969–). Volume 1 (1969) gives an historical introduction and a general review of the chemical principles of extraction metallurgy and includes a 70-page bibliography of all the books published on general and extraction metallurgy from the earliest works up to the date of publication. Each chapter ends with a comprehensive bibliography of research papers and books on the topic considered. These volumes are a particularly valuable source of information to anyone interested in extraction metallurgy.

Physical metallurgy

This branch of metallurgy is concerned with the structure of metals and alloys studied on an atomic scale by crystallography and on a larger scale by metallography; and further with the way in which these changeable structures influence many of a metal's properties, particularly strength and ductility. An elementary introduction may be obtained from *Modern Physical Metallurgy*, by R. E. Smallman (3rd edn, Butterworths, 1970), or the larger *Physical Metallurgy Principles*, by R. E. Reed-Hill (2nd edn, Van Nostrand, 1973). A much more advanced treatment of the subject is given in *Physical Metallurgy*, edited by R. W. Cahn (2nd edn., North Holland, 1971), which is a large collection of chapters by experts on their own specialisation. This book is most suitable for postgraduate and other research workers. It gives suggestions for further reading.

The methods of the crystallographer in metallurgy and some of his results are well explained in *Structure of Metals*, by C. S. Barrett and T. B. Massalski (3rd edn, McGraw-Hill, 1966), which is a well-established text for advanced students, and *X-ray Metallurgy*, by A. Taylor (Wiley, 1961).

Metals and alloys have microscopic grain structures which increase in complexity with increase in the number of added alloying elements and with use of thermomechanical treatments. An understanding of phase equilibria is necessary for the study of these microstructures, and this may be obtained from *Phase Diagrams in Metallurgy*, by F. N. Rhines (McGraw-Hill, 1956). There are books available which contain established phase diagrams. The standard work for alloys composed of two elements is *Constitution of Binary Alloys*, by M. Hansen and K. Andreko (McGraw-Hill, 1958), and its two supplements published by the same company, *Constitution of Binary Alloys, Supplement 1*, by R. P. Elliott (1965), and *Constitution of Binary Alloys, Supplement 2*, by F. A. Shunk (1969).

The techniques for preparing metals for viewing their microstructures optically or by electron microscope and interpreting them are well documented in the following books:

Electrolytic and Chemical Polishing of Metals in Research and Industry, by W. J. McGregor Tegart (Pergamon, 1956)
Interpretation of Metallographic Structures, by W. Rostoker and J. R. Dvorak (Academic Press, 1965)
Metallurgical Microscopy, by H. and S. Modin (Butterworths, 1972)
Modern Metallography, by R. E. Smallman and K. H. G. Ashbee (Pergamon, 1966)

The Structure and Properties of Alloys, by R. M. Brick, R. B. Gordon and A. Phillips (3rd edn, McGraw-Hill, 1965)
Quantitative Microscopy, edited by R. T. Dehoff and F. N. Rhines (McGraw-Hill, 1968)

Note that volume 7 of *Metals Handbook* is devoted to microstructures and volume 8 to metallography. As an aid to understanding alloy structures by personal experience, samples of most commonly used alloys together with explanatory notes and diagrams or photographs may be obtained from Metallurgical Services, Betchworth, Surrey.

There are many books published on the physics of metal deformation. *Dislocations*, by J. Friedel (Pergamon, 1964), in addition to giving a comprehensive account of the subject, gives a list of the principal works in this field (symposia, textbooks, reviews and a reference to a bibliography by others of all papers up to 1960). A smaller and more elementary work on this subject is *Introduction to Dislocations*, by D. Hull (Pergamon, 1965). A wider view of deformation and the principles for designing stronger materials is given in *Strengthening Methods in Crystals*, by A. Kelly and R. B. Nicholson (Elsevier, 1971).

An elementary introduction to the electrical and magnetic properties of metals (with a brief mention also of thermal and optical properties) is given in *The Structure and Properties of Materials: Vol. 4, Electronic Properties*, by R. M. Rose, L. A. Shepard and J. Wulff (Wiley, 1966). Many references for supplementary and advanced reading are given in addition to the ones given below.

Magnetism is primarily studied in metals; therefore they are considered in almost every book on the subject. A particularly good introduction to the magnetic properties of metals is given in *Ferromagnetism*, by R. M. Bozorth (Van Nostrand, 1951). Despite its age, this is a very useful book. Other useful references are *Electrical and Magnetic Properties of Metals*, by J. K. Stanley (ASM, 1963); *The Magnetic Properties of Metals and Alloys*, by R. M. Bozorth et al. (ASM, 1959); and *Magnetism and Metallurgy*, edited by A. E. Berkowitz and E. Kneller (2 vols., Academic Press, 1969).

The majority of metals used for their high electrical or thermal conductivity are generally used in as pure a condition as is compatible with having the necessary strength, because alloying increases strength, but decreases both forms of conduction. Only a small number of relatively pure metals or dilute alloys are used. Variation in their metallurgical constitution is small and, hence, of limited interest. However, in recent years there has been much work on the development of semiconducting and superconducting materials.

Because these properties may be favourably changed by means of metallurgical techniques, several publications have appeared on the subjects, such as *Metallurgy of Advanced Electronic Materials,* edited by G. E. Brock (Interscience, 1963); *Metallurgy of Semiconducting Materials,* edited by J. Schroeder (Interscience, 1962); *Superconductivity of Metals and Alloys,* by P. G. de Gennes (Benjamin, 1966); and The effect of metallurgical variables on superconducting properties, by J. D. Livingstone and H. W. Schadler in volume 12 of *Progress in Materials Science* (Pergamon, 1964).

The complex mechanical properties known as brittle fracture, fatigue and creep are the subject of much engineering and metallurgical research. The data relating to these problems cannot be expressed in a simple and universally applicable way. Hence, very many research papers and books have been published on these problems relating to the performance of particular metals and alloys in particular circumstances.

A general introduction to the mechanical properties of materials, which is nevertheless advanced, is given in *The Mechanical Behaviour of Materials,* edited by F. A. McClintock and A. S. Argon (Addison-Wesley, 1966). This book has chapters on the various mechanical properties each terminating in a long list of references including books devoted to that particular property.

The following are some more recent useful books:

The Mechanical Properties of Materials at Low Temperatures, by
D. A. Wigley (Plenum, 1971)
Hardness Measurements of Metals and Alloys, by H. O'Neill (2nd edn, Chapman and Hall, 1967).

Several books have been published on studies of fracture—for example, *Fracture of Structural Materials,* by A. S. Tetelman and A. J. McEvily (Wiley, 1967). *Fracture,* edited by H. Liebowitz (Academic Press, 1969–1972), is an advanced treatise of seven volumes on the engineering aspects of fracture. Volume 6 deals with metals in particular, but the other volumes are also relevant.

Continuing sources of information on fatigue and fracture are the published proceedings of the International Conference on Fracture, *Proceedings of the First International Conference on Fracture* (3 vols., Japanese Society for Strength and Fracture in Materials, 1966) and the second conference *Fracture 1969,* edited by P. L. Pratt (Chapman and Hall, 1969). A third conference was held in Munich in 1973, and the proceedings have been published in 10 volumes by the Association of German Iron and Steel Engineers.

The most recent book on this topic is *Fundamentals of Fracture Mechanics*, by J. F. Knott (Butterworths, 1973).

The most recent book on creep is *Creep Strength in Steel and High Temperature Alloys* (Iron and Steel Institute, 1973).

Fabrication metallurgy

A general introduction to fabrication metallurgy is given in *Manufacturing Properties of Materials*, by J. M. Alexander and R. C. Brewer (Van Nostrand, 1963), or the more recent *Materials and Processes in Manufacturing*, by E. P. DeGarmo (4th edn, Macmillan, 1974). The latter is descriptive rather than analytical and gives a booklist to more detailed works on materials properties and each manufacturing process. Below are some recently published books on individual processes (in addition to some volumes of *Metals Handbook* mentioned earlier).

Metal casting is considered in two recent books, *Solidification and Casting*, by G. J. Davies (Applied Science, 1973), and *Foundry Technology*, by P. R. Beeley (Butterworths, 1972). The first gives a more scientific view of the subject and includes a bibliography for further reading. The other book is much larger and deals with practical details.

An introduction to powder metallurgy is given in *Powder Metallurgy: Practice and Applications*, by R. L. Sands and C. R. Shakespeare (Butterworths, 1966), and a more detailed treatment of the subject in *Modern Developments in Powder Metallurgy*, edited by H. H. Hausner (5 vols., Plenum, 1971).

There are many recently published books on metal joining, such as *Introduction to Welding and Brazing*, by D. R. Milner and R. L. Apps (Pergamon, 1968); *The Metallurgy of Welding, Brazing and Soldering*, by J. F. Lancaster (Allen and Unwin, 1965); and *Welding Processes*, by P. T. Houldcroft (Cambridge University Press, 1966).

The Welding Institute produces a continuing series of up-to-date books on various detailed aspects of welding. The American Welding Society publishes many books on welding, including a notable six-volume series known generally as *Welding Handbook*:

Section 1   *Fundamentals of Welding*, edited by A. L. Phillips (6th edn, 1968)
Section 2   *Gas, Arc and Resistance Processes*, edited by A. L. Phillips (6th edn, 1969)
Section 3A *Welding Cutting and Related Processes*, edited by A. L. Phillips (6th edn, 1971)

Section 3B *Welding Cutting and Related Processes,* edited by A. L. Phillips (6th edn, 1971)
Section 4   *Metals and their Weldability,* edited by A. L. Phillips (6th edn, 1972)
Section 5   *Applications of Welding,* edited by C. Weisman (6th edn, 1973)

Metal finishing

An introduction to the various metal finishing methods is given in *Principles of Metal Surface Treatment and Protection,* by D. R. Gabe (Pergamon, 1972). A large number of books has been published in this field, particularly on electro-deposition of metals.

Canning Handbook of Electroplating (21st edn, Canning and Company, 1970) is a well-known practical guide. There are many books on individual metals or groups, such as *Nickel and Chromium Plating,* by J. K. Dennis and T. E. Such (Butterworths, 1972). A brief survey of many metals is given in *Modern Electroplating,* edited by F. A. Lowenheim (2nd edn, Wiley, 1963). A comprehensive account of the principles and practice of plating more than one metal, i.e. alloy plating, is given in *Electrodepostion of Alloys,* by A. Brenner (2 vols., Academic Press, 1963).

Other methods of metal coating are considered in *Metal Spraying and Flame Deposition of Ceramics and Plastics,* by W. E. Ballard (4th edn, Griffin, 1963).

Corrosion and protection

The most authoritative works on corrosion and protection published in England are *The Corrosion and Oxidation of Metals,* by U. R. Evans (Arnold, 1960), and *The Corrosion and Oxidation of Metals,* first supplementary volume, by the same author and publisher (1968). These books are very large (1118 and 496 pages, respectively) and cover a wide range of practical corrosion phenomena. The reader may wish to first consider *An Introduction to Metallic Corrosion,* by U. R. Evans (2nd edn, Arnold, 1963). An alternative introductory text is *The Fundamentals of Corrosion,* by J. C. Scully (Pergamon, 1966). Also of interest is the two volume work *Corrosion,* edited by L. L. Shreir (Newnes, 1963), a new edition of which is in preparation.

A completely different style of presentation of corrosion information is given in *Corrosion Guide,* by E. Rabald (2nd edn, Elsevier, 1968). Corrosive agents are listed alphabetically and under each name a summary is given of its interaction with metals and, hence, the suitability of various metals to contain it. This book also contains a bibliography of books on the properties of materials and the corrosion of materials and a list of journals devoted to corrosion studies.

INFORMATION ON CURRENT RESEARCH

The number of research journals carrying papers of metallurgical interest is very large because of the hybrid nature of the subject. Metals are extensively used in engineering and they are major subjects for investigation in physics and chemistry. This widely spread information is most conveniently monitored by reading journals of abstracts.

Metals Abstracts is the major international abstracting journal on metals in the English language. It has been published (jointly by The Metals Society, London and the American Society for Metals, Ohio) since 1968, when it was formed by the merger of two previously independent publications, *Metallurgical Abstracts* (IOM, 1909–67) and *Review of Metal Literature* (ASM, 1944–67). *Metals Abstracts* is published in monthly parts and *Metals Abstracts Index* is published monthly and annually.

Many other organisations publish abstracts on a limited range of subject matter or a particular metal, such as the following:

Aluminium Abstracts (Aluminium Federation)
Cast Iron Abstracts (British Cast Iron Research Association)
Cobalt Abstracts (Centre d'Information du Cobalt)
The Institution of Mining and Metallurgy (London) publishes the
   IMM Abstracts on mineral technology and extraction metallurgy.
The monthly *Journal of the Iron and Steel Institute* (London)
   contained an abstract section on ferrous metallurgy only.
Lead Abstracts (Lead Development Association)
Metal Finishing Abstracts (R. Draper, Teddington)
Nickel Abstracts (International Nickel Company, London)
Zinc Abstracts (Zinc Development Association)

If the subject matter of the research is not straightforwardly metallurgical, but could be classified as engineering or physics, etc., then it is necessary to use the abstracting journals of the other subjects. The following abstracts contain much of a metallurgical nature:

Applied Science and Technology Index (The H. W. Wilson
   Company, New York)
British Technology Index (Library Association, London)

Chemical Abstracts (American Chemical Society, Columbus, Ohio)

Engineering Index (Engineering Index Inc., New York)

Magnetism and Magnetic Materials Digest (Academic Press, New York)

Nuclear Science Abstracts (US Atomic Energy Commission, Oak Ridge, Tennessee)

Physics Abstracts (Institution of Electrical Engineers, London)

Government Reports Announcements (US Department of Commerce, Springfield, Virginia)

Bibliography of the High Temperature Chemistry and Physics of Materials (IUPAC) (Dr. M. G. Hocking, Department of Metallurgy, Imperial College, London)

Structure Reports, a Critical Guide to Papers on Crystallography of Metal Alloys and Compounds, edited by W. G. Pearson (published for the International Union of Crystallography by Oosthoek, Utrecht)

The following is a list of some of the journals which are useful in the field of metallurgy.

Acta Metallurgica
Anti-Corrosion Methods and Materials
British Corrosion Journal
British Journal of Non-Destructive Testing
Engineering Fracture Mechanics
Ironmaking and Steelmaking[4]
Journal of the Institute of Metals[1]
Journal of the Iron and Steel Institute[2]
Journal of the Less-Common Metals
Journal of Materials Science
Journal of Metals
Journal of Nuclear Materials
Journal of Physics F: Metal Physics
Materials Science and Engineering
Metal Construction and British Welding Journal
Metal Progress
Metal Science
Metallurgia and Metal Forming
Metallurgical Transactions
Metallurgist and Materials Scientist
Metals and Materials
Metals Technology[3]
Philosophical Magazine
Powder Metallurgy
Proceedings of the American Society for Testing and Materials
Transactions of the Institute of Metal Finishing
Transactions of the Institution of Mining and Metallurgy
Wear

Note. The journals 1 and 2 were replaced by journals 3 and 4, respectively, on 1 January 1974.

The following journals are available as cover-to-cover translations into English:

Automatic Welding (Automatiches-Raya Svarka)
Giessereiforschung in English (Giessereiforschung)
High Temperature (Teplofizika Vysokikh Temperature)
Industrial Laboratory (Zavodskaya Laboratoriya)
Metal Science and Heat Treatment of Metals (Metallovedenie i Termicheskaya Obrabotka Metallov)
Metallurgist (Metallurg)
Physics of Metals and Metallography (Fizika Metallov; Metallovedenie)
Protection of Metals (Zashchita Metallov)
Russian Metallurgy (Izvestiya Akademii Nank SSSR Metally)
Soviet Journal of Non-Ferrous Metals (Tsvetnye Metally)
Soviet Materials Science (Fisiko-Khimicheskaya Mekhanika Materialov)
Soviet Powder Metallurgy and Metal Ceramics (Poroshkovaya Metallurgiya)
Stal in English (Stal)
Welding Production (Svarchnoe Proizvodsto)

REVIEWS

Summaries of recent developments in research on particular topics, known as reviews, are sometimes published by research journals, but there exist some publications whose sole function is the printing of reviews. The best-known in the field of metallurgy are considered below.

The best-established periodical devoted to the whole range of metallurgy is *International Metallurgical Reviews* (The Metals Society, London and American Society for Metals, Ohio). From 1956 to 1971 inclusive this was known as *Metallurgical Reviews* and published by the Institute of Metals only. It is published in four paperback quarterly parts which combine to form the annual volume of approximately 12 independent reviews.

Research developments on the mainly physical aspects of solids are reviewed in *Progress in Materials Science* (Pergamon), originally known as *Progress in Metal Physics* for the first eight volumes. Publication is irregular, 17 volumes having been published in just over 20 years. Paperback copies of each individual review are sold as they become available and periodically some are grouped together to form a bound volume. These reviews are usually comprehensive and of a very high standard.

The other reviews in metallurgy consist of some older ones, which have not appeared recently, and a relatively new group which is not well established yet:

Advances in Materials Research, edited by H. Herman (Wiley/ Interscience, irregular)

Advances in Corrosion Science and Technology, edited by M. G. Fontana and R. W. Staehle (Plenum, irregular)

Annual Review of Materials Science (Annual Reviews Inc., annual). Annual Reviews Inc. publishes annual reviews of several other physical sciences, e.g. physical chemistry, nuclear science, which may contain metallurgical information.

Modern Materials (Academic Press) is an irregular series and 7 volumes have been published up to 1970. Each volume contains several reviews on different mainly metallurgical topics.

Progress in Applied Materials Research, edited by E. G. Stanford, J. H. Fearon and W. J. McGonnagle (Heywood) is an irregular series (volume 7 in 1967 was the last to appear). Volumes 1–3, edited by the first two authors only, were published under the title *Progress in Non-Destructive Testing.* The later issues under the new title continued to show interest in non-destructive testing.

IN THE LAST RESORT

Most metals are of commercial importance. They are manufactured or sold by large companies. If more than one company is involved in the manufacture of a particular metal, they commonly form a research or development association to further the commercial use of that metal. If a metal is produced by one company only, then all the research results and technical know-how will be concentrated in that company. In either case, those organisations form a ready source of information to the user of any metal or alloy. Some of these have been mentioned in the chapter.

A list of organisations associated with particular metals, alloys, metallurgical processes, etc., is given in *Metal Bulletin Handbook,* and lists of manufacturers and suppliers of metals and alloys are given in buyers' guides, such as *Metal Industry Handbook and Directory* (Iliffe), published annually.

28

Stress analysis

A. R. Luxmoore

INTRODUCTION

Stress analysis is a well-defined discipline which finds application in nearly all branches of engineering. The analysis of stresses and strains arises whenever a structure is designed to support some type of loading, whether it be a bridge, aeroplane, turbine, alternator or pressure vessel. It is usual to distinguish between stress analysis and structural analysis, although both are an extension of statics, with the latter being concerned with analysing the forces, bending moments, etc., in multiply connected structures, such as frameworks, and the former with the analysis of stresses within each framework member and joint. However, there is a considerable overlap between the two subjects, particularly in civil engineering.

Elementary stress analysis is often referred to as 'strength of materials', a rather misleading name which derives from a literal translation of the French name 'resistance de materiaux'. This can lead to confusion between the analysis of stresses in solid bodies and the mechanical properties of the material. Although the strength and stiffness of a material is vitally important in the final structure, the distribution of stresses is primarily a function of the shape of the structure, and stress analysis is not concerned with the study of material properties. This confusion in terminology is being resolved by the increasing use of 'solid mechanics' as a term synonymous with stress analysis, and will be found in many modern books, particularly those of American origin. At a more advanced level, the term 'continuum mechanics' (or mechanics of continua) is also used, this term being fairly explicit.

Stress analysis developed mainly in the eighteenth and nineteenth centuries, following on from the development of statics based on Newton's three laws. Mathematically, stress analysis is more complex than structural analysis, and it benefited considerably from the interest of nineteenth century mathematicians such as Airy, Cauchy, etc. The latter half of the nineteenth century saw the establishment of the theory of elasticity, the basic tool in most analyses, and textbooks, e.g. A.E.H. Love's *A Treatise on the Mathematical Theory of Elasticity,* dating from that period are still referred to in present-day work.

In the first half of this century, the analytical methods available to engineers could not solve many of the difficult geometries that were being used in the rapidly developing technology of the time, and experimental investigations became common. Model techniques were developed to a sophisticated stage, and strain measuring devices, such as electrical resistance strain gauges, proved invaluable. Photoelasticity, first suggested at the beginning of the nineteenth century by Davy as a means of analysing stresses in models, became a practical technique with the pioneering work of Coker and Filon, and the development of modern plastics, which become birefringent when stressed.

Concurrent with the development of experimental methods, the more complex problems were also being solved by numerical techniques, based on the analytical equations governing the stress distribution. The most successful technique in these early days was the finite difference method, which split the geometry into a simple mesh, and solved the equations on a piecemeal basis.

Since the 1950s, the major impact on stress analysis has been that of the computer. This has enabled numerical techniques of solution to develop at an enormous pace, and problems that were considered intractable a few years ago are being tackled now. Not only are these methods capable of solving very difficult problems, but they are also proving very economic with less difficult problems, which could be tackled by other methods. However, modern electronics is affecting every branch of engineering, and model methods of stress analysis are benefiting from automatic data collection and processing.

PHYSICAL ASPECTS OF STRESS ANALYSIS

The theory of elasticity presupposes a linear relationship between stress and strain, which is reasonably true for many structural materials loaded within a limited range, i.e. the elastic range. Outside this range, most materials suffer permanent, or plastic, deformation, and a theory of plasticity has been developed over the last 50 years.

For many years, stresses were always kept within the elastic region, which wasted a considerable portion of the material's load-carrying capability. It was left to the metallurgists to develop high-strength materials, with corresponding small plastic extensions, to gain good strength/weight ratios. Unfortunately, the reduction in ductility (plastic extension) increased the incidence of brittle fracture, whereby small flaws could propagate in a catastrophic manner at stresses within the working range of the structure. The prediction of this instability has proved a difficult problem, because of the complex stress distribution occurring round a small flaw (crack), and its modification by plastic deformation. Fracture mechanics has developed during the last two decades as a subject within those of stress analysis and material properties.

Design of structures into the plastic region has been developed by civil engineers, using limit load analysis. The structure is assumed to have plastic 'hinges' occurring at the highly stressed points, and failure occurs when sufficient hinges exist so as to allow collapse. Working loads are taken as a suitable fraction of the maximum, or limit, load. Simple plastic behaviour is assumed for the material, in order to make the calculations tractable.

Limit load analysis is restricted to static structures, as significant dynamic stresses can produce fatigue failure, which is very difficult to predict if plastic flow occurs. Fatigue is a topic in mechanical properties of materials, but it usually occurs where the stress cycle is a maximum, i.e. at stress raisers such as holes, or where vibrational modes change with changing load conditions.

Most structural materials also exhibit time-dependent effects, or creep, at some temperature range. With concrete and plastics, this occurs at ambient temperatures, but metals do not experience serious creep problems until they reach temperatures of over 30% of their melting points. Creep produces increasing deflections at constant load, and can modify stress distributions significantly. In chemical and aeronautical components, it is important to predict the effects of creep, and this can be done analytically for simple structures. For complex shapes modern numerical techniques are ideal.

Apart from problems associated with material behaviour, the theory of elasticity is limited to the type of problem it can solve analytically. It is most successful with two-dimensional stress problems, and many engineering problems reduce to conditions of either *plane stress* (for a thin, plane structure) or plane strain (a thick structure, with large constraints along the third dimension—most common). Another important class of problems involve large bending stresses compared with shear stresses, and a separate theory of plates and shells has developed concomitant with general elasticity.

PRACTICAL STRESS ANALYSIS

Most stress analysis problems are solved in industry by using standard analytical formulae that are available from a number of sources. When this formulary is not applicable, the problem is usually referred to specialist consultants with either a good mathematical background or suitable experimental equipment to solve the problem (often it is only universities that possess this type of expertise, which is of a research nature).

This approach is gradually changing with the general availability of computer programs for solving stress problems. These programs are still limited by the initial assumptions built into the program, e.g. linear elasticity, two dimensions, plate bending, etc., and specialists are still needed to operate these programs, but they do enable more accurate analysis to be performed than is possible with most standard formulary.

In Britain a great deal of engineering design is specified by publications such as the *Codes of Practice* of the British Standards Institution. These are not obligatory, but can be written into industrial contracts, and are a useful starting point for most work. They cover many aspects of design, of which stress analysis is one part. More detailed publications are issued by the American Society for Testing and Materials (ASTM), and the American Society of Mechanical Engineers (ASME), and many industrial organisations use these as guidelines. Details of the BSI publications are given in the *BSI Yearbook,* and no further reference will be made to them, as they are sufficiently well known to the industries they support, and are of little use in a general context.

Textbooks provide the major proportion of useful information for the stress analyst, with various geometries arranged in some sort of order. Journals published by the engineering institutions and other bodies contain the latest work, but as a comprehensive source of information they must be monitored regularly, or the information is missed and proves difficult to locate. In this case, review articles and abstracts are invaluable, but many relevant articles are missed because of the lack of interest by engineers in following up the latest work. It is an unfortunate fact of life that much useful information is lost in the vast morass of published literature because engineers do not have the time to look, even if they are told where to look, as is the present case. Also, it can be cheaper and more informative to do the work for oneself.

SOCIETIES

Most of the professional engineering institutions have some interest in stress analysis, particularly in aeronautical, civil, chemical, marine and mechanical engineering. In Britain their activities in the field of stress analysis are coordinated by the Joint British Committee for Stress Analysis (JBCSA), which has representatives from the following bodies:

Institute of Physics (Stress Analysis Group)
Welding Institute
British Society for Strain Measurement
Institution of Mechanical Engineers
Institution of Gas Engineers
Institution of Marine Engineers
Institute of Measurement and Control
Royal Institution of Naval Architects
Institution of Civil Engineers
Institution of Production Engineers
Institution of Electrical Engineers
Institution of Structural Engineers
Royal Aeronautical Society
Institution of Mechanical Engineers

The only specialist stress analysis society in this list is the British Society for Strain Measurement (BSSM), which publishes the journal *Strain*. The Royal Aeronautical Society and the Institution of Mechanical Engineers have considerable stress analysis interests, and the latter publishes the *Journal of Strain Analysis* on behalf of the JBCSA, as well as other journals with stress analysis interests. The RAeS runs the Engineering Science Data Unit (ESDU), which publishes numerous data sheets on stress analysis, mainly concerned with aircraft structural problems, but also of very great general interest, and there is also the journal of the RAeS, *The Aeronautical Journal*.

The American equivalent of the BSSM is the Society for Experimental Stress Analysis, which publishes the journal *Experimental Mechanics*. The interests of the American engineering institutions are similar to those of their British counterparts, but, as far as the author knows, there is no coordinating body.

The JBCSA maintains links with the International Permanent

Committee on Stress Analysis, which helps with the organisation of four-yearly international conferences. These conferences took place as follows: 1959, Delft; 1962, Paris; 1966, Berlin; 1970, Cambridge; 1974, Udino (Italy). The proceedings are useful accounts of the latest work in this field, but unfortunately they cannot be considered comprehensive.

Most of the professional institutions maintain a library service, and will provide members (and sometimes others) with a list of relevant publications on particular topics within their field.

ABSTRACTS

The relevant abstracts for stress analysis are:

Applied Mechanics Reviews (ASME)
Engineering Index (Engineering Index Inc.)
Metals Abstracts (American Society of Metals and The Metals Society)
Scientific and Technical Aerospace Reports (National Aeronautics and Space Administration)
Physics Abstracts (Institution of Electrical Engineers)
Mechanical Science Abstracts (Scientific Information Consultants Ltd)
Mathematical Reviews (American Mathematical Society)

Applied Mechanics Reviews are the most useful for the stress analyst with limited time to search the literature regularly. As their title suggests, these abstracts are limited in scope to topics in mechanics, which avoids a great deal of cross-indexing, and this makes them easy to use. They are published monthly, and contain a general review article on some aspect of mechanics. The two main subjects of interest to stress analysts are:

Rational Mechanics and Mathematical Methods, which includes the following
  sub-headings:
  Continuum Mechanics
  Analytical Methods
  Digital Computing and Programming
  Analogies and Analog Computation
Mechanics of Solids
  Dynamics
  Elasticity
  Viscoelasticity
  Plasticity
  Composite Material Mechanics
  Strings, Rods and Beams
  Membranes, Plates and Shells

Buckling
Vibrations of Solids
Wave Motion and Impact in Solids
Soil Mechanics (Basic)
Soil Mechanics (Applied)
Rock Mechanics
Material Processing
Fracture (including Fatigue)
Experimental Stress Analysis
Material Test Techniques
Structures (Basic)
Structures (Applied)
Machine Elements and Machine Design
Fastening and Joining

This list provides a very reasonable summary of the various fields of application of stress analysis, and most people should be able to identify themselves with one or more of these topics. Though very comprehensive, I do not think that these abstracts cover all possible sources of articles on stress analysis.

Engineering Index appears to provide a more comprehensive coverage of the stress analysis field, but at the price of greater complexity. As a general engineering abstracting journal, the index covers a very wide field of subjects, and the indexing leaves much to be desired. There is a great deal of cross-referencing, and the abstracts are necessarily brief. The situation is further complicated by an alphabetical subject arrangement, and small print. Rather tedious to use if one has general interests, but they provide good coverage.

The remaining abstracts have a restricted use in stress analysis, but they cover some of the topics given by the *Applied Mechanics Reviews*. *Physics Abstracts* (IEE) probably provides the next most comprehensive coverage with the following topics:

Physical Instrumentation and Experimental Techniques
Mechanical Measurements and Techniques
Pressure Measurements and Techniques
Classical Mechanics
Continuum Mechanics
Elasticity
Plasticity
Rheology
Elastic Waves, Vibrations

Metals Abstracts has a separate *Metals Abstracts Index,* published monthly with the main journal, and reference to the index under the headings 'stress analysis', etc., will elucidate the location of the articles in the main journal, where the headings are very general. The abstracted articles are restricted to those with a metallurgical connection.

A similar system is used in the *Scientific and Technical Aerospace Reports (STAR)*, published semi-monthly by NASA, which has the index at the back of the main journal. The abstracts cover all publications and reports considered pertinent to the aerospace industry.

REVIEW ARTICLES

There are many periodicals, a selection of which is listed below, which publish general review articles (at differing levels) of interest to engineers, and these often contain summaries of the latest developments in stress analysis. They are useful mainly for the practising engineer with a limited interest in the subject.

Engineer
Science and Technology
Chartered Mechanical Engineer
Engineering
Engineering and Science Review
Engineering Designer

GENERAL TOPICS

This section deals with those topics that divide stress analysis by virtue of the basic assumptions inherent in the subsequent analysis, e.g. elasticity, plasticity, plates and shells, etc. These subdivisions will be found in all fields of application of stress analysis.

The books quoted in this and the succeeding section are not the only ones available in these sections, but are those texts which, in the opinion of the author and his colleagues, are generally accepted as good source books by many stress analysts. The number of publications in this field is enormous, and a complete list is unlikely to help the serious information seeker.

Comprehensive publications

There are a number of books and journals which try to cover all aspects of stress analysis, and it is worth noting the more important of them at this stage.

Handbook of Engineering Mechanics, edited by W. Flügge (McGraw-Hill, 1962). A very useful book for the practising engineer and research worker, containing mainly theoretical analyses of many practical problems, with an introduction to the basic principles in each case.

Formulas for Stress and Strain, by R. J. Roark (4th edn, McGraw-Hill, 1965). A standard reference book of mainly elastic solutions used throughout industry and research laboratories. Lists known solutions to problems in two and three dimensions, plates and shells, buckling, etc., in a useful tabular form, and gives a brief introduction to each topic.

Foundations of Solid Mechanics, by Y. C. Fung (Prentice-Hall, 1965). An advanced mathematical treatment of complex problems in solid mechanics, including plasticity, viscoelasticity and other non-linear effects. A useful introduction to tensor notation is given by D. Frederick and T. S. Chang, in *Continuum Mechanics* (Allyn and Bacon, 1965).

Stress Concentration Around Holes, by G. N. Savin (Pergamon, 1961). An exhaustive treatment of the effects of holes in a wide variety of situations, with a good bibliography of solutions to this problem.

Stress Concentration Factors, by R. E. Peterson (Wiley, 1974). Information is provided in graphical form for a large variety of concentrating features, aimed particularly at the industrial analyst.

Journals

As there are many journals that try to cover all aspects of stress analysis, it is convenient to divide them into those with a definite civil engineering and structural bias and those concerned with mechanical engineering and allied subjects.

Civil Engineering and Public Works Review
Proceedings of the American Society of Civil Engineers. Journal of the Engineering Mechanics Division
Proceedings of the American Society of Civil Engineers. Journal of the Structural Division
Proceedings of the Institution of Civil Engineers
Structural Engineer

International Journal of Engineering Science
International Journal of Mechanical Sciences

Journal of Applied Mechanics and Technical Physics (translated
   from Russian)
Journal of Engineering Mathematics
Journal of Mechanical Engineering Science
Journal of the Mechanics and Physics of Solids
Journal of Physics D. Applied Physics
Journal of Strain Analysis
Mechanics of Solids (translated from Russian)
Nuclear Engineering and Design
Proceedings of the Institution of Mechanical Engineers
Transactions of the ASME. Journal of Applied Mechanics

Elementary solid mechanics

A great deal of useful stress analysis can be carried out with
elementary approximations of structural shapes to simple geometries,
and the following books can be used as references in this area, as well
as providing an introduction to the subject. Nearly all the analyses
are based on linear elastic behaviour, although simple plastic
behaviour is included in most of the texts, as well as elementary
accounts of buckling, plates and shells, and dynamic stresses.

Elements of Mechanics of Materials, by G. A. Olsen (2nd edn,
   Prentice-Hall, 1966)
Elements of Strength of Materials, by S. P. Timoshenko and D. H.
   Young (5th edn, Van Nostrand, 1968)
Mechanics of Materials, by F. R. Shanley (McGraw-Hill, 1967)
Mechanics of Solids and Structures, by P. P. Benham and F. V.
   Warnock (Pitman, 1973)
Strength of Materials, by G. H. Ryder (Macmillan, 1969)
Strength of Materials and Structures, by J. Case and A. H. Chilver
   (Arnold, 1971)

There are also books with a more comprehensive contents as
regards subject matter, but which still retain a fairly simple mathematical approach, and hence are useful in design, i.e. *Design for
Strength and Production,* by C. Ruiz and F. Koenigsberger
(Macmillan, 1970), and *Engineering Design,* by J. H. Faupel (Wiley,
1964).

Theory of elasticity

Elementary solid mechanics often makes simple assumptions
concerning the deformation of solids which are not strictly accurate.
In the theory of elasticity, basic compatibilty equations are employed

in differential form to ensure compatibility requirements for *small deformations only.* Basic source books on this topic are:

A Treatise on the Mathematical Theory of Elasticity, by A. E. H. Love (4th edn, Cambridge University Press, 1952)
Applied Elasticity, by C. T. Wang (McGraw-Hill, 1963)
Continuum Mechanics, by D. Frederick and T. S. Chang (Allyn and Bacon, 1965)
Mathematical Theory of Elasticity, by I. S. Sokolnikoff (2nd edn, McGraw-Hill, 1956)
Some Basic Problems of the Mathematical Theory of Elasticity, by N. I. Muckhelishvili (Noordhof, 1953)
Theoretical Elasticity, by A. E. Green and W. Zerna (2nd edn, Oxford University Press, 1968)
Theory of Elasticity, by M. Filonenko-Borodich (Foreign Languages Publishing House, 1965)
Theory of Elasticity, by S. P. Timoshenko and J. N. Goodier (3rd edn, McGraw-Hill, 1970)

Elastic stresses can also be induced by temperature variations in a body, and this topic is referred to as thermoelasticity. Two useful source books on this subject are *Theory of Thermal Stresses,* by B. O. Boley and J. H. Weiner (Wiley, 1960), and *Thermoelasticity,* by W. Nowacki (Pergamon, 1962).

The only specialist journal in this area is the *Journal of Elasticity,* but all the journals in the comprehensive publications section will include articles on this topic as well.

Plasticity

Analysis of stresses and strains in the plastic region will be found in the following:
Introduction to the Theory of Plasticity for Engineers, by O. Hoffman and G. Sachs (McGraw-Hill, 1953)
Plasticity for Mechanical Engineers, by W. Johnson and P. B. Mellor (Van Nostrand, 1962)
The Mathematical Theory of Plasticity, by R. Hill (Oxford University Press, 1950)
Theory of Perfectly Plastic Solids, by W. Prager and P. G. Hodge (Wiley, 1951)

The design of structures by limit load (or ultimate load) theory has a number of books devoted to itself. Examples are:

Plastic Analysis of Structures, by P. G. Hodge (McGraw-Hill, 1959)
Plastic Methods of Structural Analysis, by B. G. Neal (2nd edn, Chapman and Hall, 1963)
The Steel Skeleton, by J. F. Baker, M. R. Horne and J. Heyman (2 vols., Cambridge University Press, 1956)

The *International Journal of Non-linear Mechanics* and *Journal of the Mechanics and Physics of Solids* specialise in this topic, but publish material in other fields as well. See also journals in the comprehensive publications section.

Creep (viscoelasticity)

Creep is often combined with plasticity and characterised as non-linear (i.e. inelastic) behaviour, but the important difference is that creep is time-dependent, and there is viscoplastic behaviour as well as viscoelastic. Most mathematical analysis is restricted to visco-elasticity, e.g. *Mathematical Structure of the Theories of Visco-elasticity,* by B. Gross (Hermann, 1953), and *The Theory of Linear Viscoelasticity,* by D. R. Bland (Pergamon, 1960). For journals, see the comprehensive publications and plasticity sections.

Elastic instability (buckling)

Design against buckling is very important in structural analysis, and *The Stability of Frames,* by M. R. Horne and W. Merchant (Pergamon, 1965), gives an introduction to the subject.

The standard reference on this subject is *Theory of Elastic Stability,* by S. P. Timoshenko and J. M. Gere (2nd edn, McGraw-Hill, 1961).

For journals, refer to the comprehensive publications and theory of elasticity sections.

Plates and shells

An excellent introduction to theory of plates is *Elementary Theory of Elastic Plates,* by L. G. Jaeger (Pergamon, 1964).

For shells, a similar introduction is *Elementary Statics of Shells,* by A. Pflüger (2nd edn, Dodge Corporation, 1961). The standard work is: *Theory of Plates and Shells,* by S. P. Timoshenko and S. Wornosky-Kreiger (2nd edn, McGraw-Hill, 1959), and other useful design books are: *Plastic and Elastic Design of Slabs and Plates,* by

R. H. Wood (Thames and Hudson, 1961); *Stresses in Shells*, by W. Flügge (2nd edn, Springer-Verlag, 1973); *The Buckling of Plates and Shells*, by H. L. Cox (Pergamon, 1963); *The Design of Shells*, by A. Chronowicz (3rd edn, Crosby Lockwood, 1968); and *Thin-Walled Structures*, edited by A. H. Chilver (Chatto and Windus, 1967). These latter books deal also with buckling of plates and shells. A useful data book containing influence surfaces is *Influence Surfaces of Elastic Plates*, by A. Pucher (Springer-Verlag, 1964). For journals, see the comprehensive publications section.

Dynamic stresses and stress wave propagation

The effect of dynamic stresses can be split into four regions: dynamic effects which are slow compared with the speed of sound, producing stresses which can be predicted quasistatically; stresses which are induced by the structure's motion, e.g. centrifugal stresses; vibrational motions produced by natural or forced oscillations; and stress wave propagation induced by suddenly applied loads.

The first two topics are dealt with as sections in solid mechanics, and the only problem with slow cyclic loading is possible fatigue failure, considered later. With vibration, there is also the possibility of resonance occurring, and this must be avoided. There are numerous books on this subject—for example:

Mechanical Vibrations, by J. P. Den Hartog (4th edn, McGraw-Hill, 1956)
The Mechanics of Vibrations, by R. E. D. Bishop and D. C. Johnson (Cambridge University Press, 1960)
Vibration Problems in Engineering, by S. P. Timoshenko and D. H. Young (3rd edn, Van Nostrand, 1955)

An introduction to matrix methods is given by *Mechanical Vibrations: an Introduction to Matrix Methods*, by J. M. Prentis and F. A. Leckie (Longmans, 1963).

Stress wave propagation is not generally an important problem for the stress analyst, as high-intensity shock waves are rare. An interesting book giving a useful introduction is *Stress Waves in Solids*, by H. Kolsky (Dover, 1963).

Journals quoted in the comprehensive publications section will include articles on all aspects of dynamic stressing, particularly the *ASME Transactions*. Articles will also be found in physics journals, e.g. *Journal of Applied Physics* and *Journal of Physics D. Applied Physics*.

Numerical methods

The early methods, based on finite differences, are very suitable for solution using desk calculating machines, and a basic introduction is given in *Relaxation Methods,* by D. N. de G. Allen (McGraw-Hill, 1965), and *Relaxation Methods in Engineering Science,* by R. V. Southwell (Oxford University Press, 1940).

Digital computers have made the solution of large matrix arrays a simple matter, and matrix methods of structural analysis is an economic approach. Good introductions to this subject are *Matrix Methods of Structural Analysis,* by R. K. Livesley (Pergamon, 1964), and *Structural Analysis,* by R. C. Coates, M. G. Coutie and F. K. Kong (Nelson, 1972).

The finite element method of stress analysis, based on the matrix approach in structures, is described in *Finite Elements of Non-Linear Continua,* by J. T. Oden (McGraw-Hill, 1972); *Introduction to the Finite Element Method,* by C. S. Desai and J. F. Abel (Van Nostrand, 1972); *The Finite Element Method in Engineering Science,* by O. C. Zienkiewicz (2nd edn, McGraw-Hill, 1971); *Fundamentals of Finite Element Techniques for Structural Engineers,* by C. A. Brebbia and J. J. Connor (Butterworths, 1973); and *The Finite Element Method,* by K. C. Rockey *et al.* (Crosby Lockwood Staples, 1975).

Specialist journals on this topic are *Computers and Structures* and *International Journal for Numerical Methods in Engineering.*

Experimental stress analysis

Most of the textbooks in this field give details of the techniques and methods of analysis rather than accounts of actual investigations, although these are given as examples and will also be found in the journals. There are many methods available for experimentally determining stress and strain distributions, and the books listed below give details of most of them:

Elements of Experimental Stress Analysis, by A. W. Hendry (Pergamon, 1964)

Experimental Stress Analysis, by J. W. Dally and W. F. Riley (McGraw-Hill, 1965)

Experimental Stress Analysis, by G. S. Holister (Cambridge University Press, 1967)

Handbook of Experimental Stress Analysis, edited by M. Hetenyi (Wiley, 1950)

Introduction to the Theoretical and Experimental Analysis of Stress and Strain, by A. J. Durelli, E. A. Phillips and C. H. Tsao (McGraw-Hill, 1958).

Most of the journals in the comprehensive publications section publish articles on experimental methods, but there are several which specialise in this topic, including *Experimental Mechanics, Journal of Strain Analysis* and *Strain*. Physics journals also cover this topic, e.g. *Journal of Physics E. Scientific Instruments* and *Optics and Laser Technology*.

SPECIAL APPLICATIONS

As the principles of stress analysis are the same for all fields of application, it would appear unnecessary to consider topics concerned with special applications as well as the sections deriving from special assumptions dealt with under the previous heading. However, subjects in which stress analysis has important particular features not recognised previously also have their own literature dealing with these aspects, and it is for this reason that these areas (which do not include all the specialised applications of stress analysis) are mentioned here.

Aeronautics and astronautics

Apart from the popular works on elasticity, such as Timoshenko's, and the general references on stress concentrations, a great deal of information is obtained from the ESDU data sheets of the Royal Aeronautical Society, which will provide a complete list of its publications on request, and also of other publications in this field.

Relevant journals are: *Aeronautical Journal, Journal of Aircraft, Journal. American Institute of Aeronautics and Astronautics, Journal of the Astronautical Sciences, Journal of Spacecraft and Rockets* and *The Aeronautical Quarterly*.

Bioengineering

This is a relatively new field, where engineering disciplines are being applied to medicine. It involves problems such as the stress analysis of the femur, impact strength of skulls, etc., which are characterised mainly by their complex geometries and non-linear material behaviour. Serials publishing articles on stress analysis in this field are: *Advances in Biomedical Engineering, Biotechnology and Bioengineering, Journal of Biomechanics* and *Journal of Biomedical Materials Research.*

Fracture mechanics and fatigue

Crack propagation in solids is a subject of current research, as no single theory exists to explain all aspects of this phenomenon, and a considerable effort is being expended on possible solutions. One of the major problems is the difficulty of analysing accurately the stress fields at a crack tip, and this is still the subject of analytical, numerical and experimental investigations. Fatigue is an allied subject, because similar mechanisms are thought to act in each case, and final propagation of a fatigue crack is of major interest in fracture mechanics.

For practical engineering design, fatigue study can predict possible fatigue crack growth rates, which are depedent on stress levels and stress concentration factors, whereas fracture mechanics should predict the critical crack length at which catastrophic propagation occurs.

In fracture mechanics the most useful source books are those published by the ASTM and ASME:

ASME Handbook: Metals Engineering Design (McGraw-Hill, 1966)
Fracture in Engineering Materials (ASM, 1964)
Fracture Toughness Testing (ASTM Special Technical Publication No. 381, 1965)
Plane Strain Crack Toughness Testing of High Strength Metallic Materials (ASTM Special Technical Publication No. 410, 1966)

Fundamental aspects of fatigue processes are given by *Fatigue of Metals,* by P. G. Forrest (Pergamon, 1962).

For stress analysis of components subject to cyclic loads, see: *Designing Against Fatigue,* by R. B. Heywood (Chapman and Hall, 1962); *Designing by Photoelasticity,* by R. B. Heywood (Chapman and Hall, 1952); and *Random Vibration in Mechanical Systems,* by S. H. Crandall and W. D. Mark (Academic Press, 1963).

The book by Petersen on stress concentration is also of major use in fatigue design (see p. 453).

Journals specialising in fracture mechanics and fatigue are: *Engineering Fracture Mechanics, International Journal of Fracture Mechanics* and *Journal of the Mechanics and Physics of Solids,* plus those ASME, ASTM and Institution of Mechanical Engineers publications listed in the general topics section.

Pressure vessels and piping

Pressure vessels and pressurised piping are very common in the fields of aeronautical, chemical and mechanical engineering, and a great deal of literature has been produced. Much of it is concerned with the detailed design of joints and flanges between pipes and vessels, as well as the basic designs of vessels alone. Pressure vessels are basically simple structures, which become complicated by the attachment of pipes and supports. Selected texts are:

Industrial Piping, by C. T. Littlejohn (2nd edn, McGraw-Hill, 1962)
Pressure Vessel Design and Analysis, by M. B. Bickell and C. Ruiz (Macmillan, 1967)
Process Equipment Design, by L. E. Brownell and E. H. Young (Wiley, 1959)
Project Engineering of Process Plants, by H. F. Rase (Wiley, 1957)
The Stress Analysis of Pressure Vessels and Pressure Vessel Components, by S. S. Gill (Pergamon, 1970)

Pressure vessels are nearly always designed according to *British Standards* or *ASME Codes.* The relevant numbers are:

BS 1500 and *BS 1515*	(for pressure vessels)
ASME Boiler and Pressure Vessel Code	
Section *I*	(for power boilers)
Section *III*	(for nuclear power plant components)
Section *IV*	(for heating boilers)
Section *VIII*	(for pressure vessels)
ASA B31.1 & B31.3	(for pipeworks)
BS 3351	(for pipeworks)

The only specialist journal is the *International Journal of Pressure Vessels and Piping.*

Soil and rock mechanics

Soil and rock are highly inelastic materials, and the study of their deformation under load is affected by many variables, e.g. moisture content in soils. The mechanics of these materials has developed separately from other solids, where material properties can be defined fairly accurately, and this topic is not strictly a part of stress analysis as developed from the theory of elasticity. There is some overlap, and this section is presented for the sake of completeness.

A detailed account of the mechanics of soil deformation can be obtained from: *Foundations of Theoretical Soil Mechanics*, by M. E. Harr (McGraw-Hill, 1966); *Theological Aspects of Soil Mechanics*, by L. Suklje (Wiley/Interscience, 1969); *Statics of Soil Media*, by V. V. Sokolovski (Butterworths, 1960); and *Theoretical Soil Mechanics*, by K. Terzaghi (Wiley, 1943).

An account of the new developments in soil mechanics pioneered by the late Professor Roscoe at Cambridge is given in *Critical State Soil Mechanics*, by A. Schofield and P. Wroth (McGraw-Hill, 1968).

The main journals covering stress problems in soils are *Canadian Geotechnical Journal, Geotechnique* and *Journal of the Geotechnical Engineering Division, Proceedings of the American Society of Civil Engineers.*

Good source books in rock mechanics are *Elasticity, Fracture and Flow*, by J. C. Jaeger (Methuen, 1969); *Fundamentals of Rock Mechanics*, by J. C. Jaeger and N. G. W. Cook (Methuen, 1969); *Rock Mechanics and the Design of Structures in Rock*, by L. Obert and W. I. Duvall (Wiley, 1967); and *Rock Mechanics in Engineering Practice*, edited by K. G. Stagg and O. C. Zienkiewicz (Wiley, 1968). The first book in this list is a general approach to elasticity which is particularly useful for rock mechanics.

The main journals are *Geotechnique, International Journal of Rock Mechanics and Mining Sciences* and *Rock Mechanics in Engineering Geology.*

Materials technology

There are several journals primarily concerned with the uses of particular materials, and material production, that include articles on stress analysis:

Building Science
Journal of Materials Science
Magazine of Concrete Research
Materials and Structures
Metal Construction

An important application of stress analysis in this field is in the design of composite materials, particularly fibre-reinforced materials.

The principles involved in this field are outlined in:

Composite Materials, edited by L. Holliday (Elsevier, 1966)
Fibre Reinforced Materials, by G. S. Holister and C. Thomas (Elsevier, 1966)
Fibre Reinforcement, by H. Krenchel (Akademisk Forlag, 1964)
Strong Solids, by A. Kelly (Oxford University Press, 1966)

There are also several journals specialising in this topic, as well as in the general field of materials:

Composites
Fibre Science and Technology
Journal of Composite Materials

ACKNOWLEDGEMENTS

The author is indebted to his colleagues in the Departments of Civil and Mechanical Engineering, University College Swansea, for their help and co-operation, and also to Mr. J. Gammon, BP Chemicals (UK) Ltd, for help with the section on pressure vessels.

29

Hydraulics and coastal engineering

C. M. Archer

The proper control and use of water is of vital importance and is the prime function of the hydraulic engineer. In this chapter emphasis is placed on the civil engineering aspects of hydraulics, including such topics as water resources and coastal engineering, but some overlap with subjects of interest to mechanical engineers and geographers is inevitable. The treatment of water and effluent is excluded, as it is dealt with elsewhere under the general heading of public health engineering. The majority of the literature cited is British or American, although important publications from other countries have naturally been included. Generally, each citation appears once only in the section of the chapter to which it is most relevant, so that in order to obtain information on the literature concerned with a particular subject, it may be necessary to consult several sections.

There exist many establishments, institutions and organisations which produce valuable report series, annual publications, etc.; organise symposia and conferences; and provide experimental facilities and other services for the hydraulic engineer. These are listed in a later section. It is assumed that the reader has received some grounding in the basic principles and techniques of fluid mechanics and civil engineering hydraulics, but in order to make good any deficiencies in this respect, selected texts are listed at the end of the chapter. In addition, a list of texts concerned with more advanced mechanics of fluids is provided.

GENERAL SOURCES IN HYDRAULICS

Journals and books dealing with specific topics will be detailed later, but there are certain general sources of information which are relevant to many fields in hydraulics. The better handbooks provide a comprehensive treatment of the theory and practice of hydraulics, and of particular merit are *Water Resources Engineering,* by R. K. Linsley and J. B. Franzini (2nd edn, McGraw-Hill, 1972), and *Handbook of Applied Hydraulics,* by C. V. Davis and K. E. Sorensen (3rd edn, McGraw-Hill, 1969). A British counterpart to these texts is volume 2 of The Institution of Water Engineers' *Manual of British Water Engineering Practice* (3 vols., 4th edn, Heffer, 1969), which contains a list of *British Standards* relating to water. Reference should be made to current lists of British Standards Institution (BSI) publications, including *British Standards* and *Codes of Practice,* published in the *BSI Yearbook.* Another general publication is *Applied Hydraulics in Engineering,* by H. M. Morris and J. M. Wiggert (2nd edn, Ronald Press, 1972), which includes a good list of references. *Hydro-Electric Engineering Practice,* edited by G. Brown (2nd edn, Blackie, 1964), is a mammoth publication in three volumes, and volumes 1 and 2 provide a good treatment of most of the topics which come under the general heading of this chapter. An older but still relevant volume is *Engineering Hydraulics,* edited by H. Rouse (Wiley, 1950). Reference might also be made to a book which, although primarily concerned with the development of river systems, gives a very broad treatment of hydraulics; *Water Resources Development,* by E. Kuiper (Butterworths, 1965). The *Handbook of Fluid Dynamics,* edited by V. L. Streeter (McGraw-Hill, 1961), is highly recommended as this hydraulics-orientated text offers a very comprehensive treatment of both the theoretical and applied aspects of all the important topics in the field.

The important journals which provide a broad subject coverage will be mentioned here, but it should be remembered that in order to locate papers on particular topics, use must be made of the abstracting services described in the following section. The *Journal of The Hydraulics Division. Proceedings of The American Society of Civil Engineers* (ASCE) is a long-established monthly publication of high repute and contains a high proportion of the more important papers on civil engineering hydraulics and fluid mechanics. The International Association of Hydraulic Research (IAHR) produces the quarterly *Journal of Hydraulic Research,* with papers in both French and English. First issued in 1963, the journal provides a good treatment and wide coverage of theoretical and applied hydraulics. Book reviews and a list of IAHR publications are included, and advance notification is given of impending publications and forthcoming conferences, courses and symposia. The latter are of course extremely valuable, offering both sources of information and the means of exchanging ideas. A journal of similar stature is *La Houille Blanche*, a bilingual French/English publication with eight issues each year. Most of the papers are in French, some with abstracts in English. Covering hydraulics, hydroelectric engineering and fluid mechanics, including fundamental research, it similarly provides advance notification of forthcoming events. The international journal, *Water and Water Engineering* (continued as *Water Services, 1974–*), published monthly, and the British publication, *The Journal of The Institution of Water Engineers,* published eight times each year, deal with most aspects of the water cycle, and each contains useful book reviews and abstracts of current literature. The *Proceedings of The Institution of Civil Engineers* is published quarterly in two parts subtitled *Design and Construction,* and *Research and Theory,* and frequently contains papers on important theoretical aspects and practical applications in hydraulics.

Of a different nature is the review publication *Advances in Hydroscience,* edited by V. T. Chow (Academic Press, annual), which is a good source of information for advanced research and teaching. To date, eight volumes have been released and typically each volume comprises approximately five contributions by authors of high standing, setting out the current state of knowledge in various fields. A wide range of subject areas has been covered and reference to these volumes is recommended. The now biennial *Hydraulic Research in the United States* (United States Government Printing Office (USGPO) details current and recently concluded research projects in hydraulics and hydrodynamics undertaken in more than 250 laboratories in both the USA and Canada. Included is a list of the contributing industrial, university, and government laboratories.

It is sometimes necessary to obtain the precise definition of an unfamiliar term encountered in the literature on hydraulics, and in this context *Dictionary of Water and Water Engineering,* by A. Nelson and K. D. Nelson (Butterworths, 1973), *Glossary: Water and Wastewater Control Engineering* (American Public Health Association, 1969) and *Encyclopaedia of Hydraulics, Soil and Foundation Engineering,* compiled by E. Vollmer (Elsevier, 1967), would prove informative.

There exist several dictionaries designed to assist with the translation of articles on subjects in other branches of engineering, and a useful supplement to these is the publication *Technical Dictionary of Hydraulics and Pneumatics,* edited by G. Neubert (Pergamon, 1973). The equivalents of 3500 terms are given in English, French,

German and Russian, and there is a Spanish supplement. Also available is the *Dictionary of Water and Sewage Engineering*, compiled by F. Meinck and H. Mohle (Elsevier, 1963), which provides equivalent terms in English, French, German and Italian.

ABSTRACTING SERVICES

The most important British abstracting service available in the field of engineering is *Engineering Index*, published monthly, which is an international transdisciplinary publication containing abstracts arranged alphabetically by subject. The Russian *Referativnyi Zhurnal* is the most comprehensive international abstracting journal and categorises papers in science and technology into 61 sections, each abstract being written in the language of the original paper. The appropriate sections are: 17—Geofizika, 18—Geografiya, 35 —Mekhanika, 38—Nasostroenie I Kompressorostroenie, 53— Truboprovodnyi Transport and 55—Vodnyi Transport. *Applied Mechanics Reviews* comprise a monthly assessment of world literature in the engineering sciences. Two of its main sections, mechanics of fluids and combined fields, each with subsections, are relevant, and the abstracts are of a high standard.

More immediately concerned with the subject of this chapter is the valuable publication *Selected Water Resources Abstracts*, which is published semi-monthly and first appeared in 1968. Produced in America, it provides world-wide coverage of all publications in its broad field, and its well-written abstracts are arranged under 10 main subject fields, each subdivided into subject groups. The semi-monthly and annual cumulative indexes contain the usual subject and author indexes, and also organisational indexes, which give the establishment or organisation responsible for each publication. The British Hydromechanics Research Association (BHRA), in collaboration with the Hydraulics Research Station, produces *Civil Engineering Hydraulics Abstracts*, a monthly publication which was first issued in 1968 under its original title of *Channel*. A total of 550 journals are covered and the 2200 abstracts written annually are drawn from both published and unpublished sources. The abstracts are arranged under four main subject headings, with a total of approximately 30 sub-sections. A photocopy service for abstracted papers is available, and the half-yearly index issue contains subject, place name, corporate author (establishment or institution) and author indexes.

The bi-monthly *ASCE Publications Abstracts* contains abstracts of all the papers in the various *Journals* (*Hydraulics, Waterways,*

Harbours and Coastal Engineering, etc.) which comprise the *Proceedings,* and also in *Civil Engineering,* and may be helpful if ready access to these publications is not possible. In addition to the abstracts, tables of contents pages and subject and author indexes are included. *Irrigation and Power Abstracts,* published bi-monthly in English by the Central Board of Irrigation and Power (India), covers internationally a much wider range of subjects than its title suggests, and each year provides over 1000 selected abstracts from journals, books and conferences, under more than 50 subject headings. The American monthly publication *Water Resources Abstracts,* first released in 1967, places emphasis on American journals with some coverage of journals from other countries. Each issue contains 150–200 abstracts in 46 categories, but they are not arranged under subject headings.

In 1974 the Institution of Civil Engineers commenced publication of *ICE Abstracts,* which reviews papers from America and from all the European countries, including Britain. *ICE Abstracts* is published 10 times a year and replaces *European Civil Engineering Abstracts,* formerly published by the Construction Industry Translation and Information Services.

Abstracting journals dealing with specialised topics will be described later under appropriate subject headings.

When abstracting services are used the *Water Resources Thesaurus* (United States Department of the Interior, 1966) may be of assistance. This is a vocabulary for indexing and retrieving the literature of water resources research and development, and contains descriptors with good cross-references.

INDEXES AND CURRENT AWARENESS JOURNALS

Less vital than the abstracting services, but possibly convenient for quick reference over a broader range of journals, are two monthly indexes. The *British Technology Index* contains references arranged alphabetically by subject, with an author index, and includes a helpful list of the periodicals indexed. The American equivalent is the *Applied Science and Technology Index,* which cites English-language periodicals (mainly American) and offers a cumulative subject index to the year of each issue, but does not provide an author index.

The invaluable monthly current awareness journal *Hydata,* published by the American Water Resources Association, contains current tables of contents and lists of titles of the world's scientific and technical literature in the field of water resources, in its broadest sense. Indexing 50 periodicals and including selected articles from that source, it also gives titles and tables of contents of selected non-periodical literature and details of US and other patents. The bi-monthly *Fluid Mechanics Current Index,* first issued in 1972, is an American publication which cites most of the important papers on fluid mechanics, including engineering hydraulics. The citations are arranged under 17 main subject headings, each with sub-sections, and each issue contains a list of forthcoming conferences.

SPECIALISED TOPICS

It must be reiterated that because many of the specialised topics overlap and each citation is generally not repeated, then in order to obtain information on a particular topic it will usually be necessary to consult several sections of this chapter, especially that section headed 'General Sources in Hydraulics'.

Fluid mechanics

A good understanding of fluid mechanics is essential in any investigation in the realm of civil engineering hydraulics and, as mentioned previously, selected texts on fluid mechanics are listed at the end of this chapter. This section, however, will be concerned with those journals and review publications which may usefully be consulted for developments in the subject. The important *Journal of Fluid Mechanics* exists for the publication of papers on all theoretical and experimental aspects of advanced mechanics of fluids. Five volumes, each of four separate parts, are published annually, with subject and author indexes contained in the last part of each volume, and the contents include reviews of current publications. The American monthly journal *The Physics of Fluids* deals with fluid mechanics at a similarly high level over an equally broad field. A relatively new Russian equivalent is *Fluid Mechanics — Soviet Research,* which is published bi-monthly in English. The first *Annual Review of Fluid Mechanics* was produced in 1969 and each issue gives the current state of knowledge for a wide range of topics at an advanced level.

With a mechanical engineering bias are two quarterly journals which form part of the *Transactions of the American Society of Mechanical Engineers* (ASME): the *Journal of Applied Mechanics,* a good proportion of which is devoted to fairly advanced fluid mechanics, and the *Journal of Fluids Engineering,* which supplanted the *Journal of Basic Engineering* in 1973. Papers of similar interest appear in the *Proceedings of The Institution of Mechanical Engineers,* and also in *Heat and Fluid Flow,* the relatively new twice-yearly journal of the Thermodynamics and Fluid Mechanics Group of the Institution.

The irregular house journal of Armfield Engineering Ltd, *Educational Fluid Mechanics,* is mainly of interest to those concerned with teaching, but its mainly descriptive articles may be of value to others.

Hydraulic and hydrological measurements

As a summary of sources of information on methods and devices for the measurement of flow rates in open and closed channels, *Fluid Flow Measurement—A Bibliography* (BHRA, 1972) is particularly effective. It contains 2400 abstracts arranged in 53 specific subject groups, mainly by meter or method, in chronological order within each section, and includes subject and author indexes. Recent advances in the same field were described in 30 papers in *Modern Developments in Flow Measurement, Proceedings of International Conference at Atomic Energy Research Establishment (AERE), 1971,* edited by C. G. Clayton (Peter Peregrinus, 1972). For current information, reference should be made to *Fluid Flow Measurements Abstracts,* first issued by BHRA in 1974. This bi-monthly publication provides abstracts from relevant world literature, including periodicals, research reports, conference papers, books and British patent specifications. It is concerned with the complete range of flow metering and other fluids measurements.

Flow Measurement and Meters, by A. Linford (2nd edn, Spon, 1961), provides a comprehensive summary of the techniques and equipment available when the book was written, and still remains a good reference text. A similarly useful publication, although it excludes free-surface flow, is *Flow Measurement in Closed Conduits, Proceedings of Symposium at National Engineering Laboratory, 1960* (HMSO, 1962). The problems associated with unsteady flow are dealt with in a capable manner by G. P. Katys in *Continuous Measurement of Unsteady Flow* (Pergamon, 1964), translated from the Russian, which furnishes a good exposition of the relevant principles and applications.

An especially good practical manual describing methods and practices in hydrological measurement is *Stream-Gaging Procedure: A Manual* (Water-Supply Paper 888), issued by the US Department of the Interior Geological Survey (USGPO, 1962). *Stream Flow: Measurement, Records and Their Uses,* by N. C. Grover and A. W.

Harrington (Dover, 1943), outlines most of the principles fundamental to the subject and is a helpful basic reference text. More recent developments in all aspects of hydrological measurement are given in volumes 1 and 2 of *Instrumentation and Observation Techniques, Proceedings, Canadian Hydrology Symposium No. 7, Victoria, 1969* (Inland Waters Branch, Department of Energy, Mines and Resources, Ottawa, 1969).

Flow rate is but one of the many physical variables in hydraulics, and P. Bradshaw's *Experimental Fluid Mechanics* (2nd edn, Pergamon, 1970) and R. P. Benedict's *Fundamentals of Temperature, Pressure and Flow Measurements* (Wiley, 1969) together provide an interesting and highly informative introduction to the measurement of various quantities, as well as to the techniques of experimentation. The methods and principles of the measurements associated with water supply and general hydraulic engineering are detailed in the comprehensive and practical *Hydrometry,* by A. T. Troskolanski (Pergamon, 1960), translated from the Polish.

The measurement of the viscosity of fluids presents peculiar technical problems, and the theoretical and practical aspects of the subject are extremely ably covered in *Viscosity and Flow Measurement: a Laboratory Handbook of Rheology* by J. R. Van Wazer, J. W. Lyons, K. Y. Kim and R. E. Colwell (Interscience, 1963). Another relevant text is *Viscosity and its Measurement,* by A. Dinsdale and F. Moore (Chapman and Hall, 1962).

Turbulence is intrinsically a complex phenomenon and a good introduction to its quantification is given by P. Bradshaw in *An Introduction to Turbulence and its Measurement* (Pergamon, 1971). The exposition is clear and a valuable list of references is included. The latest information on new products and applications in anemometry and precision hydrometry generally may be obtained from the bi-annual house journal *DISA Information,* and a summary of the state of the art from *Fluid Dynamic Measurements in the Industrial and Medical Environments, Proceedings, DISA Conference, University of Leicester, 1972,* edited by D. J. Cockrell (2 vols., Leicester University Press, 1972).

Water resources and hydrology

These closely linked interdisciplinary topics are subjected to a great deal of research activity and, within the bounds of this chapter, a truly comprehensive description of the sources of information on the engineering aspects is not possible.

The International Association of Hydrological Sciences (IAHS) publishes the important quarterly *Hydrological Sciences Bulletin,* a bilingual French/English journal which is concerned with hydrology as an aspect of the earth sciences and of water resources. An extremely wide field is covered at a high level, and book reviews, lists of IAHS publications and details of forthcoming congresses, etc., are included. Equally valuable is the international *Journal of Hydrology,* having eight issues each year, which again deals with the complete range of hydrological problems. Two bi-monthly American journals are *Water Resources Bulletin,* produced by the American Water Resources Association (AWRA), and *Water Resources Research,* produced by the American Geophysical Union (AGU). English translations of papers selected from Russian publications in hydrology are available in *Soviet Hydrology: Selected Papers,* also issued by the AGU. The papers included in this bi-monthly journal again range over a broad field at a high level, and the only journal excepted from consideration is *Meteorologiya i Gidrologiya,* since translations of selected papers from this journal are presented in the monthly *Meteorology and Hydrology.* This latter publication also achieves a high standard, but emphasis is placed on meteorological considerations.

Reference should also be made to the *Journal of the Irrigation and Drainage Division. Proceedings of the ASCE,* because this quarterly journal often contains papers on hydrology and water resources. The antipodean approach to the field of study is given in the Australian Water Resources Council's *Water Resources Newsletter* which, published half-yearly, describes current research activities and also projects which have been undertaken.

Hydrological Research in the United Kingdom (1965–70) (National Environment Research Council (NERC), 1971) not only details the development and organisation of hydrological research in the UK and lists research projects undertaken, but also provides valuable information concerning relevant societies, institutions and other organisations, and important journals and other publications.

In recent years numerous books have been published on water resources and hydrology, and in particular on the economic and management aspects of these subjects. As a concise introduction to the basic principles and ideas involved, *Water in Britain,* by K. Smith (Macmillan, 1972), can be recommended. This descriptive text outlines applied hydrology, is interestingly informative on Britain's water resources and furnishes a useful list of references. Other introductory books, written in non-mathematical terms, are *Principles of Hydrology,* by R. C. Ward (McGraw-Hill, 1967), and *Modern Hydrology,* by R. G. Kazmann (2nd edn, Harper and Row, 1972). For information on basic hydrology with emphasis on engineering requirements, reference should be made to *Engineering Hydrology,* by E. M. Wilson (2nd edn, Macmillan, 1974); *Hydrology,* by C. O. Wisler and E. F. Brater (2nd edn, Wiley, 1965); *Hydrology for Engineers,* by R. K. Linsley, M. A. Kohler and J. L. H. Paulhus (McGraw-Hill, 1958); and *Engineering Hydrology,* by S. S. Butler (Prentice-Hall, 1957).

For more advanced reading on water resources development, two texts which have been described previously are essential: *Water-Resources Engineering,* by R. K. Linsley and J. B. Franzini (2nd edn, McGraw-Hill, 1972); and *Water Resources Development,* by E. Kuiper (Butterworths, 1965). A publication of similar standing, but a little more concerned with economic and political considerations, is *Design of Water Resource Systems,* by A. Maass *et al.* (Macmillan, 1968). If the reader is primarily interested in the economic aspects of water resources, then *Water Resources Project Economics,* by E. Kuiper (Butterworths, 1971), should be consulted. This book employs realistic problems and their solutions to deepen the reader's understanding of the principles described in the body of the text. Similar in approach are *Water Resources Systems Engineering,* by W. A. Hall and J. A. Dracup (McGraw-Hill, 1970), and *Economics of Water Resources Planning,* by L. D. James and R. R. Lee (McGraw-Hill, 1971).

The more advanced aspects of engineering hydrology are exceptionally ably covered in the comprehensive *Handbook of Applied Hydrology,* edited by V. T. Chow (McGraw-Hill, 1964). Also essential reading in this field are *Introduction to Hydrology,* by W. Viessman, T. E. Harbaugh and J. W. Knapp (Intext, 1972), and *Engineering Hydrology,* by J. Nemec (McGraw-Hill, 1972). The techniques involved in the collection and interpretation of hydrological data are clearly detailed in *Techniques of Water Resources Investigations,* by the US Geological Survey (Books 1–8, USGPO, 1968–1971).

A good introduction to computer applications in hydrology is available in *The Use of Analog and Digital Computers in Hydrology, Proceedings of Symposium, Tucson, 1966* (2 vols., IASH/UNESCO, 1969). This publication is the first in a series of 'Studies and Reports in Hydrology' issued by UNESCO, and reference to this series and also to the collection of 'Technical Papers in Hydrology' (UNESCO) is recommended. The *Guidebook on Nuclear Techniques in Hydrology* (International Atomic Energy Agency (IAEA), 1968) (Technical Reports Series, No. 91) and *Isotope Hydrology 1970, Proceedings of Symposium, Vienna* (IAEA/UNESCO, 1970) provide a similarly effective introduction to another specialised topic in hydrology.

Other texts worthy of mention concerned with water resources and hydrology are listed below:

Dynamic Hydrology, by P. S. Eagleson (McGraw-Hill, 1970)
Flood Studies in the United Kingdom (ICE, 1967)
Handbook on the Principles of Hydrology, edited by D. M Gray (Canadian National Committee for the International Hydrological Decade, 1970)
Hydrological Forecasting, by B. A. Apollov, G. P. Kalinin and V. D. Komarov (Israel Program for Scientific Translations, 1960)
Introduction to Physical Hydrology, edited by R. J. Chorley (Methuen, 1969)
River Flood Hydrology, Proceedings of Symposium, 1965 (ICE, 1966)
River Management, Proceedings of Symposium, 1966, edited by P. C. G. Isaac (Maclaren, 1967)
River Runoff: Theory and Analysis, by D. Sokolovskii (Israel Program for Scientific Translations, 1971)
Simulation Techniques for Design of Water-Resource Systems, by M. M. Hufschmidt and M. B. Fiering (Macmillan, 1966)
Streamflow Synthesis, by M. B. Fiering (Macmillan, 1967)
Synthetic Streamflows, by M. B. Fiering and B. B. Jackson (AGU, 1971 (Water Resources Monograph No. 1))
Theory of Stochastic Processes in Hydrology and River Runoff Regulation, by N. A. Kartvelishvili (Israel Program for Scientific Translations, 1969)
The Role of Water in Agriculture, Proceedings of Symposium, Aberystwyth, 1969, edited by J. A. Taylor (Pergamon, 1970)
The Water Encyclopedia—A Compendium of Useful Information on Water Resources, edited by D. K. Todd (Water Information Center, 1970)

Groundwater and wells

Groundwater is a constituent part of water resources and hydrology, and the use of wells for the abstraction of water is but one aspect of groundwater. Thus, many of the publications described in the previous section provide some information on groundwater and water wells, and in this section only those publications particularly concerned with either groundwater or wells will be considered. *Ground Water,* the American bi-monthly journal of the Technical Division of the National Water Well Association, was first published just over a decade ago and is very broad in its scope, dealing with the world's groundwater resources and the management of water, as well as with the technical aspects of well hydraulics, aquifer response, etc. Regular features include field reports, forthcoming events and book reviews. The National Water Well Association also publishes its monthly *Water Well Journal,* which gives details of new products, techniques and applications.

Reference to *Ground Water: a Selected Bibliography,* compiled by F. Van der Leeden (Water Information Center, 1971), is recommended. Relevant books, journals and general bibliographies are listed, and 1500 references classified under 32 topics in groundwater hydrology are given.

Theory of Groundwater Flow, by A. Verruijt (Macmillan, 1970), furnishes a clear exposition of the fundamental principles involved, and describes the various methods available for solving groundwater flow problems. *Physical Principles of Water Percolation and Seepage*, by J. Bear, D. Zaslavsky and S. Irmay (UNESCO, 1968) and *Theory of Fluid Flow in Indeformable Porous Media*, by V. I. Aravin and S. N. Numerov (Israel Program for Scientific Translations, 1965), offer a comprehensive advanced treatment of the subject and the former contains a good chapter on models and analogues. A similar, but more recent publication is *Dynamics of Fluids in Porous Media*, by J. Bear (Elsevier, 1972). *Water in the Unsaturated Zone: Proceedings of the Wageningen Symposium, 1966*, edited by P. E. Rijtema and H. Wassink (2 vols., IASH/UNESCO, 1969), provides a complete introduction to the specialised subject of its title. A long-established text, and still helpful for reference purposes, is *The Flow of Homogeneous Fluids through Porous Media*, by M. Muskat (McGraw-Hill, 1937), which provides both a very practical and also a good theoretical treatment of the subject. An alternative general text is *An Introduction to the Physical Basis of Soil Water Phenomena*, by E. C. Childs (Wiley/Interscience, 1969).

Advanced analytical techniques are given in *Numerical Methods in Subsurface Hydrology*, by I. Remson, G. M. Hornberger and F. J. Molz (Wiley/Interscience, 1971), and *Differential Equations of Hydraulic Transients, Dispersion, and Ground-Water Flow*, by W. H. Li (Prentice-Hall, 1972). A comprehensive treatment of theoretical and analogue methods applicable to an important type of groundwater flow problem is contained within *Seepage Through Earth Dams*, by A. A. Uginchus (Israel Program for Scientific Translations, 1966). In *Concepts and Models in Ground Water Hydrology*, by P. A. Domenico (McGraw-Hill, 1972), a rather different approach to general theoretical groundwater flow is adopted. Models and methods are grouped and classified according to a rational basis, and conservation principles, with applications, are described. The book includes a good bibliography.

A basic understanding of the more applied aspects of groundwater hydrology would be gained by reading *Ground Water Hydrology*, by D. K. Todd (Wiley, 1959); *Geohydrology*, by R. J. M. De Wiest (Wiley, 1965); and the text-manual *Introduction to Ground-Water Hydrology*, by R. C. Heath and F. W. Trainer (Wiley, 1968). *Groundwater Resource Evaluation*, by W. C. Walton (McGraw-Hill, 1970), examines all aspects of groundwater resources, including groundwater flow theory, practical well design and construction, aquifer development and management, etc., and reference to it is highly recommended. A similarly valuable publication is *Ground-Water Studies: An International Guide for Research and Practice*, edited by R. H. Brown, A. A. Konoplyantsev, J. Ineson and V. V. Kovalevsky (UNESCO, 1972), intended for the use of persons responsible for organising and conducting groundwater investigations, particularly in developing countries. An older publication still worthy of reference is *Ground Water Basin Management. Manual of Engineering Practice No. 40* (ASCE, 1963). For information concerning the specialised topic of groundwater recharge, reference should be made to *Artificial Groundwater Recharge, Proceedings of Conference, 1970* (2 vols., Water Research Association, 1971).

The theory and practice of the abstraction of groundwater is comprehensively covered in *Groundwater Recovery*, by L. Huisman (Macmillan, 1972), and an equally useful book is *Ground Water and Wells: a Reference Book for the Water-Well Industry*, by E. E. Johnson (Edward E. Johnson, 1966). Practical considerations in well design and construction are ably described with a field-orientated approach in *Water Well Technology*, by M. D. Campbell and J. H. Lehr (McGraw-Hill, 1973). For a good theoretical and practical treatment of pumping tests, *Analysis and Evaluation of Pumping Test Data*, by G. P. Kruseman and N. Ridder (International Institute for Land Reclamation and Improvement, 1970 (Bulletin No. 11)), should be consulted.

In the general field of groundwater *Elsevier's Dictionary of Hydrogeology*, by H. Pfannkuch (Elsevier, 1969), may be of assistance. It contains the English definitions of terms arranged alphabetically, together with the French and German equivalents.

Rivers, channels and sediment transport

Many of the publications previously mentioned have some relevance to the subject of this section. A journal which should be cited at this juncture is the quarterly *Journal of the Waterways, Harbors and Coastal Engineering Division, Proceedings of the ASCE,* since its scope includes all the engineering aspects of canals, rivers and channels.

A classic text which deals comprehensively with the theory and applications of open channel flow is *Open Channel Hydraulics,* by V. T. Chow (McGraw-Hill, 1959), which together with *Open Channel Flow,* by F. M. Henderson (Macmillan, 1966), and *Flow in Channels,* by R. H. J. Sellin (Macmillan, 1969), provides a complete exposition of the fundamentals of the subject. Reference may also be made to

Dynamics of Channel Flow, by V. Goncharov (Israel Program for Scientific Translations, 1964).

The particularly complex aspect of open channel flow, sediment transport, is very ably covered in *Hydraulics of Sediment Transport,* by W. H. Graf (McGraw-Hill, 1971), and a good summary is provided in 'Fluvial Sediment Transport', by J. L. Bogardi in volume 8 of *Advances in Hydroscience,* edited by V. T. Chow (McGraw-Hill, 1972). Each publication contains an extensive bibliography. Other texts which should be consulted are *Mechanics of Sediment Transport,* by M. S. Yalin (Pergamon, 1972); *Loose Boundary Hydraulics,* by A. J. Raudkivi (Pergamon, 1967); and *River Runoff: Theory and Analysis,* by D. Sokolovskii (Israel Program for Scientific Translations, 1971). For those interested in the historical development of the subject, *An Introduction to Fluvial Hydraulics,* by S. Leliavsky (Dover, 1959) and *An Approach to the Sediment Transport Problem from General Physics,* by R. A. Bagnold (USGPO, 1966) (Geological Survey Professional Paper 422–1), would repay examination.

Two essential references in the field of river engineering are *River and Canal Hydraulics,* by S. Leliavsky (Chapman and Hall, 1965), and *River Engineering and Water Conservation Works,* edited by R. B. Thorn (Butterworths, 1966). Together they provide the river engineer with a great deal of the information necessary. *Hydraulics of River Channel Closure,* by S. V. Izbash and K. Y. Khaldre (Butterworths, 1970), and *Flood Plain Management: Iowa's Experience,* edited by M. D. Dougal (Iowa State University Press, 1969) (papers presented at the Conference on Flood Plain Management, Sixth Water Resources Design Conference, Iowa State University), are useful first references for two specialised topics in river engineering.

Hydroelectric power and hydraulic structures

Water Power and Dam Construction is a long-established international journal devoted to hydroelectric power development. Published monthly, it adopts a broad approach and contains papers of a high standard on both the theoretical and applied aspects of the subject. Regular features include details of new products and abstracts of recent literature. The American viewpoint is given in the *Journal of the Power Division. Proceedings of the ASCE,* published at least twice a year; the Russian viewpoint in the monthly *Hydrotechnical Construction,* which is translated into English for the ASCE; and the Asian viewpoint in the monthly *Indian Journal of Power and River Valley Development* (published in English).

A very good summary of hydroelectric engineering and hydraulic structures generally may be obtained from four books mentioned earlier in this chapter: *Hydro-Electric Engineering Practice*, edited by J. G. Brown (3 vols., 2nd edn, Blackie, 1964); *Water Resources Engineering*, by R. K. Linsley and J. B. Franzini (2nd edn, McGraw-Hill, 1972); *Water Resources Development*, by E. Kuiper (Butterworths, 1965); and *River Engineering and Water Conservation Works*, edited by R. B. Thorn (Butterworths, 1966). Other valuable texts include *Water Power Development*, by E. Mosonyi (2 vols., 2nd edn, Akademiai Kiado, 1963); *Tidal Energy for Electric Power Plants*, by L. B. Bernshtein (Israel Program for Scientific Translations, 1965); *Design of Dams for Percolation and Erosion*, by S. Leliavsky (Chapman and Hall, 1965); *Design of Small Dams: A Water Resources Technical Publication*, by United States Bureau of Reclamation (USGPO, 1965); and the older publication *Engineering for Dams*, by P. Creager, J. D. Justin and J. Hinds (3 vols., Wiley, 1945). A useful book dealing with a particular class of hydraulic structures is *Hydraulic Energy Dissipators*, by E. A. Elevatorski McGraw-Hill, 1959).

Hydraulic machinery

In this section, those publications concerned primarily with hydraulic machinery, and in particular with pumps, are reviewed. In 1971 the BHRA first issued its bi-monthly abstracting journal *Pumps and Other Fluids Machinery Abstracts*, which contains abstracts of the world literature and British patent specifications on most types of fluids machinery. *Pumps—Pompes—Pumpen*, published monthly by the European Committee of Pump Manufacturers (Europump), is an international journal concerned with all aspects of pump applications, and is a good source of information on recent developments in the field.

For an introduction to the fundamentals of hydraulic machinery, *An Introduction to Incompressible Flow Machines*, by D. H. Norrie (Arnold, 1963), the chapter on 'Turbomachinery' in *Fluid Mechanics*, by V. L. Streeter (6th edn, McGraw-Hill, 1975), and the well-established *Centrifugal and Other Rotodynamic Pumps*, by H. Addison (3rd edn, Chapman and Hall, 1966), should be consulted. At a more advanced level, the theoretical and practical considerations in the design and use of pumps receive a comprehensive treatment in *Centrifugal Pumps*, by H. H. Anderson (2nd edn, Trade and Technical Press, 1972); *Impeller Pumps,* by S. Lazarkiewicz and A. T. Troskolanski (Pergamon, 1965); *Design and Performance of Centrifugal and Axial Flow Pumps and Compressors,* by A. Kovats (Pergamon, 1964); and *Centrifugal and Axial Flow Pumps,* by A. J. Stepanoff (2nd edn, Wiley, 1957). Equivalent texts dealing with both pumps and turbines are *Introduction to The Theory of Flow Machines,* by A. Betz (Pergamon, 1966), and *Fluid Mechanics of Turbomachinery,* by G. F. Wislicenus (2 vols., 2nd edn, Dover, 1965). Two publications dealing comprehensively with turbines alone are *Hydroturbines—Design and Construction,* by N. N. Kovalev (Israel Program for Scientific Translations, 1961), and *Hydraulic Turbines: Their Design and Equipment,* by M. Nechleba (Constable, 1957).

Good practical handbooks on the selection and application of pumps include *Pump Users Handbook,* compiled for The British Pump Manufacturers' Association by F. Pollak (Trade and Technical Press, 1973); *Pump Selection: A Consulting Engineer's Manual,* by R. Walker (Ann Arbor Science, 1972); *Guide to the Selection of Rotodynamic Pumps,* compiled by The Engineering Equipment Users Association (Constable, 1972) (EEUA Handbook No. 30); and *Pumping Manual,* compiled by the associate editors of *Pumps—Pompes—Pumpen* (3rd edn, Trade and Technical Press, 1968). Information on such technical aspects of hydraulic machinery as cavitation, vibration, pipeline resonance, etc., is available from the many publications of the Institution of Mechanical Engineers and the American Society of Mechanical Engineers. Other texts which may be found helpful are *Fluid Mechanics, Thermodynamics of Turbomachinery,* by S. L. Dixon (Pergamon, 1966); *Theory of Turbomachines,* by G. T. Csanady (McGraw-Hill, 1964); and *Pumping of Liquids,* by F. A. Holland and F. S. Chapman (Reinhold, 1966).

Water supply, irrigation and drainage

Water and waste water treatment is covered in the chapter on public health engineering, so that this facet of water supply and drainage is excluded from this section. Two relevant comprehensive journals, which have been described earlier in the section on 'General Sources in Hydraulics', are *Water and Water Engineering* (continued as *Water Services,* 1974–), and *The Journal of The Institution of Water Engineers.* The British Waterworks Association publishes the monthly journal *British Water Supply,* which provides details of recent applications and developments. The American equivalent is the well-established monthly *Journal. American Water Works Association. Aqua* is the quarterly bulletin of the International Water Supply Association, which has organised important international congresses on water supply.

The quarterly *Journal of the Irrigation and Drainage Division. Proceedings of the ASCE,* previously cited under the specialised topic of hydrology, is a valuable source of information on irrigation and drainage, and on many other related topics. Reference should also be made to *L'Irrigazione,* a bi-monthly international journal published in Italian, and *Irrigation and Power,* published quarterly in English by the Central Board of Irrigation and Power, India. The latter publication includes book reviews and gives details of current research in India.

Four texts, discussed earlier, must again be cited, since they contain good sections on water supply, irrigation and drainage. They are *Water Resources Engineering,* by R. K. Linsley and J. B. Franzini (2nd edn, McGraw-Hill, 1972); *Water Resources Development,* by E. Kuiper (Butterworths, 1965); *Handbook of Applied Hydraulics,* by C. V. Davis and K. E. Sorensen (3rd edn, McGraw-Hill, 1969); and *Manual of British Water Engineering Practice,* compiled by The Institution of Water Engineers (3 vols., 4th edn, Heffer, 1969). There follows a list of other useful books dealing with water supply and/or sewerage.

Water Supply, by A. C. Twort, *et al.* (Arnold, 1974)
Design and Construction of Sanitary and Storm Sewers, by ASCE and Water Pollution Control Federation (ASCE, 1970)
Design in Metric-Sewerage, by R. E. Bartlett (Elsevier, 1970)
Elements of Water Supply and Wastewater Disposal, by G. M. Fair, J. C. Geyer and D. A. Okun (2nd edn, Wiley, 1971)
Environmental Engineering and Sanitation, by J. A. Salvato (2nd edn, Wiley/ Interscience, 1972)
Piping Handbook, edited by S. Crocker and R. King (5th edn, McGraw-Hill, 1967)
Public Health Engineering Practice, Volume 1 Water Supply and Building Sanitation; Volume 2 Sewerage and Sewage Disposal, by L. B. Escritt (4th edn, Macdonald and Evans, 1972)
The Design of Sewers and Sewage Treatment Works, by J. B. White (Arnold, 1970)
Water Supply and Pollution Control, by J. W. Clark, W. Viessman and M. J. Hammer (2nd edn, International Textbook, 1971)
Water Supply Engineering, by H. E. Babbitt, J. J. Doland and J. L. Cleasby (6th edn, McGraw-Hill, 1962)

A broad introduction to the subject of irrigation may be obtained from *Irrigation and Water Resources Engineering,* by H. Olivier (Arnold, 1972), and detailed design information is available from

Irrigation Engineering: Canals and Barrages and *Irrigation Engineering: Syphons, Weirs and Locks,* by S. Leliavsky (Chapman and Hall, 1965). For practical details of basic agricultural drainage, reference should be made to *Erosion and Sediment Pollution Control,* by R. P. Beasley (Iowa State University Press, 1972); *Elementary Soil and Water Engineering,* by G. O. Schwab, R. K. Frevert, K. K. Barnes and T. W. Edminster (2nd edn, Wiley, 1971); *Drainage Engineering,* by J. N. Luthin (Wiley, 1966); and *Soil and Water Conservation Engineering,* by G. O. Schwab, R. K. Frevert, T. W. Edminster and K. K. Barnes (2nd edn, Wiley, 1966).

Coastal engineering

Coastal regions form a particularly vulnerable part of the environment and are the subject of an ever-increasing amount of research activity. In this section of the chapter interest is centred on the engineering aspects of coasts, estuaries, harbours and docks, and a cursory examination is made of some topics in the related subject of ocean engineering.

The *Journal of the Waterways, Harbors and Coastal Engineering Division. Proceedings of the ASCE* is an important quarterly journal which offers papers of a high standard on both the theory and practice of coastal engineering, in the broadest sense of the term. A well-established international journal, which is circulated monthly to port executives, engineers, consultants, contractors and ship owners, and which provides good descriptions of recent schemes and projects, is *The Dock and Harbour Authority.* For both of these journals annual cumulative subject and author indexes are available. *Estuarine and Coastal Marine Science,* first released in 1973, is a quarterly international journal which, although primarily concerned with ecological considerations, seems likely to contain papers of interest to coastal engineers. An informative descriptive publication, which is mainly interested in offshore petroleum activity, but which also deals with dredging and general coastal and port engineering, is the monthly *Offshore Services* (incorporating *Dredging and Coastal Engineering*). *World Dredging and Marine Construction,* a monthly publication, and *Ports and Dredging,* the house journal of IHC, Rotterdam, give practical details of new equipment and recent projects in dredging.

The monthly American journal *Sea Technology* and the newer bi-monthly international journal *Ocean Engineering* deal mainly with design, engineering and application of equipment and services in the marine environment, including submersible vehicles, but also contain some papers useful in the context of coastal engineering. The quarterly *Norwegian Maritime Research,* published in English, was first issued in 1973. A research journal of a high standard, it contains some papers on subjects of interest here, such as marine structures, mathematical models and wave behaviour. The *Journal of Physical Oceanography,* a quarterly journal first published by the American Meteorological Society in 1971, provides an advanced treatment of topics such as circulation in bays, wave frequency analysis, storm surges and turbulent mixing. Hydrographic work is an integral part of most coastal engineering problems and in this context reference should be made to the long-established bi-annual publication *The International Hydrographic Review* and its monthly supplement, *International Hydrographic Bulletin.* For information on journals concerned with advanced theoretical topics such as wave mechanics, reference should be made to the section headed 'Fluid Mechanics'.

The subject of coastal hydraulics is very ably covered in *Beaches and Coasts,* by C. A. M. King (2nd edn, Arnold, 1972) and *Coastal Engineering,* by R. Silvester (2 vols., Elsevier, 1973). All relevant phenomena are discussed both qualitatively and quantitatively, and the former publication contains a good bibliography. A similarly comprehensive text, translated from the Russian, is *Processes of Coastal Development.* by V. P. Zenkovich, edited by J. A. Steers (Oliver and Boyd, 1967), which provides a useful Russian bibliography. *Introduction to Coastline Development* and *Applied Coastal Geomorphology,* two texts again edited by J. A. Steers (Macmillan, 1971), and also two books written by J. Steers entitled *The Coastline of England and Wales* (2nd edn, Cambridge University Press, 1969) and *The Coastline of England and Wales in Pictures* (Cambridge University Press, 1960), have a strong geological and geographical bias. *Coastal Hydraulics,* by A. M. Muir Wood (Macmillan, 1969), is intended for both final year engineering students and practising engineers. Other relevant texts include the introductory *Coastal Changes,* by W. W. Williams (Routledge and Kegan Paul, 1960); the interdisciplinary *Estuaries,* edited by G. H. Lauff (American Association for the Advancement of Science, 1967); and the specialised *The Movement of Beach Sand,* by J. C. Ingle (Elsevier, 1966).

A good summary of the techniques of coast protection is provided in *Sea Defence Works: Design, Construction and Emergency Works,* by R. B. Thorn and J. C. F. Simmons (2nd edn, Butterworths, 1971). Recent developments in this field are described in *Proceedings, 13th Coastal Engineering Conference, Vancouver, 1972* (ASCE, 1973). The conferences have been held at approximately two-yearly intervals since 1950, and reference should be made to the various proceedings. An interesting descriptive approach to the subject is adopted in *The Principles of Coast Protection,* by J. Hoyle and G. King (2nd edn, J. Hoyle and G. King, 1972). *Design and Construction of Ports and Marine Structures,* by A. D. Quinn (2nd edn, McGraw-Hill, 1972), and *Dock and Harbour Engineering,* by H. F. Cornick (4 vols., Griffin, 1960–1969) are extremely valuable standard manuals providing a complete practical coverage of the field indicated by their titles. Two older texts dealing with the same subject are *Winds, Waves and Maritime Structures,* by R. R. Minikin (2nd edn, Griffin, 1963), and *The Design, Construction and Maintenance of Docks, Wharves and Piers,* by F. M. G. Du Plat Taylor (3rd edn, Eyre and Spottiswoode, 1949). *Tanker and Bulk Carrier Terminals* (ICE, 1969) and *Analytical Treatment of Problems of Berthing and Mooring Ships, Proceedings NATO Advanced Study Institute, Lisbon, 1965* (ASCE, 1970) provide information on two specialised aspects of coastal engineering.

Texts on ocean engineering include *Handbook of Ocean and Underwater Engineering,* edited by J. J. Myers, C. H. Holm and R. F. McAllister (McGraw-Hill, 1969); *Man Beneath the Sea—A Review of Underwater Ocean Engineering,* by W. Penzias and M. W. Goodman (Wiley Interscience, 1973); *Ocean Engineering Structures,* by J. H. Evans and J. C. Adamchak (MIT Press, 1969); *Dynamic Analysis of Ocean Structures,* by B. J. Muga and J. F. Wilson (Plenum, 1970); *Concrete Sea Structures, Proceedings of FIP Symposium, Tbilisi, 1972,* edited by P. V. Maxwell-Cook (Federation Internationale de la Precontrainte, 1973); and *Concrete Floating and Submerged Structures,* by R. G. Morgan (Concrete Society, 1973). The *Underwater Engineering Directory* (Spearhead, for CIRIA, 1971) contains lists of relevant UK companies, commercial and government organisations, universities and colleges, and overseas companies, and also contains a glossary. Two glossaries of note which should be mentioned here are *A Glossary of Ocean Science and Undersea Technology Terms,* edited by L. M. Hunt and D. G. Groves (Compass, 1965), and *A Glossary of Coastal Engineering Terms,* by R. H. Allen (Coastal Engineering Research Center, 1972).

Dredging is a subject of obvious importance, and a good understanding of its theory and practice may be obtained from *Hydraulic Dredging: Theoretical and Applied,* by J. Huston (David and Charles, 1970), which contains a good bibliography; *Dredging* (ICE, 1968); *Modern Dredging Practice,* by R. Hammond (Muller, 1969); and the elderly but informative *Practical Dredging,* by H. R. Cooper (Brown, Son and Ferguson, 1958).

For information on such hydrodynamic aspects of coastal engineering as waves, surges, tides, wave/structure interaction, harbour oscillation, etc., the following publications should be consulted:

Admiralty Manual of Tides, by A. T. Doodson and H. D. Warburg (HMSO, 1966)
Estuary and Coastline Hydrodynamics, by A. T. Ippen (McGraw-Hill, 1966)
Ocean Engineering Wave Mechanics, by M. E. McCormick (Wiley, 1973)
Oceanographical Engineering, by R. L. Wiegel (Prentice-Hall, 1964)
Ocean Wave Statistics, by N. Hogben and F. E. Lumb (HMSO, 1967)
Proceedings of The Symposium on Tides, (Monaco, 1967), organised by The International Hydrographic Bureau (UNESCO, 1969)
Stochastic Hydraulics: Proceedings of First International Symposium, edited by C. Chiu (University of Pittsburgh, 1971)
The Analysis of Tides, by G. Godin (Liverpool Press, 1972).
Tidal Computations in Rivers and Coastal Waters, by J. J. Dronkers (North-Holland, 1964)
Tides, by D. H. Macmillan (CR Books, 1966)
Water Waves, by J. J. Stoker (Interscience, 1957)
Wind Waves: Their Generation and Propagation on the Ocean Surface, by B. Kinsman (Prentice-Hall, 1965)

Multi-phase flow

Multi-phase flow, especially the hydraulic or pneumatic transport of solids in pipes, is becoming increasingly important in civil engineering. A good guide to the historical development of the subject may be obtained by reference to *The Hydraulic Transport of Solids in Pipes —A Bibliography,* by W. A. Thornton (BHRA, 1970). This comprehensive publication contains an interesting foreword, 1030 references arranged chronologically in specific subject groups, abstracts for most of the items, and subject and author indexes. The BHRA continued the good work by commencing publication of the quarterly *Solid–Liquid Flow Abstracts,* which contains abstracts of the world's literature on hydraulic and pneumatic transport and related topics in fluid mechanics, equipment and applications, listed under 50 subject headings. *Abstracts D: Fluid Mechanics,* published bimonthly by the National Coal Board, includes conveyance and slurry flow among its interests.

Three texts which together provide a good treatment of the fundamentals of the subject are *The Flow of Complex Mixtures in Pipes,* by G. W. Govier and K. Aziz (Van Nostrand Reinhold, 1972); *One Dimensional Two-Phase Flow,* by G. B. Wallis (McGraw-Hill, 1969); and *Fluid Dynamics of Multiphase Systems,* by S. L. Soo (Blaisdell, 1967). Reference could also be made to *Fluid Mechanics and*

Measurements in Two-Phase Flow Systems (IMechE, 1970) and *Advances in Solid–Liquid Flow in Pipes and its Application,* edited by I. Zandi (Pergamon, 1971). Some practical aspects are considered in *The Hydraulic Transport of Solids by Pipeline,* by A. G. Bain and S. T. Bonnington (Pergamon, 1970); part II of *Pumps and Blowers/ Two-Phase Flow,* by A. J. Stepanoff (Wiley, 1965); *Pumping of Liquids,* by F. A. Holland and F. S. Chapman (Reinhold, 1966); and *Piping Handbook,* edited by S. Crocker and R. King (5th edn, McGraw-Hill, 1967). *Hydraulics of Sediment Transport,* by W. H. Graf (McGraw-Hill, 1971), and *Loose Boundary Hydraulics,* by A. J. Raudkivi (Pergamon, 1967), adopt a unified approach to sediment transport and include sections on the pneumatic and hydraulic transport of solids.

Unsteady flow

It is apparent that the previously discussed specialised topics often involve transient flow, so that in each case the references cited deal with relevant aspects of this problem. In this section, however, publications primarily concerned with unsteady flow will be examined. A good concise introduction to the subject is provided in the chapter entitled 'Fluid Transients' in *Handbook of Fluid Dynamics,* edited by V. L. Streeter (McGraw-Hill, 1961). For a clear exposition of basic principles and practical aspects of concern to the hydraulic engineer, *Analysis of Surge,* by J. Pickford (Macmillan, 1969) should be consulted. Water hammer, surge tank behaviour and surges in open channels are analysed by use of graphical and finite difference methods, and a helpful list of references is provided. *Hydraulic Transients,* by G. R. Rich (2nd edn, Dover, 1963), is a similarly valuable text. A more advanced treatment over a broader range of topics is available from *Hydraulic Transients,* by V. L. Streeter and E. B. Wylie (McGraw-Hill, 1967), and *Differential Equations of Hydraulic Transients, Dispersion and Ground-Water Flow,* by W. H. Li (Prentice-Hall, 1972). In the first of these texts the computer solution of problems is emphasised, and both provide good lists of references. Two advanced texts concerned with all aspects of water hammer and surge tank function are *Waterhammer Analysis,* by J. Parmakian (Dover, 1960), and *Water Hammer and Surge Tanks,* by G. V. Aronovich, N. A. Kartvelishvili and Y. K. Lyubimtsev (Israel Program for Scientific Translations, 1970). *The Proceedings, 1st International Conference on Pressure Surges, University of Kent, 1972,* edited by H. S. Stephens and M. J. Rowat (BHRA, 1973), contains papers on most theoretical and practical aspects of pressure transients, including analytical methods, surge tanks and other methods of surge control and suppression, and two-phase flow. *Continuous Measurement of Unsteady Flow,* by G. P. Katys (Pergamon, 1964), provides a good introduction to a particular aspect of unsteady flow.

Hydraulic models

A hydraulic model is a prerequisite in many projects, and certain establishments, notably the Hydraulics Research Station, possess a wealth of expertise in this field. Most civil engineering hydraulics texts devote part of their contents to hydraulic models, but there exist some publications concerned solely with this subject. A classic reference is *Scale Models in Hydraulic Engineering,* by J. Allen (2nd edn, Longmans Green, 1952). Similar in scope, but more recent, is the useful *Hydraulic Models,* prepared by the ASCE Hydraulics Division, Committee on Hydraulic Research (ASCE, 1963). *Dimensional Analysis and Hydraulic Model Testing,* by H. M. Raghunath (Asia, 1967), explains dimensional analysis in a basic textbook manner and gives a general introduction to testing techniques with details of some applications. A text intended for postgraduate students, research workers and practising engineers is *Theory of Hydraulic Models,* by M. S. Yalin (Macmillan, 1971). It contains a fairly advanced theoretical treatment of the subject with details of practical applications. Ten papers on a particular class of hydraulic model are available in *Model Testing of Hydraulic Machinery and Associated Structures* (IMechE, 1968).

Other publications worthy of reference are:

Dimensional Analysis and Theory of Models, by H. L. Langhaar (Wiley, 1951)
Hydraulic Laboratory Practice (USBR (Engineering Monograph No. 18), 1953)
Hydraulic Model Techniques, by F. H. Allen (ICE, Vernon-Harcourt Lecture, 1959)
Hydraulics Research (HMSO) Annual Report of the Hydraulics Research Station, Wallingford
Similitude in Engineering, by G. Murphy (Ronald Press, 1950)
The Role of Models in the Evolution of Hydraulic Structures: A Symposium (1952) (Publication No. 53, Central Board of Irrigation and Power, India, 1954)

ESTABLISHMENTS, INSTITUTIONS AND ORGANISATIONS

There follows a list of establishments, institutions and organisations which produce valuable report series, design aids, statistics, annual publications, etc.; assist and encourage research and development; organise symposia and conferences; and provide experimental facilities and other services for the hydraulic engineer.

American Society of Civil Engineers
American Water Resources Association
American Water Works Association
British Hydromechanics Research Association
Coastal Engineering Research Centre. US Army Corps of Engineers
Construction Industry Research and Information Association
Delft Hydraulic Laboratory
Hydraulics Research Station
Institute of Coastal Oceanography and Tides
Institute of Hydrology
Institute of Oceanographic Sciences
Institution of Water Engineers
International Association for Hydraulic Research
International Association of Scientific Hydrology
International Water Supply Association
National Institute of Oceanography
Water Research Centre
Wimpey Laboratories Limited

ADDITIONAL TEXTS

This section contains three lists of selected texts intended to help the reader remedy any deficiencies in his knowledge or understanding of basic civil engineering hydraulics or of relevant principles and techniques of fluid mechanics. The first is a list of hydraulics textbooks of undergraduate level, and the second and third are lists of fluid mechanics, and more advanced fluid mechanics texts, respectively.

Basic civil engineering hydraulics

A Treatise on Applied Hydraulics, by H. Addison (5th edn, Chapman and Hall, 1964)
Engineering Hydraulics, edited by H. Rouse (Wiley, 1950)
Essentials of Engineering Hydraulics, by J. M. K. Dake (Macmillan, 1972)
Fluid Mechanics for Civil Engineers, by N. B. Webber (2nd edn, Chapman and Hall, 1971)

Fluid mechanics

An Introduction to Fluid Mechanics and Heat Transfer, by J. M. Kay (Cambridge University Press, 1957)
Applied Hydrodynamics, by H. R. Vallentine (2nd edn, Butterworths, 1967)
A Textbook of Fluid Mechanics for Engineering Students, by J. R. D. Francis (3rd edn, Arnold, 1969)
Elementary Fluid Mechanics, by J. K. Vennard (4th edn, Wiley, 1961)
Elementary Mechanics of Fluids, by H. Rouse (Wiley, 1946)
Engineering Fluid Mechanics, by J. E. Plapp (Prentice-Hall, 1968)
Fluid Dynamics, by R. H. F. Pao (Merrill, 1967)
Fluid Dynamics and Heat Transfer, by J. G. Knudsen and D. L. Katz (McGraw-Hill, 1958)
Fluid Mechanics, by V. L. Streeter (6th edn, McGraw-Hill, 1975)
Fluid Mechanics for Engineers, by M. L. Albertson, J. R. Barton and D. B. Simons (Prentice-Hall, 1960)
Fluid Mechanics for Engineers, by P. S. Barna (3rd edn, Butterworths, 1971)
Fluid Mechanics with Engineering Applications, by R. L. Daugherty and J. B. Franzini (6th edn, McGraw-Hill, 1965)
Foundations of Fluid Mechanics, by S. W. Yuan (Prentice-Hall, 1970)
Mechanics of Fluids, by W. J. Duncan, A. S. Thom and A. D. Young (2nd edn, Arnold, 1970)
Mechanics of Fluids, by J. W. Ireland (Butterworths, 1971)
Mechanics of Fluids, by B. S. Massey (2nd edn, Van Nostrand, 1970)
Principles of Fluid Mechanics, by W. H. Li and S. H. Lam (Addison-Wesley, 1964)

More advanced fluid mechanics

Advanced Mechanics of Fluids, edited by H. Rouse (Wiley, 1959)
An Introduction to Turbulence and its Measurement, by P. Bradshaw (Pergamon, 1971)
Boundary Layer Theory, by H. Schlichting (6th edn, McGraw-Hill, 1968)
Engineering Fluid Mechanics, by C. Jaeger (Blackie, 1957)
Essentials of Fluid Dynamics, by L. Prandtl (Blackie, 1952)
Experimental Fluid Mechanics, by P. Bradshaw (2nd edn, Pergamon, 1970)
Fluid Mechanics, by W. Kaufmann (McGraw-Hill, 1963)
Handbook of Fluid Dynamics, edited by V. L. Streeter, (McGraw-Hill, 1961)
Hydrodynamics, by H. Lamb (6th edn, Cambridge University Press, 1932)
Mathematical Models of Turbulence, by B. E. Launder and D. B. Spalding (Academic Press, 1972)
Mechanics of Fluid Flow, by P. A. Longwell (McGraw-Hill, 1966)
Modern Developments in Fluid Dynamics, edited by S. Goldstein (2 vols., Oxford University Press, 1938)
Modern Fluid Dynamics. Volume I Incompressible Flow, by N. Curle and H. J. Davies (Van Nostrand, 1968)
Momentum, Energy and Mass Transfer in Continua, by J. C. Slattery (McGraw-Hill, 1972)
Non-Newtonian Fluids, by W. L. Wilkinson (Pergamon, 1960)
Theoretical Hydrodynamics, by L. M. Milne-Thomson (5th edn, Macmillan, 1968)
The Science of Fluids, by I. Michelson (Van Nostrand Reinhold, 1970)

The Theory of Homogeneous Turbulence, by G. K. Batchelor (Cambridge University Press, 1953)

Turbulence, by J. O. Hinze (McGraw-Hill, 1959)

Turbulence Phenomena, by J. T. Davies (Academic Press, 1972)

Viscosity and Flow Measurement; A Laboratory Handbook of Rheology, by J. R. Van Wazer, J. W. Lyons, K. Y. Kim and R. E. Colwell (Interscience, 1963)

Viscosity and its Measurement, by A. Dinsdale and F. Moore (Chapman and Hall, 1962)

30

Highway, traffic and transport engineering

C. A. O'Flaherty and J. A. Lee

GENERAL

It is no longer possible to regard any or all aspects of highway, traffic and transport engineering as an isolated clearly bound academic subject. During the past 200 years the general field has progressed from its early emphasis on construction practices (eighteenth and nineteenth centuries) to the application of pre-conceived design (late nineteenth and early twentieth centuries) to design based on research and development (second quarter of the twentieth century). In more recent years (post–1950s) the engineering of transport facilities has been seen to require a knowledge not only of the physical characteristics of these facilities, but also of the nature of the demand for them, and the economic, environmental and social implications of that demand upon the community as a whole.

There are two main reasons for the impetus which has resulted in the recent renaissance of academic and practical interest in this subject area. Firstly, the sheer scale of the problem of catering for and accommodating the ever-increasing demand for mobility has emphasised the need for examining every transport proposal most carefully to ensure not only that the most economic facility (in terms of initial plus maintenance cost) is constructed, but also that it is, in fact, beneficial to the community serviced as well as to the users of the facility. Secondly, the advent and development of the computer has made possible techniques of analysis which provide the transport engineer and planner with new degrees of freedom in relation to the engineering of the transport facilities and to the problem of understanding and forecasting the need for them.

Both of these considerations have encouraged the emphasis on specialisation within the general study area, thereby making difficult the rationalisation of the vast field of literature associated with it.

HIGHWAY, TRAFFIC AND TRANSPORT TEXTBOOKS

This chapter is concerned with the use of literature in highway, traffic and transport engineering. The apparent divisibility of this subject area arises from the manner of its development within, and in association with, civil engineering. While it can be argued that the area is factually indivisible, practically it is necessary to make distinction for the purpose of classifying literature sources.

For the purpose of this text *Highway Engineering* is assumed to be the process which relates to the structural design and construction of a road facility. Generally, it involves the following engineering aspects: highway location and material surveys, soil and other highway materials engineering, pavement design and construction, drainage design and construction, highway setting-out, site management and plant usage, and road maintenance.

It will be appreciated, of course, that the design and construction of other transport facilities utilises knowledge and techniques which are also applicable to highway engineering.

In this chapter *Traffic Engineering* is assumed to deal with the movement of people and vehicles within a highway traffic system, and with determining the most effective means by which the traffic facilities can be made to accommodate the desired trips. Topics which are generally included within the framework of traffic engineering include traffic flow analysis, highway geometric design, traffic management and control, parking regulation, road lighting and road safety.

The carrying out of traffic generation and model split surveys and analyses which are often considered in the context of traffic engineering, is assumed in this text to relate more to transport planning, and thus is treated in some detail in titles listed under that heading.

Transport Engineering is in many ways the most difficult of the three terms to define, principally because its boundaries, as well as its internal content, are in a constant state of flux. To a certain extent, indeed, it may be said that the title itself is somewhat of a misnomer in that the subject is now recognised as embracing more than the 'pure' engineering of transport facilities.

Prior to the early 1950s the term 'transport engineering' was very rarely used at all. In fact, highway engineering (which included 'traffic engineering') was considered to be separate from airport engineering, railway engineering, seaport engineering, etc. Then it began to be appreciated that there was a need for an integrated approach to the design of transport facilities, and so transport engineering (or, in the United States, 'transportation' engineering) was born. This approach recognised highway engineering and traffic engineering as being specialised branches of transport engineering.

During the mid-1950s a major development took place, the implications of which were not fully appreciated for some time. This was the carrying out of the Chicago Area Transportation Study, an event which, practically, launched and gave prominence to a sophisticated form of transport planning. From this beginning, transport planning has now grown so that it is considered by many to play a dominant role within transport engineering.

Current thinking with regard to transport planning is that it not only involves the study of the demand for mobility and the manner in which that demand may be satisfied in terms of vehicles and structures, but also requires an understanding of the patterns of urban and rural development and of the physical, social and economic characteristics of the environment. The creation and implementation of transport plans can require, therefore, the application of a multiplicity of skills, not least of which are those of the engineer, economist, urban/rural planner and geographer, mathematician and statistician, and sociologist.

A by-product of this interdisciplinary approach is that it makes most difficult any precise subdivision and classification of the diffuse yet relevant literature. For practical purposes, therefore, the approach taken in this chapter is to divide the subject areas in what is hoped is a logical manner, so that the reader may have the opportunity of easily finding the references and information most appropriate to his/her requirements and knowledge. No attempt is made to provide a complete set of references, rather the intention is to provide some of the more easily available, as well as important and basic, textbook-type references published over the past decade. The divisions chosen for this purpose are as follows:

Highway engineering
    General textbooks
    Materials and physical surveys
    Pavement design, construction and maintenance
Traffic engineering
    General textbooks
    Traffic flow and geometric design
    Management and control, and parking
    Safety

Transport engineering
General textbooks
Transport history
Transport planning
Air transport and airports
Railways
Sea transport, seaports and waterways
Transport technology

RESULTS OF RESEARCH

Taking highway, traffic and transport engineering as a unit subject area, there is no doubt but that it is currently one of the most dynamic, as well as topical, areas of study. Because of the rapidity as well as the intensity of its development over the past 20 years, new knowledge is constantly being gained and old opinions discarded. Indeed, its current status may be said to be very much that of an applied research subject in the truest sense. It is essential, therefore, that the user of the technical literature should be aware of the main sources of information regarding research work.

Research information is most easily obtained from the numerous articles published in scientific/technical journals. In addition, many governmental agencies, at both the local and national level, publish their own research reports. Many of these governmental reports are most useful not only for what they factually contain, but also in that they tend to reflect governmental thinking as to what are currently important topics and policies.

A significant amount of worthwhile research is carried out in universities. As well as publishing in the technical journals, some of the more important universities also issue their own research reports, working papers, bulletins, etc.

Other groups which publish similar research reports include commercial research groups, manufacturers' associations and professional institutions.

In this chapter the sources of research information are listed under the following headings:

Periodicals
Indexes, abstracts and information services
Governmental agencies
Universities with major transport study groups
Other organisations and research groups

Before leaving this section, there are three most valuable reference sources that the writers would wish to highlight. Firstly, there are publications of the Transport and Road Research Laboratory (TRRL) in Britain. This organisation, which is the single most important source of transport research material in Britain, publishes most of its own research work. No attempt has been made in this chapter to refer to these reports under any other heading. Lists of all their publications can be obtained from:

> The Librarian
> Transport and Road Research Laboratory
> Crowthorne
> Berks.
> RG11 6AU

Secondly, there are the publications of the Transport Research Board (TRB) (formerly the Highway Research Board (HRB)) in the United States. Unlike the TRRL, the TRB does not carry out its own research work. Rather, it sponsors the preparation of a significant number of US research reports and, more important, it organises (and publishes the papers presented at) what is probably the largest annual research conference in the transport world. Lists of TRB publications are available from:

> The Director
> Transportation Research Poard
> 2101 Constitution Ave
> Washington, DC 20418

Lastly, there is a most useful bibliographic publication titled *Current Literature in Traffic and Transportation,* which is published by

> The Transportation Center Library
> Northwestern University
> Evanston, Illinois

This monthly publication, in which are listed all the main traffic and transportation research papers published in the major English-language journals, is an essential ingredient of any good transport library. (For example, the great majority of the periodicals listed in this chapter are referenced by this publication.)

STATISTICAL DATA

Statistical information as such is rarely required by either the highway or traffic engineer in his work. In contrast, the transport planner depends heavily on information of this nature in order to forecast the demand for particular forms and types of transport facilities. All national governments, and many local government and industrial groups, publish statistics on a regular basis. In addition, many valuable and detailed transport statistics are gathered for, and

published in, the various land use and transport studies that have
been carried out in urban areas in this country and abroad.
Some of the main sources of statistics relevant to the transport
scene are listed under the following headings:
General
Road transport
Air and sea transport

HIGHWAY ENGINEERING

General textbooks

Roadwork Technology, by J. H. Arnison (3 vols., Illiffe, 1967)
Highway Engineering, by R. Ashworth (Heinemann, 1966)
Highway Engineering, by V. F. Babkov and M. S. Zamakhaev (Mir Publishers, 1967)
Roads, by R. G. C. Batson and J. A. Proudlove (2nd edn, Longmans, 1968)
Method of Measurement for Road and Bridge Works, publication issued by the Department of the Environment (HMSO, 1971)
Soil Stabilisation, by O. G. Ingles and J. B. Metcalfe (Butterworths, 1973)
Road Engineering, by E. L. Leeming (3rd edn, Constable, 1952)
Highways: Vol. 2, Highway Engineering, by C. A. O'Flaherty (Arnold, 1974)
Design and Traffic Flow: Two-Lane Rural Roads (OECD, 1972)
Roads, Bridges and Tunnels, by M. Overman (Aldus Books, 1968)
Principles and Practice of Highway Engineering, by R. C. Sharma and S. K. Sharma (Asia, 1964)
Low Cost Roads: Design, Construction and Maintenance, by L. Odier, R. S. Millard, P. dos Santos and S. R. Mehra (Butterworths, 1971)
Highway Engineering Handbook, edited by K. B. Woods (McGraw-Hill, 1960)
Principles of Pavement Design, by E. J. Yoder (Wiley, 1959)

Materials and physical surveys

Soil Survey of Great Britain: Report No. 15, by the Agricultural Research Council (HMSO, 1960)
Significance of Tests and Properties of Concrete and Concrete-Making Materials: Special Technical Publication No. 169-A. (American Society for Testing and Materials, 1966)
Asphalt: Science and Technology, by E. J. Barth (Gordon and Breach, 1962)
Emulsions: Theory and Practice, by P. Becher (2nd edn, Reinhold, 1965)
Laboratory Work for Students of Construction, by S. C. Blunt and A. Cleveland (Butterworths, 1967)
Various *British Standards*, on soil, aggregate, bitumen and tar, and concrete testing
CP2001: 1957 Site Investigations (British Standards Institution)
Asphaltic Road Materials, by R. W. Hatherley and P. C. Leaver (Arnold, 1967).
Bituminous Materials: Asphalts, Tars and Pitches, edited by A. J. Hoiberg (3 vols., Interscience, 1964–1966)
Highway Materials, by R. D. Krebs and R. D. Walker (McGraw-Hill, 1971)

Deformation, Strain and Flow: An Elementary Introduction to Rheology, by M. Reiner (3rd edn, Lewis, 1969)

Pit and Quarry Textbook, by SAGA (Sand and Gravel Association) of Great Britain (Macdonald, 1967)

Bituminous Materials for Flexible Pavements (Department of Civil Engineering, University of Nottingham, 1972)

Bituminous Materials in Road Construction, by the Road Research Laboratory (HMSO, 1962)

Pavement design, construction and maintenance

Road Making Machinery, by K. Abrosimov, A. Bromberg and F. Katayev (Mir Publishers, 1965)

Street Cleaning Practice, by the American Public Works Association (2nd edn, American Public Works Association, 1959)

Street and Urban Road Maintenance, by the American Public Works Association (American Public Works Association, 1963)

Manual of Excavators, by H. Breitung (Edition Leipzig, 1968)

Various *British Standards* on bituminous surfacings

Report of the Committee on Highway Maintenance: The Marshall Report (HMSO, 1970)..

Earthmoving and Excavating Plant, by R. E. R. Hammond (CR Books, 1964)

Manual of Surface Drainage Engineering, by B. Z. Kinori (Elsevier, 1970)

Portland Cement and Asphalt Concretes, by T. D. Larson (McGraw-Hill, 1963)

Specification for Road and Bridge Works, by the Ministry of Transport *et al.* (HMSO, 1969)

Moving the Earth: The Workbook of Excavation, by H. L. Nichols (2nd edn, North-Castle Books, 1967)

Manual of Civil Engineering Plant and Equipment (2nd edn, Applied Science, 1971)

Construction Planning, Equipment and Methods, by R. L. Peurifoy (2nd edn, McGraw-Hill, 1970)

Guide to the Structural Design of Pavements for New Roads: Road Note No. 29 (3rd edn, HMSO, 1970)

Concrete in Highway Engineering, by D. R. Sharp (Pergamon, 1970)

Street and Highway Drainage, by I. R. Cole, W. A. Gelonek, J. W. Huston and W. R. Naydo (Institute of Transportation and Traffic Engineering, University of California, 1969)

Asphalt Pavement Engineering, by H. A. Wallace and J. R. Martin (McGraw-Hill, 1967)

TRAFFIC ENGINEERING

General textbooks

Traffic Engineering Practice, edited by E. Davies (2nd edn, Spon, 1968)

Traffic Engineering Handbook, edited by J. E. Baerwald (3rd edn, Institute of Traffic Engineers, 1965) (new edition in preparation)

Research on Road Traffic, by the Road Research Laboratory (HMSO, 1965)

Elementary Sampling for Traffic Engineers, by D. F. Votaw and H. S. Levinson (Eno Foundation, 1972)
Traffic Planning and Engineering, by F. D. Hobbs (Pergamon, 1974)
An Introduction to Highway Transportation Engineering, edited by D. G. Capelle, D. E. Cleveland and W. W. Rankin (Institute of Traffic Engineers, 1968)
Traffic Science, by D. C. Gazis *et al.* (Wiley, 1974)
Traffic Engineering, by T. M. Matson, W. S. Smith and F. W. Hurd (McGraw-Hill, 1955)
Fundamentals of Traffic Engineering, by N. Kennedy *et al.* (Institute of Traffic and Transportation Engineering, University of California, 1968)
Better Use of Town Roads, by the Ministry of Transport (HMSO, 1967)
Guide to Traffic Engineering Practice (National Association of Australian State Road Authorities, 1970)
Traffic Engineering: Theory and Practice, by L. J. Pignataro (Prentice-Hall, 1973)
Highways: Vol. 1. Highways and Traffic, by C. A. O'Flaherty (Arnold, 1974)
Traffic Engineering, by G. R. Wells (Griffin, 1970)
Various *Proceedings of the International Study Weeks in Traffic Engineering and Road Safety* (World Touring and Automobile Organisation (OTA), 1953–74)

Traffic flow and geometric design

A Policy on Geometric Design of Rural Highways (American Association of State Highway Officials, 1965)
The Theory of Road Traffic Flow, by W. D. Ashton (Methuen, 1966)
Proceedings of the various *International Symposia on the Theory of Traffic Flow* (Elsevier)
Highway Transition Curve Tables, by the County Surveyors' Society (Carriers Publishing Company, 1969)
Metrication—Highway Design: DOE Technical Memorandum T8/68 (Department of the Environment, 1968)
Junction Design: DOE Technical Memorandum H7/71 (Department of the Environment, 1971)
Highways Spirals, Superelevation and Vertical Curves, by H. Criswell (3rd edn, Carriers Publishing Company, 1958)
Traffic Flow Theory and Control, by D. R. Drew (McGraw-Hill, 1968)
Poisson and Other Distributions, by D. L. Gerlough *et al.* (Eno Foundation, 1971)
Mathematical Theories of Traffic Flow, by F. A. Haight (Academic Press, 1963)
Highway Capacity Manual (Highway Research Board, 1965)
Route Location and Design, by T. F. Hickerson (5th edn, McGraw-Hill, 1967)
Route Surveying and Design, by C. F. Meyer (4th edn, International Textbook, 1969)
The Layout of Roads in Rural Areas, by the Ministry of Transport *et al.* (HMSO 1968). Also *Metric Corrigendum: DOE Technical Memorandum H7/69* (Department of the Environment, 1969)
Roads in Urban Areas, by the Ministry of Transport *et al.* (HMSO, 1966)
Flows in Transportation Networks, by R. B. Potts and R. M. Oliver (Academic Press, 1972)
Queues, by D. R. Cox and W. L. Smith (Chapman and Hall, 1971)
Optimization of Transport Networks, by P. A. Steenbrink (Wiley, 1974)

Management and control, and parking

Parking of Motor Vehicles, by J. Brierley (2nd edn, Applied Science, 1972)
CP1004: [1963–1973] *Street Lighting* (9 parts, British Standards Institution)
Lorry Parking: Report of the Working Party on the Parking of Lorries (HMSO, 1971)
Traffic Control Theory and Instrumentation, by T. R. Horton (Plenum, 1965)
Multi-Storey Car Parks and Garages, by D. Close (Architectural Press, 1965)
Drugs and Driving, by G. Milner (Australasian Drug Information Services, Sydney, 1972)
Pelican Pedestrian Crossings: DOE Circular Roads 21/68 (Department of the Environment, 1968)
Zebra Crossings: DOE Circular Roads 21/68 (Department of the Environment, 1968)
Pedestrian Subways—Layout and Dimensions: DOE Technical Memorandum H2/70 (Department of the Environment, 1970)
Criteria for the Provision of Pedestrian Subways or Bridges: DOE Technical Memorandum H8/69 (Department of the Environment, 1969)
Parking in Town Centres: Planning Bulletin No. 7, by the Ministry of Housing and Local Government (HMSO, 1965)
Cars in Housing: Design Bulletin No. 12, by the Ministry of Housing and Local Government (HMSO, 1967)
Urban Traffic Engineering Techniques, by the Ministry of Transport (HMSO, 1965)
Road Pricing: The Economic and Technical Possibilities, by the Ministry of Transport (HMSO, 1964)
Traffic Signs Manual, by the Ministry of Transport *et al.* (HMSO, 1965–)
Speed Limits Outside Built-Up Areas (OECD, 1972)
Area Traffic Control Systems (OECD, 1972)
Delineating the Edge of the Carriageway in Rural Areas, by C. A. O'Flaherty (Printerhall Ltd, 1972)
Parking Design Manual (Parking and Highway Improvement Contractors Association, Los Angeles, 1968)
Proceedings of a Symposium on Area Control of Road Traffic (Institution of Civil Engineers, 1967)
Traffic Design of Parking Garages, by E. R. Ricker (Eno Foundation, 1957)
A Study of Roadway Lighting, by the Southwest Research Institute (United States Steel, Pittsburgh, 1970)
Parking Requirements for Shopping Centers: Technical Bulletin No. 53 (Urban Land Institute, Washington, 1965)
Manual on Uniform Traffic Control Devices (US Department of Transportation, 1971)
Parking Garage Operation, by R. E. Whiteside (Eno Foundation, 1961)
Zoning, Parking and Traffic, by D. K. Witherford and G. E. Kanaan (Eno Foundation, 1972)

Safety

Alcoholism and Driving, by C. J. Bridges (Thames and Hudson, 1972)
Research on Road Safety, by the Road Research Laboratory (HMSO, 1963)
Road Accidents, by D. W. Elliott and H. Street (Penguin, 1968)

A System of Reporting and Recording Traffic Accident Information, by P. A. Hall (An Foras Forbartha Teoranta, Dublin, 1967)
Human Factors in Highway Traffic Safety Research, edited by T. W. Forbes (Wiley, 1972)
Road Accidents: Prevent or Punish, by J. J. Leeming (Cassell, 1969)
The State of the Art of Traffic Safety, by A. D. Little Inc. (Praeger, 1970)
Traffic Control and Roadway Elements—Their Relationship to Highway Safety, by P. A. Mayer (US Highway Users Federation for Safety and Mobility, 1970)
Accident Proneness, by L. Shaw and H. S. Sichel (Pergamon, 1971)
A Handbook of Highway Safety Design and Operating Practices, by US Department of Transportation (US Government Printing Office, 1968)
Safer Roadside Structures, by the Texas Transportation Institute, Texas A and M. University (United States Steel, Pittsburg, 1970)
Analysis and Summary of Accident Investigations, by the US Bureau of Motor Carrier Safety (US Federal Highway Administration, 1970)
Research on Crash Barriers (OECD, 1967)
Pedestrian Safety (OECD, 1969)
Driver Behaviour (OECD, 1970)
Statistical Methods in the Analysis of Road Accidents (OECD, 1970)
Research into Road Safety at Junctions in Urban Areas (OECD, 1971)
Lighting, Visibility and Accidents (OECD, 1971)

TRANSPORT ENGINEERING

General textbooks

Urban Transit Development in Twenty Major Cities (American Automotive Foundation, 1968)
The Land-Use/Transport System, by W. R. Blunden (Pergamon, 1971)
Report of the Royal Commission on Environmental Pollution (*Cmnd. 4585*), chaired by Sir Eric Asby (HMSO, 1971)
Motorways, by J. Drake, H. L. Yeadon and D. I. Evans (Faber, 1969)
Motorways in Britain: Today and Tomorrow, by J. S. Davis (Institution of Civil Engineers, 1971)
Various proceedings of conferences organised by the *Planning and Transport Research and Computation Co. Ltd.* London
Various *Proceedings* of the *Congress of the International Union of Public Transport* (Paris)
Superhighway—Superhoax, by H. Leavitt (Doubleday, 1970)
Elements of Transport, by R. W. Faulks (2nd edn, Allan, 1969)
The Transport Problem, by C. D. Foster (Blackie, 1963)
An Introduction to Transportation Engineering, by W. W. Hay (Wiley, 1961)
Fundamentals of Transportation Engineering, by R. G. Hennes and M. Ekse (2nd edn, McGraw-Hill, 1969)
Transport Studies, by J. Hibbs (Baker, 1970)
Kitchin's Road Transport Law, by L. D. Kitchin and E. K. Wenlock (15th edn, Iliffe, 1970)
National Transportation Policy in Transition, by H. Mertins (Lexington, 1972)
The Urban Transportation Problem, by J. R. Meyer, J. F. Kain and M. Wohl (Harvard University Press, 1965)
Urban Transportation Policy, by D. R. Miller (Lexington, 1973)
Transport: Selected Readings, by D. Munby (Penguin, 1968)

500 Highway, traffic and transport engineering

Unsafe at Any Speed, by R. Nader (Grossman, 1972)
Passenger Transport—Present and Future, by C. A. O'Flaherty (Leeds University Press, 1969)..
An Analysis of Urban Travel Demands, by W. Y. Oi and P. W. Shuldiner (Northwestern University Press, 1962)
Strategy for Mobility, by W. Owen (Brookings Institution, Washington, 1964)
Transportation Engineering: Planning and Design, by R. J. Paquette, N. Ashford and P. H. Wright (Ronald Press, 1972)
The Demand for Travel: Theory and Measurement, edited by R. E. Quandt (Heath Lexington, 1970)
Highways and Our Environment, by J. Robinson (McGraw-Hill, 1971)
Wheels of Progress? Motor Transport, Pollution and the Environment, edited by J. Rose (Gordon and Breach, 1973)
Transport Design, by C. H. Stanton (Studio Vista, 1967)
Beyond the Automobile: Reshaping the Transportation Environment, by T. R. Stone (Prentice-Hall, 1971)
Transporting Goods by Road, by M. Webb (Weidenfeld and Nicolson, 1972)
Transportation and Urban Land, by L. Wingo (Johns Hopkins University Press, 1968)
Environmental Factors in Transportation Planning, by P. Weiner and E. J. Deak (Lexington, 1972)

Transport history

Highways in the Air, by the British Overseas Airways Corporation (BOAC, 1965)
London on Wheels: Public Transport in London in the 19th Century, by the British Railways Board (BRB, 1968)
Roads and their Traffic, 1750–1850, by J. Copeland (David and Charles, 1968)
The Story of London's Underground, by J. R. Day (London Transport, 1969)
A Short History of Technology from the Earliest Times to A.D. 1900, by T. K. Derry and T. I. Williams (Oxford University Press, 1970)
Transport History, 1969, by B. F. Duckham (David and Charles, 1970)
Buses, Trolleys and Trams, by C. S. Dunbar (Hamlyn, 1967)
British Transport: An Economic Survey from the 17th Century to the 20th, by H. J. Dyos and D. H. Aldoft (Leicester University Press, 1969)
A History of Transport, edited by G. N. Georgano (Dent, 1972)
Railways: An International History, by P. Hasting (Praeger, 1972)
The Story of Passenger Transport in Britain, by J. Joyce (Allan, 1967)
The Impact of Railways on Victorian Cities, by J. R. Kellett (Routledge and Kegan Paul, 1969)
Transport and Communication in Early Medieval Europe, AD 500–1100, by A. C. Leighton (David and Charles, 1972)
The Early Motor Bus, by C. E. Lee (British Railways Board, 1964)
The Horse Bus as a Vehicle, by C. E. Lee (British Railways Board, 1968)
A History of Inland Transport and Cammunication, by E. A. Pratt (David and Charles, 1970)
A Brief History of Flying: From Myth to Space Travel, by C. H. G. Smith (HMSO, 1967)

Transport planning

Economic Appraisal of Transport Projects: A Manual with Case Studies, by H. A. Adler (Indiana University Press, 1971)

British Rail after Beeching, by G. F. Allen (Allan, 1966)

Metropolitan Transportation Planning Seminars, by the American Institute of Planners (US Government Printing Office, 1972)

Developing the Transportation Plan, by the American Municipal Association (Public Administration Service, 1964)

Urban Planning Guide. Manual of Engineering Practice No. 49 (American Society of Civil Engineers, 1969)

Environmental Management: Planning for Traffic, by J. Antoniou (McGraw-Hill, 1971)

Industrial Demand for Transport, by B. T. Bayliss and S. L. Edwards (HMSO, 1970)

Transport System Planning as a Process: The Northeast Corridor Example, by H. W. Bruck, M. L. Manheim and P. W. Shuldiner (Civil Engineering Department, MIT, 1967)

Motorways in the Urban Environment, by Llewelyn-Davies Weeks Forestier-Walker and Bor, and Ove Arup and Partners (British Road Federation, 1971)

The Conurbations, by C. D. Buchanan and Partners (British Road Federation, 1969)

Introduction to Transportation Planning, by M. J. Bruton (2nd edn, Hutchinson, 1975)

Various publications of the Centre for Environmental Studies, London University, and of the Institute for Transport Studies, Leeds University

A Systems View of Planning: Towards a Theory of the Urban Regional Planning Process, by G. Chadwick (Pergamon, 1971)

Transportation Noises: A Symposium on Acceptability Criteria, edited by J. O. Chalupnik (University of Washington Press, 1971)

Urban Land Use Planning, by F. S. Chapin (University of Illinois Press, 1970)

Theory and Practice in Transport Economics, report of a symposium organised by the European Conference of Ministers of Transport (OECD, 1970)

Urban Transportation Planning, by R. L. Creighton (University of Illinois Press, 1970)

The Economic Assessment of Road Improvement Schemes: Road Research Technical Paper No. 75, by R. F. F. Dawson (HMSO, 1968)

Metropolitan Transportation Planning, by J. W. Dickey et al. (McGraw-Hill, 1975)

Environment and Change: The Next Fifty Years, edited by W. R. Gwald (Indiana University Press, 1971)

Modal Split: Documentation of Nine Methods for Estimating Transit Usage, prepared by M. J. Fertal, E. Weiner, A. J. Balek and A. F. Sevin (US Government Printing Office, 1966)

Getting the Best Roads for Our Money—The COBA Method of Appraisal (HMSO, 1972)

Road Track Costs (HMSO, 1968)

Transport Engineering Economics, by I. G. Heggie (McGraw-Hill, 1972)

The Growth and Development of 'Out of Town' Shopping Centres in the United Kingdom, by T. J. Hillier (Management Centre, University of Bradford, 1970)

Urban Mass Transit Planning, edited by W. S. Homburger (Institute of Transportation and Traffic Engineering, University of California, 1967)

Various land-use/transportation studies carried out in Britain at Aberdeen, Barnsley, Blackburn, Brighton, Bristol, Cambridge, Cardiff, Coventry, Dundee, Exeter, Grimsby, High Wycombe, Hull, Ipswich, Leicester,

London, Luton, Norwich, North Staffordshire, Nottingham, Oxford, Plymouth, Reading, Swansea, Thurrock, Torbay

An Introduction to Engineering Economics (Institution of Civil Engineers, 1969)

Reflections on Citizen Involvement in Urban Transportation Planning, by W. A. Steger (Joint Program in Transportation, University of Toronto–York University, 1972)

Various research reports issued by the Joint Program in Transportation, University of Toronto–York University, Ontario, Canada.

Regional Shopping Centres: Their Location, Planning and Design, by C. S. Jones (Business Books, 1969)

Cost-Benefit Analysis, edited by M. G. Kendall (English Universities Press, 1971)

Analytical Transport Planning, by R. Lane, T. J. Powell, and P. P. Smith (Duckworth, 1971)

Cost Benefit Analysis, edited by R. Layard (Penguin, 1972)

Planning and Transport—The Leeds Approach (HMSO, 1969)

Various publications of the Department of Planning and Transportation, Greater London Council

Principles and Techniques of Predicting Future Demand for Urban Area Transportation, by B. V. Martin, F. W. Memott and A. J. Bone (Massachusetts Institute of Technology, Report 3)

The Role of Economic Studies in Urban Transportation Planning, by J. P. Meck (US Government Printing Office, 1969)

The Preparation of Traffic and Transport Plans: Technical Memorandum No. T7 (HMSO, 1968)

Development Plans: A Manual of Form and Content, by the Ministry of Housing and Local Government (HMSO, 1970)

Report of the Panel of Enquiry on the Greater London Development Plan (The Layfield Report) (2 vols., HMSO, 1973)

Urban Traffic Noise, by B. T. Price *et al.* (OECD, 1971)

Guide for Highway Impact Studies (US Federal Highway Administration, 1972)

Predicting Highway Noise Levels: A Planning Guide for Highway Planners (Tennessee Department of Transportation, 1972)

Highway Planning Techniques: The Balance of Cost and Benefit, by G. R. Wells (Griffin, 1971)

Prediction of Traffic in Industrial Areas, by T. E. H. Williams and J. C. R. Latchford (Printerhall Ltd, 1966)

Journey-to-Work: Modal Split, by F. R. Wilson (Maclaren, 1967)

The Economics of Highway Planning, by D. M. Winch (University of Toronto Press, 1965)

Economic Analysis for Highways, by R. Winfrey (International Textbook, 1969)

Traffic System Analysis for Engineers and Planners, by M. Wohl and B. V. Martin (McGraw-Hill, 1967)

Transportation Investment Planning: An Introduction for Engineers and Planners, by M. Wohl (Heath Lexington, 1972)

Air transport and airports

Great Airports of the World, by R. Allen (2nd edn, Allan, 1968)

London's Airports, by M. Allward (7th edn, Allan, 1970)

The Case against Private Aviation, by D. Bain (Cowles, 1969)

The Boeing 747 at the Airport (Boeing Company, 1966)

British Airports and Air Traveller, 1966, edited by A. Metcalfe (Manor Publishing House, 1966)

London Airport's Traffic Study (British Airports Authority, 1967)

Australia's Two-Airline Policy, by S. Brogden (Melbourne University Press, 1968)

British Air Transport in the Seventies (Cmnd. 4018): Report of a Committee of Inquiry, chaired by Sir R. Edwards (HMSO, 1969)

Commission on the Third London Airport, Papers and Proceedings (HMSO, 1970)

The Planning and Design of Airports, by R. Horonjeff (McGraw-Hill, 1962)

Aircraft Pavement Design, edited by M. L. Hurrell (Institution of Civil Engineers, 1971)

Airports for the Future (Institution of Civil Engineers, 1967)

World Airports: The Way Ahead, edited by T. L. Dennis (Institution of Civil Engineers, 1970)

Engineering the World's Airport Passenger Terminals (Institution of Civil Engineers, 1966)

Airport Terminals (International Air Transport Association, 1966)

Various publications of the International Air Transport Association at Montreal, PQ, Canada

Various publications of the International Civil Aviation Organisation

Proceedings of the Fifth International Forum for Air Cargo (American Society of Mechanical Engineers, 1971)

A Review of the Models of the Air Traffic Control Systems, by N. Moray and L. D. Reid (Joint Program in Transportation, University of Toronto–York University, 1972)

Airfield Pavements: General Specification No. 201 (see also Nos. 203 and 204) (Ministry of Public Buildings and Works, 1965)

The Seaplanes, by H. R. Palmer (Hersant, 1965)

Air Transportation 1975 and Beyond: A Systems Approach, by B. A. Schriever and W. W. Seifert (MIT Press, 1968)

The Geography of Air Transport, by K. R. Sealy (Hutchinson, 1966)

Flight, by H. G. Stevor and J. J. Haggerty (Time–Life International, 1966)

Air Transport Economics in the Supersonic Era, by A. H. Stratford (Macmillan, 1967)

The Passenger–Aircraft Interface at the Airport Terminal, by P. C. Reese (Transportation Center, Northwestern University, 1968)

Major Commercial Airport Location: A Methodology for the Evaluation of Potential Sites, by M. S. Bambiger and H. L. Vandersypen (Transportation Center, Northwestern University, 1969)

Air Transport Policy, by S. Wheatcroft (Joseph, 1964)

Models and Their Methodology: Applications to the Studies of Flows and Resources in Air Transport, by M. Andrett and A. Bourgoyne (Institut du Transport Aerien, Paris, 1972)

Railways

The Reshaping of British Railways, by R. Beeching, *et al.* (HMSO, 1963)

Technological Change and Labor in the Railroad Industry, by W. F. Cottrell (Heath Lexington, 1970)

Modern Methods of Railway Operation, by R. Hammond (Muller, 1968)

Urban Railways and Rapid Transit, by R. Hope and I. Yearsley (IPC Transport Press, 1972)

World's Underground Railways, by F. H. Howson (Allan, 1964)

The Rapid Transit Railways of the World, by F. H. Howson (Allen and Unwin, 1971)

An Introduction to Railway Engineering, by R. A. Inglis (Chapman and Hall, 1953)

Developments in Railway Traffic Engineering (Institution of Civil Engineers, 1967)

Railway Construction and Operation Requirements for Passenger Lines and Recommendations for Goods Lines, by the Ministry of Transport (HMSO, 1967)

Civil Engineering: Railways, by B. Morgan (Longmans, 1971)

Fifty Years of Railway Signalling, by O. S. Nock (Institution of Railway Signal Engineers, 1963)

British Railways Act, 1965 (HMSO, 1965)

British Railway Track: Design, Construction and Maintenance, edited by D. H. Coombs (Permanent Way Institution, 1971)

Railway Operating Practice, by H. Samuel (Odhams, 1962)

Sea transport, seaports and waterways

Report on Small Craft Harbors. Manual of Engineering Practice No. 50 (American Society of Civil Engineers, 1969)

Report of the Committee of Inquiry into the Major Ports of Great Britain (Cmnd. 1824) (HMSO, 1967)

Report of the Committee of Inquiry into Shipping (Cmnd. 4337), chaired by Viscount Rochdale (HMSO, 1970)

Introduction to Hovercraft and Hoverports, by I. C. Cross and C. A. O'Flaherty (Pitman, 1975)

The Geography of Sea Transport, by A. D. Couper (Hutchinson, 1972)

Tanker and Bulk Carrier Terminals, edited by A. J. Savory (Institution of Civil Engineers, 1969)

Telecommunications in Ports, edited by J. N. Simmons (Institution of Civil Engineers, 1971)

Harbours Act, 1964 (HMSO, 1964)

Design and Construction of Ports and Marine Structures, by A. De F. Quinn (2nd edn, McGraw-Hill, 1972)

The Inland Waterways of England, by L. T. C. Rolt (Allen and Unwin, 1950)

A Hundred Years of Inland Transport, 1830–1933, by C. E. R. Sherrington (Cass, 1969)

Development of Ports: Improvement of Port Operations and Connected Facilities, by the UNCTAD Secretariat (United Nations Publications, 1969)

The World's Passenger Ships, by C. F. Worker (Allan, 1967)

Transport technology

The Automotive Nightmare, by A. Aird (Hutchinson, 1972)

British Railways in Transition, by D. H. Aldcroft (Macmillan, 1968)

Switzerland: Its Railways and Cableways, Mountain Roads and Lake Steamers, by C. J. Allen (Allan, 1967)

Alternatives to the Internal Combustion Engine, by R. U. Ayres and R. P. McKenna (Johns Hopkins University Press, 1972)

The Technology of Urban Transportation, by D. S. Berry, G. W. Blomme, P. W. Shuldiner and J. H. Jones (Northwestern University Press, 1966)

The Glideway System: A High-speed Ground Transportation System in the Northeastern Corridor of the United States (MIT Press, 1966)

Project METRAN: An Integrated, Evolutionary Transportation System for Urban Areas (MIT Press, 1966)

Containerisation International Yearbook, compiled by the staff of Containerisation International (National Magazine Company, annual)

Various technical reports (Nos. 71104–71107) on *Pedestrian Conveyors*, by A. C. Browning (Royal Aircraft Establishment, Farnborough, 1972)

Tramways of the World, by J. Joyce (Allan, 1965)

Cars for Cities, by the Ministry of Transport (HMSO, 1967)

Intertown Public Transport Alternatives for Canberra (National Capital Development Commission, Canberra, 1975)

Passenger Transport: Present and Future, by C. A. O'Flaherty (Leeds University Press, 1969)

New Movement in Cities, by B. Richards (Studio Vista, 1966)

Vertical Transportation: Elevators and Escalators, by G. R. Strakosch (Wiley, 1967)

Passenger Conveyors, by J. M. Tough and C. A. O'Flaherty (Allan, 1971)

New and Novel Transportation Systems: Planning Principles, Operating Characteristics, and Costs (Institute of Transportation and Traffic Engineering, University of California, 1970)

The Hybrid Vehicle Concept for Short Range Inter-City Transport, edited by M. A. Williamson (Transportation Research Institute, Carnegie-Mellon University, 1970)

Proceedings of the Carnegi-Mellon Conference on High Speed Ground Transportation (Transportation Research Institute, Carnegie-Mellon University, 1969)

Transport Modes and Technologies for Development (United Nations, 1970)

RESULTS OF RESEARCH

Periodicals

Accident Analysis and Prevention
African Transport Review
Air Cushion Vehicles
Air Line Pilot
Air Transport World
Airport Forum
Airport World
Airports International
American Highways
American Road Builder
Astronautics and Aeronautics
Australian Road Research
Australian Transport
Automotive Engineering
Automotive News
Autostrade
Aviation Week
Behavioural Research in Highway Safety
Better Roads
Bulletin of Economic Research

Buses
Care on the Road (RSPA)
Cargo Handling and Shipbuilding Quarterly
Chartered Institute of Transport Journal
Chemistry and Industry
Civil Engineering
Civil Engineering (ASCE)
Containerisation International
Defense Transportation Journal
Dock and Harbour Authority
Ekistics
Electric Railway Society Journal
Environment and Planning
Environmental Studies
Flight International
Freight
Highway Engineer
High Speed Ground Transportation Journal
Highway and Urban Mass Transportation
Highway Focus
Highway User
Highways and Road Construction
Hovering Craft and Hydrofoil
ICAO Bulletin
Institute for Rapid Transit Digest
Interavia
International Bulletin (International Federation of Pedestrians)
International Freighting
International Railway Journal
ITA (Institut due Transport Aerian) *Bulletin*
Japanese Railway Engineering
Journal of Air Traffic Control
Journal of Regional Science
Journal of Safety Research
Journal of the American Institute of Planners
Journal of the Institute of Rail Transport
Journal of the Institution of Municipal Engineers (continued as *Chartered Municipal Engineer*)
Journal of the Israel Shipping Research Institute
Journal of the Public Road Transport Association
Journal of the Royal Town Planning Institute
Journal of the Urban Planning and Development Division. ASCE Proceedings
Journal of the Waterways, Harbors and Coastal Engineering Division. ASCE Proceedings
Journal of Transport Economics and Policy
Journal of Transport History
Maritime History
Modern Railroads
Modern Railways
Modern Tramway
National Cooperative Highway Research Program Results Digest
National Ports Council Bulletin
National Safety News
New Concepts in Urban Transportation

New Zealand Road Safety
Operational Research Quarterly
Operations Research
PTJ (Passenger Train Journal)
Parking
Passenger Transport
Ports and Harbors
Public Roads
Rail International
Rail News (British Rail)
Railroad History
Railway Gazette International
Railway Technical Research Institute. Quarterly Report (Japanese National Railways)
Railway Transportation
Railway World
Regional Studies
Road Tar
Road and Road Construction
Rural and Urban Roads
Seaports and the Shipping World
Shell Aviation News
Surveyor and Public Authority Technology
Texas Transportation Researcher
Traffic Administration
Traffic Engineering
Traffic Engineering and Control
Traffic Management
Traffic Quarterly
Traffic Safety
Traffic World
Transactions of the Institute of Marine Engineers
Transport and Communications Bulletin for Asia and the Far East
Transport Economics
Transport History
Transportation
Transportation Engineering Journal. ASCE Proceedings
Transportation Journal
Transportation Planning and Technology
Transportation Research
Transportation Research Board Special Report
Transportation Research Circular
Transportation Research News
Transportation Research Record
Transportation Science
UITP Revue (International Union of Public Transport)
Waterways Journal

Indexes, abstracts and information services

American Society of Civil Engineers Publications Abstracts (bi-monthly)
Applied Science and Technology Index (monthly)
Australian Government Publications (monthly and annual)

Australian National Bibliography (monthly)
British Technology Index (monthly)
Canadian Government Publications (monthly and annual)
Central Road Research Institute (New Delhi) *Abstracts* (irregular)
Current Literature in Traffic and Transportation (monthly)
Engineering Index (monthly and annual)
Fast Announcement Service of the National Technical Information Service (weekly)
Library Bulletin of the Department of the Environment (bi-weekly)
International Survey of Current Research and Development on Roads and Road Transport (International Road Federation, Washington; annual)
Geotechnical Abstracts (monthly)
Government Publications Issued During ... (HMSO, monthly)
Highway Research Information Service (HRIS) Abstracts (Quarterly)
Highway Safety Literature (weekly)
Monthly Catalog, US Government Publications (monthly)
Statistical Theory and Methods' Abstracts (quarterly)
Technical Road Notes of the National Association of Australian State Road Authorities (annual)
Transportation: Current Literature (weekly)
Transportation Research Abstracts (monthly)
Transportation Research Board (TRB) Publications (annual)
International Union of Public Transport (UITP) Biblio-Index (quarterly)
Urban Transportation Research and Planning: Current Literature (bi-weekly)

Government agencies

British

British Airports Authority (London)
British Rail (London)
British Waterways Board (London)
British Standards Institution (London)
Building Research Station (London)
Department of the Environment (London)
Dock and Harbour Authority (London)
Greater London Council (London)
National Ports Council (London)
Transport and Road Research Laboratory (Crowthorne, Berks.)
Various Passenger Transport Authorities

American

American Association of State Highway Officials (Washington, DC)
Association of American Railroads (Washington DC)
Civil Aeronautics Board (Washington, DC)
Department of Transportation (Washington DC)
National Bureau of Standards (Washington DC)
National Highway Traffic Safety Administration (Washington, DC)
National Safety Council (Washington, DC; Chicago)
National Transportation Safety Board (Washington, DC)
Transportation Research Board (Washington, DC)
US Bureau of Public Roads (Washington, DC)

US Bureau of Economics (Washington, DC)
US Federal Highway Administration (Washington, DC)
Various State Highways Departments

Other

An Foras Forbartha (Dublin)
Australian Road Research Board (Melbourne)
Central Road Research Institute (New Delhi)
Commonwealth Bureau of Roads (Melbourne)
Commonwealth Railways (Melbourne)
Commonwealth Scientific and Industrial Research Organisation (CSIRO, Melbourne)
Department of Shipping and Transport (Canberra)
Indian Roads Congress (New Delhi)
International Bank for Reconstruction and Development (IBRD, Washington, DC)
International Union of Public Transport (UITP) (Brussels)
Laboratoires des Ponts et Chaussees (Paris)
National Association of Australian State Road Authorities (Sydney)
National Institute of Road Research (Pretoria)
National Research Council (Canada)
New Zealand Road Research Unit (Wellington)
Organisation for Economic Cooperation and Development (OECD) (Paris)
Société Nationale des Chemins de fer Francais (Paris)
Standards Association of Australia (Brisbane)
United Nations (New York)
Various State Highways Departments (Australia)

Universities

Universities undertaking research in highway, traffic and transport engineering are legion, so many as to make it impossible to mention all. Perhaps some of the principal establishments are the following:

British: Birmingham, Cranfield Institute of Technology, Leeds, University College London, Newcastle, Salford, Southampton
American: California, Carnegie-Mellon, Massachusetts Institute of Technology, Michigan, Northwestern, Pennsylvania State, Texas, Stanford
Others: Melbourne and New South Wales (Australia); Queens, Toronto, York and Waterloo (Canada)

Other organisations

British

Association of Public Lighting Engineers (London)
British Parking Association (London)
British Road Federation (London)

British Tar Industry Association (London)
Cement and Concrete Association (London)
Institution of Civil Engineers (London)
Institution of Highway Engineers (London)
Institution of Municipal Engineers (London)
Road Haulage Association (London)
Royal Town Planning Institute (London)

American

Air Line Pilots Association (Washington, DC)
Air Transport Association of America (Washington, DC)
American Association of Asphalt Paving Technologists (Denver)
American Concrete Institute (Detroit)
American Institute of Planners (Washington, DC)
American Society of Photogrammetry (Falls Church, Virginia)
American Society of Planning Officials (Chicago)
American Trucking Association (Washington, DC)
Eno Foundation for Transportation (Saugatuck, Connecticut)
Highway Users Federation for Safety and Mobility (Washington, DC)
Illuminating Engineering Society (New York)
Institute of Traffic Engineers (Washington, DC)
Society of Automotive Engineers (New York)
Transit Association (Washington, DC)
Transportation Association of America (Washington, DC)

Other

Australian Asphalt Pavement Association (Sydney)
Institut du Transport Aerien (Paris)
Institution of Engineers, Australia (Canberra)
International Air Transport Association (Montreal)
International Civil Aviation Authority (Canada)
International Railway Congress Association (Brussels)
International Road Federation (Geneva and Washington)
New Zealand Institute of Engineers (Wellington)
Permanent International Association of Road Congresses (PIARC) (Paris)

STATISTICAL DATA

General

Transit Fact Book (American Transit Association, New York, annual)
Transportation Annual (Distribution Age, Philadelphia, annual)
Annual Abstract of Statistics (HMSO, London, annual)
International Travel Statistics (International Union of Official Travel Organisations, Geneva, annual)
Annual Bulletin of Transport Statistics for Europe (United Nations, New York, annual)

Statistical Yearbook (United Nations, New York, annual)
Monthly Bulletin of Statistics (United Nations, New York, monthly)
Statistical Abstracts of the United States (US Government Printing Office, Washington, DC, annual)
Transport Statistics in the United States (US Government Printing Office, Washington, DC, annual)

Road transport

Automobile Facts and Figures (Automobile Manufacturers Association, Detroit, annual)
American Trucking Trends (American Trucking Association, Washington, DC, annual)
Basic Road Statistics (British Road Federation, London, annual)
Highway Expenditures, Road and Motor Vehicle Statistics (International Road Federation, Geneva and New York, annual)
World Highways (International Road Federation, Geneva and New York, monthly)
World Road Statistics (International Road Federation, Geneva and New York, annual)
Highway Statistics (US Government Printing Office, Washington, DC, annual)
Highway Statistics (HMSO, annual)

Air and sea transport

World Air Transport Statistics (International Air Transport Association, Montreal, annual)
Handbook of Airline Statistics (US Government Printing Office, Washington, DC, annual)
Statistical Handbook of Civil Aviation (US Government Printing Office, Washington, DC, annual)
World Port Index (US Naval Oceanographic Office, Washington DC, annual)

31

Public health engineering

G. K. Anderson

The general professional title 'public health engineer' is now so wide-ranging in its meaning that an individual cannot possibly study or practise in depth all of those aspects to which the title refers. Public health engineering literature is also now produced at such a rate as to make it impossible for the individual to keep pace with it, let alone appreciate all of the implications of the books and papers published in the field.

It has been found necessary here to be selective in noting specific examples of sources of information and also to divide the subject into groupings into which the public health engineer will normally specialise. It is therefore hoped that by doing this anyone wishing to carry out a literature search in his specialised area will be able to begin the investigation using only a limited number of specific but valuable sources of information. Only texts published in English are quoted, but this is not intended to imply that there are not many sources available in other languages.

Public health engineering has its origins rooted mainly in the nineteenth century among those scientists and laymen who were able to see the need for cleaning up the air, water and land around them, but who at that time did not have the technology (or legislative power) available to see that their demands were carried out. One outcome of this was that the work of the public health engineer fell between that of the chemical engineer and the civil engineer and as a result has often been an art rather than a science. It is only over the past 10 years or so that he has organised himself into a recognised professional in his own right and with this has come a vast improvement in the quality of literature specifically produced for and by the public health engineer.

EXAMINATION AND STANDARDS

Since most of the work is associated in some way or the other with
the protection of the environment, it is essential that the quality of
waters, wastewaters, air, etc., be measured by recognised, reproducible, standard techniques. A number of textbooks are available
which enable an analyst to carry out standard examination procedures which will be identical with those procedures carried out by
other analysts in a variety of widespread laboratories. In addition,
these so-called 'standard methods' indicate to others the exact procedure used for data comparison purposes. In water and wastewater analysis the most widely used book is that by the American
Public Health Association, *Standard Methods for the Examination
of Water and Wastewater* (13th edn, APHA, 1971). For similar
analysis in saline waters *A Practical Handbook of Sea Water
Analysis,* by J. D. H. Strickland and T. R. Parsons (Bulletin 167,
Fisheries Research Board of Canada, 1968) has proved invaluable.

For monitoring air quality, the Institute for Air Pollution Training
has produced *Introduction to Air Pollution Control* (US Department Health, Education and Welfare, Manual No. 422, 1969), which
covers all of the recommended techniques for air quality measurement.

Further analytical techniques may be found in the following
textbooks:

Analysis of Raw, Potable and Waste Waters, by the Department
 of the Environment (HMSO, 1972)
Chemistry for Sanitary Engineers, by C. N. Sawyer and P. L.
 McCarty (2nd edn, McGraw-Hill, 1967)
Pollution Microbiology, by M. S. Finstein (Dekker, 1972)
The Bacteriological Examination of Water Supplies, by the Ministry
 of Housing and Local Government (HMSO, 1969)

With respect to the setting, designing for and implementation of
standards, it may be necessary to study Acts of Parliament, Bills of
Congress, etc., in order to determine those standards applicable, but
a number of references may be used as guidelines. For example, it
is strongly recommended that the *5th Royal Commission Report on
Sewage Disposal* (HMSO, 1908) be studied, since this forms the
backbone to most subsequent effluent quality criteria, and the *1st*
and *3rd Reports of the Royal Commission on Environmental
Pollution* (HMSO, 1971 and 1972) also indicate allowable levels of
receiving water quality. Possibly the most comprehensive work in
the field of water and wastewater quality is a text edited by J. E.
McKee and H. W. Wolf and produced by the State Water Resources

Control Board of California, *Water Quality Criteria* (Publication 3-A, Resources Agency of California, 1971), in which over 3800 references are cited on the subject.

Other recommended texts relating to standards are:

European Standards for Drinking Water (World Health Organisation, 1970)

International Standards for Drinking Water (2nd edn, World Health Organisation, 1963)

GENERAL TEXTBOOKS

As stated before, the subject of public health engineering is so wide that few modern textbooks attempt to cover the whole field, and in this section are included those books which cover a number of areas. Each of the books referred to contains a comprehensive reference list and may therefore be used as a starting point in a review of a particular subject. Two books attempted to cover the whole accepted range of subjects, the first being that of E. B. Phelps, entitled *Public Health Engineering* (Wiley, 1948), which was followed by *Public Health Engineering*, by P. C. G. Isaac (Spon, 1953).

More recently, books have tended to cover only a specific area, and in that of water and wastewater *Water and Wastewater Engineering*, by G. M. Fair, J. C. Geyer and D. A. Okun (2 vols., Wiley, 1968), appears to be the most comprehensive, with the details of hydraulics being most satisfactorily covered in *Handbook of Applied Hydraulics*, edited by C. V. Davis and K. E. Sorenson (3rd edn, McGraw-Hill, 1969).

Additional textbooks of a general nature include:

Engineering in Public Health, by H. E. Babbitt (McGraw-Hill, 1952)

Municipal and Rural Sanitation, by V. M. Ehlers and E. W. Steel (McGraw-Hill, 1965)

Water Pollution Microbiology, by R. Mitchell (Wiley/Interscience, 1972)

Water Science and Technology, by T. H. Y. Tebbutt (John Murray, 1973)

World Health, by F. Brockington (Pelican, 1960)

GENERAL PERIODICALS

There are many journals in which one may find papers and research reports covering all aspects of public health engineering. The *Journal of the Institution of Public Health Engineers* seeks to be as broad as possible. However, the list of periodicals is vast and it would be of more value to itemise these rather than discuss the content of each individual journal. The title of the journal may indicate the special area of interest, but notwithstanding this, one is able to find an extremely varied subject matter in each.

American Journal of Public Health
Biochemical Journal
Chemical Engineering
Chemistry and Industry
Environmental Health
Environmental Science and Technology
Industrial and Engineering Chemistry
Journal of Applied Bacteriology
Journal of Applied Chemistry and Biotechnology
Journal of Bacteriology
Journal of General Microbiology
Journal of the American Chemical Society
Proceedings of the American Society of Civil Engineers. Journal of the Sanitary Engineering Division
Proceedings of the Institution of Civil Engineers
Process Biochemistry
Public Works Journal
Royal Society of Health Journal
Surveyor
Transactions of the Institution of Chemical Engineers
Water Research

Since extensive searches of the literature are often time-consuming, frequently without reward, the service of abstracting may be of considerable value and in this section should be included *Selected Water Resources Abstracts, Public Health Abstracts* and, of even more general nature, *Engineering Index.*

WATER POLLUTION

Textbooks

Although it is often difficult to decide exactly which category a particular textbook comes into, there are certain which may be considered to deal basically with water pollution and its many facets. Possibly the most widely used of the many books available are the three volumes by L. Klein, *River Pollution Vol. 1 Chemical Analysis* (Butterworths, 1959), *River Pollution Vol. 2 Causes and Effects* (Butterworths, 1962) and *River Pollution Vol. 3 Control* (Butterworths, 1966). *Applied Stream Sanitation,* by C. J. Velz (Wiley, 1970), covers the subject from a more mathematical aspect, and a number of books deal with the biological sciences relationship with water pollution, among these being:

Biochemical Ecology of Water Pollution, by P. R. Dugan (Plenum, 1972)

Biology and Water Pollution Control, by C. E. Warren (Saunders, 1971)

Microbial Aspects of Pollution, by G. Sykes and F. A. Skinner (Academic Press, 1971)

P.C.G. Isaac edited the proceedings of a symposium on *River Management* (Maclaren, 1967) which covered the control of pollution, the effects of discharges and the methods by which a river could be utilised to realise its full potential in its relationship with man and the environment.

With the ever-increasing use of modelling as a tool for water resources management has come a number of articles and books on the subject, but the most useful appears to be by the Institution of Water Engineers, who published the proceedings of a *Symposium on Advanced Techniques in River Basin Management: the Trent Model Research Programme* (IWE, 1972). It may be useful to have access to a text covering the legal situation in water pollution, and this may be found in *The Law of Rivers and Watercourses*, by A. S. Wisdom (2nd edn, Shaw, 1971).

Journals

No satisfactory journal exists which deals exclusively with water pollution as such, but there are many which touch on the subject either in detail or briefly. These will be found in other sections of this chapter.

WASTE TREATMENT

Textbooks

Possibly nowhere in public health engineering is a particular subject covered by so many books as is waste treatment, although in the opinion of the author no completely satisfactory book has been published. The most widely used book for design purposes is no doubt that by L. B. Escritt entitled *Sewerage and Sewage Disposal. Calculations, Design and Specifications* (3rd edn, CR Books, 1965). This is a most exhaustive text attempting to provide the designing engineer and student with a comprehensive outline of the subject matter in an empirical manner. L. G. Rich in his *Unit Processes of Sanitary Engineering* (Wiley, 1963) covers the material from a more academic viewpoint, while a detailed study is given of the biological treatment units by W. W. Eckenfelder Jr. and D. J. O'Connor in *Biological Waste Treatment* (Pergamon, 1961). Many public health engineers have little or no training in either chemistry or microbiology, but this may be rectified by studying *Microbiology for Sanitary Engineers,* by R. E. McKinney (McGraw-Hill, 1962), and *Chemistry for Sanitary Engineers,* by C. N. Sawyer and P. L. McCarty (2nd edn, McGraw-Hill, 1967).

The advent of SI units has led to a number of design books in either SI or metric units of which *Sewers and Sewage Works — Metric Calculations and Formulae,* by L. B. Escritt (Allen and Unwin, 1971), is a useful text.

A particular field of waste treatment is that devoted to industrial wastes, where many books are available. *The Treatment of Industrial Waste,* by E. B. Besselievre (2nd edn, McGraw-Hill, 1969), relates this very broad subject in a simple but adequate way. Although little waste treatment is carried out where marine disposal is practised, this section would be incomplete without reference to the most quoted and comprehensive book on the subject, *Waste Disposal in the Marine Environment,* edited by E. A. Pearson (Pergamon, 1959).

Other major textbooks in waste treatment include:

Design in Metric—Sewerage, by R. E. Bartlett (Elsevier, 1970)
Sewage Treatment, by R. L. Bolton and L. Klein (2nd edn, Butterworths, 1971)
Sewage Treatment, by K. Imhoff and G. M. Fair (2nd edn, Wiley, 1956)
Sewage Treatment Plant Design (ASCE, 1959)
Sewerage and Sewage Treatment, by H. E. Babbitt and E. R. Baumann (8th edn, Wiley, 1958)
The Design of Sewers and Sewage Treatment Works, by J. B. White (Arnold, 1970)
Water Quality Engineering for Practicing Engineers, by W. W. Eckenfelder (Barnes and Noble, 1970)

Journals and abstracts

The *Journal of the Water Pollution Control Federation* is devoted exclusively to waste treatment and effluent disposal and as such offers a wide range of papers on the subject at a very high level. Its counterpart in the United Kingdom, *Water Pollution Control,* tends to be concerned more with sewage treatment practice and as such makes a valuable contribution to the field. Recent research in waste treatment is documented in the *Proceedings of the International Conference on Advances in Water Pollution Research* and a similar conference specialising in industrial wastes results in a most comprehensive annual, *Proceedings of the Industrial Water and Waste Conference* at Purdue University. *WRC Information* (formerly *Water Pollution Abstracts*) covers all aspects of water pollution, but considerable space is devoted to waste treatment.

The following list of journals includes those in which a considerable percentage of papers are devoted to wastewater, its effects, disposal and treatment:

Abstracts on Hygiene
Effluent & Water Treatment Journal
Environmental Science and Technology
Industrial Water Wastes
Progress in Industrial Microbiology
Public Health Bulletin
Water and Sewage Works
Water and Waste Treatment
Water and Wastes Engineering

WATER TREATMENT

Textbooks

With the world demand for water expected to double within the next 20 years, the lack of high-quality available water is rapidly becoming one of the major problems of both the developed and underdeveloped world. The water engineer is faced with providing water of sufficient quantity which is wholesome and safe from health hazards. Many disciplines are concerned in this area, including the civil engineer, chemical engineer, biologist, chemist, etc., and it would not be possible to list here all of those texts and journals in the various special fields. The explosion in terminology has created many difficulties in communication, and an attempt to overcome this has been made by A. Nelson and K. D. Nelson in their *Dictionary of Water and Water Engineering* (Newnes–Butterworths, 1973). The Institution of Water Engineers have also tried to bring together all of the disciplines in the three volume *Manual of British Water Engineering Practice* (4th edn, Heffer, 1969), which is regularly updated. For a specialist chemistry book produced for engineers, *Chemistry for Sanitary Engineers,* by C. N. Sawyer and P. L. McCarty (2nd edn, McGraw-Hill, 1967), has proved to be invaluable. There are many texts on hydrology, but the most suitable for water engineers are possibly *Introduction to Hydrometeorology*, by J. P. Bruce and R. H. Clarke (Pergamon, 1966), and *Groundwater Hydrology*, by D. K. Todd (Wiley, 1959). A book which covers a very wide range from irrigation and hydrology to treatment of water is that by R. K. Linsley and J. B. Franzini entitled *Water Resources Engineering* (McGraw-Hill, 1964), and in the high-quality water field *Industrial Water Treatment Practice*, by P. Hamer, J. Jackson and E. F. Thurston (Butterworths, 1961), gives an excellent background of sophisticated treatment techniques.

The most widely used book in the water treatment design field is possibly *Water Supply*, by A. C. Twort *et al.* (2nd edn, Arnold, 1974), which gives an adequate description of the basic unit processes. Other water treatment books found to be of value are listed below:

Engineering Water Quality Management, by P. H. McGauhey (McGraw-Hill, 1968)
Water Supply and Sewerage, by E. W. Steel (4th edn, McGraw-Hill, 1960)
Water Supply Engineering, by H. E. Babbitt, J. J. Doland and J. L. Cleasby (6th edn, McGraw-Hill, 1962)
Water Treatment and Examination, edited by W. S. Holden (Churchill, 1970)
Water Treatment Plant Design, by the American Society of Civil Engineers *et al.* (American Water Works Association, 1969)

Journals

Many of the journals cited in other sections contain papers on water treatment and allied subjects, notably the *Proceedings of the Institution of Civil Engineers* and the *Proceedings of the American Society of Civil Engineers,* but more specialised coverage of the subject is given in *Journal of the Institution of Water Engineers* and *Journal of the American Water Works Association.* Other journals of immediate value to water engineers are *Water Treatment and Examination* and *Water and Wastes Engineering.*

AIR POLLUTION

Textbooks

The subject of air pollution appears to be handled more adequately in American textbooks than in those of UK origin. The two most widely used books on the subject are *Air Pollution*, edited by A. C.

Stern (3 vols., 2nd edn, Academic Press, 1968), and *Air Pollution Handbook,* by P. C. Magill, F. R. Molden and C. Arkley (McGraw-Hill, 1956). These both cover the causes, effects and controls of air pollution and are thus complete in themselves. F. N. Frenkiel and P. A. Sheppard in their book *Atmospheric Diffusion and Air Pollution* (Academic Press, 1959) deal with the subject from a more meteorological point of view, while A. Gilpin in *Control of Air Pollution* (Butterworths, 1963) is concerned mainly with methods of avoiding and controlling air pollution.

Journals

Air pollution is a fashionable subject and is thus researched into and discussed in a very wide range of professional journals. For example, many medical journals contain articles on the effects of air pollution, while trade journals and engineering journals often discuss the control aspects. The most comprehensive journal dealing exclusively with the subject is the *Journal of the Air Pollution Control Association* with a bias towards the control side. *The Proceedings of the National Incinerator Conference* are largely devoted to the relationship between pollution and incineration, whereas the *American Industrial Hygiene Association Journal* is concerned more with health aspects and deals with both the enclosed industrial environment and the general atmospheric environment. An abstracting service is provided by the *Air Pollution Control Association Abstracts* and will prove invaluable to those carrying out research in the subject. Some other journals are listed below which contain, or abstract articles on air pollution:

Atmospheric Environment (formerly *International Journal of Air and Water Pollution*)
Industrial and Engineering Chemistry
Journal of Applied Meteorology
Journal of Industrial Chemistry
Journal of the American Chemical Society
Meteorological and Geoastrophysical Abstracts

SOLID WASTES

Textbooks

Probably more than in any other field of public health engineering, the collection, treatment and disposal of solid wastes or refuse has been treated as an art-form rather than a science. For this reason a wide range of engineering disciplines has been associated with the problem, with few personnel receiving any specialised training in the subject. A number of textbooks and manuals have been produced, of which that by F. Flintoff and R. Millard entitled *Public Cleansing* (Maclaren, 1968) is an excellent text covering the field in general. Specialised topics have been well provided for in certain areas of solid waste disposal. R. E. Bevan's *Controlled Tipping of Refuse* (Institute of Public Cleansing, 1967) provides a full account of controlled tips (sanitary landfill). Although incineration of refuse receives much criticism from an environmental point of view, it is still widely practised and there is a tendency for more municipalities to use the system. A widely used text is that by R. C. Corey entitled, *Principles and Practices of Incineration* (Wiley, 1969), in which most of the major incineration systems are outlined. A recent major report by the Department of the Environment entitled *Refuse Disposal, Report of the Working Party on Refuse Disposal* (HMSO, 1971) gives a comprehensive view of the current situation in the UK. Other texts which serve as a useful adjunct to solid waste disposal study and design are:

Municipal Cleansing Practice, by F. Flintoff (Contractors Record, 1950)
Municipal Refuse Disposal, by the American Public Works Association (3rd edn, Public Administration Service, 1970)
Sanitary Landfill (ASCE, 1959)

Journals

No major professional journal is devoted solely to solid wastes in general, but one may find many excellent articles and papers in such journals as *Compost Science,* which concentrates on the practice of composting. Solid waste collection, treatment and disposal are all covered in *Public Works.* In the UK it may be necessary to study such journals as the *Proceedings of the Institution of Civil Engineers, Surveyor* or *Process Biochemistry* for specialised papers on current research and design criteria. Other useful sources of information are the journals and proceedings of a number of specific industries, and in particular the *Journal of Technical Associations of the Pulp and Paper Industry* (known as *TAPPI*). The following is a list of other major journals which contain solid waste material:

American City
Chemical and Engineering News

Industrial and Engineering Chemistry
Journal of Industrial Wastes
Mechanical Engineering
Paper Industry Journal

OTHER SOURCES OF INFORMATION

It would be unwise to assume that only textbooks and technical journals contain information relating to public health engineering. There are two other prime sources of information in the United Kingdom which must be considered: firstly, those universities which have specialist departments in public health engineering and, secondly, the research establishments responsible to the government (DOE).

Universities

The following is a list of universities which have major departments carrying out research and organising symposia, and consequently publishing reports, bulletins and symposia proceedings:

Birmingham University
Imperial College London
Southampton University
University College London
University of Newcastle-upon-Tyne
University of Strathclyde

Research establishments

A vast amount of literature is available in the form of annual reports, bulletins and research reports from the following establishments:

Hydraulics Research Station
Warren Springs Laboratory
Water Research Centre-Medmenham Research Station
Water Research Centre-Stevenage Research Station

32

Structural engineering

R. M. Birse

The scope of structural engineering is taken to include analysis, design and supervision of construction of most buildings or structures where more than nominal loading has to be carried over moderate or longer spans; industrial structures of all kinds from television transmission masts to piers and jetties; and all types of bridges, multi-storey parking garages and so on.

What is generally known as the construction industry is not, perhaps, so well defined as structural engineering, but it certainly includes the building industry as well as a wider range of civil engineering. Nevertheless, any information which is intended for the construction industry is likely to have considerable relevance to structural engineering.

Some other topics closely related to structural engineering are also considered in this chapter in the wider context of civil engineering, the over-all coverage consisting of four sections: structural engineering; fire engineering; construction materials; and construction management and plant.

It may be useful to mention here that, as in other branches of engineering, several organisations are engaged in studies of the information needs of the construction industry, one of the foremost being the Construction Industry Research and Information Association (CIRIA), whose publications include:

Computerised Information for the Construction Industry
(Proceedings of Conference, June 1969)
It Pays to be Informed
(Proceedings of Conference, October 1969)

Abstracts Survey and Recommendations
(Report of the Working Party, 1970)
Processing and Storing Information for Building Design
(Report R31, May 1971)
International Sources of Information
(CIRIA Information Liaison Group (CILG), 1972)
Facsimile Transmission of Information and Data for the Construction Industry
(Report R35, 1971)

The Royal Institute of British Architects (RIBA) published in 1968 *The Organisation of Information in the Construction Industry,* which contains papers on the use of CI/SfB for general and project information.

Both the Building Research Establishment (BRE) and the Transport and Road Research Laboratory (TRRL) are much concerned with information for broad sections of the construction industry as well as for their own use. Reports on this subject include:

Information Retrieval at the Building Research Station, by D. P. Delany and H. H. Neville (BRS Current Paper 28/69, 1969)
RRL Computer-aided Technical Information Service, by P. E. Mongar (RRL Report LR 177, 1968)

These reports provide very useful summaries of two contrasting approaches to the storage and retrieval of information, and they would repay study by anyone who might make use of the information services available to enquirers at BRE and TRRL.

Three publications which attempt to perform the very valuable service of pointing an enquirer in the right direction, so that he can see where the answer may be found, should be mentioned here.

The CIRIA *Guide to Sources of Information* (2nd edn, 1974) lists potential sources of information in the UK, ranging from government departments and research laboratories through professional institutions, building centres and libraries, to trade associations and industrial research laboratories. Some indication is given of the extent to which assistance is normally available from each source, and any series of publications are listed briefly.

The *Redland Guide to the Construction Industry* (formerly *House's Guide to the Building Industry*), edited by J. H. Cheetham *et al.* (Redland Ltd., annual), has established itself as an invaluable 'annual guide to the recommendations, regulations and statutory and advisory bodies of the construction industry'. The first part of the *Guide* is a 'Directory of the Construction Industry' somewhat similar in scope to the CIRIA *Guide*. Then follow sections dealing with:

Major Reports on Construction and Planning
Legislation on Construction and Planning
Education and Training
Communication in Construction (including select lists of reference and periodical publications)
Current Topics—including accidents, computers (with an excellent select bibliography), metrication, research, the environment, and winter construction
Guide to the Building Regulations

Altogether, the *Redland Guide* is without doubt the most comprehensive reference 'source' publication covering the construction industry as a whole. The *Construction Industry Handbook,* edited by R. A. Burgess *et al.* (2nd edn, MTP Construction, 1973), which is described more fully in later sections of this chapter, contains a directory of 'Information Sources in the Construction Industry', which includes many British organisations together with a number of European and international bodies such as CIB (International Council for Building, Research, Studies and Documentation), CEB (European Committee for Concrete), FID (International Federation for Documentation), FIP (International Federation of Pre-Stressing), ISO (International Organisation for Standardisation), and so on.

STRUCTURAL ENGINEERING

Structural engineering brings many of its practitioners into close contact with other professions such as architecture and building, but publications in these fields have been included only where they are likely to be of direct interest. References to soils and foundations will be found in Chapter 33 on *Soil Engineering*. Stress Analysis is covered in Chapter 28.

Abstracting and indexing services, bibliographies

Finding the answer to a particular question and keeping up to date in one's particular field of research or practice are two of the most frequent objectives of all users of technical information. In a search for recent papers on a particular topic, much time can be saved by making use of abstracting and indexing services, some of which cover several hundred source journals, entries being indexed under some system of subject classification. Information sources of this kind covering the whole field of engineering are described in Chapter 11.

An indexing periodical confining itself to civil engineering is *Articles in Civil Engineering (ACE)*, published by the University of Bradford Library, which is essentially a current awareness service since there are no cumulations or indexes. Articles from more than 100 journals (about 10 in foreign languages) and a selection of technical reports are classified under 44 subject headings, which include structural analysis, structural members, structures, buildings, engineering materials, etc. *ACE* aims to include articles which would be of interest to research workers and designers as well as those engaged in civil engineering construction.

The *Building Science Abstracts (BSA)*, published monthly through HMSO by the Building Research Establishment (BRE) of the Department of the Environment (DOE), are naturally research-orientated, though not narrowly so. Over 300 journals are abstracted, almost one-third of them in a language other than English (abstracts are printed in English), and entries are classified into 20 subject headings, about half of which are directly related to structural engineering, materials and construction. In addition to articles from journals, some books, reports and a selection of manufacturers' literature are also included.

The abstracts are printed as index cards of A6 size, four to a page single-sided. Each abstract bears a unique identification number, but no classification other than the DOE subject category, and no list of keywords. It is envisaged that the cards might be collected (as pages) under the subject divisions, or filed (as cards) under authors' names, in any standard subject classification system such as UDC or SfB, or under keywords selected by the user. A list of the principal journals together with a comprehensive subject and name index is published annually, the latter arranged alphabetically by computer in the form of a cyclic title and author index.

Another publication of the Department of the Environment, through its library service, is *Current Information in the Construction Industry* (fortnightly), which lists in Part I forthcoming courses, conferences and exhibitions, and in Part II selected additions to the DOE Library including books, pamphlets and periodical articles. The abstracts in Part II are classified under UDC subject headings, and selected keywords (usually not more than three or four) are printed for each abstract although there is no keyword subject index. Every 6 months the information in Part II is consolidated and published as *Construction References*, again classified under UDC subject headings, with alphabetical author and subject indexes.

There is an abridged alphabetical key to the UDC classification in each issue of *Construction References,* which will be of assistance to anyone not at home with numerical systems. The more comprehensive *Information on Building* includes both numerical and alphabetical schedules based on UDC of terms relevant to the construction industry in the broadest sense.

These publications of the DOE are intended to form a current awareness and reference system for the Department's design staff, including architects, civil engineers, electrical engineers, heating and ventilation engineers, management staff, mechanical engineers and structural engineers.

Most abstracting and indexing journals aim to publish their entries as soon as possible after the appearance of the original article, and all four of the above journals regularly achieve a time-lag of only 1–4 months for British publications, up to 6 months or more for foreign publications.

The newly published *ICE Abstracts* is a joint venture with the American Society of Civil Engineers, whose publications are fully abstracted along with those of the American Concrete Institute, the Prestressed Concrete Institute (USA) and the Water Pollution Control Federation (USA). In addition, abstracts (in English throughout) are taken from almost 100 journals published in 20 European countries, about three-quarters of the original articles being in languages other than English, though more than two-thirds of these are in French, German, Italian or Spanish. The *ICE Abstracts* are published 10 times a year in A5 format, entries being indexed in each issue under keywords, but without any numerical classification such as UDC being used.

An abstracting system that is different in many ways is the *Geodex Structural Information Service,* produced by Geodex International Inc., of Califorina. It includes abstracts of articles from periodicals, papers from conferences and symposia, and doctoral theses from the USA and Canada. Only about 40 English-language journals are regularly scanned, but each abstract is indexed in the special Geodex optical coincidence card index under at least five and sometimes as many as twelve or more keywords, selected from a list developed by Geodex.

This system, though slightly cumbersome in operation, has several distinct advantages over, say, the cyclic title index of the *Building Science Abstracts:*

(1) Geodex can use any reasonable number of appropriate keywords, whereas *BSA* is confined both in number and scope to words actually occurring in the title of the article.

(2) The optical coincidence system enables searches to be conducted on a broad or narrow front as required, since the index cards (each relating to only one keyword) may be used in any number or combination.

On the other hand, of course, it is obvious that compilation of the Geodex abstracts and selection of the keywords for each one is a skilled and time-consuming task, whereas the cyclic title index of the *BSA* is computer-compiled. This contrast may be partly responsible for the much longer time-lag with Geodex, usually between six and twelve months by the time the quarterly up-dating packages are received in the UK.

Another abstracting journal worthy of mention is *Canadian Building Abstracts,* published monthly by the National Research Council of Canada, Division of Building Research. Entries are printed separately in English and French, classified under UDC numbers, and special emphasis is naturally placed on such subjects as frozen ground engineering, ice and snow, and timber construction.

The CIRIA Report *Abstracts Survey and Recommendations,* already referred to is a very valuable guide to these and other abstracting services—it lists no fewer than 34 in the UK and 23 in other countries, some of which are published in more than one language. For example, *Bygglitteratur,* the Norwegian Building Abstract Service, appears in English as well as the Scandinavian languages.

Specialised subject bibliographies can be of particular value to both research workers and practising engineers, and there are several useful series readily available in the UK.

The Department of the Environment produces bibliographies on a wide range of subjects, and its abstracting current awareness journal, *Current Information in the Construction Industry,* not only lists DOE bibliographies as they appear, but also includes those from the Building Research Establishment, the Cement and Concrete Association and various other sources in the UK and overseas.

A useful list of over 80 bibliographies on subjects relevant to the construction industry appears under the heading 'Bibliography' in the CIRIA *Index of Technical Publications* (2nd edn, 1970), and, of course, lists may also be obtained of the bibliographies available from the organisations mentioned here.

One bibliography that merits special mention, in view of the current attention focused on the subject, was published in 1973 jointly by the American Society of Civil Engineers (ASCE) and the International Association for Bridge and Structural Engineering

(IABSE) on *Tall Buildings*. It includes brief summaries of the articles listed.

Bibliographies on various aspects of structural engineering are available from the Institution of Structural Engineers, and a particularly interesting service is being developed by the Transport and Road Research Laboratory with the aid of its computer-based information storage and retrieval system. Since each reference stored is indexed under a number of selected keywords, a bibliography in the form of a computer print-out can be produced for any of the keywords singly or in combination, and the search can be confined to the previous 12 months only, or extended to cover 5 years or more as required.

Lastly in this section should be mentioned the 'review' type of publication such as *Progress in Construction Science and Technology*, edited by R. A. Burgess *et al*. (Medical and Technical, No. 1, 1971, No. 2, 1973), to which a selection of 'state-of-the-art' reviews is contributed by a leading authority on each topic. A feature of such articles is often a comprehensive and carefully selected bibliography, which can form an excellent starting point for a literature search in that area.

The *Construction Industry Handbook,* described in detail later, also features a number of review articles of this type.

Journals

Pride of place goes naturally to *The Structural Engineer,* published monthly by the Institution of Structural Engineers, which in addition to technical papers includes professional and industrial news items (new materials, equipment, components, etc.), book reviews and lists of library accessions, and occasional special features on topics such as piling or demolition. Synopses of the papers are printed A6 size in the same issue, classified under UDC. It is worthy of mention that the Institution's library holds some 5000 books related to structural engineering, and subscribes to about 100 journals covering a wide field.

The corresponding publication in the United States is the *Journal of the Structural Division. Proceedings of the American Society of Civil Engineers.*

There are few other journals of note covering the whole field of structural engineering, and *only* structural engineering. However, many journals in engineering, civil engineering, highway engineering, applied mechanics and other more distant disciplines will from time to time contain papers on structural topics.

It should be noted that a number of journals (e.g. *Concrete,*

Magazine of Concrete Research, Construction Steelwork) regularly include papers on aspects of the properties, fabrication, etc., of a particular constructional material as well as on structural design, detailing and construction in that material. Such journals have been included either in this section or under the appropriate material in a later section (occasionally in both) depending on their emphasis.

A selection of journals relevant to Structural Engineering is listed below.

American Society of Civil Engineers (ASCE):
  Transactions (abstracts of papers in *Proceedings* and *Civil Engineering*)
  Proceedings: Journal of the Engineering Mechanics Division (and the journals of other divisions)
  Civil Engineering (ASCE)

Civil Engineering (controlled circulation) (formerly *Civil Engineering and Public Works Review*) (Morgan Grampian)

Engineering (Engineering, Chemical and Marine Press)

Engineering Institute of Canada:
  Transactions
  Engineering Journal
  Divisional Papers

Institution of Civil Engineers:
  Proceedings. Part 1—Design and Construction
  Proceedings. Part 2—Research and Theory
  New Civil Engineer

Institution of Engineers (Australia):
  Journal
  Civil Engineering Transactions

International Association for Bridge and Structural Engineering:
  Publications (Switzerland)
  International Civil Engineering (translations into English): (International Civil Engineering Publications, Jerusalem)

  Earthquake Engineering and Structural Dynamics (Wiley). The journal of the International Association for Earthquake Engineering.

International Journal of Mechanical Sciences (Pergamon)

International Journal of Solids and Structures (Pergamon)

Building Research and Practice (London and Paris). The journal of the International Council for Building Research, Studies and Documentation (CIB).

Building Science (Pergamon)

Computers and Structures (Pergamon)

British Corrosion Journal (British Joint Corrosion Group, London)

Materials and Structures (RILEM) (France)

American Concrete Institute:
Journal
Supplements
Special Publications

Concrete (Concrete Society, London)

Indian Concrete Journal (Cement Marketing Company of India, Bombay)

Magazine of Concrete Research (C and CA, London)

Precast Concrete (C and CA, London)

Prestressed Concrete Institute Journal (USA)

Portland Cement Association (USA): *Research and Development Bulletins*

Construction Steelwork—Metals and Materials (Portal Press: controlled circulation). The journal of the structural steel industry.

Acier–Stahl–Steel (Belgium—editions in English and five other languages)

Almost all of the above journals are scanned regularly for relevant articles and papers by at least two of the abstracting services mentioned at the beginning of this chapter, and none is omitted by all of them.

The implication is clear, then, re-emphasising the message of Chapter 11. Make the greatest possible use of the secondary sources of information, for they can spotlight the two, ten or twenty articles out of perhaps 100 journals that could be relevant to one particular topic of interest to yourself.

In addition to the published general lists of periodicals mentioned in Chapter 3, there are several useful specialised lists of the periodicals taken by libraries, such as those of the DOE, C and CA, and other organisations within the construction industry. Such lists are usually brought up to date regularly and if available to enquirers they provide a very good survey of the worthwhile periodicals in a particular field.

Foreign language journals, translations

An engineer wishing to (or driven to) extend his search for knowledge into languages other than English faces considerably greater problems. There must be few who would subscribe to a foreign-language journal if they had not sufficent fluency at least to get the gist of interesting articles. The structural engineer who is also a linguist has a wide field to choose from, of which the following is a very brief selection.

Belgium:	*Précontrainte—Prestressing*
Denmark:	*Ingeniøren*
France:	*Annales des Ponts et Chaussées*
	Beton Armé
	Construction
	Societé des Ingenieurs Civils de France: Bulletin
Germany:	*Bauingenieur*
	Beton und Stahlbetonbau
	Stahlbau
Italy:	*Giornale del Genio Civile*
	Prefabbricare
Netherlands:	*Cement*
	Ingenieur
Norway:	*Betongen Idag*
Poland:	*Inzynieria i Budownectwo*
Russia:	*Beton i Zhelezobeton*
Spain:	*Hormigon y Acero*
Sweden:	*Acta Polytechnica*
	Byggnads Ingenjoren
	Nordisk Betong

The majority of engineers must, however, rely on abstracts in English to assess the importance of papers in other languages, and seek translations of those they wish to study.

ACE—Articles in Civil Engineering regularly scans some ten or twelve foreign-language journals, but by far the widest coverage is provided in the *Building Science Abstracts,* extending to almost 100 journals. This is partly due to the participation of the Building Research Establishment in the abstract exchange scheme administered by CIB, the International Council for Building Research, Studies and Documentation, based in Rotterdam. Abstracts of foreign-language papers are given in English in *BSA,* together with a note of any available translation of the whole paper. It has already been noted that the *ICE Abstracts* covers some 100 foreign-language journals, including more than half of these listed above, all abstracts being printed in English.

Translations are of course particularly expensive to produce, and the work is normally undertaken only when some research organisation such as the Cement and Concrete Association considers it justified, no comparable paper having been published in English. There is, however, at least one journal which consists solely of '. . . articles translated (into English) from the world's leading publications about structural problems': *International Civil Engineering* is published monthly in Israel, and a typical issue contained translations from Belgium, Israel, USSR, Holland and Spain.

Among the organisations producing translations of interest to structural engineers are the following:

Building Research Establishment *Library Translations.*
Department of the Environment (Property Services Agency (PSA) Library Service). A list of translations with abstracts is available as *Construction Information in Translation,* covering the years 1963–1973.
Construction Industry Research and Information Association
Cement and Concrete Association

New translations are noted in the *BLLD Announcement Bulletin* of the British Library, Lending Division, and in the DOE's *Current Information in the Construction Industry (CICI).*

Reports, theses, conference proceedings

Many thousands of reports and hundreds of theses on topics related to structural engineering appear every year from all corners of the globe, so clearly they cannot be noted individually.

Some of the most useful series of reports originating in the UK are as follows.

Building Research Station:
Current Papers
BRE Digests
Princes Risborough Laboratory (formerly Forest Products Research Laboratory):
Technical Notes
Timberlab Papers
Transport and Road Research Laboratory:
Road Notes

Road Research Technical Papers
TRRL Reports
Construction Industry Research and Information Association:
Reports
Technical Notes

In addition, most of the industrial research associations issue reports on their work, so it is fortunate indeed that there are two excellent guides to this mass of information:

CIRIA *Index of Technical Publications* (2nd edn, 1970)
DOE *Annual List of Publications* (list dated 1971, published 1973)

The CIRIA *Index* lists approximately 4000 publications from 175 sources, arranged in alphabetical subject order. About 70 of the 175 sources are directly related to structural engineering, and under the main subject heading of 'Brickwork', for example, well over 100 references are listed under almost as many detailed sub-headings.

The DOE list contains all publications of the Department itself and its research establishments issued during 1971. It is the first such list to be produced by DOE and it is intended to be an annual publication in future.

In fact, of course, all the government research laboratories as well as many industrial research associations issue *Annual Reports*, which are often very informative, with summaries of current research projects and sometimes notes on new work started. In addition, *Annual Reports* will generally include lists of publications and articles contributed by members of staff to other journals during the year.

It should be noted at this point that the HMSO monthly *Government Publications Issued During . . .* is essential as a current awareness service in that area where it is particularly important to keep up to date.

Selected reports from Britain and the USA are indexed in *ACE*, and for general updating on recently issued reports there are two principal sources:

(1) The British Library, Lending Division: *BLLD Announcement Bulletin*, a monthly guide to British reports, translations and theses (subject classifications tend to be broad because of the over-all coverage of the *Bulletin*).

(2) Department of Industry. Technology Reports Centre: *R & D Abstracts*, a fortnightly journal of abstracts of science and technology reports.

General sources of information on reports, theses and conference proceedings will be found under the appropriate heading in the earlier chapters.

The organisation of conferences, congresses, and symposia is now a highly developed 'industry' (as is, it is said, the writing of papers for presentation at the same) and many national and international bodies are involved. Such events can be, of course, extremely valuable in themselves as opportunities for interchange of information for those able to attend, and the collected papers and discussions are usually published in a volume of proceedings at a later date.

Some of these of most recent interest to structural engineers are listed below.

Institution of Civil Engineers:
  Prestressed Concrete Pressure Vessels (1968)
  Safety on Construction Sites (1969)
  Hazards in Construction (1972)
  Mechanisation for Road and Bridge Construction (1972)
  Steel Box Girder Bridges (1973)

Institution of Structural Engineers:
  (with University College of South Wales and Monmouthshire)
    Developments in Bridge Design and Construction (1971)
  (with University of Sheffield) *Joints in Structures* (1972)

American Society of Civil Engineers:
  Structural Plastics—Properties and Possibilities (1969)
  Fifth Conference on Electronic Computation (1970)
  Planning and Design of Tall Buildings (1973)

International Association for Bridge and Structural Engineering:
  Seventh Congress, Rio de Janeiro (1964). (Themes include methods of calculation; structural steels, means of connection; steel bridges.)
  Eighth Congress, New York (1968). (Themes include safety; tall buildings; thin-walled structures; dynamic loads.)
  Ninth Congress, Amsterdam (1972). (Themes include interaction problems; long-span roofs; elevated highways and viaducts; tall slender structures.)
  Reports of Working Commissions—
  Concepts of Safety of Structures and Methods of Design (1969)
  Design of Concrete Structures for Creep, Shrinkage and Temperature Changes (1970)
  Mass-produced Steel Structures (1971)

International Association for Shell and Spatial Structures:
Folded Plate and Prismatic Structures (1970)
Shell Structures and Climatic Influences (1972)

Cement and Concrete Association:
The Structure of Concrete and its Behaviour Under Load (1965)
Fifth International Congress of the Precast Concrete Industry (1966)
Sea-Dredged Aggregates for Concrete (1968)
(with Fédération Internationale de la Précontrainte—FIP) *Steel for Prestressing* (1968)
(with FIP) *Sixth Congress of the FIP, Prague* (1970)
(with FIP) *Concrete Sea Structures* (1972)

Concrete Society:
The Design of Prestressed Concrete Bridge Structures (1967)
First International Congress on Lightweight Concrete, London (1968)
Design for Movement in Buildings (1969)
Drawing and Detailing by Automated Procedures (1970)
The Science of Admixtures (1970)
Advances in Concrete (1971)

American Concrete Institute:
Torsion of Structural Concrete (1968)
Concrete Bridge Design (1969)
Concrete for Nuclear Reactors (3 vols., 1970)
Impact of Computers on the Practice of Structural Engineering in Concrete (1972)

Réunion Internationale des Laboratoires d'Essais et de Recherches sur les Matériaux et les Constructions (RILEM):
Static and Dynamic Testing of Full Scale and Model Structures (1969)
Concrete and Reinforced Concrete in Hot Climates (1971)
Experimental Analysis of Instability Problems (1971)
(with ASTM and CIB) *The Performance Concept in Buildings* (1972)

A list of all RILEM symposia to date was published in *Materials and Structures* No. 32, March/April 1973.

British Constructional Steelwork Association:
Structural Steelwork (1966)
Steel Bridges (1969)

Iron and Steel Institute:
Design in High Strength Structural Steels (1969)

DOE Building Research Establishment:
Prospects for Fibre Reinforced Construction Materials (1972)
Construction Industry Research and Information Association:
The Modern Design of Wind Sensitive Structures (1970)
British Ceramic Research Association:
Second International Brick Masonry Conference (1970)
The Plastics Institute:
Plastics in Building Structures (1965)

Relevant conference proceedings are noted as published in most of the abstracting and indexing journals and in some libraries' accession lists. In addition, there is the monthly *Index of Conference Proceedings Received,* published by the BLLD.

Legislation, standards and units, metrication

On the legislative level the work of the structural engineer is governed principally by one or other of the following.

England and Wales: *The Building Regulations: Statutory Instruments, 1972 No. 317 and 1973 No. 1276*
London: *London Building Acts and Constructional By-Laws* (Greater London Council, 1972)
Scotland: *The Building Standards (Scotland) (Consolidation) Regulations 1971, and Amendment Regulations 1973*

The Scottish Regulations are supplemented by a series of *Explanatory Memoranda* (HMSO) and for England and Wales there is a *General Guidance Note to the Building Regulations* (HMSO, 1973) as well as a useful book *Guide to the Building Regulations,* by A. J. Elder (Architectural Press, 1972).
There are of course many other Acts which may apply to particular structural engineering works, ranging from the Town and Country Planning Act 1971, through the Clean Air Acts 1956 and 1968, the Factories Act 1961, the Fire Precautions Act 1971 and the Licensing Act 1964, to the Public Health Acts 1936 and 1961. The very helpful DOE *Guide to Statutory Provisions* (HMSO, 1973) summarises over 100 of the more important statutory provisions affecting the design and construction of buildings. It includes a comprehensive subject index in which all the Acts relating to fire, food, harbours or hotels, for example, are listed under the appropriate subject heading.
The DOE *Guide* referred to above does not, however, attempt to

deal with any of the differences separating the Scottish and English statutory authorities and legal systems. For a concise exposition of these differences, with valuable references in each of its 12 sections, the *AJ Legal Handbook,* edited by E. Freeth and P. Davey (Architectural Press, 1973), should be consulted.

It may on occasion be very useful to be able to refer to a summary of an Act if the document itself is not readily available, and in this respect section 10 'Legislation on Construction and Planning' in the *Redland Guide to the Construction Industry,* which summarises more than 20 Acts within 60 pages, can be invaluable. In the same way, section 14 'Guide to the Building Regulations' acts as a useful conspectus of this far from simple document as it applies to England and Wales.

In the preparation and revision of a necessarily wide range of standards and codes of practice, the construction industry is well served by the British Standards Institution (BSI), and the *BSI Yearbook* provides a complete list of all BSI publications in numerical order, with brief abstracts of all standards and codes, and an alphabetical subject index.

Sectional Lists are also available on building, codes of practice, iron and steel, etc., and it is worth noting that a useful survey of the standards relating to any particular field will be found in the relevant code of practice, where all other codes and standards referred to in it will be listed.

Codes and standards are often of necessity both complex and concise, and in many cases their interpretation and implementation in structural design benefits from (if it does not actually depend on) some supplementary explanation and instruction. In the past this was usually done by the publication of a 'guide' more or less independently of the code itself; for example,

Explanatory Handbook on the BS Code of Practice for Reinforced Concrete CP 114 (1957), by W. L. Scott, W. Glanville and F. G. Thomas: (2nd edn, C and CA, 1965)

A Guide to the BS Code of Practice for Prestressed Concrete CP 115, by F. Walley and S. C. C. Bate (C and CA, 1961)

The Structural Use of Timber (CP 112: 1965), by P. O. Reece and L. G. Booth (Spon, 1967)

At least one recent guide, however, has been developed concurrently with the code itself by members of the drafting committee, so that the two documents are fully compatible and complementary: *Handbook on the Unified Code for Structural Concrete (CP 110: 1972),* by S. C. C. Bate, W. B. Cranston *et al.* (C and CA, 1972).

This code has been published by BSI in three parts, and together with the *Handbook* the total cost is almost £26, providing a new twist to the phrase 'a wealth of information'.

Standards are of course a particularly vital area in which to keep up to date, and the monthly *BSI News* is by far the best way of doing this, although some journals do note the issue of new standards and codes from time to time. There are many other overseas bodies comparable to BSI, the American Society for Testing and Materials (ASTM) being one of the most useful in often providing standards where none yet exists in BSI. *ASTM Standardisation News* is the counterpart of the *BSI News* in giving full details of new *ASTM Standards* and *Tentatives* on publication. Many foreign standards can be purchased in this country through BSI.

BSI has also taken a leading part in the change to the metric (SI) system throughout the construction industry, a process which is now virtually complete in advance of many other British industries. Some of the documents which paved the way to this achievement are:

PD 6030: 1967 Programme for the Change to the Metric System in the Construction Industry
PD 6031: 1968 Use of the Metric System in the Construction Industry
BS 3763: 1970 The International System of Units (SI)
PD 5686: 1972 The Use of SI Units

A particularly useful booklet on metric units comes from the National Physical Laboratory: *Changing to the Metric System: Conversion Factors, Symbols and Definitions,* by P. Anderton and P. H. Bigg (4th edn, HMSO, 1972)

The Department of the Environment has published a *Metric Bibliography* (1971) and a series under the general title 'Metrication in the Construction Industry':

No. 1 *Metric in Practice* (1970)
No. 2 *Calculations in SI Units* (1970)
No. 3 *Craftsmen's Pocket Book* (1970)
No. 4 *Metric Reference Book* (1971)

Product information

Up-to-date information on materials, products and components is no less vital to the structural engineer than the latest code of practice or textbook, but it is much more difficult to deal efficiently with trade literature arriving in an engineer's office at an average rate of seven items every day, four of which will be retained for some time at least. A thousand items a year to be filed (or not filed), and months later, perhaps, searched for and located (or not located) when the need arises.

Little wonder that a fascinating and very informative report from the Department of the Environment, Directorate of Research and Information entitled *Commodity Information for the Construction Industry—a Survey of Supply and Demand* (HMSO, 1971), on which the above generalised figures are based, concludes that more than £50m is spent in the construction industry every year on the production, distribution, storage and retrieval of commodity information. The report also notes that, on average, trade and technical journals were received in engineers' offices at the rates of 55, 30, and 10 per month, respectively, for large, medium and small offices.

Most professional and technical journals include at least some information on new products in semi-editorial format, quite apart from advertising displays. Among those most likely to be received in structural engineers' offices are *The Structural Engineer, New Civil Engineer, Consulting Engineer, Civil Engineering, Building Research and Practice* and the DOE's monthly *Construction*.

The Building Centres in London and elsewhere, though used more extensively by architects than engineers, should not be overlooked as sources of product information, more especially since they now participate in the project known as *Facsimile Information Network Development* (FIND), whereby photocopies of technical or trade literature can be reproduced via the national telephone system at the centre nearest to an enquirer's office for collection or onward transmission by post.

Independent appraisals of materials and products are undertaken by the Agrément Board and by the Greater London Council (GLC), the results being published as Agrément Certificates and in the GLC's monthly *Development and Materials Bulletin*, respectively. Advice may also be sought from the Building Research Establishment through its Building Research Advisory Service.

The CIRIA *Guide to Sources of Information* (1974) lists several hundred organisations, many of which are trade associations of an advisory nature, such as the Aluminium Federation, the Brick Development Association, the British Ready Mixed Concrete Association, and many others. There is an alphabetical subject index listing the appropriate sources of information under such varied headings as bridges, bricks and bird nuisance prevention.

Finally, it should be noted that there is now a *BS4940: 1973*

Recommendations for the Presentation of Technical Information about Products and Services in the Construction Industry.

General reference sources, handbooks

Sources of general information on structural engineering will often belong primarily to civil engineering, building or the construction industry as a whole, but no attempt has been made here to classify them in that way. Instead, they have been grouped under the functional headings of guides, technical references, and directories and catalogues.

Guides

The most comprehensive publication of this kind is the *Redland Guide to the Construction Industry,* edited by J. H. Cheetham (Redland Ltd., annual), already referred to in some detail at the beginning of this chapter, which is particularly useful in being revised annually, and extended in scope in response to demand.

Other 'guides' that may be found helpful with some types of enquiry include:

Institution of Civil Engineers: *List of Members*
Institution of Structural Engineers: *Year Book and Directory*
Federation of Civil Engineering Contractors: *Handbook*
Association of Consulting Engineers: *The Consulting Engineers Who's Who and Year Book*

Technical references

Since the handy *Structural Engineer's Data Book,* by D. A. Cresswell and J. H. G. King (3rd edn, Pitman, 1966) is so out-of-date in its present edition, it is fortunate that after an interval of 14 years the excellent *Civil Engineer's Reference Book,* edited by L. S. Blake (3rd edn, Newnes–Butterworth, 1975), has been completely revised and reset in a single volume. The aim of the book is stated to be a concise presentation of the fundamentals of the theory and practice of all branches of civil engineering, and there must be few of its 42 sections that do not relate in some way to structural engineering. The bibliographies and references at the end of each section are on the whole particularly helpful, though they vary from a scant 3 items to more than 3 pages of carefully selected references.

The many handbooks of American origin are of limited value on this side of the Atlantic, even though the size of their market must give them a clear advantage with this type of expensive publication. *The Standard Handbook for Civil Engineers,* edited by F. S. Merritt (McGraw-Hill, 1968), and the earlier *Civil Engineering Handbook,* edited by L. C. Urquhart (4th edn, McGraw-Hill, 1959), made few concessions to British civil engineering practice, and must be appreciably less useful since the advent of metrication here.

Perhaps, in an age of rapid technological advance, the species 'compendium' is in danger of extinction like the mammoth and the dinosaur, too large and ponderous to survive in a changing world. *The Construction Industry Handbook,* edited by R. A. Burgess *et al.* (2nd edn, MTP Construction, 1973), does not set out to be a comprehensive work of reference, but it does provide some useful and up-to-date information on: building in the EEC; analysis of the construction industry (construction company finances and accounts); reviews and developments (six selected topics including smoke control in buildings, and clay products); the metric system; properties of building materials; environmental design data; and information sources in the construction industry.

Since it is often in a relatively unfamiliar and distant area of knowledge that a structural engineer may need some facts and figures, the wide scope of *Kempe's Engineers Year-Book* (2 vols., Morgan Grampian, annual) makes it a particularly valuable reference source. The contents of its 90 chapters range from mathematical tables, thermodynamics, and mechanical handling to gearing, motor vehicles, fuels, explosives and legal notes for engineers; together with the more closely related theory of structures, plastics, fire protection, reinforced and prestressed concrete, design of steel structures, bridges and bridgework, and so on.

Each year some of the chapters in *Kempe* are substantially rewritten and brought up to date, although inevitably the complete cycle of revision occupies several years and metrication, for example, seems to have made relatively little headway up to now.

Some other more specialised technical reference volumes are:

Reinforced Concrete Designer's Handbook, by C. E. Reynolds (8th edn, Viewpoint, 1974)
Concrete Engineering Handbook, edited by W. S. Lalonde and M. F. Janes (McGraw-Hill), 1961
Steel Designers' Manual (4th edn, CONSTRADO/Crosby Lockwood, 1972)
Structural Steel Designers' Handbook, edited by F. Merritt (McGraw-Hill, 1972)

Manual of Steel Construction (7th edn, American Institute of Steel Construction, 1970)

Timber Construction Manual, by the American Institute of Timber Construction (Wiley, 1966)

ACI Manual of Concrete Practice (American Concrete Institute, 1973)

Firms and product information

Trade directories and similar annual publications likely to be found useful by the structural engineer include:

Architects Standard Catalogues (6 vols., Standard Catalogue Company)

B & C J Directory (IPC Building & Contracts Journals)

Concrete Year Book (Cement and Concrete Association)

UK Trade Names (Kompass)

Sell's Building and Civil Engineering Trades List (Business Directories)

Specification (2 vols., Architectural Press)

Laxton's Building Price Book (Kelly's Directories)

Spon's Architects' and Builders' Price Book

Kelly's *Directories*

There are also a number of commercial information services which for an annual fee will provide and regularly update a set of files on products, materials, etc., as part of an office information system. Such services include *Barbour Index, Wears-Milne, Pindex, BLIS (Building Library and Information Services),* and the *NBA and 'BUILDING' Commodity File* compiled by the National Building Agency and published by *The Builder.*

Textbooks, monographs, etc.

There are so many textbooks and monographs on various aspects of structural engineering that any list must be very selective. The general principle has been to include some of the most useful recent titles under each heading, since any worthwhile books or papers published in earlier years will almost certainly be referred to or at least listed in a bibliography.

Book reviews should not be neglected as a means of assessing recently published literature. Among the journals in which reviews appear regularly are *The Structural Engineer, New Civil Engineer, Civil Engineering* and *Building Science.*

Structural engineering in general

Institution of Structural Engineers:
  Aims of Structural Design (1969)
  Stability of Modern Buildings (1971)
  Structural Engineering, by R. N. White, P. Gergley and R. G. Sexsmith (4 vols., Wiley, 1972)

Structural theory and analysis

Theory of Structures, by S. P. Timoshenko and D. H. Young (2nd edn, McGraw-Hill, 1965)

Energy Theorems and Structural Analysis, by J. H. Argyris and S. Kelsey (Butterworths, 1960)

Matrix Methods of Structural Analysis, by R. K. Livesley (Pergamon, 1964)

Matrix and Digital Computer Methods in Structural Analysis, by W. M. Jenkins (McGraw-Hill, 1969)

Structural Analysis, by R. C. Coates, M. G. Coutie and F. K. Kong (Nelson, 1972)

Dynamics in Engineering Structures, by V. Kolousek (translated by R. F. McLean and J. S. Fleming) (Butterworths, 1973)

Plastic Theory of Structures, by M. R. Horne (Nelson, 1971)

Computer Programs for Structural Analysis, by W. Weaver (Van Nostrand, 1967)

The Finite Element Method in Engineering Science, by O. C. Zienkiewicz (2nd edn, McGraw-Hill, 1971)

Concrete structures

Oscar Faber's Reinforced Concrete, by J. Faber and F. Mead (2nd edn, Spon, 1961)

Structural Concrete, by R. P. Johnson (McGraw-Hill, 1967)

Statistical Theory of Concrete Structures, by M. Tichy and M. Vorticek (Irish University Press, 1972)

Ultimate Load Analysis of Reinforced and Prestressed Concrete Structures, by L. L. Jones (Chatto and Windus, 1962)

Yield-line Analysis of Slabs, by L. L. Jones and R. H. Wood (Chatto and Windus, 1967)

Yield-Line Formulas for Slabs, by K. W. Johansen (C and CA, 1972)

Reinforced and Prestressed Concrete in Torsion, by H. J. Cowan (Arnold, 1965)

Concrete Bridge Design, by R. E. Rowe (Applied Science, 1962)

Steel structures

Design of Steel Structures, by B. Bresler, T. Y. Lin and J. B. Scalzi (2nd edn, Wiley, 1969)

Steel Structures Design and Behaviour, by C. G. Salmon and J. E. Johnson (Intertext, 1972)

Thin Walled Steel Structures, edited by K. C. Rockey and H. V. Hill (Crosby Lockwood, 1969)

Steel Space Structures, by Z. S. Makowski (Michael Joseph, 1965)

Timber structures

An Introduction to Timber Engineering, by H. J. Andrews (Pergamon, 1967)

Timber – an Introduction to Structural Design, by D. Beckett and P. Marsh (Surrey University Press, 1974)

Loadbearing brickwork and blockwork

Design of Loadbearing Brickwork in SI and Imperial Units, by W. Thorley (Heinemann, 1970)

Structural Masonry, by S. Sahlin (Prentice-Hall, 1971)

Structural plastics

Structural Design with Plastics, by B. S. Benjamin (Van Nostrand, 1969)

Pneumatic structures

Principles of Pneumatic Architecture, by R. Dent (Architectural Press, 1971)

Air Structures, by C. Price and F. Newby (HMSO, 1971)

FIRE ENGINEERING

The Institution of Fire Engineers was founded in 1924 to 'promote the science and practice of fire extinction, prevention and engineering'. The fact that only in 1973 was the first Chair of Fire Engineering established in a British university reflects the general complacency of other professions and the public at large to what amounts to an annual 'disaster' involving the loss of 1000 lives and almost £250m worth of property.

Fire engineering embraces many diverse disciplines in science, technology, medicine and sociology, and the writer is indebted to Professor D. J. Rasbash of the Department of Fire Safety Engineering in the University of Edinburgh for assistance in the preparation of this section.

Abstracting and indexing services, bibliographies

The National Academy of Sciences in Washington DC publishes *Fire Research Abstracts and Reviews* three times per annum. In each issue entries are classified under 14 subject headings, and there is an alphabetical subject and author index. Cumulated indexes are published annually with the *Abstracts,* and in a separate volume every 5 years.

Both *Building Science Abstracts* and the DOE's *Current Information in the Construction Industry* (also *Construction References*) include abstracts of articles on fire engineering topics of particular concern to architects and structural engineers.

The Joint Fire Research Organisation (JOFRO) set up by the DOE and the Fire Offices' Committee publishes annually its *Library Bibliography No. 5: References to Scientific Literature on Fire.* Entries are classified under seven subject headings and there are subject and name indexes.

Journals

Institution of Fire Engineers Quarterly
JOFRO (Joint Fire Research Organisation) Quarterly
British Fire Services Association Journal
Fire Prevention. The journal of the Fire Protection Association (FPA).
Fire Prevention Science and Technology (FPA, London)
Fire Surveyor. Publication of the Fire Surveyors Section of the Incorporated Association of Architects and Surveyors.

Fire (Unisaf House, Tunbridge Wells). The journal of the British Fire Services.

Fire International (Unisaf House, Tunbridge Wells)

Fire Protection Review (Benn)

Fire Journal (National Fire Protection Association (NFPA), Boston)

Fire News (NFPA, Boston)

Fire Technology (NFPA, Boston). Sponsored by the Society of Fire Protection Engineers, a section of NFPA.

Combustion and Flame (USA). The journal of the Combustion Institute.

Journal of Fire and Flammability (USA)

Foreign-language journals, translations

Of the many foreign-language journals, one may be singled out for special mention, the *VFDB Zeitschrift*, the journal of Vereinigung zur Förderung des Deutschen Brandschutzes, the German Fire Protection Association.

Translations are few and far between, although occasionally the Building Research Establishment or the Department of the Environment may translate an article of special interest. It should be noted, however, that articles appearing in *Fire International* are printed in English, French and German in parallel columns.

Reports, theses, conference proceedings

Much of the information given in the earlier section on reference sources is also relevant here. Some useful series of reports on fire engineering topics are listed below:

Joint Fire Research Organisation:
  Fire Research Technical Papers
  Fire Notes
  Fire Research Digests
  Fire Research (Annual Report of the Fire Research Station)

Fire Protection Association:
  Planning for Fire Safety in Buildings
  Fire Safety Data Sheets
  Technical Information Sheets

Home Office:
  Fire Precautions Guides (No. 1 'Fire Precautions in Town Centre Redevelopment') (HMSO, 1972)

Proceedings of recent conferences and symposia include:

Joint Fire Research Organisation:
Fire and Structural Use of Timber in Buildings (1967)
Behaviour of Structural Steel in Fire (1968)
Movement of Smoke on Escape Routes in Buildings (1969)
Fire-Resistance Requirements for Buildings—a New Approach (1971)

American Society for Testing and Materials:
Special Technical Publication STP No. 422. Proceedings of Symposium on (a) Restraint in Fire Tests and (b) Measurement and Control of Smoke in Building Fires (1967)

University of Edinburgh:
Fire Safety of Combustible Materials (1975)

CIB/BRE Fire Research Station:
The Control of Smoke Movement in Building Fires (1975)

Legislation, standards and units, metrication

The information given under this heading in the earlier section is of equal relevance to fire engineering, but it should be noted that in the United States an important series of *Fire Standards* is published by the National Fire Protection Association.

Product information

The principal sources of information on the properties of construction materials and components in relation to the effects of fire are the publications of the Fire Research Station and the Agrément Board, together with the considerable amount of data on the fire resistance of different materials and structural elements which is contained in the Schedules to the Building Regulations.

The Fire Research Station report, *Sponsored Fire Resistance Tests* (HMSO, 1960), describes the results of tests carried out on a number of proprietary floor and roof systems, and several of the series of *Fire Notes* are devoted to similar topics.

F1. Fire Resistance of Floors and Ceilings, by G. I. Bird (HMSO, 1961)

F4. Tests on Roof Constructions Subjected to External Fire (HMSO, 1970)

F9. 'Surface Spread of Flame' Tests on Building Products, by F. C. Adams and Barbara F. W. Rogowski (HMSO, 1966)

The Agrément Board's *Certificates of Assessment* of new products will include reference to their fire-resisting properties where appropriate to the nature of their intended use.

General reference sources, handbooks

Among the relatively small number of general reference sources in fire engineering the following are likely to be found useful:

Dictionary of Fire Technology (Institution of Fire Engineers, 1958)
Fire Protection Handbook, edited by G. H. Tryon (13th edn, NFPA Boston, 1969). A monumental tome of over 2000 pages. Full coverage of US fire engineering practice is given.
A Complete Guide to Fire and Buildings, edited by E. W. Marchant (Medical and Technical, 1972). Fourteen chapters contributed by specialist authors, with useful lists of references throughout.

Textbooks, monographs, etc.

Fire Safety in Buildings: Principles and Practice by G. J. Langdon-Thomas (A. and C. Black, 1972)
Fire and Timber in Modern Building Design, by L. A. Ashton (Timber Research and Development Association (TRADA), 1970)
Fire Performance of Timber; a Literature Survey (TRADA, 1972)
The Role of Structural Fire Protection, by Merrett Cyriax Associates (C and CA, 1969)
Fire and Buildings, by T. T. Lie (Applied Science, 1972)
'Principles of Structural Fire-resistance' by H. L. Malhotra. Paper No. 8 in *Progress in Construction Science and Technology*, edited by R. A. Burgess *et al.* (No. 1, Medical and Technical, 1971)
'Reinstatement of Fire Damaged Building Structures' by E. Dore and W. S. Watts. Paper No. 6 in *Progress in Construction Science and Technology*, edited by R. A. Burgess *et al.* (No. 2, Medical and Technical, 1973)
'Smoke Control in Buildings' by E. W. Marchant in *The Construction Industry Handbook*, edited by R. A. Burgess *et al.* (2nd edn, Medical and Technical, 1973)

CONSTRUCTION MATERIALS

Abstracts and bibliographies

The abstracting and indexing services described at the beginning of this chapter all include construction materials in their coverage, and there are only a few abstracting journals devoted solely to construction materials.

The *British Ceramic Abstracts* are issued monthly by the British Ceramic Research Association, with a comprehensive coverage of clays and the manufacture, properties and uses of clay products. More than 200 journals are regularly abstracted, about 30 of them being concerned with structural uses of clay products.

Concrete Abstracts are issued bi-monthly in the USA, covering the whole field of concrete technology, design and construction in just over 100 journals, all but 10 of them published in the USA and including such esoteric titles as *Dixie Contractor* and *Rocky Mountain Construction*.

The *Journal of the American Concrete Institute (ACI)* prints under the heading of 'Technical Reviews' what are in effect abstracts of articles on concrete from many sources, unclassified but with a list of keywords for each entry. The *Journal* also includes abstracts of new ACI *Special Publications*.

The Cement and Concrete Association's (C and CA) *Magazine of Concrete Research,* published quarterly, contains a list of papers and books on cement and concrete received in the C and CA library, covering articles, books, conference proceedings, reports, translations, patents and standards, etc., from a wide range of sources. In a typical issue almost 200 such items were listed under sixteen subject headings.

The Cembureau annual publication *Compilation of Research in Progress; Plain and Reinforced Concrete* lists briefly the current research projects of over 300 institutions in almost 30 countries.

Specialised bibliographies on various aspects of construction materials are published from time to time by the Department of the Environment, the Building Research Establishment and the Cement and Concrete Association, and noted in several of the current awareness periodicals mentioned above.

Journals

Journals dealing with the whole range of construction materials are rather rare. Among them, *Materials and Structures* has already been mentioned. *Building Materials* is aimed more at the architect than the structural engineer. The publications of the American Society for Testing and Materials (ASTM) are naturally slanted towards testing and evaluation, i.e. the *Journal of Testing and Evaluation* and the irregular *Special Technical Publications (STP)* series published as separate items. It is worth noting that the *Building Science Abstracts* search at least six journals on corrosion and a further two or three on materials science for articles relevant to construction materials.

Concrete and its constituents are probably best served of all materials by journals and other sources of information, several of which have been noted under concrete structures in the first part of the chapter. Cement itself is the concern of *Cement, Lime and Gravel* (London) and *Cement Technology* (C and CA, London), while the broader field of cement and concrete technology is covered by the *Journal* and other publications of the American Concrete Institute, as well as *Cement and Concrete Research* (Pergamon/ ACI), *Concrete*, the *Magazine of Concrete Research* (C and CA, London) and the *Research and Development Bulletins of the Portland Cement Association* in America.

These and other journals on cement and concrete are covered fairly thoroughly by all the abstracting and indexing journals, but again the most extensive coverage of other materials is provided by the *Building Science Abstracts*.

Useful source journals include:

Acier Stahl Steel (Belgium)
Metal Construction and British Welding Journal
  (Welding Institute, London)
Construction Steelwork—Metals and Materials (Portal Press: controlled circulation)
Timber Trades Journal and Woodworking Machinery (Benn)
Forest Products Journal (USA)
American Ceramic Society:
  Bulletin
  Journal
British Ceramic Society:
  Proceedings
  Transactions and Journal
Europlastics Monthly
Journal of Composite Materials
Composites
Glass Technology

Foreign-language journals, translations

Among the foreign-language journals from which articles are frequently abstracted, and sometimes translated into English are:

Beton und Stahlbetonbau
Nordisk Betong
Zement–Kalk–Gips
Stahlbau

Many others are included in the lists of journals abstracted by *BSA, CICI, ICE Abstracts* and *ACE*, and in the lists of journals taken by specialist libraries.

Reports, theses, conference proceedings

The reports issued by the government research laboratories and by CIRIA, as listed in the earlier section, deal from time to time with construction materials as well as structural engineering topics. Reports emanating from other bodies such as the industrial research associations will naturally be within their particular area of interest, and many of them are listed under the appropriate subject heading in the CIRIA *Index of Technical Publications* (2nd edn, 1970).

General guidance on sources of information on reports, theses and conference proceedings is given in Chapters 3 and 5, and the selected conference proceedings listed in the earlier section are wholly or partly relevant to various construction materials.

Standards and units, metrication

To complement the *Codes of Practice* already mentioned, there are many *British Standards* listed in the *BSI Yearbook,* specifying the composition, properties, dimensions, methods of test and other essential characteristics for a very wide range of construction materials, forming a vital link in the chain of communication from designer to contractor and manufacturer.

For the structural engineer the change to the metric (SI) system has been largely a paper revolution (though hardly less traumatic on that account), while the manufacturers and distributors of materials and products for the construction industry have had many special problems to face both during and after the transition period. It is worth remembering, therefore, that on the subject of metrication there are many very informative manuals, booklets and brochures available from the manufacturers and trade associations listed in the CIRIA *Guide to Sources of Information* (1974) and other directories.

General reference sources, handbooks

For general reference sources and handbooks, construction materials are singularly badly served by existing literature. Reference books and handbooks on materials are very often compiled with the mechanical or production engineer in mind, sometimes the electrical or chemical engineer and occasionally the architect, but the writer knows of only one recent publication aimed at civil and structural engineers, i.e. *The Construction Industry Handbook,* edited by R. A. Burgess *et al.* (2nd edn, MTP Construction, 1973). One section of this handbook is devoted to a compilation of data on a selection of construction materials, their properties being listed in accordance with the principles of the CIB *Master Lists for Structuring Documents Relating to Buildings, Buliding Elements, Components, Materials and Services* (CIB Report No. 18, 1972). Unfortunately, concrete, for example, is only listed under types of cement or products such as concrete bricks, while structural steels are not mentioned.

For detailed information on the properties of a particular material, therefore, the structural engineer will frequently have to refer to one of the more specialised sources of information listed in the next section, or perhaps look over the shoulder, as it were, of his colleagues in another discipline:

Metals Reference Book, edited by C. J. Smithells (5th edn, Butterworths, 1976)
Handbook of Engineering Materials, edited by D. F. Miner and J. B. Seastone (Wiley, 1955)
Encyclopaedia of Engineering Materials and Processes, edited by H. R. Clauser *et al.* (Van Nostrand Reinhold, 1964)

Textbooks, monographs, etc.

General textbooks on materials are little better. An approach to the properties of materials through atomic structure and bonding, phase diagrams and imperfections, seems to lead inevitably to an emphasis on metallic and organic materials with cements and concrete relegated to a few pages or in extreme cases only a few lines, and other important structural materials such as brickwork omitted altogether.

The one notable exception is *The New Science of Strong Materials,* by J. E. Gordon (Pelican, 1973), which makes quite fascinating reading out of such topics as cracks and dislocations, mice in gliders, and glass fibres, and at the same time succeeds in conveying much useful and practical information on the peculiarities of many materials, and is therefore highly recommended for all structural engineers.

Textbooks and other sources of information on building materials, on the other hand, written in most instances primarily for architects, are not without interest and value to the structural engineer. Among many such publications are:

Dictionary of Building Materials, by W. Kinniburgh (C. R. Books, 1966)

Materials for Building, by L. Addleson (Iliffe, 1972)
> Vol. 1 *Physical and Chemical Aspects of Matter and Strength of Materials*
> Vol. 2 *Water and its Effects: 1*
> Vol. 3 *Water and its Effects: 2*

Further volumes will deal with the effects on buildings of fire, heat and sound vibration.

Building Materials Technology, by L. A. Ragsdale and E. A. Raynham (2nd edn, Arnold, 1972). Deals with a wide range of materials in the context of their use in building construction.

Chemistry of Building Materials, by R. M. E. Diamant (Business Books, 1970). Reasonably full, though not entirely accurate, coverage of most construction materials, with a useful bibliography for each chapter.

When we turn to some particular materials, however, the scene is very different and the problem sometimes becomes one not of searching for the few but of selecting from the many.

Cement and concrete

The Chemistry of Cement and Concrete, by F. M. Lea (3rd edn, Arnold, 1970)

Properties of Concrete, by A. M. Neville (2nd edn, Pitman, 1973)

Concrete Materials and Practice, by L. J. Murdock and G. F. Blackledge (4th edn, Arnold, 1968)

Concrete Mix Design, by F. D. Lydon (Applied Science, 1972)

Concrete Corrosion, Concrete Protection, by I. Biczok (3rd edn, Akademia Kaido, Budapest, 1972)

Creep of Concrete: Plain, Reinforced and Prestressed, by A. M. Neville (North-Holland, 1970)

Lightweight Concrete, by A. Short and W. Kinniburgh (2nd edn, Applied Science, 1968)

Bricks and brickwork

Elements of Loadbearing Brickwork, by D. Lenczner (Pergamon, 1972

Timber

Timber, by H. E. Desch (5th edn, Macmillan, 1973)

Stone

Stone for Building, by H. O'Neill (Heinemann, 1966)

Plastics

Plastics in the Modern World, by E. G. Couzens and V. E. Yarsley (Pelican, 1968)

CONSTRUCTION MANAGEMENT AND PLANT

The title of this section, if it were not to be a paragraph in itself, could hardly begin to convey the variety and scope of civil engineering construction—contract law; business management and economics; specifications and quantities; estimates and tenders; contract supervision; construction management and plant; site safety . . . to name but a few.

Within the limited space available, therefore, only a selection of the most essential references can be given, together with some 'pointers' to more specialised and comprehensive sources of information.

Abstracts and bibliographies

Although there are no abstracting services specialising in construction management and plant, most of those previously described include these subjects in their general coverage of civil engineering or the construction industry.

ACE—Articles in Civil Engineering does not give a separate subject heading to construction, but relevant articles are included under appropriate headings such as bridges, soil mechanics and foundations, docks and harbours, road construction, etc. About 20 of the journals scanned by *ACE* are at least partly devoted to engineering construction and plant.

The *ICE Abstracts* provides an unusually wide coverage of British, American and foreign-language journals dealing with civil engineering construction. Entries are indexed under relevant keywords.

The *Building Science Abstracts* include sections headed construction industry, operations, and component manufacture and assembly, while the DOE's *Construction References* are indexed more accurately but somewhat less compactly under such UDC headings as:

347.4	Contract law
621.879.2	Excavating plant
624.15	Foundation engineering
658	Management
693.546.3	Guniting

It must be borne in mind, however, that none of these abstracting services sets out to deal comprehensively with construction management and plant. They will all be selective to a degree, in line with their own particular approach, whether theoretical or practical.

As might be expected, there are not many bibliographies in this field of engineering, but some of the Building Research Station's *Library Bibliographies* are relevant, e.g. *No. 183 Tower Cranes, No. 186 Winter Building* and *No. 189 Critical Path Analysis.*

To supplement the relatively few bibliographies, reference must be made once again to the CIRIA *Index of Technical Publications* (2nd edn, 1970), where a great deal of useful information may be found under such headings as access, accidents, builders' plant, building organisation, contracts, cost analysis, economics, estimating, explosives, and so on, right through to work study, working rules and workplaces.

Journals

In addition to the journals listed below, most of which can be said to specialise in aspects of construction management and plant, many others will from time to time publish articles in this field, which can most easily be tracked down through one or another of the abstracting periodicals.

Accidents (Department of Employment, HMSO). Covers accidents and safety precautions at factories, offices, shops, docks and construction sites.
Building Technology and Management. The journal of the Institute of Builders.
Construction (Department of the Environment)
Construction Methods and Equipment (McGraw-Hill)
Construction Plant and Equipment (incorporating Muck Shifter and Cranes) (Morgan Grampian: controlled circulation)

Contract Journal (IPC Building & Contract Journals)
Contractor and Plant Manager (Westbourne Publications, Surrey: controlled circulation)
Proceedings of the ASCE. Journal of the Construction Division
Tunnels and Tunnelling (Morgan Grampian)

The 'controlled circulation' journals, like the many others in that category, are distributed free of charge to persons in approved positions of responsibility, which must include most chartered engineers.

Two at least of the many journals listed in other sections of this chapter should be mentioned again here. *Civil Engineering* in its new format gives a fairly well balanced coverage to construction as well as to theory, design and research. Its monthly feature 'Insite' includes abstract-like announcements of new equipment, plant, products, services and trade literature of potential interest to civil, structural and municipal engineers and contractors. The Institution of Civil Engineers' weekly news magazine *New Civil Engineer* complements the Institution's *Proceedings* with its content of feature articles and announcements dealing with construction machines, materials and methods together with items on new construction legislation, company finances, court cases and enquiries, and other topics of current interest.

Foreign language journals, translations

Of the very large number of foreign language journals published, only three have been listed as typical of those regularly scanned by one or more of the principal UK abstracting and indexing services. Should a wider survey be necessary, the list of journal holdings of an appropriate library (BRE, C and CA, etc.) may be consulted, together with the more general sources mentioned in Chapter 4.

Belgium: *Technique des Travaux*
France: *Travaux*
Germany: *Bautechnik*

Reports, theses, conference proceedings

Apart from those mentioned elsewhere in this chapter, an important series of reports on engineering construction methods and plant is produced by the Military Engineering Experimental Establishment (MEXE) at Christchurch. While some of these reports are obviously classified with restricted, if any, outside circulation, some may be obtained on application to MEXE.

In addition to the conference proceedings previously listed, the Institution of Civil Engineers has published a number of reports dealing with construction work:

Guide to Specifying Concrete (1967)
Safety in Civil Engineering (1968)
Safety in Sewers and at Sewage Disposal Works (1969)
Safety in Wells and Boreholes (1972)

Naturally, too, the Construction Industry Research and Information Association has initiated research into many problems of direct concern to designers and contractors, the results in many cases being issued as a report:

The Pressure of Concrete on Formwork (Report No. R1, 1965)
Wind Flow in an Urban Area (Report No. R10, 1967)
Placing Concrete in Deep Lifts (Report No. R15, 1969)
Effect of Site Factors on the Load Capacities of Adjustable Steel Props (Report No. R27, 1971)
Civil Engineering Bills of Quantities (Report No. R34, 1971)

Many reports on the construction industry have been published by various government departments, and it may be useful to refer to the *Sectional List No. 61 Building,* issued and periodically revised by HMSO. Detailed summaries, presenting the gist of almost 20 major reports on construction and planning within about 50 pages, can be found in section 9 of the *Redland Guide to the Construction Industry.*

Among what might be called the 'minor league' of reports and similar series, there are at least three that deserve special mention in the context of construction.

Department of Employment: *Safety, Health and Welfare Booklets,* e.g. Nos 6A to 6F 'Safety in Construction Work', No. 25 'Noise and the Worker', No. 47 'Safety in the Stacking of Materials'
Department of the Environment: *Advisory Leaflets*
Cement and Concrete Association: *Man on the Job Leaflets*

These series are intended as guides to good practice in many different building and construction operations, primarily for '. . . the builder, foreman and craftsman'. The engineer who buys them to hand over his foreman, however, would do well to glance at the contents first.

Legislation, Standards and Units, Metrication

The only additions to the references given earlier are the following publications concerned with civil engineering contracts:

Institution of Civil Engineers:
General Conditions of Contract (5th edn, ICE, 1973)
Standard Method of Measurement of Civil Engineering Quantities (ICE, 1972)

It should be noted that the Institution also publishes special versions of the *Conditions of Contract* for overseas and international contracts.

Product information

As with the materials, products and components covered in the earlier section, information on new items of construction plant is not hard to find in the advertisement pages of most construction journals. Independent appraisals are rare, although they can be found from time to time in the editorial pages of the same journals, and in some of the reports produced by MEXE and TRRL.

For once the CIRIA *Index of Technical Publications* (2nd edn, 1970) is almost too reticent on the breadth of its coverage by using the heading 'builders' plant' to include items such as air locks, excavators, locomotives and snow ploughs as well as the more common compressors, concrete mixers, hoists, lorries, pumps, ropes and so on.

General reference sources, handbooks

Hudson's Building and Engineering Contracts, by I. N. Duncan Wallace (Sweet and Maxwell, 1970)
Handbook of Construction Management and Organisation, edited by J. B. Bonny (Van Nostrand, 1972)
Handbook of Heavy Construction, edited by J. A. Havers and F. W. Stubbs (2nd edn, McGraw-Hill, 1971)
Manual of Civil Engineering Plant and Equipment, edited by J. M. Paxton (Vol. 1, 2nd edn, Applied Science, 1971) (Vols. 2 and 3 in preparation)
Contractors' Plant Association Handbook (The Association, annual)

British Construction Equipment—Building, Quarrying, Civil Engineering (9th edn, Federation of Manufacturers of Construction Equipment and Cranes, 1973)

Textbooks, monographs

The Construction Industry of Great Britain, by J. R. Colclough (Butterworths, 1965)
Civil Engineering Procedure (Institution of Civil Engineers, 1971)
Engineering Law and the ICE Contracts, by M. W. Abrahamson (2nd edn, Applied Science, 1969)
Civil Engineering Specification, by I. H. Seeley (Macmillan, 1968)
Civil Engineering Quantities, by I. H. Seeley (SI edn, Macmillan, 1971)
Estimating for Building and Civil Engineering Works, by Spence Geddes (5th edn, Newnes–Butterworth, 1971)
An Introduction to Engineering Economics (Institution of Civil Engineers, 1969)
Building Economy: Design, Production and Organisation, by P. A. Stone (Pergamon, 1966)
Industrial Relations in Construction, by W. S. Hilton (Pergamon, 1968)
Civil Engineering: Supervision and Management, by A. C. Twort (2nd edn, Arnold, 1972)
Construction Management in Principle and Practice, edited by E. F. L. Brech (Longmans, 1971)
Construction Management, by I. Atkinson (Elsevier, 1971)
Construction Planning, Equipment and Methods, by R. L. Puerifoy (2nd edn, McGraw-Hill, 1970)
Critical Path Methods in Construction Practice, by J. M. Antill and R. W. Woodhead (2nd edn, Wiley, 1970)
Network Planning in the Construction Industry, by W. Jurecka (Maclaren, 1969)
Cost Control in the Construction Industry, by J. Gobourne (Butterworths, 1973)
Builders' Plant and Equipment, by G. Barber (2nd edn, Butterworths, 1973)
Plant Hire for Building and Construction, by H. T. Mead and G. L. Mitchell (Butterworths, 1972)

33

Soil engineering

N. E. Simons

Soil engineering is the branch of engineering science dealing with the behaviour of soil when subjected to stresses or to the action of percolating water. It forms a rational basis for foundation engineering and for earth dam design. It includes the development and application of quantitative methods in the design of foundations and earth-retaining structures and for examining the stability of natural soil slopes and earth or rockfill dams.

During the past 25 years, the available literature has increased at a quite remarkable rate, reflecting the increasing interest which is being paid to the subject by the engineering profession, and, equally, the outstanding advances in knowledge which have been made. Many eminent research workers and engineers have applied themselves to soil engineering and the names of Terzaghi, Bishop, Bjerrum, Casagrande, Peck, Skempton and Rowe, come readily to mind. Their writings form a most valuable source of information.

Because of the vast amount of literature available, it is inevitable that the reviewer has had to exercise personal judgement when presenting the selected items which follow, and informative publications have had to be omitted on grounds of space.

The various sources of information have been considered under the headings of books, codes of practice, journals, and the proceedings of conferences and symposia. Finally, an information retrieval system is briefly described.

Rock mechanics and mining engineering are considered to be outside the scope of this chapter.

BOOKS

Many excellent reference books exist and it has been difficult to prepare a selective list. The subject has been subdivided into six sections, each of which has been covered by a small number of authoritative works.

Basic principles in soil mechanics

From Theory to Practice in Soil Mechanics: Selections from the Writing's of Karl Terzaghi (Wiley, 1960). Provides an ideal means to demonstrate the prerequisites and techniques for successfully practising soil mechanics.

Fundamentals of Earthquake Engineering, by N. M. Newmark and E. Rosenblueth (Prentice-Hall, 1971). Presents a very thorough and current understanding of all the basic principles and techniques related to structures subjected to earthquake and random oscillatory stresses.

Fundamentals of Soil Mechanics, by D. W. Taylor (Wiley, 1965). Written primarily as a textbook for undergraduate and graduate students, but also of interest to practising engineers and specialists in soil engineering.

Principles of Soil Mechanics, by R. F. Scott (Addison-Wesley, 1963). Aimed at providing a textbook for both undergraduate and graduate students in soil mechanics, and is also of interest to research workers and practising engineers.

Soil Mechanics, by T. W. Lambe and R. V. Whitman (Wiley, 1969). Designed as a text for an introductory course in soil mechanics and contains many numerical examples and problems. While written principally for the student, should also be of value to the practising engineer as a reference document.

Soil Mechanics, by T. H. Wu (Allyn and Bacon, 1966). The major topics covered are water movement through soils, elastic deformation of soil masses, and failure in soil masses.

Soil Mechanics for Road Engineers, by the Department of Scientific and Industrial Research, Road Research Laboratory (HMSO, 1952). Presents the collective experience of a large body of people concerned with the use of soils and soil engineering techniques in road construction and maintenance.

Theoretical Soil Mechanics, by K. Terzaghi (Wiley, 1943). The classic fundamental textbook of theoretical soil mechanics.

Seepage

Flow of Homogeneous Fluids Through Porous Media, by M. Muskat (J. W. Edwards, 1946). Gives a comprehensive treatment of the flow of homogeneous fluids through porous media, treated in four parts:
Foundations
Steady state flow of liquids
Non-steady state flow of liquids
Flow of gases through porous media

Groundwater and Seepage, by M. E. Harr (McGraw-Hill, 1962). Provides the engineer with an organised analytical approach to the solutions of seepage problems.

Theory of Groundwater Flow, by A. Verruijt (Macmillan, 1970). Presents the fundamentals of the theory of groundwater flow, and the most effective methods for solving groundwater flow problems occurring in civil engineering practice.

Theory of Ground Water Movement, by P. Y. Polubarinova-Kochina (Princeton University Press, 1962). The main content consists of an exposition of the mathematical theories of the flow of groundwaters.

Field and laboratory testing

Laboratory Testing in Soil Engineering, by T. N. W. Akroyd (Soil Mechanics Ltd, 1957. Describes procedures which have evolved in the laboratories of Soil Mechanics Ltd over a number of years.

Soil Testing for Engineers, by T. W. Lambe (Wiley, 1951). Describes laboratory procedures for the more common soils tests and includes numerical examples.

Sub-Surface Exploration and Sampling of Soils for Civil Engineering Purposes, by M. J. Hvorslev, (US Waterways Experimental Station, 1949). Gives a comprehensive description of soil sampling equipment and a discussion of the basic principles of soil sampling.

The Measurement of Soil Properties in the Triaxial Test, by A. W. Bishop and D. J. Henkel (2nd edn, Arnold, 1962). Restricted to a treatment of the triaxial test alone, and of the ways of meeting the various problems which arise in its use in the laboratory.

Engineering geology

A Geology for Engineers, by F. G. H. Blyth and M. H. de Freitas (6th edn, Arnold, 1974). Essentially a textbook of geology, with emphasis on engineering applications.

Geology and Engineering, by R. F. Leggett (2nd edn, McGraw-Hill, 1962). Provides general discussions on the application of geological principles to civil engineering practice, illustrated by examples drawn from over 20 different countries.

Principles of Engineering Geology and Geotechnics, by D. P. Krynine and W. R. Judd (McGraw-Hill, 1957). Designed as a textbook for civil engineering and advanced geology students and as a reference work for practising civil engineers and engineering geologists.

Foundation engineering

Civil Engineering Reference Book. Vol. 2, edited by J. Comrie: 'Soil Mechanics' by H. Q. Golder and A. C. Meigh; 'Site Investigation' by R. Glossop and I. K. Nixon; 'Foundations and Earthworks' by Guthlac Wilson, revised by E. O. Measor (Butterworths, 1961). A concise guide to good practice and to recent methods and theories.

Design Manual: Soil Mechanics, Foundations, and Earth Structures, Navdocks DM-7 (Department of the Navy, Bureau of Yards and Docks, Washington 25 DC, 1962). Covers the engineering application of soil mechanics to the design of all foundations and earth structures and contains many useful design charts and tables.

Foundation Design and Construction, by M. J. Tomlinson (3rd edn, Pitman, 1975). Provides a manual of foundation design and construction methods for the practising engineer, and incorporates recent developments in the application of soil mechanics science to foundation engineering.

Foundation Engineering, edited by G. A. Leonards (McGraw-Hill, 1962). Combines the knowledge and experience of recognised authorities into a single text with sufficient continuity and scope to be useful to both students and practising engineers.

Foundation Engineering, by R. B. Peck, W. E. Hanson and T. H. Thornburn (2nd edn, Wiley, 1974). Develops and utilises knowledge of soil mechanics, and presents a vast store of practical engineering information essential to the foundation engineer.

Foundation Instrumentation, by T. H. Hanna (Trans. Tech. Publications, 1973). Presents a case for the role of field measurements in the solution of foundation engineering problems, treats the principles of measurement and gives details of available instruments.

Foundations, by A. L. Little (Arnold, 1961). Divides the subject into shallow and deep foundations, the latter group being subdivided into those which are piled and those which are not. Much of the book is concerned with case records.

Foundations, Retaining and Earth Structures, by G. P. Tschebotarioff (2nd edn, McGraw-Hill, 1973). An introduction to the theory and practice of design and construction. Considerable space has been devoted to records of failures, of field measurements and to other experimental evidence.

Soil Mechanics in Engineering Practice, by K. Terzaghi and R. B. Peck (2nd edn, Wiley, 1967). In three parts, part A dealing with the physical properties of soils, part B with the theories of soil mechanics and part C with the art of getting satisfactory results in earthwork and foundation engineering at a reasonable cost.

A Short Course in Foundation Engineering, by N. E. Simons and B. K. Menzies (IPC, 1975). Covers the methods currently available for predicting failure loads and deformations for shallow and piled foundations.

Embankment dams

Design of Small Dams (United States Department of the Interior, Bureau of Reclamation, 1965). Presents instructions, standards and procedures for use in the design of small dams.

Earth and Earth Rock Dams: Engineering Problems of Design and Construction, by J. L. Sherard, R. J. Woodward, S. F. Gizienski and W. A. Clevenger (Wiley, 1963). Presents the main elements of current practice, experience and opinion on the design and construction of earth and rockfill dams.

Embankment-Dam Engineering. Casagrande Volume, edited by R. C. Hirschfeld and S. J. Poulos (Wiley, 1972). Contains 11 papers dealing with various aspects of the design and construction of earth dams.

'Embankment Dams', by O. Elsden, H. G. Keefe and A. W. Bishop: Chapter IX in *Hydro-electric Engineering Practice. Vol. 1. Civil Engineering,* edited by J. Guthrie Brown (Blackie, 1964). Gives an excellent background to the design and construction of embankment dams.

CODES OF PRACTICE AND SPECIFICATIONS

Virtually every country has its own recommended codes of practice relating to the various aspects of soil engineering, and these naturally should be consulted when working abroad. Only the relevant British codes and specifications, however, are included in this section.

Road Note 29. A Guide to the Structural Design of Pavements for New Roads (HMSO, 1970)

Civil Engineering Code of Practice No. 2:1951 Earth Retaining Structures (Institution of Structural Engineers)
CP 2003:1959 Earthworks (British Standards Institution)
CP 2004:1972 Foundations (British Standards Institution)
BS 1377: 1975 Methods of Test for Soils for Civil Engineering Purposes (British Standards Institution)
CP 2001:1957 Site Investigations (British Standards Institution)
Specification for Road and Bridge Works (HMSO, 1969)

JOURNALS

In 1948 the Institution of Civil Engineers, London, commenced the publication of *Géotechnique,* intended to fulfil four purposes: to promote international collaboration between workers in soil mechanics and related sciences; to publish papers on specialised aspects of these subjects; to encourage the pursuit of engineering geology; and to make the results of research available to the practising civil engineer. *Géotechnique,* published quarterly, has a circulation approaching 4000 and is generally considered to be the outstanding journal in this field. Other journals of particular relevance to engineers are:

Canadian Geotechnical Journal. Published quarterly by the National Research Council of Canada.
Earthquake Engineering and Structural Dynamics. The journal of the International Association for Earthquake Engineering, published quarterly by Wiley.
Engineering Geology. Published quarterly by Elsevier, Amsterdam.
Ground Engineering. Published bi-monthly by Foundation Publications Ltd.
Quarterly Journal of Engineering Geology. Published quarterly by The Geological Society of London.
Transportation Research Abstracts. Published monthly by the Transportation Research Board, National Academy of Sciences.
Transportation Research Record. Published irregularly by the Transportation Research Board, National Academy of Sciences.
Journal of the Geotechnical Engineering Division. Proceedings of the American Society of Civil Engineers. Published monthly by ASCE.
Soils and Foundations. Published quarterly by the Japanese Society of Soil Mechanics and Foundation Engineering.

The Journal of the Boston Society of Civil Engineers, published quarterly, from time to time includes important articles on soil engineering. These papers have been made available in a special series in three volumes, entitled 'Contributions to Soil Mechanics', covering the periods 1925–1940, 1941–1953 and 1954–1962.

In addition to the above, many research establishments issue publication series describing the results of their work and some of these are listed below:

Building Research Establishment, Watford, Herts.
Danish Geotechnical Institute, Copenhagen
Norwegian Geotechnical Institute, Oslo
Swedish Geotechnical Institute, Stockholm
Transport and Road Research Laboratory, Crowthorne, Berks.

PROCEEDINGS OF CONFERENCES AND SYMPOSIA

In 1936, the first conference of the International Society of Soil Mechanics and Foundation Engineering was held at Cambridge, Massachusetts. Since then, both international and regional conferences have taken place at regular intervals and the published proceedings of these meetings form an invaluable source of information. The various conferences are listed below.

International Conferences

1st, Harvard, 1936
2nd, Rotterdam, 1948
3rd, Zurich, 1953
4th, London, 1957
5th, Paris, 1961
6th, Montreal, 1965
7th, Mexico City, 1969
8th, Moscow, 1973

Regional Conferences

African

1st, Pretoria, 1955
2nd, Lourenco Marques, 1959
3rd, Salisbury, 1963
4th, Cape Town, 1967
5th, Luanda, 1971
6th, Durban, 1975

Asian

1st, New Delhi, 1960
2nd, Tokyo, 1963
3rd, Haifa, 1967
4th, Bangkok, 1971

Australia–New Zealand

1st, Melbourne, 1952
2nd, Christchurch, 1956
3rd, Sydney, 1960
4th, Adelaide, 1963
5th, Auckland, 1967
6th, Melbourne, 1971

European

1st, Stockholm, 1954
2nd, Brussels, 1958
3rd, Wiesbaden, 1963
4th, Oslo, 1967
5th, Madrid, 1972

Pan-American

1st, Mexico City, 1960
2nd, Sao Paulo, 1963
3rd, Caracas, 1967
4th, Puerto Rico, 1974

At the present time, conferences and symposia are regularly held all over the world at frequent intervals, discussing topical and local aspects of soil engineering.

Meetings held in the UK

Pore Pressure and Suction in Soils (Butterworths, 1961)
Grouts and Drilling Muds in Engineering Practice (Butterworths, 1963)
Chalk in Earthworks and Foundations (Institution of Civil Engineers, 1966)

Large Bored Piles (Institution of Civil Engineers, 1966)
Ground Engineering (Institution of Civil Engineers, 1970)
In Situ Investigations in Soils and Rocks (British Geotechnical Society, 1970)
Behaviour of Piles (Institution of Civil Engineers, 1971)
Field Instrumentation in Geotechnical Engineering (Butterworths, 1973)
Settlement of Structures (Pentech Press, 1974)
Diaphragm Walls and Anchorages (Institution of Civil Engineers, 1974)
Off-shore Structures (Institution of Civil Engineers, 1974)

Proceedings of Speciality Conferences organised by the Soil Mechanics and Foundation Division of the American Society of Civil Engineers

1960 University of Colorado. Research Conference on Shear Strength of Cohesive Soils
1964 Northwestern University. Design of Foundations for Control of Settlement
1966 University of California. Stability and Performance of Slopes and Embankments
1968 Massachusetts Institute of Technology. Placement and Improvement of Soil to Support Structures
1970 Cornell University. Lateral Stresses in the Ground and Design of Earth Retaining Structures
1972 Purdue University. Performance of Earth and Earth Supported Structures

American Society for Testing and Materials Special Technical Publications

No. 322 *Field Testing of Soils* (1962)
No. 351 *Symposium on Soil Exploration* (1964)
No. 361 *Laboratory Shear Testing of Soils* (1964)
No. 377 *Compaction of Soils* (1965)
No. 392 *Instruments and Apparatus for Soil and Rock Mechanics* (1965)
No. 399 *Vane Shear and Cone Penetration Resistance Testing of In-Situ Soils* (1966)
No. 444 *Performance of Deep Foundations* (1969)
No. 450 *Vibration Effects of Earthquakes on Soils and Foundations* (1969)

No. 477 *Determination of the In-Situ Modulus of Deformation of Rock* (1970)
No. 479 *Special Procedures for Testing Soil and Rock for Engineering Purposes* (5th edn, 1970)
No. 483 *Sampling of Soil and Rock* (1971)

This is by no means a complete list of all ASTM publications relating to soil engineering, but it is believed to contain the more important and fairly recent works.

International Commission on Large Dams

The objects of the International Commission on Large Dams are to encourage improvements in the design, construction, maintenance and operation of large dams by bringing together information thereon, and by studying questions relating thereto.

These objects are obtained principally by holding international congresses, the transactions of which contain valuable papers covering all aspects of the subject.

Transactions of International Congress on Large Dams

1st, Stockholm, 1933
2nd, Washington, 1936
3rd, Stockholm, 1948
4th, New Delhi, 1951
5th, Paris, 1955
6th, New York, 1958
7th, Rome, 1961
8th, Edinburgh, 1964
9th, Instanbul, 1967
10th, Montreal, 1970
11th, Madrid, 1973

GEOTECHNICAL ABSTRACTS AND GEODEX RETRIEVAL SYSTEM

Geotechnical Abstracts is a service of literature documentation for soil mechanics, foundation engineering, rock mechanics and engineering geology, published monthly by the German National Society of Soil Mechanics and Foundation Engineering. Each issue contains

144 abstracts of literature published worldwide in more than 500 journals and other sources. Two editions are available: a cardboard edition for use with simple card files, and a paper edition for use with retrieval systems. The abstracts are arranged in each issue according to the International Geotechnical Classification System (IGC), thus permitting the user to use the abstracting journal for scanning of items of interest as well as to organise the abstracts in a simple card file.

The *Geodex Soil Mechanics Information Service* is a high-speed research and geotechnical information retrieval system for abstracts of literature published in journals and the proceedings of conferences and symposia, and is based on *Geotechnical Abstracts*. It uses some 300 punched cards, known as keyword cards or feature cards, and holes have been punched in each card through the code numbers of all abstracts pertaining to the particular keyword named on the card. A typical abstract may have its code number punched on eight or more different cards, thus enabling users to find the abstract again from all of these different viewpoints. The same simple technique makes it possible to search the entire file for papers pertaining only to a group selected keywords, thus quickly narrowing the search to papers of specific interest.

34

Land surveying

J. R. Smith

The profession of land surveying, or the science of measurement, covers a vast field of topics, and because of the confusion this often causes, some definitions will form a useful starting point.

Although the term 'surveying' in all its various connotations means the measurement of particular quantities, some of which may seem rather nebulous, from the land surveyor's point of view he is, in general, trying to measure three quantities—angles, distances and heights— although these may come in such disguises as the dimensions of the earth, the variation of gravity or the movements of the face of a dam. However, by combining them it is possible to either produce a map or plan, or alternatively, from designs on a plan, to set the location of a structure on the ground.

It is convenient to classify the subject under seven headings, but it must be borne in mind that these will inevitably overlap. The headings are: historical aspects; geodetic surveying; topographic surveying; large-scale and engineering survey; cadastral surveying; hydrographic surveying; and a general section. These will each be defined in their respective sections, but first of all some mention of the relevant professional institutions in the United Kingdom is required.

INSTITUTIONS

The principal professional body in the United Kingdom which covers all these fields of surveying (and many others besides) is the Royal Institution of Chartered Surveyors, 12 Great George St, Westminster, London SW1P 3AD, although out of its total membership of some 30 000 fewer than 1000 come into the category of qualified (chartered) land surveyors. The land surveyors division of this Institution has a small Technical Advisory Group which helps in answering queries of a technical nature so long as they do not entail the giving of a professional opinion. The members of this Technical Group are all professional land surveyors and so are able to more readily supply answers to the majority of queries than could a library staff with a general background.

As well as this Technical Advisory Group, the Institution naturally has an extensive library of land survey books, including an historical collection dating from 1538. This collection contains copies of most of the important early writings on surveying such as:

This Boke Sheweth the Manner of Measurynge, by R. Benese (1538)
Pantometria, by L. Digges (1591)
The Surveiors Dialogue, by J. Norden (1610)
The Description and Use of the Sector, by E. Gunter (1623)
The Compleat Surveyor, by W. Leybourn (1653)
Geodaesia, or the Art of Surveying, by J. Love (1688)

These, and the 100 other pre-1900 volumes are only available for reference purposes. Similarly, the extensive set of land survey theses (about 400) covering all aspects of the profession are only available on a reference basis. Lists of both the theses and historical collection can be obtained from the Librarian at the above address.

Other relevant societies are:

The Photogrammetric Society, Department of Photogrammetry and Surveying, University College, Gower St, London WC1. While holding regular meetings, this Society has no permanent accommodation and as a result its library material is held at the RICS. Its organ is the *Photogrammetric Record*.

The Society of Survey Technicians (SST). This body only came into being in the late 1960s, but is already making itself felt in industry. It had a newsletter in the form of the *Surveying Technician*, which has recently become a journal. The address of the Society is Aldwych House, Aldwych, London WC2B 4EL.

The Hydrographic Society is even newer than the SST but will be bringing together many interests of the sea. It publishes the *Hydrographic Journal*. As yet it has no permanent offices, but its secretary is Lt. Cdr. A. Ingham RN (Retd), Department of Surveying, NE London Polytechnic, Forest Rd, London E17.

HISTORICAL

Among the modern volumes that are essential reading for anyone investigating historical aspects of the profession are:

Chartered Surveyors: Growth of a Profession, by F. M. L. Thompson (Routledge and Kegan Paul, 1968)
Early Science in Oxford, by R. W. T. Gunther (14 vols., Oxford University Press, 1923–1945)
Historical Metrology, by A. E. Berriman (Dent, 1953)
History of the Retriangulation of Great Britain, by the Ordnance Survey (HMSO, 1967)
Mathematical Practitioners of Hanoverian England 1714–1840, by E. G. R. Taylor (Cambridge University Press, 1966)
Mathematical Practitioners of Tudor and Stuart England 1485–1714, by E. G. R. Taylor (Cambridge University Press, 1967)
The Art of Navigation in Elizabethan and Early Stuart Times, by D. W. Waters (Hollis and Carter, 1958)
The Roman Land Surveyors, by O. A. Dilke (David and Charles, 1971)

A series of articles in the *Australian Surveyor,* by A. P. H. Werner entitled 'A Calendar of the Development of Surveying' (**21**, No. 3, September 1966 to June 1968) gives innumerable references and clues to landmarks in all aspects of the profession.

The *Philisophical Transactions of the Royal Society* is also useful source material in this context and an example of this is 'On the Attraction of the Himalaya Mountains and of the Elevated Regions Beyond Them Upon the Plumb Line in India' by J. H. Pratt (*Phil. Trans. R. Soc.* **145** (1855) and **149** (1859)).

GEODETIC SURVEYING

Geodesy is the science of measuring the size and shape of the earth, although the term is used in a much broader context. For nearly 2000 years scientists have been attempting to measure the earth's dimensions. After starting as being thought of as a plane, it progressed through a sphere and spheroid to triaxial ellipsoid and geoid. Nowadays, observations on to earth satellites are providing much useful information. It embraces the survey of continents and the use of satellites to tie continents together. The measurement of the variation of gravity throughout the world, on both land and water, is a specialised science within itself. The term 'geodetic accuracy' is also sometimes used in surveys of small areas if a very high degree of refinement is aimed at in the observations, such as in the setting out of precision machines and reactors.

The standard works of reference in English for all aspects of geodesy are *Geodesy,* by G. Bomford (3rd edn, Oxford University Press, 1971), and the classical coverage in *Geodesy,* by A. R. Clarke (Oxford University Press, 1880). However, various continental volumes are often referred to, and these include:

Géodésie Genérale, by J. J. Levallois (4 vols., Edition Eyrolles, 1969)

Handbuch der Vermessungskunde, by W. Jordan, O. Eggert and M. Kneissl. This is in many volumes and selected portions have been translated into English by the Army Map Service, Washington 1962)

Lehrbuch de Geodäsie, by C. F. Baeschlin (Orell Füsslie Verlag, 1948)

Complementary to these are books and booklets on specific aspects, and among those of most use are:

Adjusting Calculations in Surveying, by I. Hazay (Akadémiai Kiadó, 1970)

A History of the Mathematical Theories of Attraction and the Figure of the Earth, by I. Todhunter (Cambridge University Press, 1873; Dover, 1962)

Coordinate Systems and Map Projections, by D. H. Maling (Philip, 1973)

'Figure of the Earth' in *Encyclopaedia Metropolitana,* Vol. 5 (1845)

Geodesy, by W. M. Tobey (Geodetic Survey of Canada, 1928)

Geodetic Reports of the Survey of India, published over many years prior to World War II.

Map Projections for Geodesists, Cartographers and Geographers, by P. Richardus and R. K. Adler (North-Holland, 1972)

Physics of the Earth's Interior, by B. Gutenberg (Academic Press, 1959)

Survey Adjustment and Least Squares, by H. Rainsford (Constable, 1957)

Textbook of Field Astronomy, by C. A. Biddle (HMSO, 1958)

The Earth and its Gravity Field, by W. Heiskanen and F. A. Vening-Meinesz (McGraw-Hill, 1958)

The Earth, its Origin, History and Physical Constitution, by H. Jeffreys (5th edn, Cambridge University Press, 1970)

Theory of Errors and Generalised Matrix Inverses, by A. Bjerhammar (Elsevier, 1973)

Theory of the Adjustment of Normally Distributed Observations, by J. M. Tienstra (Elsevier, 1956)

Various publications of the former US Coast and Geodetic Survey, such as the *Manual—Triangulation, Computation and Adjustment,* by W. F. Reynolds (Special Publication 138, US Department of Commerce, 1934)

However, Bomford's *Geodesy* is by far the most useful of all these because of its bibliography of 531 entries.

TOPOGRAPHIC SURVEYING

Topographic surveys are small-scale surveys of large areas. The scale is usually 1:50 000 or less. Such surveys are concerned with the basic features of nationwide areas and in some cases even more than one country. From a framework of triangulation, trilateration or traverse with sides of up to 50 km or so, a breakdown gives a more dense coverage of control points. Today such points often form the ground control for detail mapping by air survey methods.

Among the most respected textbooks in this category are:

Introduction to the Algebra of Matrices, by E. H. Thompson (Hilger, 1969)

Practical Field Surveying and Computations, by A. L. Allan, J. R. Hollwey and J. B. Maynes (Heinemann, 1968)

Plane and Geodetic Surveying, by D. Clark, revised by J. E. Jackson (Vol. 2, Constable, 1973)

Principles and Use of Surveying Instruments, by J. Clendinning and J. G. Olliver (3rd edn, Van Nostrand, 1972)

Principles of Surveying, by J. Clendinning and J. G. Olliver (3rd edn, Van Nostrand, 1972)

After a break of many years, the Ordnance Survey has recently restarted its series of professional papers, each of which describes a survey project, test or piece of apparatus in great detail.

Photogrammetry, or the use of both aerial and terrestrial photographs in the compilation of maps, has developed rapidly since World War I, and there is a great deal of literature available, but mostly in the form of papers rather than textbooks. The most comprehensive references are the *Manual of Photogrammetry,* edited by M. M. Thompson (2 vols. 3rd edn, American Society of Photogrammetry, 1966), and the *Manual of Colour Air Photography* (American Society of Photogrammetry, 1968).

Other works include:

Air Photography Applied to Surveying, by C. A. Hart (2nd edn, Longmans, 1943)
Elementary Air Survey, by W. K. Kilford (3rd edn, Pitman, 1973)
Elementary Photogrammetry, by D. R. Crone (Arnold, 1963)
Photogrammetry, by B. Hallert (McGraw-Hill, 1960)
Physical Aspects of Air Photography, by G. C. Brock (Longmans, 1952)

LARGE-SCALE AND ENGINEERING SURVEY

Engineering survey is a rather loose expression covering all forms of large-scale (about 1:2500 and larger) other than those that can be described as cadastral or hydrographic. It includes such projects as surveys for roads, railways, dams, housing estates, precision machine location, tunnelling, etc. The accuracy required in engineering surveys can vary from the approximate or coarse to the highest attainable.

There are many textbooks that go to varying depths into the subject, and among the most useful are:

Basic Metric Surveying, by W. S. Whyte (Butterworths, 1969)
Desk Calculators, by J. R. Smith (Crosby Lockwood Staples, 1973)
Electromagnetic Distance Measurement, by C. D. Burnside (Crosby Lockwood Staples, 1971)
Surveyor's Guide to Electromagnetic Distance Measurement, edited by J. J. Saastamoinen (University of Toronto, 1967)
Fundamentals of Survey Measurement, by M. A. R. Cooper (Crosby Lockwood Staples, 1974)
Modern Theodolites and Levels, by M. A. R. Cooper (Crosby Lockwood Staples, 1971)
Optical Distance Measurement, by J. R. Smith (Crosby Lockwood Staples, 1970)
Plane and Geodetic Surveying, by D. Clark, revised by J. E. Jackson (Vol. 1, Constable, 1969)
Surveying, by A. Bannister and S. Raymond (3rd edn, Pitman, 1972)
Surveying for Young Engineers, by S. W. Perrott, revised by A. L. Allan (3rd edn, Chapman and Hall, 1970)
Surveying Problems and Solutions, by F. A. Shepherd (Arnold, 1968)

The above volumes by Crosby Lockwood Staples form a series of single-topic volumes under the general title of *Aspects of Modern Land Surveying* and further volumes are in preparation.

In recent years a disturbing trend has been noticed in textbooks aimed at the civil engineering and construction fields of surveying. Many books have been written by non-surveyors and, in general, they leave a great deal to be desired and in some instances display a complete lack of understanding of the subject. Such a trend is very regrettable and the particular texts should be avoided and further similar writings actively discouraged.

CADASTRAL SURVEYING

This is the surveying of a particular area of land and the production of a plan of it for the purpose of ownership and the registration of the property on the National Register system of the particular country concerned. It is, therefore, particularly interested in the demarcation and location of boundaries and the scale employed is usually 1:2500 or larger.

At present there is little written in textbook form specifically on cadastral survey, although many of the field operations are covered by books in the previous section. For many years the standard text for examination purposes on the land registration aspect has been *Land Registration,* by E. Dowson and V. L. O. Sheppard (HMSO, 1956), but this is not a very readable volume and a successor has been recently commissioned as *The Registration of Title to Land,* edited by S. R. Simpson.

The lack of literature is due to some extent to there being no cadastral survey operating in the UK because of the use of general boundaries for registration purposes as opposed to fixed boundaries in many continental countries. With entry to the Common Market, interest in cadastral surveying is likely to increase because any engineering projects in countries operating with fixed boundaries will become involved in the 'pincushion' of concrete markers, as it has been described.

HYDROGRAPHIC SURVEYING

As the name implies, this covers any survey related to areas of water. In particular, it includes coastal surveys, offshore islands, estuaries and bays as well as encompassing the requirements for offshore oil rigs and deep-draught tankers. Besides land features bordering the water, a prime aspect is the measurement and location of soundings.

The standard work of reference is the *Admiralty Manual of Hydrographic Surveying* (2 vols., HMSO, 1965 and 1970) supplemented by *Radio Aids to Maritime Navigation and Hydrography*

(2nd edn, International Hydrographic Bureau, 1965). A recent book worthy of mention is *Hydrography for the Surveyor and Engineer*, by A. E. Ingham (Crosby Lockwood Staples, 1974).

GENERAL

The proceedings of conferences and congresses provide a very useful source of information on all aspects of land surveying. Next to periodicals they are the best means of keeping up to date with the rapidly changing scientific scene. Their one drawback is that, on occasions, papers are written for conferences simply for the sake of writing papers, or perhaps as a means of the author being able to get finance for going to the conference as opposed to the true worth of papers in providing details of new techniques and developments. In the same vein is the writing of papers for journals solely with a view to their usefulness in getting promotion at a later date, but it is only the reader who can sort the wheat from the chaff.

Regular gatherings are held of the Conference of Commonwealth Survey Officers in Cambridge and the *Proceedings* are published by HMSO. The International Federation of Surveyors (FIG) meet in different countries every 3 years and all their papers are published as *Archives*. In between the three-yearly meetings the FIG have yearly meetings of their organising committees, again in different countries, and these countries often hold an international survey conference to coincide with the committee meeting. For instance, in 1972 the committee met in Israel and a week-long conference was held in Tel Aviv at the same time. The proceedings of such conferences are similarly produced by the country concerned. The theme of the Tel Aviv conference was 'The Computer and the Surveyor' and the bound papers form one of the first comprehensive volumes on this fast-developing aspect of the profession.

In a similar way to the FIG, the International Society of Photogrammetry (ISP) publishes the papers of their regular congresses as *International Archives of Photogrammetry*. In the more specialised field of geodesy the body concerned is the International Association of Geodesy (IAG), whose General Assemblies and National Reports are published as *Travaux de L'IAG*, by the Bureau Central de L'Association Internationale de Géodésie at 39ter, Rue Gay Lussac, 75005, Paris.

Complementary to these international gatherings and their proceedings are various national meetings for which papers and/or proceedings are available. Among the countries doing this are: America, with the American Society of Surveying and Mapping, which holds annual congresses (1972 saw the 32nd such meeting); Australia, with the Institution of Surveyors, Australia (the papers of their 15th Survey Congress held in Newcastle in 1972 are available in bound form); South Africa, where their Institute of Surveyors' 4th National Survey Conference was held in 1972 and the papers published.

REFERENCE WORKS

Admiralty List of Radio Signals (HMSO, irregular)
Admiralty Tidal Handbooks (Nos. 1–3, HMSO, 1958, 1960, 1964)
Apparent Places of Fundamental Stars (International Astronomical Union, annual)
Constants, Formulae and Methods Used in the Transverse Mercator Projection, by the Ordnance Survey (HMSO, 1954)
Ordnance Survey Maps. A Descriptive Manual, by J. B. Harley (HMSO, 1975)
Photogrammetric Guide, by J. Alberta and W. Kreiling (Wickmann, Karlsruhe, 1975)
The Astronomical Ephemeris (HMSO, annual)
The Nautical Almanac (HMSO, annual)
The Star Almanac for Land Surveyors (HMSO, annual). This is specially designed for the surveyor and includes 'Ephemeris of the Sun'. 'Apparent Places of Stars', 'Circumpolar Stars', 'Pole Star Tables' and 'Refraction Table'.
Various projection tables such as the two manuals recently placed on public sale by the US Army, *Universal Transverse Mercator Grid* (TM5–241–8), which describes the worldwide UTM system and lists applicable formulae, and *Universal Transverse Mercator Grid Tables* (TM5–241–11), for latitudes 0°–80°; Clarke 1866 spheroid (metres), coordinates for $7\frac{1}{2}$ minutes intersections. These are obtainable from The Superintendent of Documents, US Government Printing Office, Washington DC 20402.

With the advent of desk calculating machines with built-in trigonometric and logarithmic functions, the requirement for tables of these functions is decreasing, although as yet it cannot be said that such tables are obsolete. A wide selection exists, but among those of particular use to the land surveyor are the following:

5 figure log tables. *Fünfstellige Vollständige Logarithmische und Trigonometrische Tafeln*, by F. G. Gauss (Wittwer, 1956)
6 figure log tables. *Logarithmisch-Trigonometrische Tafeln fur alte (Sexagesimale) Teilung mit Sechs Dezimalstellen*, by C. Bremiker (Wittwer, 1950)

7-figure log tables. *Seven Place Logarithmic Tables of Numbers and Trigonometrical Functions,* by G. F. Vega (Hafner, 1957)
6-figure natural tables. *Sechsstellige Tafel der Trigonometrischen Funcktionen,* by J. Peters (Dümmler, 1971)
8-figure natural tables. *Eight Figure Tables of Trigonometric Functions for Every Second of Arc,* by J. Peters (Chelsea Publishing Company, 1965)

Other useful references are:

Handbook of Mathematical Functions, by M. Abramowitz and I. A. Segun (Dover, 1964)
Highway Transition Curve Tables (Metric), by The County Surveyors Society (Carriers Publishing Company, 1969)
Tacheometric Tables for the Metric User, by D. T. F. Munsey (Technical Press, 1971)

DICTIONARIES

Definitions of Terms Used in Geodetic and Other Surveys, by H. C. Mitchell (Special Publication 242, US Department of Commerce, 1948)
Multilingual Dictionary of Technical Terms in Cartography (International Cartographic Association, 1973)
Multilingual Dictionary of the International Federation of Surveyors (Elsevier, 1963)
Photogrammetric Dictionary (Elsevier, 1961)
Slownik Geodezyjny W 5 Jezykach-Polskim, Rosyjskim, Niemieckim, Angielskim, Francuskim, by W. Sztompke (Panstwowe Przedsiebiorstwo Wydawnictw Kartograficznych, Warszama, 1954)
Standard Definitions of Terms Used in Photogrammetric Surveying and Mapping (National Mapping Council of Australia, 1963)

MISCELLANEOUS REFERENCE WORKS

The following publications are particularly useful:

Education for the Profession, by A. L. Allan (International Federation of Surveyors, 1968, revised 1974). This booklet provides background information about the education of surveyors in the FIG countries as a means of enabling international discussions on educational topics to proceed with the minimum of misunderstanding of the various educational systems.

Organisations for Land Surveying, by J. R. Hollwey (International Federation of Surveyors, 1968). This booklet provides background information about the various organisations which prepare maps and plans and which carry out survey operations in the different countries of FIG.

The Surveyors Profession in Commonwealth Countries. A Handbook of Professional Societies is published by the Commonwealth Association of Surveying and Land Economy (CASLE), 12 Great George St, London SW1P 3AD.

CARD INDEXES

International Bibliography of Photogrammetry. Since 1958 the International Institute for Aerial Survey and Earth Sciences (ITC), Kanaalweg 3, Delft, Netherlands, has published a card index on the UDC system covering all aspects of photogrammetry. It is in English, French and German with both a subject and author index.

International Geodetic Bibliography. This was originally published in bound form covering 3 years at a time, but as from 1961 the data have become available in card form as *International Geodetic Documentation* and as a monthly review, *Bibliographica Geodaetica*. In four languages—English, French, German and Russian—both are published in two groups—Group 1 Geodesy and Group 2 Surveying—by Zentralstelle für Internationale Dokumentation der Geodäsie at the Geodetic Institute der TU, Momsenstrasse 13, Dresden 8027, East Germany.

Both of these card index systems are supported by various countries undertaking to supply abstracts from all the relevant journals in their language or, more particularly, in their country.

PERIODICALS AND ABSTRACTING JOURNALS

Over 100 journals from 30 countries, in 15 different languages, contain items of interest to the surveyor, and this range is now expanding even more with his increased incursions into the field of electronics. Of these journals, many have a cross-section of papers from all the different aspects of survey, although there are some which concentrate on particular facets. For example, *Bulletin Géodésique* and the *Photogrammetric Record*. Today, however, there is such an overlap of interests that none of the branches of land survey can operate in isolation from the others. For instance, photo-grammetry is becoming of increasing importance in the survey control of motorway projects and geodetic techniques are used in the measurement of engineering deformations and the location of precision machinery. With computers and electronic calculators in abundance, the mathematical concepts that used to be reserved solely for the geodesist can now be introduced into the adjustment of engineering schemes.

Among the languages involved in the range of journals, interesting items can be found in German and Italian through Russian and Serbo-Croat to Turkish. Luckily the mathematics is the same in almost every language, so that often it is possible to make some sense of an article in a foreign language without having a working knowledge of that language.

Abstracting journal

Geodetic Abstracts. A quarterly publication of the US Army Corp of Engineers, it contains abstracts from many foreign-language publications, particularly European. The circulation list appears to be limited to official bodies, and the RICS is on this list. The abstracts in the American journal *Surveying and Mapping* are from the same source as this publication. The choice of abstracts covers a wide field, but they are often of a research character.

Periodicals

Canadian Surveyor. This is a quarterly publication covering a wide range of interests from cadastral through engineering to geodesy, although there often seems to be a bias of papers towards the photo-grammetric side. Its contents are of a high standard and worthy of perusal by anyone doing research as well as those simply trying to keep up to date. It has two sections, the body being technical papers, while a yellow-page supplement has chatty items of Institution news as well as reviews and smaller items. It is the organ of The Canadian Institute of Surveying, Box 5378, Station 'F', Ottawa, Canada, K2C 3JI.

Geodesy and Aerophotography. This is an American translation from the Russian and is very much a researcher's tool. It is an expensive bi-monthly, but nevertheless good value, particularly for those interested in the higher mathematical aspects of surveying. It is wholly technical and covers all aspects of surveying from satellite orbits to structural deformations and cartography. One unfortunate but unavoidable aspect is that is usually appears about 18 months after the Russian original. It is published by the American Geophysical Union, 2100 Pennsylvania Ave, NW Washington, DC 20037.

International Hydrographic Review. This is a large volume appearing every 6 months as the organ of The International Hydrographic Bureau, Ave President J. F. Kennedy, Monte Carlo, Monaco. While all the papers obviously have an hydrographic flavour, they are of a good technical standard and often of interest to those other than hydrographic officers. It is essential reading on any course which covers the marine aspects of surveying. A more frequent publication from the same source is the *International Hydrographic Bulletin,* but this does not contain papers—only short technical news items.

Journal of the Surveying and Mapping Division. Proceedings of the ASCE. Of much interest from the United States is this journal, which is one of some ten similar divisional journals produced by the American Society of Civil Engineers. It contains only technical papers and the discussion relevant to them, and these cover all aspects of survey, although the photogrammetric content is limited. It can be a useful journal for both general and research purposes. It appears every 2 months and contains a card index system listing the keywords of the paper together with abstracts. The editorial offices are at 345 East 47th St, New York, NY 10017.

New Zealand Surveyor. As yet this is but a smaller version of the *Australian Surveyor* with only a limited number of papers of a research nature. It is a six-monthly publication from Box 831 Wellington, New Zealand.

Photogrammetric Engineering and Remote Sensing. This is the monthly organ of the American Society of Photogrammetry, 105 N. Virginia Ave, Falls Church, Virginia 22046.

Photogrammetric Record. This is the organ of the Photogrammetric Society and appears twice a year. It is very well produced and invariably has articles of a high technical quality. The editorial address is: Department of Photogrammetry and Surveying, University College, Gower St, London WC1.

South African Survey Journal. A slimmer volume than the previous ones, this appears with alternate pages usually in English and Afrikaans. It has both technical and institutional sections, and although there are a limited number of the former, past issues have contained a scattering of important articles, particularly with an engineering flavour. Address is c/o Department of Land Surveying, University of Natal, Durban.

Surveying and Mapping. This is the journal of the American Society of Surveying and Mapping. Published quarterly, it has a technical section and a social section. The technical content is of variable quality and at times there seems to be an abundance of social material. It is probably of limited research value, although the abstracts of foreign-language journals are very useful. The technical items themselves cover all aspects of the profession and the predominant theme is that related to engineering survey work. ACSM, Suite 430, Woodward Building, 733 15th St NW Washington DC 20005.

The Australian Surveyor. Not such a lengthy publication as the *Canadian Surveyor* and until a few years ago not of the same technical standard. However, with a revamping of the whole journal around 1970 the standard is rising and many of the papers are worthy of study. Unlike the *Canadian Surveyor,* there is much less photogrammetry and more papers with an engineering bias. The address is Box 4793 G.P.O., Sydney, NSW 2001, Australia.

The Chartered Surveyor. This is the monthly journal of the Royal Institution of Chartered Surveyors. By virtue of having to cater for other types of surveyor besides the land surveyor, the land survey content is limited to an occasional paper and a monthly column of land survey notes. These notes contain abstracts of technical articles from other journals, information on new equipment and literature, and reports of conferences. As from September 1973 a quarterly supplement is devoted to land, hydrographic and mineral surveying. The address for both is 12 Great George St, Westminster, London SM1P 3AD.

The Survey Review. This is one of the most useful of English-language journals and it has been published as a quarterly continuously since 1931. Its contents cover all aspects of the profession, with contributors from all over the world. Although its circulation is only about 1400, this is spread over some 80 countries. Recently it has been thought to be somewhat top-heavy with highly mathematical papers, but steps are in hand to try and remedy this, the problem, as always, being to get prospective authors to put pen to paper. The editorial address is: Department of Photogrammetry and Surveying, University College, Gower St, London WC 1.

So far, all the journals mentioned have been in English, but there are a number of foreign-language journals that should be regularly scanned. While most of these are only in the parent language, there are some with English abstracts. As has already been stressed, the mathematics of such articles is fairly understandable in any language and as such foreign journals should not be immediately rejected out of hand as being unintelligible. The following are the more important of such publications.

Allgemeine Vermessungs Nachrichten. The monthly journal of the German survey profession, it contains much useful technical material but there are only infrequent English translations or abstracts. Covers all topics. Address: Herbert-Wichmann-Verlag, Rheinstrasse 122, PO Box 210729, 7500 Karlsruhe 21, West Germany.

Bolletino di Geodesia e Scienze Affini. A quarterly publication from Italy with English summaries. The text can be in any of several languages including English, but the majority are in Italian. It covers all aspects of the profession to a good technical standard. From: Dell' Instituto Geografico Militaire, Via C.Battisti 10, Firenze, Italy.

Bulletin Géodésique. This is the quarterly journal of the International Association of Geodesy, with most of the papers in English, although contributions are acceptable in French or German. As the name implies, it is very much devoted to the higher aspects of geodesy and is usually very mathematical. Nevertheless, some of these geodetic topics are occasionally relevant to precise engineering survey. Photogrammetry and cadastral surveying do not feature in this publication. From Bureau de L'Association Internationale de Géodésie, 39ter Rue Gay Lussac, 75005, Paris.

Fotogrammetriska Meddelanden. A specialised photogrammetric journal from Scandinavia which has many articles in English. Of more use as a research tool than for the solution to general problems. Institutionen för Fotogrammetri, Kungl. Tekn. Högskolan, Stockholm 70.

Geodetski List. Another somewhat difficult journal to digest since the language is Serbo-Croat. It is a quarterly covering all topics. Savez Geodetskih Inzenera I Geometara, S.R.Hrvatske, Zagreb, Berislaviceva 6, Jugoslavia.

Geometre. The monthly French survey journal (not to be confused with a Belgian one of similar title) covers other forms of survey as well as that of prime interest here. It is somewhat similar to *The Chartered Surveyor,* although its percentage of land survey papers is probably greater. There are often items of good technical content and it has a section devoted to problems posed for answer by the readers. It does not have English summaries. Published at 40 Avenue Hoche, Paris 8e.

Przeglad Geodezyjny. A monthly Polish journal with brief English abstracts. Has good technical material, often on the engineering side. Longer abstracts would be beneficial with such a language as Polish. Export & Import Enterprise, 'Ruch', ul Wilcza 46, Warzawa.

All the above periodicals are available for reference at the RICS library. Other material worthy of regular perusal are the infrequent publications of various International Survey Institutions. Examples are:

Department of Geodetic Science, Ohio State University, USA
Department of Surveying, University of New South Wales, Kensington, NSW 2033
Division of National Mapping, Canberra, Australia
ITC, Kanaalweg 3, Delft, Netherlands
Netherlands Geodetic Commission, Delft
Ordnance Survey (Professional Papers) Romsey Road, Maybush, Southampton
Deutsche Geodetic Kommission, 9 Munchen 22, Marstallplatz 8, West Germany

Further very useful sources, particularly with reference to new instrumentation, are the journals from the various instrument manufacturers, i.e.

Jena Review, VEB Carl Zeiss, Jena, East Germany
MOM Review, Hungarian Optical Works, Budapest XII, Csorsz Utca 35/43
Sureveying News (Vermessungs Informationen), VEB Carl Zeiss, Box 190, Jena
Wild Reporter, Wild Heerbrugg Ltd, CH-9435, Heerbrugg, Switzerland

BIBLIOGRAPHY

Technical Information for Surveyors, by H. L. Rogge and M. J. Roberts (International Federation of Surveyors (FIG), 1968)
Commission 3 papers, FIG Congress, Wiesbaden (1971)

35

Mining and mineral technology

J. McFarlane

The widely varying nature and scale of operations of the different sections of the mineral industry, coupled with the wide range of scientific, technological and management concepts which are necessarily involved in some of the larger-scale and sophisticated operations, mean that a complete survey of literature would embrace a substantial proportion of the UDC classification system used in libraries. Inspection of the inside rear cover of the *IMM Abstracts,* published bi-monthly by the Institution of Mining and Metallurgy, will demonstrate the extent of the field covered in this chapter. In practice, a simplified classification can be adopted for this chapter in preference to the UDC system, in which allied subjects may be found to be sometimes separated because the UDC covers the whole field of science and technology.

An extension of this classification would lead into the fields related to mining and mineral technology, such as geology, engineering (civil, mechanical, electrical), pure science (mathematics, physics, chemistry), and the technologies using mineral raw materials, such as metallurgy, chemical engineering, ceramics, fuel technology. These subjects have close affinities with our restricted field of the present chapter and overlap is inevitable. A literature search in any of the subjects listed in the classification above will lead into the literature of these related fields.

THE OCCURRENCE OF AND SEARCH FOR MINERALS

The most important developments in this field during the last two decades have been in the field of indirect prospecting. The use of geophysical and geochemical techniques in the search for minerals has become an accepted part of mineral operations in all parts of the world. The literature reflects this spread of the techniques, and many case studies in prospecting have appeared in print since 1950.

General and economic geology

Textbooks

There have never been large numbers of textbooks on general or economic geology, possibly owing to the fact that in many parts of the world geology is regarded as a minority subject in schools. The most interesting development in geological textbooks has been in the use of colour photography. High-quality colour printing is no longer prohibitively expensive and the three-dimensional effect of colour is of great benefit in geological illustrations. A short list of selected geological texts is given below, but special mention should be made of the series produced by the Open University. These have appeared during the last few years and represent a very modern approach to the subject at very reasonable cost. The value for money approaches that of the government publications, and the treatment of the subject is authoritative, modern and interesting. The Institute of Geological Sciences (formerly the Geological Survey of Great Britain) has for many years produced relatively lowly priced publications covering most parts of Britain in the form of *Memoirs* describing the areas covered by their one-inch scale maps. Another very useful series by the Institute is the *British Regional Geology* covering the country in about fifteen slim volumes. In the economic field, the Institute of Geological Sciences has recently begun a series of pamphlets on individual minerals and the British resources of them. These are worthy successors to the earlier wartime series on minerals.

Periodicals

These originate mostly from professional bodies and scientific societies since there is a relatively small market for commercial publications. The few most important exceptions are listed below with the principal publications produced by learned societies in the geological field. The main function of the latter is to publish the scientific papers read before the societies by various authors and research workers. Although there are 'fashions' in the popularity of certain parts of the subject which change as years go by, it is true to say that the annual output by individual societies presents a cross-section of the various interests within the societies and it is often difficult to find a paper on a particular subject. Fortunately each society normally produces an annual index which helps in a literature search. It is clearly a great advantage to receive the publications of a society as they appear, which is one of the benefits of membership.

In recent years there has been an increase in the number of societies catering for specialists within particular areas of geological science. The proceedings of such bodies contain material covering smaller fields in greater detail, which is very useful in a literature search. Some of the older bodies, such as the Geological Society of London, have followed this trend by publishing not simply a general *Quarterly Journal*, but also a *Quarterly Journal of Engineering Geology*. The Society also produces one or more *Special Memoirs* or other publications on indivdual subjects in each year. These are available to members at reduced cost.

Prospecting

Textbooks

Some admirable general textbooks in this area of study have appeared during the last 40 years. These vary from small general texts for the non-specialist who needs to know a little about the techniques and their uses, to the large tome intended to be a working manual for the prospector. Because the indirect methods are of relatively recent origin, the textbooks in this area are more numerous and more up to date and can be taken as good starting points for more detailed studies. As the subject has developed, however, it has become more difficult to produce an authoritative textbook to cover the whole subject, and recourse has to be made to the periodical specialist publications.

Periodicals

Several specialist societies have been formed in this area, such as the European Association of Exploration Geophysicists, the Association of Exploration Geochemists and the Geochemical Society. The bulk of the current advanced material in these fields appears in the transactions of these bodies, but a few papers occasionally appear in the publications of the more general societies and professional bodies. Physical- and chemical-orientated bodies should not be forgotten as possible sources of information in this field.

Evaluation of mineral deposits

There have been great strides forward in this field in recent years mainly because the use of computers of various sizes has extended the possibilities of calculation and shortened the time needed for trying out theories in evaluation. Naturally much has been written by the workers involved, although not so much as would have been desired, owing to the commercially confidential nature of the work. There is some overlap with the management literature, as, for example, in the reports on conferences on the use of computers in the mining industry.

So far, no recent textbook on the subject has appeared and obviously, earlier works of some forty or fifty years vintage take no account of computer-developed techniques. The most useful information can only be obtained from the published proceedings of conferences and courses run by professional bodies and universities in the USA, South Africa, Canada and the United Kingdom.

Selected publications (geological)

Abstracts and indexes

Abstracts of North American Geology (formerly *Geoscience Abstracts*)
Bibliography and Index of Geology
Geotitles Weekly
Mineralogical Abstracts
NCB Abstracts C Coal and Mining Geology

Periodicals

Earth Science Review
Economic Geology
Engineering Geology
Geochimica et Cosmochimina Acta
Geoexploration
Geological Magazine
Geophysical Prospecting
Geophysics
International Journal of Mathematical Geology

Journal of Engineering Geology
Journal of Geochemical Exploration
Journal of the Geological Society
Journal of Geology
Journal of Petrology
Mineralogical Magazine
Mining Geology
Proceedings of the Geologists Association
Scottish Journal of Geology
Transactions of the Institution of Mining and Metallurgy

Books

Use of Earth Sciences Literature, edited by D. N. Wood (Butterworths, 1973

Guide to Geologic Literature, by R. M. Pearl (McGraw-Hill, 1951)

A Guide to Information Sources in Mining, Minerals and Geosciences, edited by S. R. Kaplan (Interscience, 1965)

Memoirs of the Geological Survey of Great Britain, Institute of Geological Sciences (HMSO, various dates)

British Regional Geology, Institute of Geological Sciences (HMSO, various dates)

Statistical Summary of the Mineral Industry, Institute of Geological Sciences (HMSO, annual)

Geologist's Handbook (NCB, 1960)

An Introduction to Geological Structures and Maps, by G. M. Bennison (2nd edn, Arnold, 1969)

The Geological History of the British Isles, by G. M. Bennison and A. E. Wright (Arnold, 1969)

Geological Maps and their Interpretation, by F. G. H. Blyth (Arnold, 1965)

Manual of Field Geology, by R. R. Compton (Wiley, 1962)

Understanding the Earth, by I. G. Gass, P. J. Smith and R. C. Wilson (Artemis, 1970)

Structural and Field Geology, by J. Geikie (6th edn, Oliver and Boyd, 1953)

Principles of Geology, by J. Gilluly, A. C. Waters and A. O. Woodford (3rd edn, Freeman, 1968)

Principles of Stratigraphy, by A. W. Grabau (2 vols., 3rd edn, Dover, 1924)

Elements of Structural Geology, by E. S. Hills (2nd edn, Chapman and Hall, 1971)

Principles of Physical Geology, by A. Holmes (Nelson, 1965)

A Field Guide to Rocks and Minerals, by F. H. A. Pough (3rd edn, Constable, 1970)

Introduction to Geology, by H. H. Read and J. Watson (2nd edn, Macmillan, 1970)

Fundamentals of Geology, by J. J. W. Rogers and J. A. S. Adams (Harper and Row, 1966)

The Science of Geology, by D. A. Robson (Blandford, 1968)

Geological Maps, by B. Simpson (Pergamon, 1968)

Rocks and Minerals, by B. Simpson (Pergamon, 1968)

Geology and Scenery in England and Wales, by A. E. Trueman (Penguin, 1972)

Textbook of Petrology: Vol. 1 Petrology of Igneous Rocks, Vol. 2 Petrology of the Sedimentary Rocks, by F. H. Hatch (12th edn and 6th edn, Chapman and Hall, 1961 and 1971)

Precious Stones and Minerals, by H. Bank (Warne, 1970)

Economic Mineral Deposits, by A. M. Bateman (2nd edn, Wiley, 1967)

Geology of the Industrial Rocks and Minerals, by R. L. Bates (Constable, 1969)

Ore Microscopy, by E. N. Cameron (Wiley, 1961)

Manual of Mineralogy, by E. S. Dana (18th edn, Wiley, 1971)

Microscopic Identification of Minerals, by E. W. Heinrich (McGraw-Hill, 1965)

Chemical Methods of Rock Analysis, by P. G. Jeffrey (Pergamon, 1970)

Identification of Mineral Grains, by M. P. Jones and M. G. Fleming (Elsevier, 1965)

Minerals and Rocks in Colour, by J. F. Kirkaldy (2nd edn, Blandford, 1968)

Gems and Gem Materials, by E. H. Kraus and C. B. Slawson (5th edn, McGraw-Hill, 1947)

Mineral Deposits, by W. Lindgren (McGraw-Hill, 1933)

Rocks and Mineral Deposits, by P. Niggli (Freeman, 1954)

Ore Deposits, by C. F. Park and R. A. Macdiarmid (2nd edn, Freeman, 1970)

Introduction to the Mineral Kingdom, by R. M. Pearl (Blandford, 1966)

Elements of Mineralogy, by F. Rutley (26th edn, Murby, 1970)

Determination Tables for Ore Microscopy, by C. Schouten (Elsevier, 1965)

Elements of Optical Mineralogy, by A. N. Winchell (4th edn, Wiley, 1951)

Physical Methods in Determinative Mineralogy, edited by J.
Zussman (Academic Press, 1967)
Atomic Absorption Spectrometry in Geology, by E. E. Angino and
G. K. Billings (Elsevier, 1967)
Nuclear Techniques for Mineral Exploration (International Atomic
Energy Agency, 1971)
The Geology of Sand and Gravel, by S. H. Beaver (Sand and Gravel
Association, 1968)
A Geology for Engineers, by F. G. H. Blyth and M. H. de Freitas
(6th edn, Arnold, 1974)
Blast Vibration Analysis, by G. A. Bollinger (South Illinois
University Press, 1971)
Basic Geology for Engineers, by J. Bundred (Butterworths, 1969)
North Sea Oil—the Great Gamble, by B. Cooper and T. F. Gaskell
(Heinemann, 1966)
Introduction to Geophysical Prospecting, by M. B. Dobrin (2nd edn,
McGraw-Hill, 1960)
Mineral Resources, by P. T. Flawn (Rand McNally, 1966)
Coal: its Formation and Composition, by W. Francis (2nd edn,
Arnold, 1961)
Geophysical Exploration, by F. W. Dunning (HMSO, 1970)
General Principles of Geochemical Prospecting, by I. I. Ginzburg
(Pergamon, 1960)
Interpretation Theory in Applied Geophysics, by F. S. Grant and
G. F. West (McGraw-Hill, 1965)
Applied Geophysics for Engineers and Geologists, by D. H. Griffiths
and R. F. King (Pergamon, 1965)
Geochemistry in Mineral Exploration, by H. E. Hawkes and J. S.
Webb (Harper and Row, 1962)
Geophysical Exploration, by C. A. Heiland (Hafner, 1963)
Electrical Methods in Geophysical Prospecting, by G. V. Keller and
F. C. Frischknecht (Pergamon, 1966)
Statistical Analysis of Geological Data, by G. S. Koch and R. F.
Link (2 vols., Wiley, 1970 and 1971)
Geological Prospecting and Exploration, by V. M. Kreiter (Central
Books, 1968)
Introduction to Statistical Models in Geology, by W. C. Krumbein
(McGraw-Hill, 1965)
Principles of Geochemistry, by B. Mason (3rd edn, Wiley, 1966)
The Mineral Resources of the Sea, by J. L. Mero (Elsevier, 1964)
Handbook of Subsurface Geology, by C. A. Moore (Harper and
Row, 1965)
Development and Exploitation of Oil and Gas Fields, by I. Muravyov
(Central Books, 1965)

Geophysical Prospecting for Oil, by L. L. Nettleton (McGraw-Hill, 1940)

Mining Geophysics, by D. S. Parasnis (2nd edn, Elsevier, 1973)

Fault and Joint Development in Brittle and Semi-Brittle Rock, by N. J. Price (Pergamon, 1966)

Folding and Fracturing of Rocks, by J. G. Ramsay (McGraw-Hill, 1967)

Geological Aspects of Mining, by J. Sinclair (Pitman, 1958)

Gemstones, by G. F. H. Smith (13th edn, Methuen, 1958)

Coal Mining Geology, by I. A. Williamson (Oxford University Press, 1967)

World Petroleum Congress, 8th, Moscow, 1971 (Vols. 2 and 3, Applied Science, 1971)

Mineral Resources, by K. Warren (Penguin, 1973)

THE EXTRACTION OF MINERALS (MINING)

The extraction of any useful material from the earth's crust, with the possible exception of water, is covered in this section. Sand and gravel operations off the coast of East Anglia, coal mining at a depth of a kilometre in the Ruhr, gold mining at depths of several kilometres in South Africa and copper mining by surface pit in New Guinea are examples enough to show the diversity of the subject covered and the problems posed in following the literature.

Textbooks

The constant and inevitable development of the technology concerned with the extraction of mineral wealth from the earth's crust plus the diversity referred to above mean that it is not easy to produce a general textbook which is sufficiently detailed and up to date to help the reader to gain an over-all impression of the modern mineral industry. A well-written, adequately illustrated textbook would take several people two or three years to produce and after a few more years would probably be thought to be out of date. In some cases a whole mineral industry in a country can be developed and exhausted within three decades. In addition, there is a relatively small market for general mining books, as there are probably only about 400 mining engineers qualifying each year throughout the world. The best way to obtain general information on the mineral industry is to consult books written on individual types of mines, e.g. coal-mining or gold-mining. These often deal with the history of the mining areas as well as the geology and mining techniques.

So far as mining general descriptions are concerned, most available textbooks are out of date and need to be supplemented by reference to more recent publications. Examples are given below of some of these general descriptions.

The situation with respect to textbooks on the separate subjects listed in the classification is rather better. Mining technology impinges on other disciplines in its specialist branches, and publishers have sought to satisfy the various composite markets. Examples are in ground control studies, where surveyors, civil engineers, geologists, physicists and mathematicians are active in research and development. Thus, it is best to look at each of the listed sections of mining technology, since the outside associations differ in each section.

Development of mines

The driving of tunnels and shafts in order to gain access to mineral deposits has progressed from the use of explosives to break the rock and shovels to load it for transportation away from the site, to the use of machines which continuously cut into the rock and load into a conveying system behind the machine. These rapid advance techniques are of interest to civil engineers engaged in road tunnelling, water tunnelling and similar activities. There are several textbooks on this subject, the best of which are the collected contributions to conferences which have taken place in Britain, North America and South Africa. These volumes represent the experience of experts throughout the world who are engaged in this fascinating work.

Mining methods

Surface mining techniques have developed greatly both in extent and in size of operation during the last two decades, and now account for over half the world's mineral production. The development of equipment which has enabled this increase in scale to take place has stimulated interest in this field and there are several relatively recent textbooks which can be recommended as giving a balanced picture of the situation. Particular mention should be made of two volumes published by the American Institute of Mining Metallurgical and Petroleum Engineers: *Surface Mining,* edited by E. P. Pfleider (1968), and *Case Studies in Surface Mining,* edited by H. L. Hartman (1969). These are very well produced and will probably remain as standard works for some years.

General works on underground methods are not so recent or well written, with the exception of the volumes by S. D. Woodruff entitled *Methods of Working Coal and Metal Mines* (3 vols., Pergamon, 1966), which do cover stratified and non-stratified minerals, although the former is more completely described. Most textbooks on underground methods are restricted in scope or out of date, and only *Introduction to Mining,* by B. Stoces (2 vols., Lange, Maxwell and Springer, 1958), is likely to keep its wide appeal. An interesting volume by C. H. Fritzsche and E. L. J. Potts entitled *Horizon Mining* (Allen and Unwin, 1954) covers mine development and methods used under particular circumstances. This was of great importance at the time of publication, but the principles enunciated in the book are of more lasting significance and are worthy of note.

In order to read accounts of the latest mining methods it is usually necessary to consult papers written about individual operations and published in the periodicals of the professional bodies.

Ground control and rock mechanics

This has been a subject of intense activity during the last 30 years since it has been recognised as having a profound effect on the safety and economics of mining operations. There have been numerous textbooks published covering the whole field or selected areas within it. These have ranged from the highly mathematical to the purely practical and empirical so that it is necessary to select carefully in order to obtain a treatment at the right level with the appropriate approach for a particular purpose. Fortunately most of the authors have made their approach clear in the introductory remarks or even in the title.

Mine transportation

This section of mining technology can be regarded as that branch of mechanical engineering applied to mining. Several pre-war text-books dealt with the subject comprehensively, but developments during the last 30 years have rendered these books out of date. Later works have tended to deal with parts of the subject such as winding engines, winding ropes, locomotives, conveyors, hydraulic transport, trucks, etc. Many of these publications have been written by specialists employed by companies who manufacture the equipment. It should not be forgotten that very detailed information about individual machines and equipment can be obtained from the brochures produced by the manufacturers. These are often of very high standard, particularly as regards illustrations. One of the problems in this field is that equipment is constantly being developed and textbooks therefore tend to have a limited useful life except for the industrial historian. However, a book dealing with the principles of system design as opposed to actual equipment design is likely to remain useful for much longer. Such a book is moreover more useful to the student in understanding the purpose and limitations of mining equipment.

Mine environment

The problems of providing reasonable working conditions in and around surface and underground mining installations have gradually become regarded as an essential element in mine management. Even so there are relatively few textbooks which deal comprehensively with the problem of ventilation, lighting, dust, noise, heat and industrial diseases related to mining operations. Several texts deal with ventilation, but the remaining fields have a very restricted literature. Most of the useful detailed information can be obtained by consulting official publications, such as those of the Safety in Mines Research Establishment, who have built up a world-wide reputation for the high quality of their work and of their published reports.

Reference should also be made to the mining legislation, which in most countries in the world lays down minimum standards of practice in the provision of safe and healthy working conditions in mines. Mine surface installations may also be covered by general industrial legislation in some countries. The International Labour Office in Geneva has also tried to spread the better standards of more advanced mining countries to the rest of the world and has published a great deal in this field.

Mine management and economics

The mining industry has been in the forefront of developments in the management field, particularly where mining operational and financial units have been large and where, consequently, the benefits of scientific management have been most evident. Textbooks in this area relating particularly to mine management are few in number and these are now somewhat dated. Reference to more general works on management and economics will enable much more modern ideas to be appreciated, but it should be borne in mind that the mineral industry has special problems in both these fields and due allowance should be made for these special circumstances.

Mention has already been made of the development of computer methods in mineral deposit evaluation. Similar extensions have been made in the field of mineral economics and mine planning. Marketing trends and open-pit planning are examples of the sort of study which have been mechanised by the use of computers. Most of the reports on this work have appeared as contributions to the periodicals published by the professional bodies.

Surveying

The surveying of mines is very important from the operational and safety point of view and in turn the existence of mining activities has been a stimulus for survey operations in developing countries. General textbooks in surveying are widely available and the choice is largely dependent on the taste of the reader. Some take a basic trigonometrical approach, whilst others concentrate on instrumental work. In recent years the development of self-reducing tacheometers, self-levelling levels, electronic distance measurement and gyroscopic attachments for orienting theodolites have led to a need for new textbooks, which has rapidly been satisfied. These more recent books have obvious advantages over the earlier standard works.

Textbooks in mine surveying are very limited in number. Particular mention should be made of *Surveying Practice,* issued by the National Coal Board (1973). While this is not a textbook in the usual sense it gives details of procedures, standards and conventions of symbols which must be of value to any mine surveyor.

Periodicals

The principal periodicals concerned with mineral extraction operations are those produced by the professional bodies, with a few notable commercial publications making a significant contribution. Among the latter are *Colliery Guardian, Mining Journal* and *Mining Magazine, Mine and Quarry* and *Quarry Management and Products,* formerly *Quarry Managers Journal.* These are general magazines dealing with particular industries, the latter two being published in association with learned societies. The *International Journal of Rock Mechanics and Mining Sciences* is a good example of a magazine for the specialist research worker. Other commercial technical magazines published in the United States, Canada and Europe cover general mining matters, often on an international scale. *Engineering and Mining Journal* (USA), *Canadian Mining Journal* and *Glückauf* (Germany) are good examples of these. All of these commercial publications carry technical news sections, new equipment descriptions and personnel items as well as technical articles containing descriptions of installations, equipment, innovations or new ideas.

The professional bodies concerned with the mineral industries of the world include the Institution of Mining Engineers (UK) (coal), Institution of Mining and Metallurgy (world metal mining), Institute of Quarrying (UK), the American Institute of Mining Engineers, the Canadian Institute of Mining and Metallurgy, the South African Institute of Mining and Metallurgy and the Australian Institute of Mining and Metallurgy. There are also, of course, similar bodies in other countries publishing in their own languages, often giving a synopsis of the contents in English.

Generally speaking, these publications consist of technical papers read and discussed at meetings of the bodies so that the subjects covered range across the mineral industry and even occasionally outside it. It is thus very difficult to locate a paper on a particular subject without reference to the annual index which most journals contain. Some of the bodies such as the Institution of Mining Engineers have produced general indexes covering ten- or twenty-year periods. These are usually arranged with subject and author sections.

The Institution of Mining and Metallurgy has tried to reduce the amount of searching required in two different ways. Firstly, the papers are published in three groups which appear in separate monthly issues on a three-month cycle. Thus, all geological papers appear in four issues during the year, as do the mining papers and the mineral processing papers. Secondly, the Institution offers an abstracting service. All books and articles on subjects related to the mineral industry are summarised briefly with very thorough cross-referencing. The *IMM Abstracts* appear six times per year and probably are the most comprehensive available for the industry. The classification used has already been referred to and appears in *Figure 35.1.*

Access to the publications of a professional body is an obvious advantage of membership and, in addition, many institutions have libraries and a postal loan service for members. A similar loan service and a personal visit facility is available at the British Library, Lending Division at Boston Spa, Yorkshire. A microcopying service is also available for certain borrowers or material.

Entries in *IMM Abstracts* are arranged according to the Universal Decimal Classification (UDC). The English edition of the schedules may be purchased, as parts of B.S.1000, from the British Standards Institution, 2 Park Street, London, W.1, England.

Universal Decimal Classification — Summary

Generalities	0	Drifting	622.26
Scientific theories	001.5	Mining methods (surface)	622.271
Bibliographies	011/016	Mining methods (underground)	622.272/.275
Periodicals	05	Working by boreholes	622.276
Associations, Congresses	06	Support, linings	622.28
Social Sciences	3	Mining specific minerals	622.3
Statistics	31	Carbonaceous	622.32/.33
Economics	33	Metalliferous	622 34
Taxation	336.2	Building stones	622.35
Production, Costs	338	Non-metallics	622.36
Legislation, Public Administration	34/35	Precious stones	622.37
Education	37	Ventilation	622.41/.46
Trade	38	Illumination	622.47
Mathematics	51	Drainage	622.5
Physics	53	Haulage, hoisting	622.6
Chemistry	54	Mineral Processing	622.7
Mineralogy	549	Specific minerals	622.7–32/–37
Geology	55	(arranged as mining specific minerals)	
Prospecting	550.8	Control, efficiency, etc.	622.7.01/.09
Physical geology	551.1/.4	Sorting	622.72
Stratigraphy	551.7	Comminution	622.73
Petrology	552	Sizing, screening	622.74
Economic Geology	553	Classification	622.75
Description and properties of ores	553.1	Gravity concentration	622.762/.764
Origin	553.2	Flotation	622.765
Metalliferous deposits	553.3/.4	Dense media separation	622.766
Building stones	553.5	Leaching	622.775
Non-metals	553.6	Electrostatic separation	622.777
Mineral springs	553.7	Magnetic separation	622.778
Gems	553.8	Roasting, drying, briquetting	622.79
Carbonaceous deposits	553.9	Storage, transport of concentrates	622.799
Medical Sciences	61	Hazards, Accidents	622.8
Industrial hygiene	613.6	Dust, explosions, fires	622.81/.82
Accidents: prevention, protection	614.8	Rock pressure	622.83
Engineering (General)	62	Accidents, safety measures	622.86
Materials testing	620.1	Civil Engineering	624
Corrosion	620.19	Ports, Dams	627
Economics of energy	620.9	Public Health Engineering	628
Power, Plant, Mechanical Engineering	621	Management	65
Mining	622	Chemical Technology	66
Economics	622.013	Metallurgy	669
Surveying	622.1	Processes and equipment	669.02/.09
Exploration and development	622.22	Ferrous metals	669.1
Machine mining, rock drills	622.233	Non-ferrous metals	669.2/.8
Hydraulic mining, dredging	622.234	Instrumentation	681
Blasting	622.235	Computers	681.3
Exploration and well drilling	622.24	Biography	92
Shafts, sinking	622.25	History	93

Figure 35.1   IMM abstracts classification

Before leaving mining literature, mention should be made of the publications of companies producing mining equipment or providing specialist services. Apart from the periodical house magazines of the larger companies, which only occasionally have articles of substantial technical content, many of these companies are happy to supply copies of the brochures describing their products, as was mentioned earlier. These are of great interest to the practising engineer and give information on the newest equipment and techniques. Finally, general information on mining operations and details of financial results can be obtained from the annual reports of companies who operate mines, whether these are privately or publicly owned. Some of these reports are very detailed and some are profusely illustrated.

Selected publications (mining)

Abstracts

Engineering Index
Geomechanics Abstracts (formerly *Rock Mechanics Abstracts*)
IMM Abstracts
International Petroleum Abstracts (formerly *Institute of Petroleum —Abstracts*)
NCB Abstracts A—Technical Coal Press
Petroleum Abstracts
Safety in Mines Abstracts

Periodicals

Canadian Mining and Metallurgical Bulletin
Canadian Mining Journal
Chamber of Mines Journal
Chartered Surveyor
Coal Age
Coal News
Colliery Guardian
Engineering and Mining Journal
Geologie en Mijnbouw
Geotechnique
Glückauf
International Journal of Rock Mechanics and Mining Sciences
Journal of Mining and Geology
Journal of the Institute of Petroleum
Journal of the South African Institute of Mining and Metallurgy
Mine and Quarry
Mining Congress Journal
Mining Engineer
Mining Engineering
Mining Journal
Mining Magazine
Mining Technology

Mining Ventilation
Photogrammetria
Photogrammetric Record
Pit and Quarry
Proceedings of the Australasian Institute of Mining and Metallurgy
Quarry Managers Journal (continued as *Quarry Management and Products*)
Review de l'Industrie Minerale
South African Mining and Engineering Journal
Steel Times
Survey Review
Transactions of the Institution of Mining and Metallurgy
Tunnels and Tunnelling
US Bureau of Mines. Reports of Investigations. Information Circulars
Western Miner
World Mining

Books

Dictionary of Mining, by A. Nelson (Butterworths, 1964)
Glossary of Mining Geology, by G. C. Amstutz (Elsevier, 1971)
Prospecting and Operating Small Gold Placers, by W. F. Boericke (2nd edn, Wiley, 1936)
Hydraulics Applied to Underground Mining Machinery, by W. Buchanan (Pitman, 1966)
Offshore Drilling and Rigs, edited by L. Draper (Royal Institution of Naval Architects, 1971)
The Strength, Fracture and Workability of Coal, by I. Evans and C. D. Pomeroy (Pergamon, 1966)
Metal Mines of Southern Wales, by G. W. Hall (G. W. Hall, 1971)
Coal Mines Rescue and Firefighting, by J. D. Jenkins and J. W. Waltham (Griffin, 1956)
The Modern Technique of Rock Blasting, by U. Langfors and B. Kihlstrom (2nd edn, Wiley, 1967)
Elements of Mining, by R. S. Lewis and G. B. Clark (3rd edn, Wiley, 1964)
Novel Drilling Techniques, by W. C. Maurer (Pergamon, 1968)
Mining and Minerals, by G. A. Northedge and E. N. Davies (Pergamon, 1968)
Examination and Valuation of Mineral Property, by R. D. Parks (4th edn, Addison-Wesley, 1957)

Mining Engineers Handbook, by R. Peele (3rd edn, Wiley, 1941)

Working of Mineral Deposits, by G. Popov (Central Books, 1971)

Economics for the Mineral Engineer, by E. J. Pryor (Pergamon, 1958)

Quarrying in Somerset (Somerset County Planning Department, 1971)

Underground Lighting in Mines, Shafts and Tunnels, by A. Roberts (Technical Press, 1958)

Coal Mining Economics, Organisation and Management, by J. Sinclair (2nd edn, Pitman, 1967)

Planning and Mechanised Drifting at Colleries, by J. Sinclair (Pitman, 1963)

Quarrying, Open-Cast and Alluvial Mining, by J. Sinclair (Elsevier, 1969)

Winning Coal, by J. Sinclair (Pitman, 1960)

Prospecting for Gemstones and Minerals, by J. Sinkankas (Van Nostrand Reinhold, 1971)

Mine Ventilation, by A. Skochinsky and V. Komarov (Central Books, 1969)

Advanced Coal Mining, by B. M. Vorobjev and R. T. Deshmukh (2 vols., Asia, 1967)

Methods of Working Coal and Metal Mines, by S. D. Woodruff (3 vols., Pergamon, 1966)

Heat Exchange Calculations for the Ventilation of Dead End Galleries in Deep Coal Mines, by A. N. Yagel'skii (Israel Program for Scientific Translations, 1960)

Guide to the Coalfields, edited by E. G. Corbin (Colliery Guardian, annual)

Mining Year Book (W. R. Skinner, annual)

Manual of Excavators, by H. Breitung (Interbook, 1968)

General Excavation Methods, by A. B. Carson (McGraw-Hill, 1961)

Ore Mining and Materials Handling (Iron and Steel Institute, 1963)

Earth Moving and Excavating Plant, by R. Hammond (Maclaren, 1964)

Drilling of Rock, by K. McGregor (Maclaren, 1967)

Mining Subsidence, edited by A. Thompson (Institution of Civil Engineers, 1962)

Engineering Geology and Rock Mechanics, by N. Duncan (2 vols., L. Hill, 1970)

Engineering Properties of Rocks, by I. W. Farmer (Spon, 1968)

Fundamentals of Rock Mechanics, by J. C. Jaeger and N. G. W. Cook (Chapman and Hall, 1971)

Principles of Engineering Geology and Geotechnics, by D. P. Krynine and W. R. Judd (McGraw-Hill, 1957)

Rock Mechanics in Engineering Practice, by K. G. Stagg and O. C. Zienkiewicz (Wiley, 1968)

The Finite Element Method in Engineering Science, by O. C. Zienkiewicz (2nd edn, McGraw-Hill, 1971)

Tunnelling in Rock, by E. E. Wahlstrom (Elsevier, 1973)

Mechanics of Bulk Materials Handling, by N. Brook (Butterworths, 1971)

Materials Handling Equipment, by N. Rudenko (2nd edn, Central Books, 1970)

Conveyors and Related Equipment, by A. Spivakovsky and V. Dyachkov (Central Books, 1966)

Underground Electric Haulage, by L. Szklarshi (Pergamon, 1969)

Manual of Modern Underground Haulage Methods for Mining Engineers, by H. R. Wheeler (Griffin, 1946)

Coal Mining, by A. R. Griffin (Longmans, 1971)

British Mining Fields, by J. E. Metcalfe (Institution of Mining and Metallurgy, 1969)

Alluvial Prospecting and Mining, by S. V. Griffith (2nd edn, Pergamon, 1960)

British Coal: a Review of the Industry, by J. Platt (Lyon, Grant and Green, 1968)

Mining: an International History, by J. Temple (Benn, 1972)

Practical Field Surveying and Computation, by A. L. Allan *et al.* (Heinemann, 1968)

Photogeology and Regional Mapping, by J. A. E. Allum (Pergamon, 1966)

Elementary Photogrammetry, by D. R. Crone (Arnold, 1963)

Introduction to Geodesy, by C. E. Ewing and M. M. Mitchell (Elsevier, 1970)

Survey Adjustments and Least Squares, by H. F. Rainsford (Constable, 1957)

Tacheometry, by F. A. Redmond (Technical Press, 1951)

Surveying Problems and Solutions, by F. A. Shepherd (Arnold, 1968)

Engineering Surveying, by W. Schofield (2 vols., Butterworths, 1972 and 1974)

Metalliferous Mine Surveying, by F. Winiberg (5th edn, Mining Publications, 1966)

Surveying Practice (NCB, 1973)

Project Surveying, by P. Richardus (Elsevier, 1974)

THE BENEFICIATION OF MINERALS (MINERAL PROCESSING)

Studies relating to the preparation of mineral products for marketing either as finished products or as raw material for further processing, such as smelting, have been given various labels, such as mineral dressing, mineral processing, mineral technology, minerals engineering or applied mineral science. The boundaries between these subjects and other related fields are ill-defined and overlap is inevitable. The direct extraction of minerals, such as salt by borehole, can be regarded as mining or mineral processing; the preliminary treatment of ores of some metals before smelting can be regarded as mineral processing or metallurgy; the blending or agglomeration of minerals prior to use in the chemical or fuel industry may also be regarded as spanning the boundaries.

Clearly, the mineral beneficiation processes are based on chemistry, physics, geology and mathematical principles, so that a survey of literature in this field must begin with reference to those parts of these sciences which have a direct application to the problems of mineral processing. It is not proposed to cover the whole field of the sciences, since this is dealt with elsewhere in the series.

In recent years the problem of waste disposal has become more important for industry as the awareness of environmental conservation has grown. Because the mineral industry has always been involved with this problem, it has been in the forefront of the technology of waste disposal and of the reduction of pollution.

Another important trend in the mineral industry which has had a profound effect on the technology and economics of mineral beneficiation is the need to work mineral deposits with lower useful mineral content or grade. Although this has resulted in some increases in market price for the various mineral concentrates produced from such low-grade deposits, there is a limit to such increases and the mineral technologist has had to find economical methods of beneficiation in order to operate within the marketing constraints. Thus, automatic processes and non-manual control of processes have been of prime concern to the industry, and this is reflected in the literature as well as the technology of the industry.

Textbooks

Most of the general textbooks in this field belong to the earlier years of this century and as such have limited value at the present time as a guide to the activities in mineral beneficiation. The technology has advanced so much during the last 40 years that much of the equipment now used was either unknown or too expensive to use when these texts were written. As in the case of mining textbooks, the period elapsing between the commissioning of a well-written and well-illustrated book covering the whole processing field and the time when technology has advanced to cause the book to be out of date, along with the limited size of the student and specialist markets, has meant that most publishers have not been tempted to produce authoritative textbooks. Most of the works which have appeared have been of the specialist type aimed at a more restricted field. These are obviously of less wide appeal, but at least they tend to have a longer useful life.

The publications which give a general view of the activities in mineral processing are those produced by the professional bodies and learned societies recording the proceedings of general conferences which have been held at various centres throughout the world. Many of the contributions are of a specialist nature, but usually there are some 'state-of-the-art' papers which survey the developments within a particular subject or country over the past few years. The Institutions of Mining and Metallurgy in the Commonwealth countries and in the USA are particularly active in this respect, and special mention should be made of the *Proceedings of the Commonwealth Mining and Metallurgical Congress,* which is held every 4 years in one of the countries of the Commonwealth.

It has been noted already that specialist textbooks in mineral processing are more numerous. Admirable volumes have appeared on a wide variety of subjects, such as surface chemistry, crushing and grinding, fine particle measurement and identification, control of processing systems and the mineralogy and processing of individual minerals or groups of minerals, such as clays. It would be impossible to compile a complete list of these specialist volumes, and the list given at the end of the chapter selects only examples of the more important. A thorough examination of library catalogues or an inspection of *IMM Abstracts* will yield many more titles of interest in this field.

In addition to the publications produced by the professional bodies and by commercial publishers, there are some government publications which are worthy of mention. Occasional special volumes have been commissioned by bodies such as the late Department of Scientific and Industrial Research on particular topics, such as crushing and grinding. These are both authoritative and cheap, but are unfortunately infrequent. Within the field of mineral processing, many items of equipment and procedures are covered by *British Standards.* These should be consulted to give details of 'weights and measures' standards and safety specifications in relation to plant and equipment.

Periodicals

There are very few commercially published magazines produced within the strictly limited field of mineral beneficiation. Some of the mining magazines do carry articles on mineral processing and should be consulted. Similarly, there are few professional bodies limiting their interest to mineral processing and usually technical material appears alongside mining articles. In recent years the Institution of Mining and Metallurgy has published the mineral beneficiation papers separately, so that reference to such material is now rather more convenient.

A recent development in periodical publication is the use of a commercial magazine, *Mine and Quarry,* as the official organ of the Minerals Engineering Society in much the same way as *Iron and Coal Trades Review* at one time published the proceedings of the now defunct National Association of Colliery Managers. A similar relationship exists between *Quarry Management and Products* and the Institute of Quarrying.

It should not be forgotten that an appreciable amount of significant material in mineral processing is published in periodicals produced by scientific societies and professional bodies concerned primarily with related subjects. The proceedings of bodies such as the Institute of Fuel, Institute of Metallurgy, Institution of Chemical Engineers, the Metals Society (formerly the Iron and Steel Institute), the Society for the Chemical Industry, the Chemical Society and the Institute of Physics are all of occasional interest, and an inspection of the titles of papers published by these bodies can often yield useful leads to information. Many practitioners in mineral beneficiation find membership of one or other of these societies very useful for reasons which have already been mentioned in respect of the main mineral-based professional institutes.

Selected publications (mineral processing)

Abstracts

Chemical Abstracts: 53 Mineralogical and geological chemistry
54 Extractive metallurgy
55 Ferrous metals and alloys
56 Non-ferrous metals and alloys
57 Ceramics

Fuel Abstracts and Current Titles
Industrial Diamond Review-Abstracts
Metals Abstracts

Periodicals

Industrial Minerals
International Journal of Mineral Processing
Minerals Processing
Mineral Science and Engineering
Transactions of the Institution of Mining and Metallurgy

There are also quarterly publications, mainly produced by the
development organisations, for most metals, e.g. tin, lead.

Books

Mineral Processing (Proceedings of the 7th International Congress),
    edited by V. N. Arbiter (Gordon and Breach, 1966)
Principles of Mineral Dressing, by A. M. Gaudin (McGraw-Hill,
    1939)
Carbonisation of Coal, by J. Gibson and D. H. Gregory (Mills and
    Boon, 1971)
Identification of Mineral Grains, by M. P. Jones and M. G. Fleming
    (Elsevier, 1965)
Economics for the Minerals Engineer, by E. J. Pryor (Pergamon,
    1958)
Mineral Processing (Proceedings of the 6th International Congress),
    edited by A. Roberts (Pergamon, 1965)
Mineral Use Guide, by R. H. S. Robertson (Macmillan, 1960)
Coal Preparation and Power Supply at Collieries, by J. Sinclair
    (Pitman, 1962)
Handbook of Mineral Dressing, by A. F. Taggart (Wiley, 1945)

CONCLUSION

It will be noted that consideration of the literature concerned with
mineral beneficiation has been dealt with as a whole and not in
individual sections as was the case with the mining literature, or
with the geological literature. This was felt advisable in view of
the fragmented nature of the literature concerned with mineral

processing. The technology has always been regarded as either a branch of chemical engineering or metallurgy or as a necessary evil attached to mining technology. This is reflected in the organisation of the professional institutes. As a result there has been no channelling of technical publications in this field and the investigator has to seek widely for his material.

No apology is offered for repeating the earlier comments on the advantages of membership of a scientific society or professional institute. By receiving the regular publications one is informed of activities within the industry and also a collection of back numbers of these journals is the nucleus for a private technical library. In addition, one has access to the library of the society either by personal visit, or by postal loan service. Special publications are also available, often at reduced rates, and, of course, attendance at meetings and conferences is the right for a member. Additional facilities such as the *IMM Abstracts* also are of inestimable value to anyone engaged in work within the industry.

Index

Abbreviations
  dictionaries of, 143–144
  journal titles, 32
Abridgments of patent specifications,
    82–83
Abstracting and indexing journals, 5,
    151–165 (*see also* under specific
    subjects)
  lists of, 163
Abstracting journals
  coverage, 153
  delay, 153
  indexing of, 153
Abstracts
  definition, 152
  types, 152
Active networks, 219
Aerodynamics, 279–281
Aeronautics and astronautics, 276–300
  abstracting services, 297
  aerodynamics, 279–281
  aerospace materials, 289–290
  aircraft structures, 287–289
  astronautics, 295–296
  avionics, 295
  conferences, 296
  gas dynamics, 281–284
  journals, 299–300
  mechanics of flight, 292
  performance literature, 293
  propulsion (air breathing engines),
    284–285
  propulsion (rockets), 285–287
  reports, 297–298
  reviews, 298–299

Aeronautics and astronautics
                           continued
  stability and control, 294–295
  stress analysis, 459
Aerospace materials, 289–290
Air-breathing engines propulsion,
    284–285
Air conditioning, 402–404
Air pollution, 337–338, 519–520
Air transport, 502–503
Aircraft
  structures, 287–289
  V/STOL, 290–292
American National Standards, 101
American National Standards Insti-
    tute, 101
American Society for Testing and
    Materials, 101 (*see also* ASTM)
Amplifiers, 219–222
  operational, 219
Analogue and hybrid computation,
    255–257
  high-speed analogue computing, 257
  hybrid computation, 256–257
Analogue/digital conversion, 251–252
Applied electronics, 216
Applied Mechanics Reviews, 155, 395,
    450–451, 467
Applied Science and Technology Index,
    163, 331, 340, 441, 468, 507
Artificial intelligence, 233, 253
Aslib, 10
Aslib Commonwealth Index of Un-
    published Translations, 46
Aslib Index to Theses . . ., 39

618 *Index*

Minerals, 588–595
abstracting services, 591
deposits, 591
geology, 589–590
journals, 589–592
prospecting, 590–591, 594
textbooks, 592–595
Mining, 595–605
abstracting services, 600–602
books, 603–605
development of mines, 596
journals, 599–603
mine environment, 598
mine management and economics,
598–599
mine transportation, 597, 598, 605
mining methods, 596–597, 603–
604
rock mechanics, 597, 604–605
surveying, 599, 605
textbooks, 595–599, 603–605
Modelling, chemical engineering, 318
Models, hydraulic, 486
*Monthly Catalog of United States
Government Publications,* 65
Multi-phase flow, hydraulics, 484–485

National Bureau of Standards, 102
National Engineering Laboratory,
126
National Physical Laboratory, 126
National standards, 94–95
Noise, communications, 230
Non-ferrous metals, 432–433
Non-governmental international or-
ganisations, 135
Non-linear control systems, 264–265
Nuclear Science Abstracts, 65, 72, 155,
316, 442

Ocean engineering, 483
Operational amplifiers, 219
Optical coincidence cards, 14, 192
Organisations
intergovernmental, 133–135
international, 130–133
non-governmental international,
135–137

*Pandex Current Index to Scientific and
Technical Literature,* 162, 166
Parking, traffic engineering, 498
Patent office, 79–82
Patent search, 78–82

Patents, 4, 76–92 (*see also* under
specific subjects)
bibliography, 89–92
Canada, 84, 88, 90–91
Classification Key, 80–81
European, 86
Germany, 85, 91–92
Great Britain, 76–83, 89
infringement in Britain, 77–78
international, 86–87, 92
statistics, 87–88
USA, 83–84, 88, 90
Patent specifications, abridgments,
82–83
Pavements, 496
Peek-a-boo cards, 14, 192
Performance, aircraft, 293
Periodicals, *see* Journals
Personal indexes, 184–200
computers, 197–198
Physical electronics, 215–217, 222
Physical metallurgy, 436–439
Physical standards, 94
Physics, solid state, 222
Planning, production, 349
Plant, construction, 555–560
Plasticity, 455–456
Plastics, 554
structures, 545
Plates, 456–457
Pneumatic structures, 545
Post-coordinate indexing systems,
18–19
Power, direct conversion, 205–206
Power generation, 204–205
and conversion, 396–398
Power system analysis, 208–210
Power systems, electrical, 201–213
Pre-coordinate indexing systems, 16–
18
Pressure vessels, 461
Primary literature, 3–5
Proceedings in Print, 38
Process control, 268
Process plant, 318
Product information, 115–123 (*see
also* under specific subjects)
collection of catalogues, 119–121
commercial services, 121
microfilm systems, 122–123
Production engineering, 339–355
abstracting services, 340–341
conferences, 343–344
dictionaries, 345–346

Date Due